Design process improvement

A review of current practice

John Clarkson, MA, PhD, MIEE, CEng
Claudia Eckert, MSc, PhD

Engineering Design Centre, Department of Engineering, University of Cambridge,
Trumpington Street, Cambridge, CB2 1PZ, UK

British Library Cataloguing in Publication Data
Design process improvement : a review of current practice
1. Engineering design 2. Engineering design — Management
I. Eckert, Claudia, 1965– II. Clarkson, John, 1961–
620′.0042
ISBN 185233701X

Library of Congress Cataloging-in-Publication Data
Design process improvement : a review of current practice / edited by John Clarkson and Claudia Eckert.
p. cm.
Includes bibliographical references and index.
ISBN 1-85233-701-X (alk. paper)
1. Engineering design. I. Clarkson, John, 1961– II. Eckert, Claudia, 1965–
TA174.D4682 2004
620′.0042—dc22 2004052526

ISBN 1-85233-701-X
Springer Science+Business Media
springeronline.com

Typesetting: Electronic text files prepared by editors
Printed in the United States of America
69/3830-543210 Printed on acid-free paper SPIN 10889838

Design process improvement

A review of current practice

Edited by
John Clarkson and Claudia Eckert

With 345 Illustrations

Springer

Foreword

Industry and Academia have worked together for many years on research topics for their mutual benefit at varying levels of engagement on a wide variety of topics, a large proportion of which have been in the pursuit of technology to improve products or manufacturing methods. There is a growing awareness within industry that improvements in people and processes are just as vital for the health of the business.

Good design is fundamental to product success. Consequently there is a great incentive to improve. An understanding of the good practices that are being employed in some quarters and recognition of the opportunities that exist is a good starting point for improvements. There is a view that Industry is not involved in formulating or incorporating the outcomes of design research and Academia is not addressing the needs of Industry.

This book addresses the issues. It provides a wide-ranging review of current academic design research that has been carried out in close cooperation with Industry. It reveals current practices that would be of benefit to those concerned with design in its widest context. It also highlights areas where further research would be beneficial.

This is a timely contribution in the drive to understand and improve the design process, which is so vital to business success.

G E Kirk, RDI
Chief Design Engineer,
Engineering Fellow,
Rolls-Royce Civil Aerospace

Preface

The process is important! I learned this lesson the hard way during my previous existence working as a design engineer with PA Consulting Group's Cambridge Technology Centre. One of my earliest assignments involved the development of a piece of laboratory automation equipment for a major European pharmaceutical manufacturer. Two things stick in my mind from those early days – first, that the equipment was always to be ready for delivery in three weeks and, second, that being able to write well structured Pascal was not sufficient to deliver reliable software performance. Delivery was ultimately six months late, the project ran some sixty percent over budget and I gained my first promotion to Senior Engineer.

At the time it puzzled me that I had been unable to predict the real effort required to complete the automation project – I had genuinely believed that the project would be finished in three weeks. It was some years later that I discovered Kenneth Cooper's papers describing the Rework Cycle and realised that I had been the victim of "undiscovered rework". I quickly learned that project plans were not just inaccurate, as most project managers would attest, but often grossly misleading, bearing little resemblance to actual development practice.

I was proud of my well-structured subroutines founded on a well-structured programme architecture. So why did the equipment that had worked perfectly in the laboratory behave so badly when installed on the customer's site? I had fixed numerous 'bugs' prior to delivery in response to a barrage of performance tests and observed the customer using the equipment without fault. However, I now know that it is generally accepted that as many bugs that you find prior to delivery you are likely to leave in the code.

A few years later I was asked to lead a team to develop an automatic control system for a prototype firefighter training simulator to be delivered to the Royal Navy. This was a hugely complex undertaking with many seemingly conflicting requirements, the need to use unproven technology to create the simulated environment, and a typically robust fixed-price contract. Despite these challenges we delivered a system with over five hundred inputs and outputs on time, to budget and with only four minor software errors reported in over ten years of use.

John Clarkson
Reader in Engineering Design,
Director, Cambridge Engineering Design Centre

I have often reflected on the differences between these two projects in order to understand the reasons behind their widely differing outcomes. The 'products' were evidently different. Whilst the simulator development was undoubtedly more complex than the earlier automation project, it was commercially and technically far more successful. The key, it appeared, was the design process.

At the start of the simulator project I realised that I would personally have to demonstrate the integrity of the control system – a complex mix of hardware and software to control propane gas burners, smoke generators and fans in an enclosed, dark space – before it could be adopted by the Navy instructors. The simulator was to be designed to provide a safe, controllable, realistic and environmentally friendly training environment for the trainees and their instructors. Given the failure of my earlier programming attempts, these requirements focused my attention on acquiring the necessary skills to complete such a project.

I undertook training in structured software analysis and development (SSAD) techniques and applied this to the task at hand with liberal doses of risk identification, analysis and control, driven by a rigorous Ministry of Defence quality master plan. Half of the project was spent understanding the requirements, ten percent on coding the software, and the remaining forty percent testing the resulting system.

I learned many things from this experience. First, that active risk management is critical to the successful development of a complex product. However, on its own it is insufficient. There is a need to identify and characterise critical components or systems within a product and to actively manage the interfaces between them. This provides a reference point for communication of the product architecture and its asssociated functionality, and a basis for continuous risk management. Second, that risk management and interface management have also to be integral to the management of the design process. The risk of not achieving project goals can be as important as the risk of not achieving techinical goals. There is equally a need to identify and characterise critical activities within a design process and to manage the interfaces between them.

My fascination for the design process has grown from these early experiences and inspired much of the recent research in the

Engineering Design Centre at the University of Cambridge. There is always room for improvement in design. Maybe there is a need for a better product, or for a better, more effective and economic, design process – the late delivery of new products has been shown to be the single largest contributor to the loss of company profits in the UK. Our own experience of working with automotive, aerospace and healthcare companies has shown that effective communication, management of change and process planning are all essential ingredients for an effective product development process.

This book aims to develop an understanding of these issues as a means to facilitate design process improvement. Part I contains a series of review articles written by a team of international experts on models of design, perspectives on design, design practice and design management. Part II provides an introduction to the wealth of academic research on these topics by presenting the activities of research centres from around the world.

This book is for:

- design managers and business leaders who want to improve their company design procedures
- designers who want to know how to design more efficiently
- researchers who want to explore the field of design process improvement.

It is intended to be used as a reference resource to provide an introduction to the key issues currently facing design practitioners and researchers – further reading is suggested through the many references to other resources. In addition, I hope that the descriptions of the research centres will encourage design practitioners to explore the benefits of collaboration with the research community. Together there is much potential for understanding and improving the design process as a means to ensuring continued customer satisfaction and commercial success – the process is important!

Contents

Introduction

Design issues

Models of design

Perspectives on design

Design practice

Design management

Design research

Index and authors

Acknowledgements

The editors would like to thank all those who have helped bring this book about, in particular, the contributors of the review chapters. Their efforts and patience were greatly appreciated. Thanks are due to the UK Engineering and Physical Sciences Research Council for its funding of the Cambridge Engineering Design Centre, of which this book is a major output. Also, Rolls-Royce, BAE SYSTEMS and Airbus UK who, as partners to the research programme, have been ardent supporters of our development of new process improvement practices.

Thanks are also due to the Cambridge team of Jerome, Tim, Yolanda, Brendan, Tido, Seena, David, Tomas, Anja, Rene, Labi and Esther for contributing ideas and proof-reading many of the chapters in this book. Special thanks must go to John who captured many of the images used in the book, to Anthea for reading and re-reading numerous proofs, and to Mari for her tireless enthusiasm for drawing diagrams and her attention to detail in the latter stages of production.

Introduction

The reality of design

Claudia Eckert and John Clarkson
University of Cambridge

Had Percy Pilcher not died following a glider crash on 2 October 1899 he might have been the first man to fly a powered aeroplane. He had nearly completed the design for the first triplane, a design potentially more advanced than the Wright Brothers' plane at that time.

Pilcher had a background in mechanical engineering through a naval apprenticeship, but his real obsession was flight. The science of flight was in its infancy and in order to learn as much as he could, Pilcher designed and built a series of gliders in order to gain flying experience. The first glider, named the *Bat* was completed in 1895. This was followed in quick succession by the *Beetle*, the *Gull* and the *Hawk* (Figure 1) which first flew in early 1896. Pilcher experimented with a variety of control strategies and mastered the art of moving his body to correct for disturbances in flight.

He also travelled to Germany to visit Otto Lilienthal who had based his own gliders on detailed studies of birds. Both recognised that only powered flight could keep man in the air and, since steam engines were too heavy, Pilcher opted for the newly developed internal combustion technology and proceeded to design his own 4 hp engine. However, he encountered a dilemma – the added weight of an engine would increase the flying speed unless the wing area was also increased, and any increase in wing area would render the glider uncontrollable by weight-shift alone. Pilcher, presented with this unfortunate trade-off, opted for a higher flying speed in order to maintain the optimal wing area and continued development of the Hawk (Jarrett, 2001).

When he was about to despair, Pilcher received an unsolicited letter from Octave Chanute, one of America's leading aviation pioneers, who had read about the Hawk. Chanute suggested that instead of using one big wing he should stack several smaller ones on top of one another to achieve greater surface area but with a lightweight, robust design. This was the breakthrough Pilcher had needed and, despite initial scepticism, he proceeded to design the *Duck*, a quadruplane based on Chanute's design. Later he would design and build a triplane in order to demonstrate the potential of powered flight.

Pilcher arranged a trial flight of his powered glider for 30 September 1899 in order to attract much needed funding. However, when his engine failed he flew his glider instead, crashed and was all but forgotten by history. The Wright Brothers, learning from Lilienthal, Pilcher and Chanute, focused their efforts on developing an adequate method of control. This led to their first powered flight on 17 December 1903. When a replica of Pilcher's powered glider was built in 2003, it maintained controlled flight for longer than the Wright Brothers' first flight.

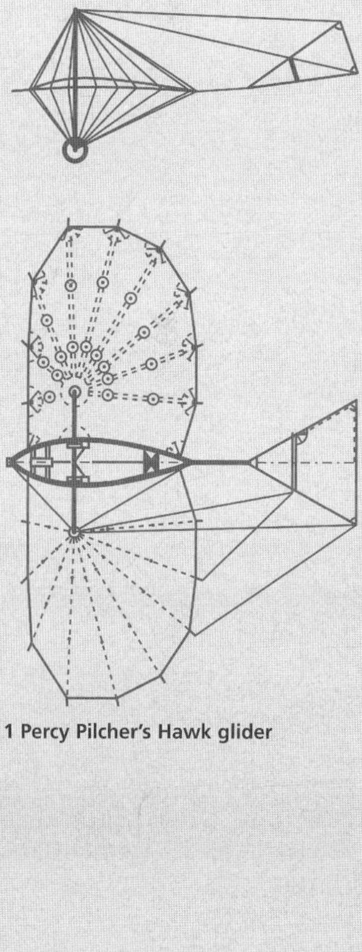

1 Percy Pilcher's Hawk glider

Increasing complexity of design

Percy Pilcher's story illustrates not only some of the common challenges of design, but also how design processes have changed in the last 100 years. He was a person who understood the theory in his field as well as anybody. He corresponded by letter with Chanute, and otherwise worked predominantly on his own – he did not need to produce detailed descriptions or models of his work.

Most designs are created by evolution from existing designs.

Pilcher's Hawk glider was based on Lilienthal's designs (Jarrett, 1987), which were inspired by birds' wings. His plane incorporated radical new ideas that were derived from the basic principles of physics, yet the fundamental technical challenges that he faced are still as pertinent as ever: aerodynamics, weight, material properties, engine efficiency, controllability etc. However, in early aviation history the trade-offs inherent in aircraft design had yet to be discovered and viable product architectures needed to be established. As a result, Pilcher built prototypes to test his designs and ultimately died because one such design proved unreliable.

Today, an aircraft is designed by over ten thousand people, each with very specific technical knowledge, and built by people with a wide range of different skills, so that an individual engineer can have only a cursory understanding of some of the contributing fields. Designs are developed using modern computer-aided design (CAD) systems and many specialised software packages, enabling extensive virtual and physical testing.

Safety has now become paramount with detailed design, verification, validation and documentation taking up over 75% of an aircraft development programme. In addition, most designs are created by evolution from existing ones, and only refer to the basic principles of physics if established design procedures fail. Major changes originate from major technological breakthroughs, such as the development of swept wings or jet engines, and key technical trade-offs have become sufficiently well understood that aircraft have settled into a number of generic designs.

2 Modern aircraft have hundreds of thousands of parts
© Airbus

Aircraft have evolved from humble beginnings a little over one hundred years ago to become highly complex products. They have hundreds of thousands of parts, designed, produced, assembled and maintained all over the world. This complexity brings its own challenges (see Chapter 7, Complexity). Modifying one product to generate the next is a way of coping with such complexity – starting from an existing design guarantees known properties of the product and reduces the overall effort required for design and manufacture (see Chapter 10, Engineering change).

In a safety-critical industry, such as aerospace, certification requirements tend to make companies very conservative in their response to change, with a preference for utilising existing critical parts wherever possible. Consumers are also often reluctant to purchase products that are too radical because they do not want to carry the risk of being 'guinea pigs' unless they have a need that cannot be met with existing technology. The skills required to operate and maintain products also cause bias towards incremental development, because sudden changes may require expensive retraining as well as the need to maintain dual platforms while the old design is phased out.

Designers' own thought processes can also be conservative (see Chapter 8, Thinking and representing in design). Many think about new designs with reference to existing ones, using mental representations of both physical embodiments and functions and performance factors; see Schön (1988), Oxman (1990) and Eckert and Stacey (2001) for discussions of the roles of types of design element and examples of design thinking. However, this locks them into tacit assumptions about the structure of the new design that are very difficult to escape – a phenomenon known to psychologists as fixation (Purcell and Gero, 1996).

Designs are integrated systems, with many physical, functional and behavioural links between the different parts. Issues such as vibration, noise or electromagnetic susceptibility (EMS) connect seemingly distant parts, so that small design modifications can have huge effects. For example, small variations in manufacturing can affect the natural frequency of an assembly making it susceptible to unwanted vibration (see Chapter 10, Engineering change).

Designs are integrated systems, with many physical, functional and behavioural links between the different parts.

The challenge for many designers is to maintain an adequate overview of a complex emerging product and its equally complex design process. Thinking by referring to existing products helps, since a new design inherits properties from the existing designs. External representation can also help designers to gain an overview of a product, but no matter how good the techniques are such representations can only capture part of the design.

The scope of design

In everyday language, as well as in professional literature, *design* is described in two distinct ways – by reference to the process of design or to the product that has been designed. The former, also known as *designing*, is often described as an iterative process in which the need, or problem, is understood as the solution is generated and evaluated (see Chapter 1, Models of designing). In this sense, designing occurs all the time in everyday activities – when

3 Design occurs in engineering and in many other domains as diverse as textiles, software and architecture

we plan our weekend, prepare a meal, look for the best route home. It is also present in many diverse professions – when a solicitor puts a case together, when a mathematician thinks about a proof or a teacher plans a lesson. The alternative view sees design as all the activities that lead from the first idea for a product to its pre-production prototype. Design then includes a wide range of tasks with very different underlying cognitive processes.

The *design process* may also have two distinct definitions. The generic procedure set out by a company to pursue design is often referred to as "the process", as is the actual sequence of activities that take place during design. The aim of this book is to assist primarily with improving the actual process, helping to make it easier and more efficient, rather than to prescribe a range of generic procedures.

Design across domains

Design occurs in engineering and in many other domains as diverse as textiles, software and architecture. Intuitively we recognise these design activities as similar, but it is often difficult to say what actually is similar and what is different. Academia has only recently turned its attention to this issue, and a research programme has been established to bring together designers from different domains together to talk about their design processes (Eckert *et al.*, 2004). Initial findings indicate that, although the products are different, the processes of their creation are in many ways similar – however, within each process the emphasis that is given to particular activities varies greatly between domains. For example, in the design of safety-critical products, verification and validation play a very important role in the evaluation of product safety, whereas in graphic design the product is evaluated subjectively by the designer.

Although products are different, the processes of their creation are in many ways similar.

There is also an increasing overlap in the activities of different domains. For example, a wider range of textiles are being used in the automotive industry, where styling now affects the geometric properties of many engineering products. As a result, the differences between design activities across domains can often be less than within the same domain.

Products

Engineers design everything from screwdrivers to an aircraft (Ulrich and Eppinger, 2003). Only at a very abstract level can the design process of these products be the same (see Chapter 1, Models of designing). However, design research typically has looked at what is general in design processes, rather than focusing on providing interesting insights into specific processes.

The timescale of an engineering process can range from a few days for simple components to a decade for complex innovative projects such as an aircraft, with upwards of ten thousand person-years of design effort going into a single product. As we discuss later, the drivers that influence different design processes can vary enormously. Very mature mass produced products are likely to be predominantly driven by price, whereas other, lower volume products may differentiate themselves through functionality. The core expertise required to design these two types of product can be very different.

At present we have little theoretical understanding of how products affect the processes by which they are designed, and *vice versa*, and how they are influenced by the designer and user, let alone how these relationships influence each other.

With different disciplines come different ways of working, thinking and talking about design.

Activities

The design of a modern product, such as a car, requires the collaboration of a multi-disciplinary team (see Chapter 9, Communication in design). Typically, mechanical engineers work with styling experts, electrical engineering and materials specialists to come up with the design of the product. In the later stages, manufacturing personnel become involved to influence the design process or, as still too often happens, to transform the design into one that can be manufactured. With these different disciplines come different ways of working, thinking and talking about design, each engaging at different stages of the typical design process (Figure 4).

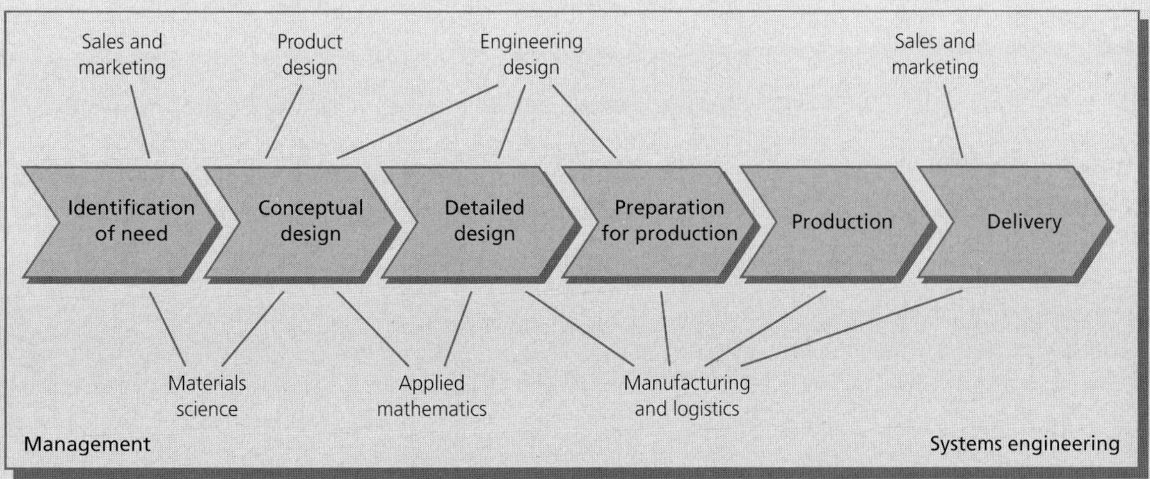

4 Disciplines engaged during the design process and their predominant focus

5 The design of a modern product, such as a telephone, requires the collaboration of a multi-disciplinary team

The process will usually begin with the *sales and marketing* team identifying the need for a new product. Such needs are likely to result from extensive market and commercial analysis or, more rarely, from a response to a direct customer request. For example, in the aerospace sector customers can specify their individual engine requirements. Design consultancies and some specialist component suppliers may depend entirely on their customers to define the product requirements. However, more usually, sales and marketing analyse the market, looking carefully at competitors' products and development activities. They establish trends and try to ascertain what their target customers want through the use of focus groups and surveys. Marketing and sales personnel are often non-engineers and while marketing rhetoric often involves figures, the language describing engineering requirements is vague. For example, a sports car must be "sportive" or "masculine", terms that need translating into technical requirements (see Chapter 4, Requirements engineering).

Design processes require *management* on many levels. The organisation needs to be managed strategically (see Chapter 17, Product portfolio management) at the product introduction process level (see Chapter 16, Integrated new product development) and at the design process level (see Chapter 2, Design planning and modelling). At the same time, individuals need to be managed so that they are supervised and encouraged, a role engineers (as managers) are often ill prepared to take on. Often they have to fall back on their own natural ability to cope with the people side of their job (see Chapter 5, Human resources).

In recent years managers have also increasingly been expected to manage the risk of non-compliance to specification (product risk) and the risk of non-compliance to the project plan and budget (process risk), thus limiting the commercial risk associated with the product development process (see Chapter 11, Risk in the design process).

Implicitly or explicitly, many design activities involve systems engineering or systems thinking.

Implicitly or explicitly, many design activities involve *systems engineering* or systems thinking (see Chapter 3, Systems engineering), where a classical systems view breaks a system down hierarchically and initially sees the subsystems as black boxes, with a specified functionality. Only once the functionality and role of those subsystems are clearly understood and their interfaces are fully defined are they broken down further. Mechanical engineers are rarely trained in systems thinking and can find it difficult, while electrical and software engineers are often more comfortable with such an approach.

The tasks of *engineering designers* typically include the conceptual and detailed design of individual components or subsystems. They are provided with requirements and constraints they need to translate into a viable design and usually have an existing component or design they are starting from. Conceptual design can be a holistic process, very much akin to an artistic design process, where the designers pull diverse information together, consider multiple trade-offs and synthesise several solutions. Others use automated tools to create and evaluate novel ideas (see Chapter 6, Artificial intelligence for design process improvement).

6 Car engines are designed to fit within the bonnet profile that is defined prior to a styling freeze

Many engineers also have exceptional visualisation abilities which help them to 'see' and 'manipulate' parts. This often draws them early into detailed design of components by working on specific solutions in their minds. If they have not considered all the options, this approach can lead to errors and design iteration. At the other end of the process, many engineers find it difficult to stop when the design is 'good enough' rather than when it is 'optimal'. This can put them at odds with their managers, who need to plan the process.

A substantial part of most engineering processes is design analysis, such as weight calculations or stress analysis. The more complex and safety critical the design becomes, the more this is necessary. Design analysis is essentially *applied mathematics*. Designers learn algorithms and are trained to recognise the situations in which they need to be applied. They use software tools for many of the calculations, but need to identify input information at the right level of maturity at each stage of the process. This identification process requires an in-depth understanding of the product, its constraints and the design process. Consequently, analysis tasks are very different from synthesis tasks and require different skills to undertake them.

The development of novel materials requires the skills of a *materials scientist*, who experiments with new combinations of materials, properties and manufacturing processes. Often the selection of a material with the right properties is enough (Ashby, 1999), but at other times new developments are necessary, either within the company or in conjunction with a supplier. Understanding the behaviour of materials under the expected operating conditions is vital in any design, but it can be hard to predict for novel materials or applications.

At the other extreme of the design process lies *product design*, where the skill of the designer lies in understanding customer needs and market trends, and translating them into a physical form that conforms to the required brand image. Product designers often draw inspiration from designs and objects far outside the realm of the product they are working on – for example, designers of sports cars and trainers draw inspiration from each other.

The skill of the product designer lies in understanding customer needs and market trends, and translating them into a physical form that conforms to the required brand image.

In many companies design engineers still have surprisingly little contact with *manufacturing engineers*, a legacy of ancient boundaries. The latter, whose experience is often practice-based, are expected to be mindful of potential problems with manufacturing, assembly and maintenance. They need to understand the capabilities of their company and its suppliers, and to anticipate likely areas of difficulty and identify potential cost savings.

In addition to the above groups, every company has logistics experts, administrators, personnel advisors (see Chapter 5, Human resources), etc. All these groups have their own professional skills and mental predisposition, and their professional focus is most often directed towards their specific skills rather than the product they are working on.

Table 7 provides some examples of the differences between the professional groups discussed above, in terms of the typical descriptions they use and the uncertainties they face in their work. The different groups typically focus on different degrees of detail and make decisions in different ways. Their different mindsets lie at the heart of many of the communication problems within organisations and can lead to inefficiencies in the product development process. Conversely, people who have worked on the same aspects of problems share a common understanding of their tasks and are likely to use similar mental and social constructs.

7 Differences between disciplines. Examples of the variation in the disciplines that may be involved in the design of a complex product, such as a car or an aircraft.

Discipline	Description	Uncertainty	Degree of detail	Distance to goal	Decision making
Sales and marketing	High-level product characteristics	Company capability, market, world events	Low	Known	Selection of alternatives
Management	Tasks, deadlines, budgets	Resources, technical risk	Low	Known	Constraint resolution
Engineering design	Sketches, drawings, CAD systems	Technical, missing information	High	Not known	Function and fit
Applied mathematics	Formulae, simulation models	Underlying specifications	High	Known	Objective evaluation
Materials science	Formulae, empirical models	Physical performance	High	Known	Physical evaluation
Systems engineering	High-level diagrams	Influencing factors	Low	Not known	Weighting
Product design	Sketches, models, mood boards	Context, emergent properties	Low, high	Not known	Subjective
Manufacturing and logistics	CAM systems, MRP systems	Known types	Very high	Known	Potentially objective

Pressures on design

Companies are looking for an appropriate design process to achieve the desired product within given time, cost and resource constraints. However, this in itself is not enough. The process needs to assure the long-term success of the company, taking into consideration the needs of other concurrent and future projects. This involves developing the product within the constraints of available resource, whilst also developing and nurturing skills in the company. In addition, the relationship with suppliers and customers needs to be maintained. All these factors constitute 'classical' constraints (Figure 8), which apply to all design processes (see Chapter 12, Design for X). In addition, new pressures are always arising.

Classical constraints

The product is the strongest constraint on its own design process. In addition, each design process is constrained by time and cost (Figure 8). Tendering agreements define timescales and budgets and often impose harsh penalties for late delivery. Internal projects are also likely to be constrained by budgets. These constraints may vary, but no project will have an unlimited budget.

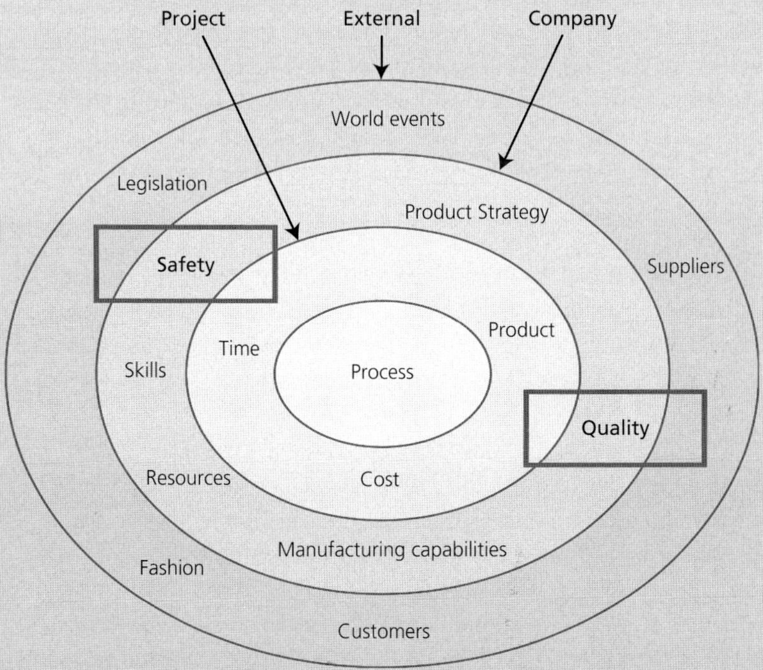

8 Constraints on the design process

Research by McKinsey for the British Department of Trade and Industry (DTI), looking at a cross-section of UK industry (DTI, 1991), has estimated that on average, if a project is shipped 6 months late, the percentage loss in after-tax profit will be over 30% of that expected with delivery on time. If production costs are 9% too high, a loss of about 20% may be expected. However, a project that finishes on time, but overruns its development costs by 50%, is predicted to incur a loss of less than 5%.

Product quality and safety are also fundamental drivers for most engineering processes (see Chapter 14, Quality management). This is particularly visible in industries with a strong regulatory influence, such as healthcare or aerospace. In these cases quality and safety are the key drivers of rigorous validation and verification processes which can consume a substantial part of the engineering budget. This is noticeably different from, for example, fashion design, where the products are inherently safer and the manufacturing quality of the product can be tested by simple means and the design judged by the designer in the light of their understanding of the market.

The company in which a design process is carried out also adds its own constraints to the process. Typically, companies want to make best use of their available skills and make strategic decisions on how to develop them. This will lead at times to inexperienced people being trained alongside those that are more experienced, or conversely to the most appropriate person not being available for a particular project. A product development strategy (see Chapter 17, Product portfolio management) can also place demands and restrictions on specific projects, such as the need to carry the cost for a new development or reuse parts that were developed for other projects.

The company in which a design process is carried out adds its own constraints.

Some companies design products in a way that makes best use of their existing or intended manufacturing resources and are highly reluctant to subcontract parts that could be produced in house. Others have a policy of competitive tendering, where internal departments have to bid against external candidates. No matter where a product is designed, the way in which it is produced constrains the design process.

There are also many external factors constraining the process. Customers generate constraints by directly setting requirements for the product or by displaying a need. Similarly, suppliers set constraints through cost, quality and availability of their parts, systems and services. Legislation also plays a strong part, for example, new emission standards drive the development of engines (Jarratt *et al.*, 2003), and safety directives drive the development of medical equipment and devices (Ward and Clarkson, 2004). In addition, fashions change and not only determine the appearance of a product, but also the technology people like to use. All design processes are directly or indirectly affected by the outside world. Market forces and the health of the economy influence the sales, and hence profitability, of a company. While many companies are acutely aware of the importance of successful processes, they often do not know how to achieve them.

New challenges

Engineering companies will face a number of new challenges in the future. With rapid technological development and strong international competition, all companies now must design better products faster and more efficiently. Many time and cost factors relating to a design process are fixed; e.g. testing requirements may be set by external regulatory authorities. In practice, this means that companies must improve the effectiveness of their design processes by maximising the use of their available knowledge base (see Chapter 13, Engineering knowledge management) to avoid making mistakes – and to plan more effectively (see Chapter 2, Design planning and modelling). Already many large organisations have so called 'right first time' policies, aiming to produce a fully functioning first prototype.

Many companies also face an ever increasing diversification of their markets, and are driven to produce a greater variety of products (see Chapter 10, Engineering change). In the developed world, customers have ever more specialised requirements. While true mass customisation of even moderately complex products is still a long way off, the desire for personalised products is increasing. Personalisation changes the economics of manufacturing. Products made to order are more likely to meet the customer's requirements and warehouse time is kept to a minimum. At the same time, there is a great need for cheap mass-produced products in the developing world, where the legislatory requirements may be less stringent, but where customers are nonetheless likely to have sophisticated tastes and demands.

Increased awareness of environmental issues and the need to make technology more sustainable are leading to increased legislation. For example, new emissions requirements for the aerospace industry raise the possibility that current engine technology will not be able to be adapted to meet the new targets. Such a step change in an already complex product will be an enormous challenge for the entire aerospace industry and will only be achievable through greater integration of design effort across the supply chain.

9 Aircraft engine manufacturers are moving towards selling "power by the hour" rather than engines – they will be paid for the hours that their engines are operating and carry the responsibility when it does not
Reproduced with the kind permission of Rolls-Royce plc

The nature of the supply chain is also changing. Traditionally, where parts could not be bought from stock, companies passed complete descriptions of parts to suppliers, requiring many design decisions to be made before an order can be placed. This often led to a delay in the design process or imposed an artificial order on design activities. As a result an alternative approach has emerged, where companies are building up close relationships with preferred suppliers with whom they are sharing design information that has yet to be finalised. The aim is to foster trust along the supply chain and, consequently, suppliers are increasingly selling services rather than parts.

New manufacturing capabilities are emerging that will change design processes considerably.

New manufacturing capabilities are emerging that will change design processes considerably. For example, 'rapid manufacturing' allows the creation of almost any shape in plastic or metal without requiring expensive tooling. This has the potential to revolutionise all such parts, since they can be adapted on a part-by-part basis to ensure that mass customisation is a real option. Many restrictions on part geometry will also be removed, allowing the manufacture of single, multi-feature parts where previously assemblies of components were necessary. Rapid manufacturing is already making prototyping significantly easier, as parts can be made directly from their design descriptions. A new manufacturing paradigm is emerging, moving away from some of the characteristics of design outlined in this chapter so far. The pressure to design as much as possible by modification and to minimise the number of prototype iterations may be reduced.

Maintaining the right balance

All companies have to operate within a number of constraints, but their influence on the product development process will vary from product to product, project to project, company to company, and with time. Often one factor is dominant. One year it might be quality, the next the environment, then safety and so on. In each case considerable effort will be directed towards addressing the dominant constraint without, where possible, compromising performance with regard to other constraints.

Design process improvement may be viewed as a multi-objective optimisation process in which a solution is sought that can satisfy all constraints as well as is possible (Figure 10). As with all multi-objective optimisation problems, it pays companies to find a robust solution that copes with the inevitable uncertainties of design processes, rather than finding a perfect solution that only satisfies a very narrow range of conditions.

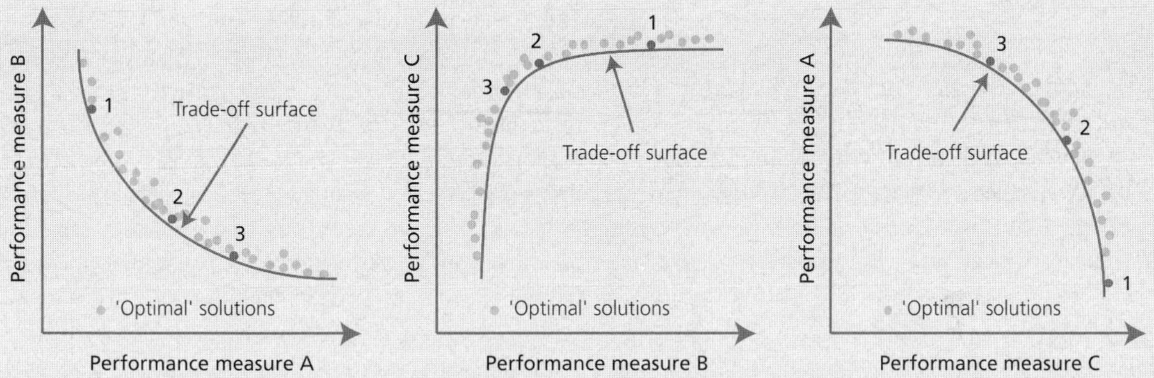

10 An example of the trade-offs possible for optimising multiple performance measures

In practice, companies often focus on improving a specific performance measure whilst trying to maintain a balance across a broad range of measures. For example, there may be a drive to improve reliability whilst maintaining existing delivery and cost targets. Such initiatives rely on defining suitable metrics to measure 'before' and 'after' performance, which is easier where such performance metrics are readily and objectively measurable.

This approach works less well for the intangible issues faced during the design processes, such as planning (see Chapter 2, Design planning and modelling) or communication (see Chapter 9, Communication in design). Problems in these areas are likely to have multiple causes and performance is hard to measure. For example, how well organisations communicate can ultimately only be measured by how successful they are in delivering the product on time.

Many process-improvement techniques have been developed for manufacturing processes, where performance is measurable and is often routinely measured. It is important that companies exercise caution when endeavouring to apply these techniques to design process improvement.

Influences on design processes

Most companies are subject to similar constraints. However, some cope with them better then others. When one looks at the design process beyond the abstract process models, the details are different in each company. The product is likely to be the major influence on the process. Although, even if the product is very similar, different teams within an organisation are likely to use different processes.

Design process improvement requires an understanding of how design processes work and what influences their behaviour.

Design process improvement requires an understanding of how design processes work and what influences their behaviour. This section looks at 'drivers', which are the major factors influencing the processes and 'characteristics', which are their observable properties (Figure 11).

A design process is a complex entity which can be looked at from two viewpoints: the actions that are carried out and the observable behaviour. Both, though unlikely to be a perfect match, can be viewed as a network with causal connections between the elements. One action leads to the next and no action is initiated unless motivated by another action or an external driver. As with all networks, if some parts are changed then others are affected. If some actions go wrong then others are affected. Hence, understanding a process means understanding these causal connections.

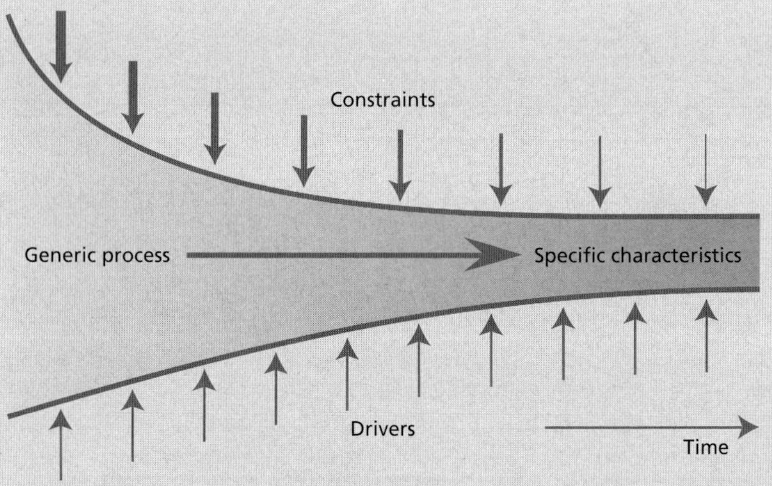

11 Constraints, drivers and their influence on shaping the specific characteristics of a design process

Drivers

The constraints described in Figure 8 are in turn major drivers of the design process. However, this list alone does not explain what can be observed about a particular process. For that it is necessary to explore the drivers and their relative influence on it. Such drivers arise from the generic properties of the product, the industry sector in which it is applied or properties of the organisation, rather than the individual process. For example, all products must be safe, but some industries, such as aerospace, healthcare and automotive, produce safety-critical products. Here safety-criticality becomes a major driver of the design process and all such processes share some common characteristics.

Table 12 shows some examples of drivers and the behaviour that may follow from them. For example, in a safety-critical product the designs are more likely to be developed as modifications to existing products in an effort to minimise the risk of product failure (see Chapter 11, Risk in the design process). Therefore, companies making such products are likely to put effort into understanding the implications of any changes and may handle these changes very carefully. Safety-critical products also require rigorous evaluation to demonstrate their 'fitness for purpose'. This will involve a range of review, verification and validation activities, typically leading to a longer design and development process. Such processes will need careful planning to maximise the effectiveness of the evaluation process, since the cost of evaluation has to be balanced against the benefit of finding errors early or, conversely, providing evidence of correct operation.

Drivers	Consequences
Artistic	→ designers might see themselves as artists → believe that design is not rational → unwilling / unable to explain their designs → not taught to solve problems → tacit and analogous problem solving → difficulties in solving numerical problems → designers unsatisfied when they have to work to tight constraints → technical subjects not understood / seen as inferior → visual appearance of the product very important → tension between visual and technical design / designers
Fashion dependent	→ designers need to study fashion context → design needs to be timely → designed to tight deadlines for shops → success of design depends on properties of other designs → designers are paid for their ability to understand the fashion context → designs are hard to justify → mood board culture → style and physical appearance of designs is important → design has to be looked at, worked on shortly before it can be launched → rework
Safety critical	→ designs are modifications of working designs → change needs to be understood and handled carefully → designs are tested rigorously and frequently → longer design process → design process needs careful planning → certification requirements → designers are more willing to employ analytical or formal methods
Certification	→ new design handled as modification to existing designs →modular approach preferred → innovation only when required → different countries specify different rules → versions of designs or single overspecified design

12 Examples of drivers and their consequences

The consequential behaviour that arises as a result of the drivers (Table 12) is not always evident, but it is likely to be. In addition, the consequences are not independent, and neither are the drivers. However, thinking of drivers and their consequences helps to see the causal connections between aspects of process behaviour and offers a potential explanation for problems. The direct consequences of drivers are difficult to change and design process improvement, if it is to be successful, must embrace and support them rather than fight against them.

Dominant drivers

The presence of particular process drivers contributes to the definition of the characteristics of an individual process. Generally, when there are multiple drivers it is difficult to assess which one causes which behaviour. However, usually it is possible to identify one driver as dominant and in a well managed process the drivers are often explicitly prioritised. Rankings may change in some companies over time, in others they will not. Consider, for example, the following cases taken from the aerospace, engineering and healthcare sectors.

GKN Westland Helicopters produce highly specialised naval and air-sea rescue helicopters. Their products include the EH101, which was originally developed for the UK military. Subsequent customers have each had their own specific needs, but, given the desire to procure a proven product, order a variation of the EH101. Hence, in this case customisation is the dominant driver. However, changes to any part of the helicopter can have knock-on effects across the entire craft (Clarkson *et al.*, 2001), and so to assure financial success the company must assess these changes before accepting any new contract and then execute them efficiently (Figure 13). Consequently, customisation is the core business and the core challenge for GKN Westland Helicopters.

The design of modern diesel engines is heavily influenced by environmental legislation (Jarratt *et al.*, 2003), with stringent targets for noise and emissions that are different for the USA, Europe and the developing world. This leaves companies with a choice: either they standardise their products to meet the toughest requirements and sell them to all their customers; or they diversify their range and produce a large number of different products to suit different markets. One such company, Perkins, has opted for diversification, developing new engines while still producing products based on 30 year old designs. Consequently, Perkins design and manufacture engines to meet orders ranging in size from one unit, to support existing customers, to tens of thousands of units. While they aim to meet the needs of their many customers

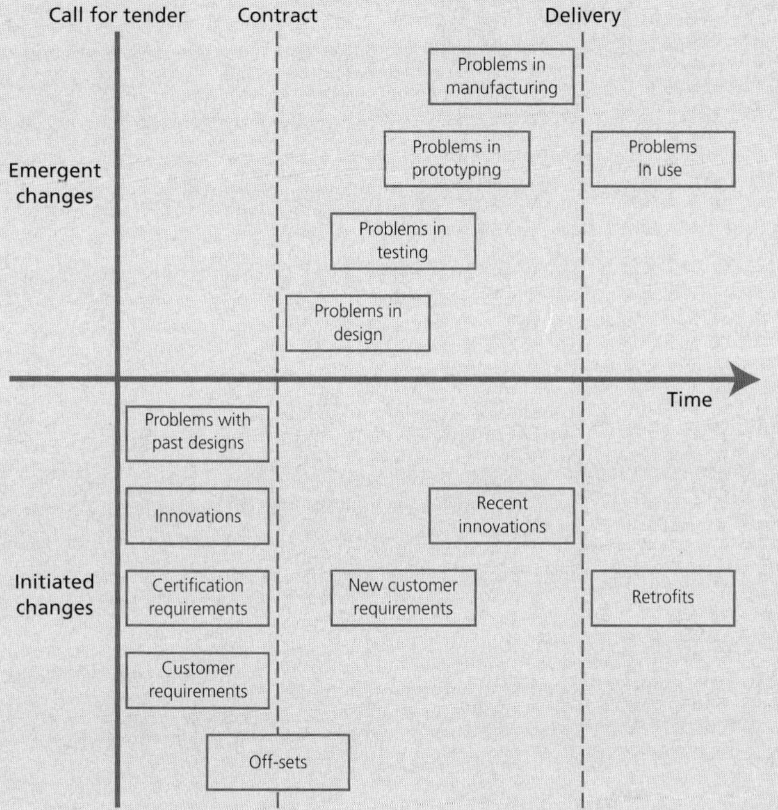

13 Sources of change in design leading to product customisation
© Springer-Verlag

through standard configurations, they also need to carry out a large number of small adaptations, resulting in thousands of different engines. Evidently, recent changes to legislation represent the core challenge for Perkins.

Medical equipment has traditionally been purchased on the basis of a minimum specified level of functionality and price. This has increasingly led to a proliferation of different equipment to provide the same basic functions in different hospitals or even on the same ward. Such diversity contributes, along with a number of other factors, to a significant level of adverse incidents associated with hospital care. A recent study has proposed that a shift towards products that are designed with enhanced levels of safety in mind could help reduce the incidence of such errors (Clarkson *et al.*, 2004). Such a change requires the adoption of a user-centred systems-based approach by equipment suppliers, purchasers and users. As a result, patient safety is emerging as a new driver for medical equipment design.

Characteristics

As a process unfolds it is possible to observe its behaviour which may exhibit certain characteristics, such as the iterations that occur, the way communication is conducted or the way in which teams are divided. Characteristics can be positive or negative. For example, the open exchange of information can be a positive characteristic of a process, whereas the misinterpretation of ambiguous information is likely to be a negative characteristic.

Process characteristics are not independent and there can be causal connections between them. For example, in processes with open exchange of information, misinterpretation of communication is rarer. However, processes nonetheless exist where both characteristics are seen to be present.

Patterns instead of models

Design processes are not deterministic. As a result, they are all different unless looked at in very abstract terms. However, common drivers can lead to similar characteristics in different processes. These similarities can be recognised as patterns of designing, describing elements of process behaviour.

Patterns of designing

Design processes are not deterministic. As a result, all design processes are different unless looked at in very abstract terms.

Patterns of designing describe aspects of a design process that may be shared with other processes (as illustrated in Figure 14). The 'star' represents all of design – a concept with indeterminate boundaries. Different colours indicate different types of pattern, for example, patterns in problem-solving or in communication. Each design process can be described with reference to the patterns that occur within it. These patterns of designing may also occur in other processes, with or without any overlap. For example, two processes might not share any patterns, yet still both may be valid. This makes it possible to describe very different processes using similar terminology.

Researchers have long been concerned with describing design by contrast to what is not design (Love, 1979). If they look at patterns of designing they no longer need to do this. Whether a pattern or an activity is considered to be "design" is neither here nor there – if it is relevant, it should be considered.

The characteristics that some instances of designing share with others come in clusters – different instances of designing can have a lot in common because they share powerful determinants of the form of the designing process. Clusters of consistently shared characteristics form the patterns. If we observe such a cluster, then we can hypothesise that the shared characteristics are linked by causal relationships, or are all symptoms of some as yet un-

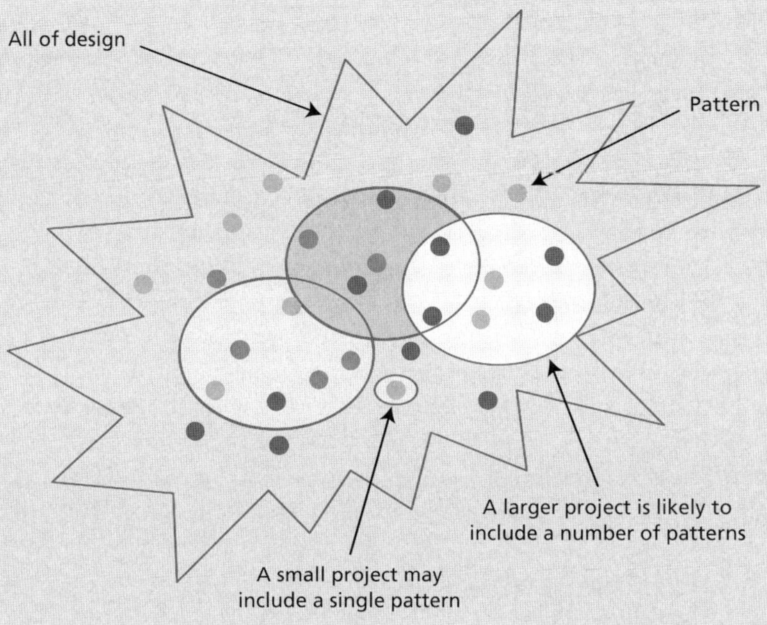

14 Patterns of designing

recognised underlying cause. We can test such a hypothesis in two ways: by looking at other design processes to see whether the presence of some of these attributes predicts the presence of other attributes; and by trying to construct and test theories of how the attributes are causally related.

Patterns of designing can be detected on all the levels at which it can be analysed and described – time, number of participants, the portion of the whole artefact being considered, and the activities that are the units of analysis. Moreover, given sufficiently rich observations of design processes, we can look for patterns comprising features at different levels of description. However, identifying similarities that can be represented as patterns of designing is not trivial, since this involves finding appropriate abstractions of observable phenomena.

Note the avoidance of the term design pattern; this refers to an abstractly formulated solution to a recurring problem, together with a description of the type of problem it fits and the consequences of using it. This is an idea introduced into architecture by Christopher Alexander (Alexander *et al.*, 1977) and widely adopted in software engineering (Gamma *et al.*, 1995). The same notion has long been implicit in engineering practice, and 'design by modification', as discussed earlier, is based on the use of solution patterns.

An example of a pattern

15 Large projects are carried out with large teams often in different locations and having different sets of experiences
Reproduced with the kind permission of Rolls-Royce plc

Patterns can exist and be described at any level of detail. For example, a 'high'-level pattern might relate to the size of the design team. This in itself can vary dramatically from one project to another. For example, some processes are solitary activities. Graphic design usually involves one individual interacting with clients, carrying out the design tasks and planning the process accordingly. In a similar way, engineers on a large project may work in relative isolation.

Other processes are carried out in small teams of collocated people, as in textile companies or small engineering firms. In such cases everybody is to some extent aware of the others' skills, tasks and needs, and communication is influenced by understanding of people as much as understanding the tasks. Small teams typically have a manager, who directs the team but negotiates tasks with individuals.

Large projects, be they in software, construction or large-scale engineering, are carried out with large teams often in different locations and having different sets of experiences. Large teams need to be carefully managed to facilitate the integration of tasks since the interface between the sub-teams is often one of the main sources of failure and uncertainty in the design process. The measures that companies can take to assure the smooth running of the process are similar in all large projects.

Other patterns can be much more detailed and specific. For example, consider a company of 40 people collocated in attractive premises. Most people were well informed, via a personal network of informal conversations, about the activities of others. However, everybody resented not being provided with information officially. They had to "find out for themselves". Management was aware of poor communication and had as a result become very cautious about releasing information officially, preferring formal communication through the heads of groups. The result was unmotivated staff and resentment between subgroups. This scenario illustrates an anecdotal connection (pattern) relating bad management to communication problems.

A pattern is not just any 'chunk' of a process, but something that is recognised as a recurring event and hence has a meaning beyond the single instance.

Patterns as causal stories

A pattern is not just any 'chunk' of a process, but something that is recognised as a recurring event and hence has a meaning beyond the single instance. It must also have a persuasive causal structure that links observed behaviour. Whilst it is never possible to fully explain every causal link in a complex context like design, what is offered must add insight and be persuasive. A pragmatic approach is to identify patterns that are useful and help practitioners. In this way it should be possible to build up a library of patterns, each consisting of a brief description, an illustrative story and a formal description.

Patterns can exist and be described at any level of detail.

The consistent presence (or absence) of a group of characteristics in a range of design processes will not be persuasive as a pattern of designing until there is a causal explanation of the way in whih these characteristics are related. Thus, the next step, given a possible pattern, is to hypothesise one or more plausible explanations as to how the characteristics share common causes, or are linked by a chain of causes and effects. Instances of designing can then be scrutinised for supporting (or conflicting) evidence. Hypothesising causal relationships helps to formulate more focused questions about what is really happening in an episode of designing.

The function of models and patterns

Traditional models of design processes try to provide an abstract description of general design processes at varying levels of detail. Professional designers understand such processes and their corresponding activities. They provide designers and design managers with a common vocabulary for a generic design process and their own specific process. At the same time they provide reminders of what should be accomplished at or before certain points in any process. For example, the Pahl and Beitz (1995) model has a list of generic tasks associated with each design phase. The more detailed these models are, the more they serve as a checklist for the design process. Specific process models are often assembled using the conventions and terminology of the generic models. A further function of the generic models is to aid visualisation of the design process. However, such models are likely, in practice, to provide few insights into the process.

In contrast, the primary function of patterns of designing is to provide insights into the actual process by supplying potential causal explanations for observed phenomena. Patterns can help companies to see the root of their problems (or successes) by linking behaviours to drivers. Once the root cause has been identified, it is often fairly straightforward to change (or maintain) a process to ensure future success. By finding the same pattern in another process, companies can also investigate how similar problems (or successes) have been dealt with (or achieved) in the past.

Traditional models of design processes try to provide an abstract description of general design processes at varying levels of detail.

In practice, designers and design managers often try to cope with their problems by talking to other people, or by gaining insight from books and magazines describing similar situations. While companies can compare their products with those of their competitors, they generally have very little access to the process experiences of others except through anecdotal evidence or via the transfer of staff between organisations. Patterns are a way to give them a sense of what is likely to occur in a process and what aspects of this they can change.

Successful processes

The aim of any attempt at design process improvement is to make a process more effective and efficient in order to ensure that a sufficiently good product will be developed on time and on budget. However, it is not easy to define a successful process in isolation from all those to which it is connected (for example, processes in other projects or within supplier companies).

A simple design project?

Academic literature on design often focuses on individual design processes, *i.e.* single projects isolated from other projects. These can occur in some special circumstances, but are unusual. One exception can be found in design consultancies where designers work on one project after another, often with little connection between them. They put a team together for each project, depending on the technical and managerial skills required, plan the design process and deliver this solution to the best of their abilities. Their skills lie in generic problem solving and designing to meet the needs of their customers.

Small engineering companies, too, occasionally design individual products for niche markets in relative isolation from other products or design processes. However, such examples of individual design processes are rare. Design is far more likely to take place as part of a complex web of related products and design processes. A successful process will then be one that builds on past experience and takes appropriate account of other co-existing ones.

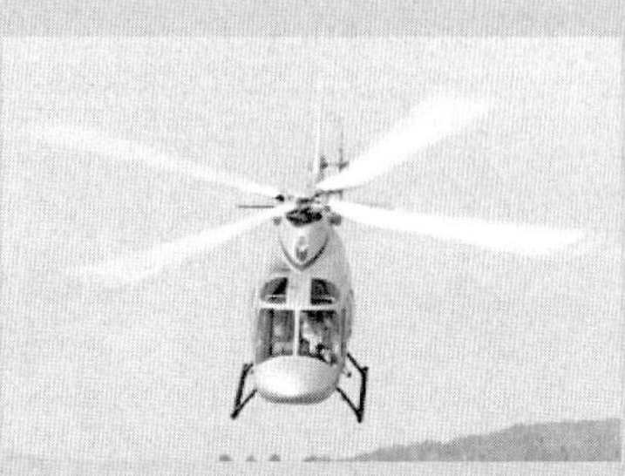

16 A successful process will be one that builds on past experience
© AgustaWestland

Multiple intertwined processes

Most engineering design processes are intertwined with other engineering processes, embedded in other business processes in the organisation, and linked to a number of supplier companies' processes. For example, Figure 17 illustrates some of the processes that have been observed in a world-leading supplier of diesel engines.

At any one point in time several different, but related, engines are being designed, requiring continuous management of resources and prioritising of activities. At the same time, the company has to implement hundreds of small

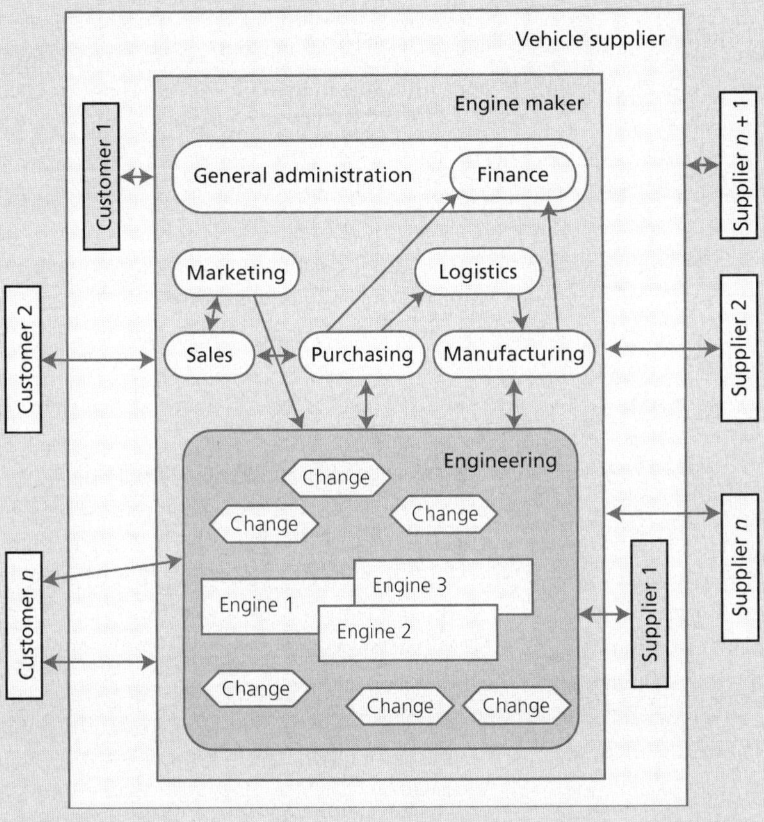

17 An example of intertwined processes

changes to engines to meet new requirements. On each project, engineers are working with many other departments in the organisation.

On the basis of past sales figures, the marketing department works out likely sales of the new products and provides high-level specifications. The sales group establishes how much can be charged for the product and looks for customers. Using sale predictions, the purchasing departments work in close co-operation with engineering to procure parts and subsystems from suppliers. Logistics are responsible for assuring the smooth flow of these parts through the company in support of manufacturing.

Production experts work with designers to understand how the new product can be made, ideally an on-going collaboration throughout the design process. Cost engineers evaluate the new products to see how they could be modified to be cheaper. All these functions are supported by further administrative functions which include finance, personnel, warehousing, etc.

18 Part of a bigger picture

The engine maker is a subsidiary of a vehicle supplier and shares some activities with its parent company. The vehicle supplier is moving towards a global purchasing strategy, so that it selects suppliers for key components such as fuel pumps for all its engine-making companies. Over the years, the vehicle supplier has attempted to concentrate research in the USA, but the engine supplier has regained some of its core research. Staff are moved to and from the USA and project teams can span two continents.

The vehicle supplier is both a supplier and a customer to the engine maker, whereas the latter has many other suppliers for 'off-the-shelf' and specially made components. For some suppliers the engine maker is a large customer, who can influence them significantly. For others, who are key suppliers to the automotive industry, the engine maker is a small customer with leverage.

The engine maker has customers of all sizes. Some buy thousands of engines, while others, like makers of specialised vineyard diggers, buy only a few engines a year. Every customer and every supplier relationship ties the engine maker's design processes to other design processes, putting time, resource and cost constraints on them. Most of these interactions are routine and do not cause problems. However, they influence design decisions and impact upon the planning of the design process.

All these intertwined processes are quite different in character and require different levels of support. For example, logistics tend to use a standard process that they follow under all circumstances, and hence they refine their process until it meets as many situations as possible. This is typical of repeatable processes with known, predictable tasks that may be managed using workflow techniques (see Chapter 15, Workflow for design) and requirements engineering research (see Chapter 4, Requirements engineering). However, design processes are quite different, being less certain in their outcome.

Errors in design

Design processes are determined by multiple intertwined factors ranging from the characteristics of the product to the capabilities of the organisation. This is particularly evident when design goes catastrophically wrong, resulting in tragic loss of life or the financial collapse of a company.

Failure is often due to a number of problems occurring in the same design.

Ultimate failure is often due to a number of problems occurring in the same design, so that safety margins are eroded and vital warning signs are missed. A famous example is the series of crashes of the de Havilland Comet airliner (Walker and Henderson, 1999). Originally conceived during World

War II, the Comet was developed as a showcase of British engineering in the austere post-war years with an ambitious target – the Comet was jet propelled, generating massively greater thrust than a piston-engined aircraft. It also needed to carry more fuel. Hence, strength and lightness were all important and it was decided to build the fuselage from thin aluminium sheets. A cruising altitude of 40 000 feet was required for efficient operation of the engines, at which level the cabin was pressurised to a pressure equivalent to an altitude of 8 000 feet. This resulted in a pressure differential of 56 kPa across the fuselage wall – twice the value previously used.

De Havilland had earlier developed a special manufacturing process to glue the windows in place with this kind of fuselage. However, for the Comet a riveting process was chosen, because tooling for gluing was too difficult for the square windows. Although the manufacturer had conducted many tests on parts of the aircraft, they did not have the time to conduct sufficient pressure tests on a complete prototype. Three aircraft crashed within the first year of service. Analysis of the wreckage pointed to fatigue-induced cracks in the fuselage as the cause of the accidents, resulting from high stress concentrations at the corners of the square windows. When an improved Comet was launched four years later, the American competition had caught up and Boeing dominated the world market for decades to come.

The Comet crashes were ultimately the result of structural failure following from evaluation failure. This, in turn, was evidently the result of a failure of project management (Table 19).

Design process should be tailored to the product under development, the competence of the design team and the aspirations of the users.

Type	Causes
Structural failure	A mismatch between structural and/or material properties and the system (functional) requirements, for example, cracking, corrosion, creep, melting etc., can lead to failure. Such problems are aggravated in innovative designs, when not all the technical properties of the emerging design can be known.
Evaluation failure	Testing can fail to pick up potential failures when the future use cases are inadequately understood or the test set-up is incorrectly designed.
Interaction failure	Mistakes in human/system interaction can occur both during design, for example, unexpected changes or flawed decision making, and during use, for example, potentially dangerous sequences of actions or poor inspection procedures.
Management failure	Inadequate management of the design process can lead to late design changes, inappropriate expectations of the design team, and the inability to keep up with technical innovation.
Marketing failure	Inadequate assessment of market needs can lead to designs that are 'ahead of their time' or good ideas that fail to meet the user's needs through poor execution.

19 Potential causes for dramatic design failure (adapted with permission – James, 2004)

What makes a process successful?

When things go wrong it is usually possible to point to the factors that have contributed to the failure. However, since there are many factors that influence the outcome of a design process, it is much harder to describe what makes a process successful. 'Success' factors are often interrelated, and little effort is typically expended to explore the causal links between them and performance measures for a project. Most companies are reluctant to experiment with their design process to gather such data, and the majority make no attempt to learn from their successes. Postmortems are usually reserved for the projects that fail.

Success or failure is clearly linked to the abilities of the design team, their understanding of the users' needs for a product, the architecture of the product, and the process by which it is realised. In addition, the interplay between these four factors is critically important. For example, the design process should be tailored to the product under development, the competence of the design team and the aspirations of the users (Figure 20). Knowledge of the product, process and users is also vital if the design team is to learn from past successes and failures.

Success in design is also more likely if designers and design managers are aware of: *models of design* that can be used to describe the design process; *perspectives on design* from which to view the design process; good *design practice* to improve product and process performance; and a range of approaches to ensure effective and efficient *design management*. The chapters that follow, comprising the first part of this book, address these issues in more detail.

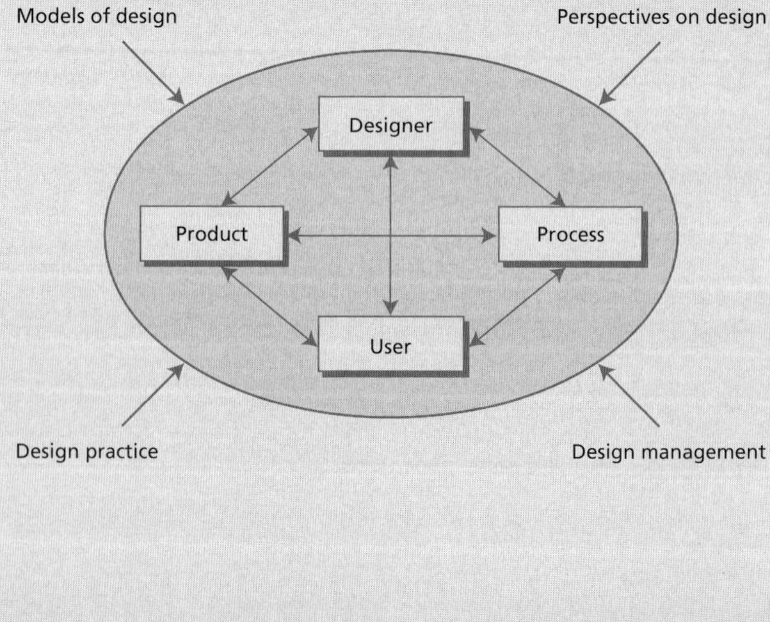

20 The interplay between product, process, designers and users

Conclusions

Over the past 20 years design research has focused on methods for delivering more efficient processes and support tools for design. The German methodological tradition, exemplified by Hubka (1982), Ehrlenspiel (1995) and Pahl and Beitz (1995), has had enormous influence on design practice. It presents design as a rational process that goes through a number of stages, and informs designers what they need to have achieved by the end of each stage. This has led to the use of gateways and checklists, which are a major means of managing design processes. The majority of engineers and designers are exposed to these models at university, which in turn leads to them being taken up more widely in industry.

The current challenge to the design research community is to provide designers with a wider range of methods and tools to support specific activities within the design process and to improve its overall co-ordination. However, while some excellent progress has been made in this respect, a common theoretical understanding of design remains elusive and there are a number of fundamental questions that remain unresolved (see Table 21 for some examples). Future progress will depend upon the combined research efforts of industry and academia, focused on a mix of immediate practical challenges and longer term research to improve our understanding of fundamental design processes.

Potential areas for future design process research
• Learning from success and failure • Handling complexity in design • Defining new paradigms for process planning and management • Transferring practice between design domains • Understanding the influence of social, psychological and organisational issues • Training of engineers for the realities of industrial practice • Managing product, process and rationale information through the whole life-cycle • Providing practical tools for multi-objective, multi-criteria optimisation • Understanding the relationship between products and processes • Integrating product, process and user knowledge • Modelling hierarchical and dynamic processes and products • Integrating information and knowledge capture into the design environment • Harnessing the power of artificial intelligence in design

21 Some challenges for future design research

References

Alexander C, Ishikawa S, Silverstein M *et al.* (1977) A pattern language: towns, buildings, construction. Oxford University Press

Ashby MF (1999) Materials selection in mechanical design. Butterworth-Heinemann

Clarkson PJ, Simons CS, Eckert CM (2001) Predicting change propagation in complex design. ASME DETC'01, Pittsburgh, PA, USA

Clarkson PJ, Buckle P, Coleman R *et al.* (2004) Design for patient safety: a review of the effectiveness of design in the UK Health Service. Journal of Engineering Design, 15(2): 123–140

DTI (1991) Managing product creation – a management overview

Eckert CM, Stacey MK (2001) Designing in the context of fashion – designing the fashion context. DTRS'01, Delft, The Netherlands

Eckert CM, Blackwell A, Bucciarelli L *et al.* (2004) What designers think we need to know about their processes: early results from a comparative study. Design 2004, Dubrovnik, Croatia

Ehrlenspiel K (1995) Integrierte Produktentwicklung: Methoden für die Prozessorganisation, Produkterstellung und Konstruktion. Hanser

Gamma E, Helm R, Johnson R, Vlissides J (1995) Design patterns. Addison Wesley

Hubka V (1982) Principles of engineering design. Butterworth

James MN (2004) Failure as a design criterion. http://www.plymouth.ac.uk/si. Reproduced with permission of MN James

Jarratt TAW, Eckert CM, Clarkson PJ (2003) Environmental legislation as a driver of design. ICED'03, Stockholm, Sweden

Jarrett P (1987) Another Icarus: Percy Pilcher and the quest for flight. Smithsonian Institution Press

Jarrett P (2001) Percy Pilcher and the challenge of flight. NMS Publishing

Love SF (1979) Planning and creating successful engineering designs. Van Nostrand

Oxman RE (1990) Prior knowledge in design: a dynamic knowledge-based model of design and creativity. Design Studies, 11(1): 17–28

Pahl G, Beitz W (1995) Engineering design: a systematic approach. Springer

Purcell AT, Gero JS (1996) Design and other types of fixation. Design Studies, 17(4): 363–383

Schön DA (1988) Designing: rules, types and words. Design Studies, 9(3): 181–190

Ulrich KT, Eppinger SD (2003) Product design and development. McGraw-Hill
Walker T, Henderson S (1999) The first jet airliner: the story of the de Havilland Comet. Scoval
Ward JR, Clarkson PJ (2004) An analysis of medical device-related errors: prevalence and possible solutions. Journal of Medical Engineering and Technology, 28(1): 2–21

Design issues

In the first part of this book each chapter stands on its own as the authors' view of the particular topic, complete with its own set of references and examples - yet they are ordered to allow for those who wish to read more widely. In particular, the early chapters on models of design provide a useful introduction to the later chapters on perspectives on design, design practice and design management.

- **Models of design**
  - Models of designing
  - Design planning and modelling
  - Systems engineering
- **Perspectives on design**
  - Requirements engineering
  - Human resources
  - Artificial intelligence for design process improvement
  - Complexity
  - Thinking and representing in design
- **Design practice**
  - Communication in design
  - Engineering change
  - Risk in the design process
  - Design for X
- **Design management**
  - Engineering knowledge management
  - Quality management
  - Workflow for design
  - Integrated new product development
  - Product portfolio management
  - The transfer of methods into industry

Models of design

Chapter 1

Models of designing

David Wynn and John Clarkson
University of Cambridge

Many authors have proposed theories, models and methods in their search to explain or improve upon aspects of design practice. This field of literature, commonly known as design methodology, is primarily concerned with:

> *...the study of how designers work and think; the establishment of appropriate structures for the design process; the development and application of new design methods, techniques and procedures; and reflection on the nature and extent of design knowledge and its application to design problems.*
> (Cross, 1984)

Despite the extensive research undertaken since the 1950s, there is no single model which is agreed to provide a satisfactory description of the design process (Bahrami and Dagli, 1993). Likewise, there is no 'silver bullet' method which can be universally applied to achieve process improvement. Instead, most methods have a well-defined and often relatively narrow focus, ranging from the generation of mechanism concepts (e.g. Pahl and Beitz, 1996) through to the management of project risk (e.g. Baxter, 1995). Even so, implementation and use of such methods is often problematic, as experienced by Bucciarelli (1996).

There is no 'silver bullet' method which can be universally applied to achieve process improvement.

In this chapter, some popular approaches to the design process are presented and their practical relevance is discussed. Throughout the chapter, a classification framework is developed to support the discussion and to relate the diverse range of forms exhibited by these models.

Classifying models of designing

Design is well known as an ill-structured and pernicious problem (Rittel and Webber, 1984); it is difficult to describe the design process satisfactorily, and it is an equally challenging task to describe the relationships between models concerned with its various aspects. Many classification schemes have been used to frame discussions of such literature, including those of discipline, nationality of origin, and the historical development of form. Reflecting many other aspects of design research, however, such frameworks seem as diverse and difficult to relate as the models they describe.

The following sections briefly discuss three classification schemes which we believe are useful in highlighting issues of practical relevance. These schemes are the interrelated dimensions of: stage vs activity-based models; problem vs solution-oriented literature; and abstract vs analytical vs procedural approaches. Preceding the discussion of the models themselves, these schemes are described in greater depth in the following sections.

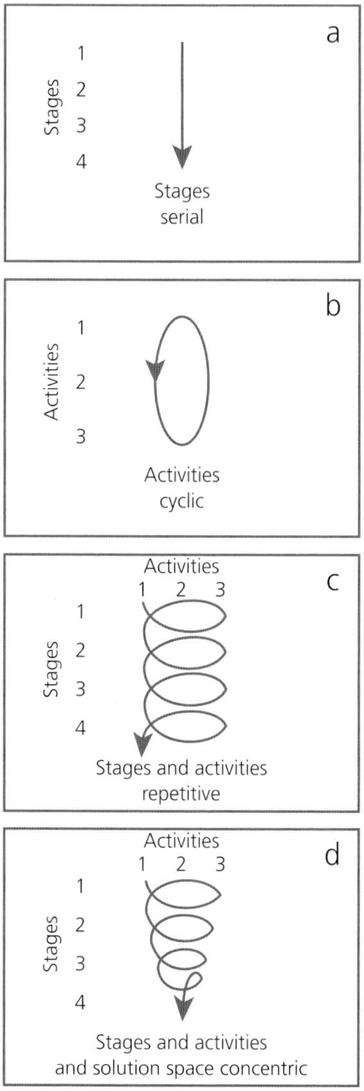

1.1 A typology of design models (reproduced with permission of Blessing, 1994)

Stage-based vs activity-based models

According to Blessing (1994), models of designing may be classified using the four categories shown in Figure 1.1. This framework is based on the earlier theorising of Hall (1962), who proposed a two-dimensional perspective of project development in which the phase-based structure of the project lifecycle lies orthogonal to the iterative problem-solving process which takes place within every phase. Asimow (1962) further developed this theory, transferring Hall's ideas from the domain of systems engineering to that of design. Asimow described the essentially linear, stage-based chronological structure of the project as the morphological dimension of the design process, and the highly cyclical, rework-intensive activities characteristic of the designer's day-to-day activities as the problem-solving dimension.

Blessing refers to those models concerned with Asimow's morphological and problem-solving dimensions as stage- and activity-based respectively (Figure 1.1a and b). She also notes the existence of combined models (Figure 1.1c) which prescribe well-structured, iterative activities within each stage (e.g. Hubka, 1982); by comparison, purely stage-based models indicate only the possibility of rework using feedback loops between stages (e.g. French, 1999). Some combined models illustrate convergence on a design solution (Figure 1.1d) by using progressively more concrete activities in each stage (e.g. Evans, 1959). It will be seen that models with a stage-based component are more useful in practice than their purely activity-based counterparts.

Solution-oriented vs problem-oriented literature

Another commonly used scheme places literature into either of the following two categories, according to the strategy the author proposes is used to reach the design goal (e.g. Lawson, 1980; Birmingham *et al.*, 1997):

- *Solution-oriented*, in which an initial solution is proposed, analysed and then repeatedly modified as the design space and requirements are explored together
- *Problem-oriented*, in which the emphasis is placed upon abstraction and thorough analysis of the problem structure before generating a range of possible solutions.

Observing graduate students of architecture and science asked to solve a simple problem, Lawson (1980) concluded that the strategy chosen in practice is determined by training and background; designers preferred the more creative 'try it and see' solution-oriented approach, while the scientifically trained focused on unravelling the problem before attempting to synthesise solutions. Lawson went on to describe the interlinked and subjective nature of problem specifications and design solutions, a persuasive argument supported by many other authors (e.g. Jones, 1970; Cross, 1994), and concluded that real design problems cannot be solved in a purely problem-oriented fashion. In fact, it is generally recognised that completing a design requires application of both of these strategies at one point or another, according to the individual nature of each problem the designer encounters (Frost, 1992). In conclusion, it may be seen that stage-based models typically adopt a problem-oriented strategy, whereas activity-based models may be either problem- or solution-oriented in nature.

Abstract vs procedural vs analytical approaches

The focus of this chapter is on the relevance or applicability of literature to the problem of improving the effectiveness of a design project. With this in mind, a third set of categories is proposed here to form the framework used in this chapter (Figure 1.2):

- *Abstract approaches*, which are proposed to describe the design process at a high level of abstraction. Such literature is often relevant to a broad range of situations, but does not offer specific guidance useful for process improvement.
- *Procedural approaches*, which are more concrete in nature and focused on a specific aspect of the design project. They are less general than abstract approaches, but more relevant to practical situations.
- *Analytical approaches*, which are used to describe particular instances of design projects. Such approaches consist of two parts: a representation used to describe aspects of a design project, such as the design structure matrix or DSM (Steward, 1981); and techniques, procedures or computer tools, which make use of the representation to understand better or improve the process of design; see Browning (2001).

To relate this new typology to the previous discussion, abstract models are usually activity-based in nature – although this is by no means clear in many cases – and thus may adopt either a problem- or solution-oriented strategy. Conversely, procedural models are problem-oriented in nature and always contain a stage-based component. Analytical models of designing are fundamentally different and are described in Chapter 2; the remainder of this chapter further develops the framework depicted in Figure 1.2 via a more detailed discussion of abstract and procedural approaches.

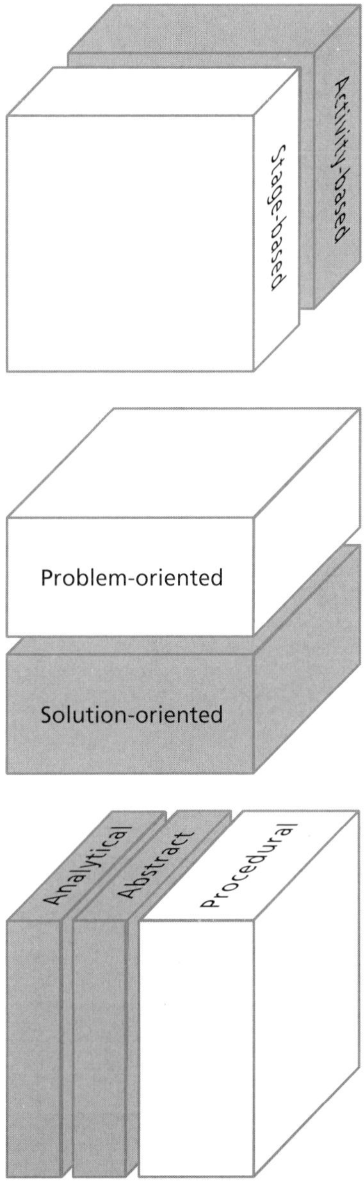

1.2 A classification of design process literature

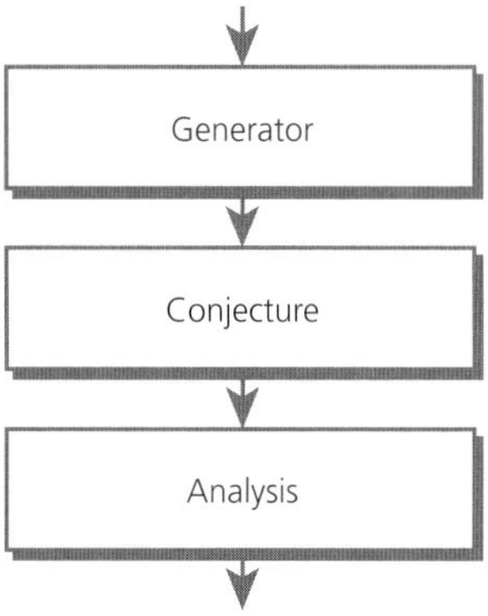

1.3 Darke's model of the problem-solving process in architectural design
Reprinted with permission from Elsevier

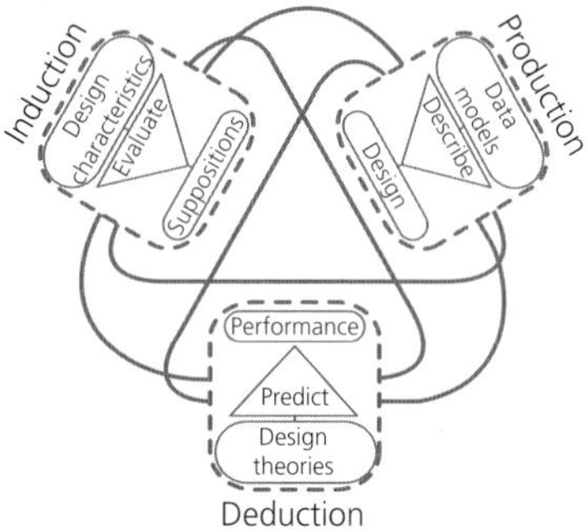

1.4 March's PDI model of reasoning in design (March, 1984)

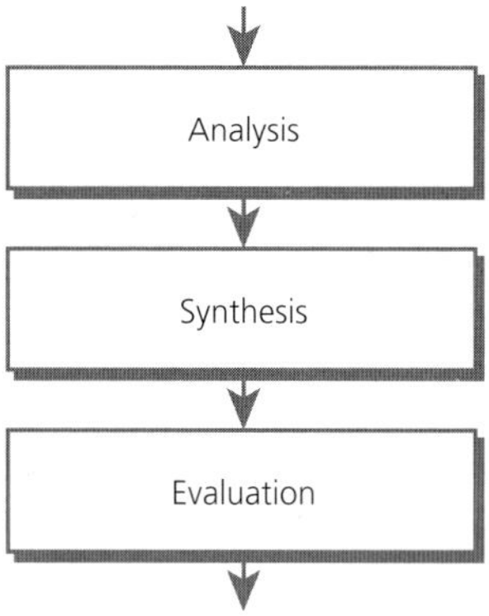

1.5 Jones' model of the design process

Abstract approaches

A commonly held theory of designing is that designers can and should resist bringing their own preconceptions to bear on a problem. In opposition to this problem-oriented perspective, Hillier *et al.* (1972) proposed the conjecture-analysis theory to reflect the belief that a designer would pre-structure a problem in order to solve it; that is, that existing knowledge and previous experiences would be used to influence the nature of the solution. This concept forms the basis of the solution-oriented models of design, which are usually considered to be more realistic descriptions of the designer's thought process than their problem-oriented counterparts.

One example of a solution-oriented model is that given by Darke (1979) following observations of architectural design practice (Figure 1.3). Darke argues that the designer does not start by studying an explicit list of problem factors and objectives to be met by the design, but rather tries to reduce the set of possible solutions to a smaller class which is more manageable. To achieve this, a subset of the objectives is chosen, based on prior experience of similar problems and subjective judgement. Darke terms this subset the primary generator, consideration of which leads to a possible solution or conjecture being produced. This enables further clarification of the design requirements, against which the solution is tested and further improvements are made.

March (1984) proposed a particularly interesting solution-oriented model of reasoning in design, termed the production–deduction–induction (PDI) model. Drawing on the philosophy of Peirce (1923), March argued that the two conventionally understood forms of reasoning, *i.e.* deduction and induction, are only able to describe the evaluative and analytical aspects of design respectively. He proposed that Peirce's third type of reasoning, termed abductive or productive, is responsible for the essential creative activities. From this he developed the triple activity model shown in Figure 1.4.

In the first phase, of productive reasoning, the designer draws on the vague problem statement and his or her existing knowledge to conceive a candidate solution. In the second phase, deduction, based on understanding of key physical principles, is used to analyse or predict the system behaviour. In the third phase, inductive reasoning is used to identify possible means of improving performance by altering certain aspects of the design, leading to the production of a better solution. In common with other solution-oriented models, the highly cyclic nature of design is given primary emphasis.

In contrast, the problem-oriented models are essentially linear, as typified by the description given by Jones (1963), in which the design process comprises the three stages of analysis, synthesis and evaluation (Figure 1.5). The initial analysis stage involves consideration of the problem and its structuring into a set of objectives. Synthesis involves the generation of a range of solutions, and evaluation involves the critical appraisal of the solutions against the objectives. A similar model is proposed by Ehrlenspiel (1995) for the domain of problem solving in systems analysis, illustrating the divergence of the design space as solutions are generated, and convergence during evaluation and selection of concepts (Figure 1.6).

Cross (1994) proposed a four-stage variant (Figure 1.7), in which the designer first explores the ill-defined problem space before generating a concept solution. This is then evaluated against the goals, constraints and criteria of the design brief. The final step is to communicate the design specification either for manufacture or integration into a more complex product. Since generation does not always result in a satisfactory solution, Cross includes a feedback loop between the evaluation and generation stages.

These problem-oriented models are based on the premise that the engineer is capable of formulating a solution-neutral problem statement, and propose that the final design should be more dependent upon logical deduction than prior experience. This assumption, common to all problem-oriented literature, forms the basis of the procedural design models introduced below.

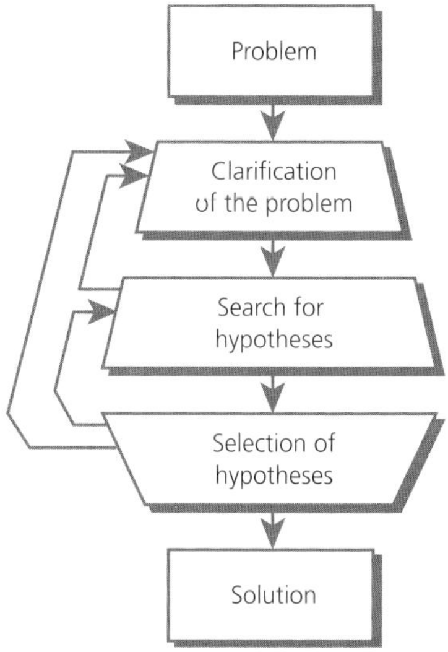

1.6 The procedural cycle for systems analysis (Reproduced with permission – Ehrlenspiel, 1995)

Discussion

All the theories and models discussed above provide high-level, generic descriptions of design practice; as such, abstract approaches do not explain the process of designing in detail. They are characterised by a small number of stages or activities and do not describe the specific steps or techniques which might be used to reach a solution. The practical applicability of such approaches is described rather colourfully by Lawson as:

> *...about as much help in navigating a designer through his task as a diagram showing how to walk would be to a one year old child...*
>
> *...Knowing that design consists of analysis, synthesis and evaluation will no more enable you to design than knowing the movements of breaststroke will prevent you from sinking in a swimming pool.*
>
> (Lawson, 1980)

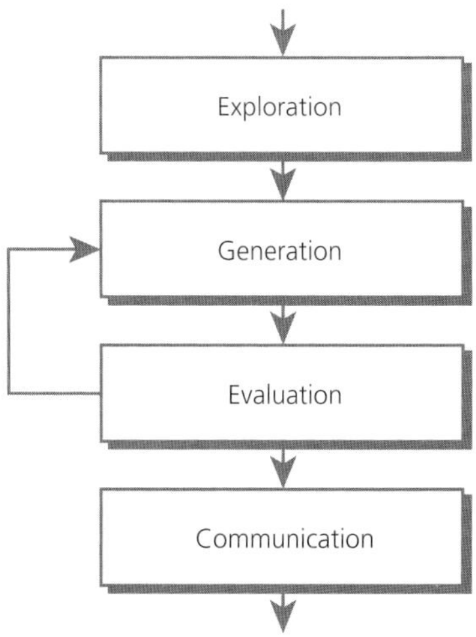

1.7 Cross's model of the design process (Cross, 1994)

1.8 A classification of procedural approaches to designing

Procedural approaches

Procedural approaches are more concrete in nature than the abstract theories and models discussed above, typically incorporating a larger number of phases and focusing on a specific audience and/or industry sector. Such literature is commonly categorised as follows (Finger and Dixon, 1989):

- *Descriptive*, resulting from investigation into actual design practice. Processes and procedures observed in industry form the basis of texts which are used primarily for teaching, training and research purposes. The abstract theories and models introduced in the previous section are descriptive in nature.
- *Prescriptive*, which are distillations of best practice intended to improve effectiveness or efficiency in some aspect of the design project. Such procedures are usually targeted towards a particular audience (for example, student, design engineer or manager) and domain (for example, industrial or mechanical design).

Prescriptive approaches recommend or prescribe guidelines, stages or techniques which, if implemented correctly, are thought to improve performance in specific aspects of the product or project. To illustrate, a procedure may be intended to improve product reliability, or to improve visibility of the design process to its participants.

Hubka (1982) expresses a commonly held view by recommending such procedures when searching for solution concepts in order to cover a wider search space, and also suggests that following a systematic approach can be particularly beneficial in all review and revision activities.

Archer (1965) proposes that systematic approaches are particularly useful under one or more of three conditions: when the consequences of being wrong are grave; when the probability of being wrong is high (for example, due to lack of prior experience); and/or when the problem is complex, characterised by many interacting variables.

Although widespread, the classification of approaches as descriptive or prescriptive is of limited practical use, since both descriptive and prescriptive aspects may be found in most literature. To illustrate, consider the following distinction of scope in procedural approaches:

- *Models*, which refer to a description or prescription of the morphological form of the design process.
- *Methods*, which prescribe systematic procedures to support the stages within a model.

Most procedural literature combines collections of prescriptive methods, such as brain-storming, synectics, functional analysis or morphological combination, with a problem-oriented model to illustrate the context of each method. Furthermore, it will be seen in the following section that models and methods are often intertwined, with the stages of each model being dependent upon the methods from which it is composed. The pernicious classification problem of prescriptive vs descriptive and model vs method will not be discussed further in this chapter; instead, approaches are classified by focus, with literature falling between the following two extremes:

- *Design-focused*, which supports the generation of better products by the application of prescriptive models and methods to the design process (e.g. Pahl and Beitz, 1996).
- *Project-focused*, which advocates approaches to support or improve management of the design project, project portfolio or company (e.g. Hales, 2004).

A summary of the emerging classification for procedural approaches is given in Figure 1.8, which shows the dimensions of descriptive vs prescriptive, design-focused vs project-focused and model vs method. The following sections introduce some well-known procedural models and methods, beginning with models focused on the process of design.

Design-focused literature

Most procedural models present design as a series of stages, each of which is visited only once by the ideal process. A different perspective is offered by Evans (1959), who proposes a combined stage and activity model concentrating on the iterative nature of the design process. Noting that one of the most fundamental problems of design lies in making trade-offs between many interdependent factors and variables, Evans' model argues that design cannot be achieved by following a linear process. He demonstrates this using the example of bridge design, where the structure must be chosen to support the dead weight of the material, but the weight is not known until the structure has been defined. According to Evans, such interdependencies are characteristic of design, a view later supported by Eppinger *et al.* (1994) and many others.

To solve these issues, an iterative procedure is adopted; early estimates are made and repeatedly refined as the design progresses, until such time as the mutually dependent variables are in accord. Based upon this principle, Evans proposes a prescriptive model for ship design, shown in Figure 1.9. The radial lines show the aspects of the ship – the interdependent variables – which must be chosen for the design to be complete. As the project progresses,

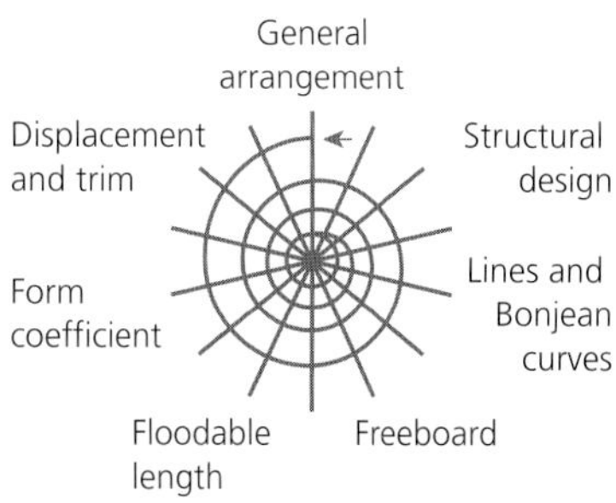

1.9 The design spiral (adapted from Evans, 1959)
Reprinted by permission from the American Society of Naval Engineers

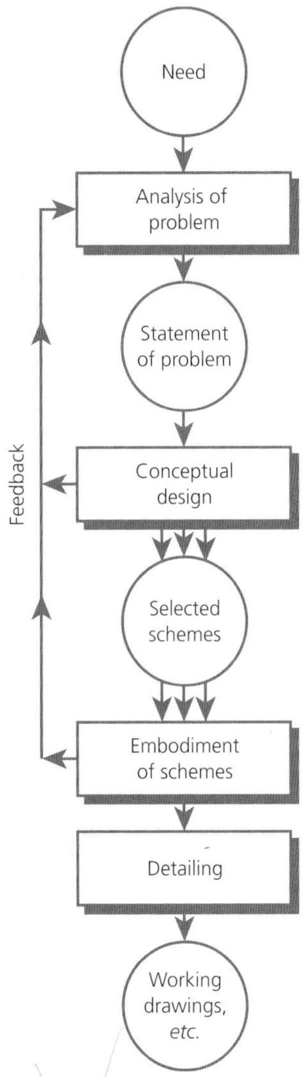

1.10 The design process (Reproduced from French, 1999)
© Springer-Verlag

these variables are gradually refined by repeated attention until the ultimate, balanced solution is reached. At each iteration the manoeuvring room for each variable decreases as the interdependencies are gradually resolved, smaller modifications are required, and different methods may be applied to each problem. Evans notes that the effort required to improve the design increases as the solution converges, and that more and more resources may be applied as the project moves towards completion.

A typical example of the more common stage-based models was proposed by French. The model, shown in Figure 1.10, is based on design practice observed in industry. It consists of four stages (French, 1999):

- The process begins with the observation of a market need, which is then analysed, leading to an unambiguous problem statement. This takes the form of a list of requirements which the product must fulfil.
- During the conceptual design phase several concepts are generated, each representing a set of physical principles for solving the problem. These schemes are transformed into a more concrete representation to allow assessment and comparison. The resulting concepts are evaluated and one or more are chosen to form the basis of the final solution.
- The chosen architecture is then solidified in the embodiment phase, where the abstract concept is transformed into a definitive layout.
- Finally, the remaining details are added to remove all ambiguity from the solution, allowing the release of instructions for manufacture.

French accounts for the non-linear activity sequences highlighted by Evans by describing his model as hierarchical in nature; in other words, a project may encompass several stages of the model according to the varying completeness of each aspect of the design.

Perhaps the most well known of the stage-based models was proposed by Pahl and Beitz (1996) for mechanical design (Figure 1.11). Each of the four prescribed phases consists of a list of working steps which they consider to be the most useful strategic guidelines for design. They propose that following their prescribed steps ensures nothing essential is overlooked, leading to more accurate scheduling and resulting in design solutions which may be more easily reused.

Although many more design-focused models may be found in the literature, Cross and Roozenburg (1992) describe how most have converged upon the general form proposed by authors such as Pahl and Beitz and French. Other examples may be found in the work of Dym and Little (2000), Ullman (2003), Pugh (1991) and Roozenburg and Eekels (1995).

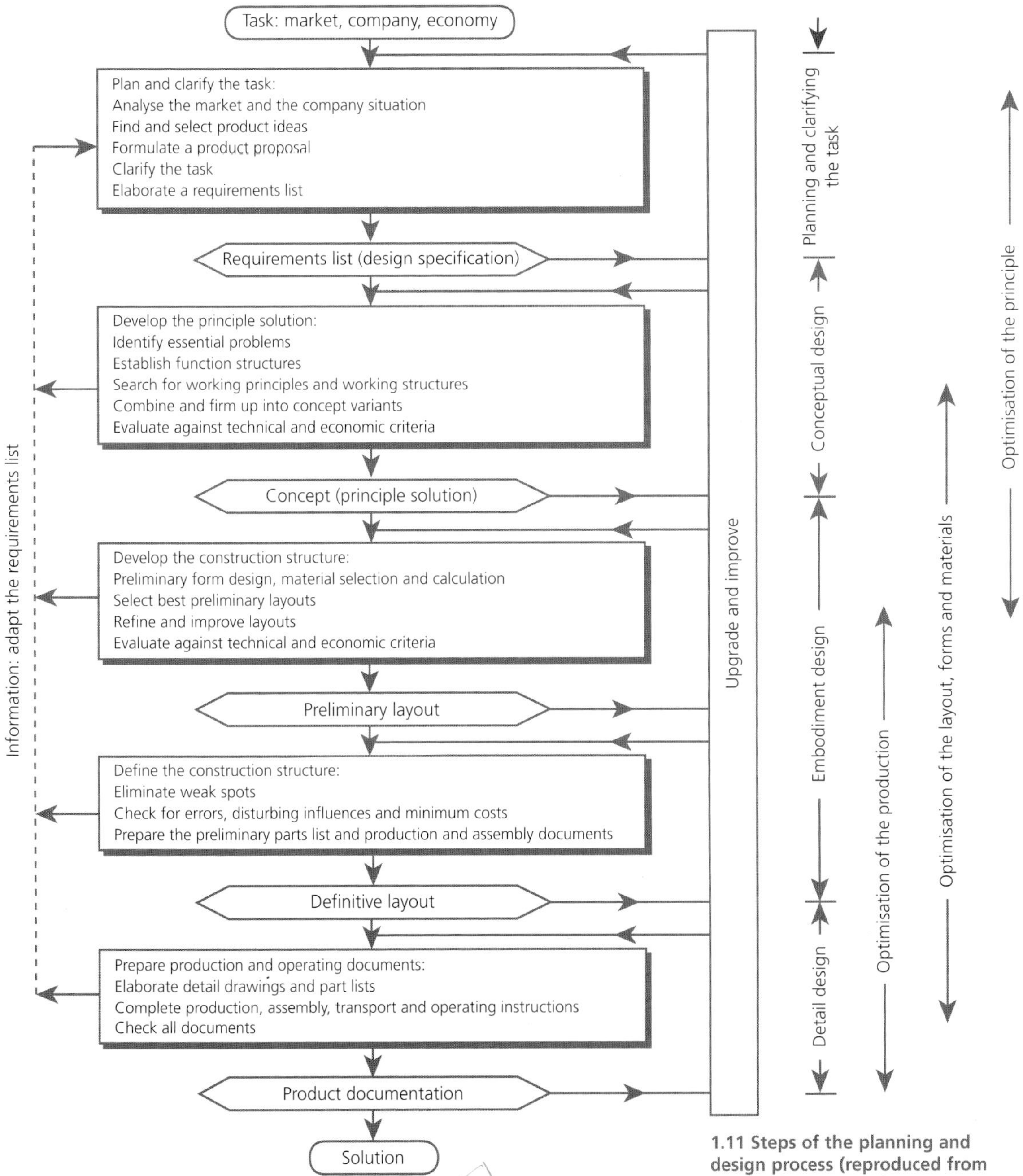

1.11 Steps of the planning and design process (reproduced from Pahl and Beitz, 1996)
© Springer-Verlag

In common with many other authors, Pahl and Beitz (1996) believe that the most challenging problem in design – or the most resistant to solution by systematic methods – lies in making the creative leap between problem definition and solution concept. In the mechanical design literature, a common feature of methods attempting to support this step is an emphasis placed on understanding the relationship between the function and form of physical structures and mechanisms.

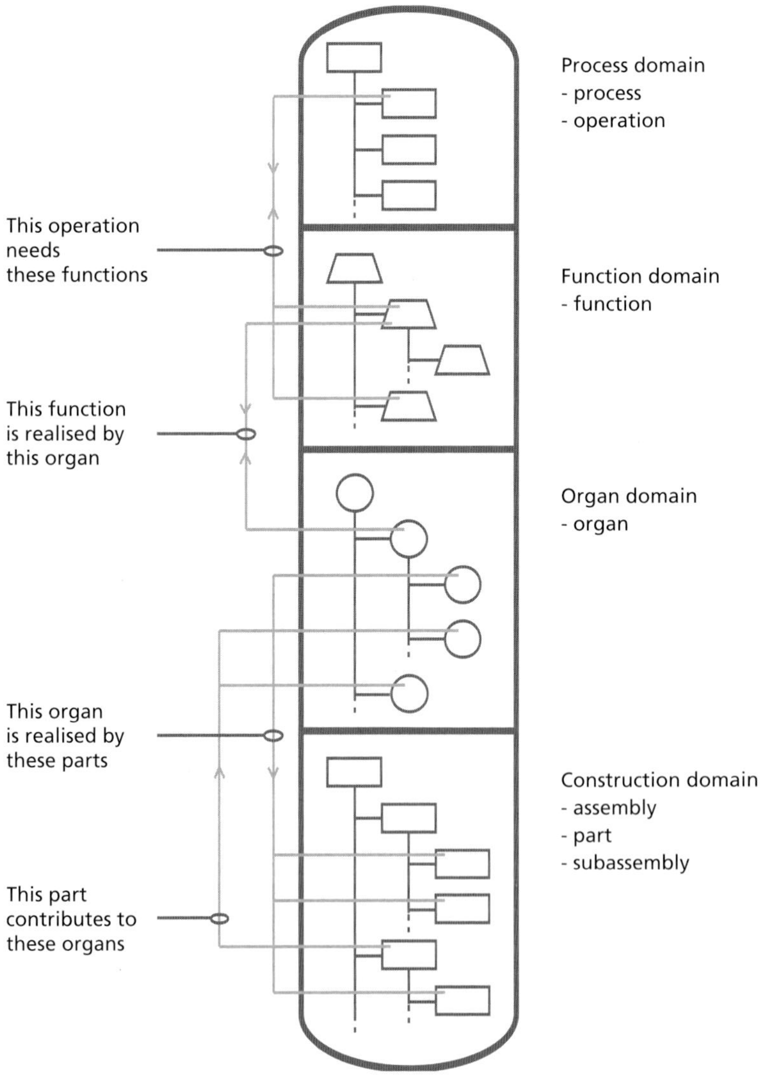

1.12 Product model or chromosome derived from the theory of domains (Reproduced from Andreasen, 1992)

Most authors advocate the establishment of a function structure, in which various physical principles such as 'friction' or 'leverage' are combined to solve the design problem (develop principle solution, in Pahl and Beitz's terminology). These functions are an abstract formulation of the solution, independent of possible realisations such as 'roller bearing'. Defining the interactions of functions in terms of transformations of energy, material and signal flows provides an intermediate step between problem and concept solution. Once a function structure is in place, many authors recommend the systematic consideration of a number of possible means of realising each function, using morphological matrices and other combinatorial methods to ensure that a wide range of possible design concepts is covered.

Hubka (1982) further developed this functional decomposition method by proposing the consideration of 'organs' as a part of the mechanical design process. Organs, by analogy to biological systems, describe the complex relationship between functions and their realising components. Hubka describes organs as the groups of physical components that perform collections of functions, where the essential features of each organ are those spaces, surfaces or lines which represent the localities where the necessary effects take place. Andreasen and co-workers take a similar approach in their theory of domains (Andreasen *et al.*, 1997), in which technological artefacts are represented from the four structural perspectives of technological principle (process domain), functions, organs, and physical parts (Figure 1.12) (Andreasen, 1992). Further rationalising the step between the product requirements and synthesis of the physical form, Andreasen (1980) proposes that parts may be derived from organs, which in turn are derived from functions and physical principles (Figure 1.13).

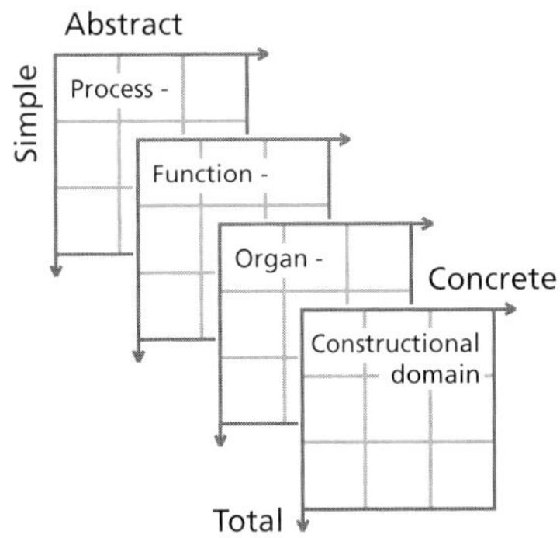

1.13 Design of a mechanical system involves gradually building up models and determining data in four domains (Reproduced with permission – Andreasen, 1980)

Many proponents of systematic methods agree that design problems require a creative and intuitive step which cannot be made by following prescribed activities. Their methods reflect this belief by supporting but never prescribing the creative leap itself, which remains the preserve of the human designer. Pahl and Beitz (1996), for example, stress that their methods are intended to "encourage creativity, and at the same time drive home the need for objective evaluation of the results". Other authors, such as Kusiak, (1999) take an alternative approach based on the systems theory derived from the work of authors such as Simon (1996), who argue that the complex behaviour of systems such as mechanical products is not a holistic property, but rather arises from the rational interactions between subsystems. Furthermore, Simon argues that such systems are nearly decomposable – in other words, may be treated as essentially hierarchical in nature. Applying this reductionist perspective to machine systems, Hubka (1982; Hubka and Eder, 1988) proposes that they may be generated by following a sequence of entirely rational operations. His method describes the product to be one element interacting with several others, such as users and operators, which together form a technical process which is designed to solve the identified need.

Discussion

The design-focused models introduced above are strongly focused on the technical aspects of solving design problems, describing or prescribing the steps thought necessary to progress from problem to solution. Some such authors propose models which are general in nature and have been applied to many design disciplines – the general form of Evans' spiral model, for example, is still in use after more than four decades in diverse fields from ship design (Rawson and Tupper, 1994) to computer software design (O'Donovan, 2004). Others are more strongly focused on product structure and are thus less relevant outside the target discipline.

The methods accompanying each model are intended for use by engineers and designers to support the execution of individual design steps. They typically concentrate on the early stages of the design process – Roozenburg and Cross (1991) went so far as to question the existence of detailed procedures for the embodiment and detail design stages. As with models, methods may be dependent or independent of discipline; whereas morphological combination is of limited use in the design of non-mechanical products such as micro-processors, brainstorming and requirements analysis are applicable in most situations. According to Pugh (1991), successful product design is subject to the integration of such general design methods with traditional engineering expertise.

In practice, the applicability of such models and methods is limited by their product-focused perspective, which implies that the key difficulty in a design project lies in finding solutions to the technical problems. In reality, however, even the simplest design process is a highly complex socio-technical activity requiring a much broader range of skills, from marketing to human resource management. This is highlighted by Ulrich and Eppinger (2003), who describe how bringing a screwdriver to market takes six individuals and a period of 12 months; more complex products, such as passenger aircraft, require the organisation of tens of thousands of man-years' effort. Furthermore, many authors describe how most complex design projects place strong limitations on early concept design, with constraints such as existing product platforms and legislative requirements often predetermining the form of the solution (Pugh, 1991; see Figure 1.14). In such circumstances many concept design methods are of limited use, and the primary difficulty design companies face lies in the integration of diverse methods, disciplines, tools and personnel (e.g. Andreasen and Hein, 1987).

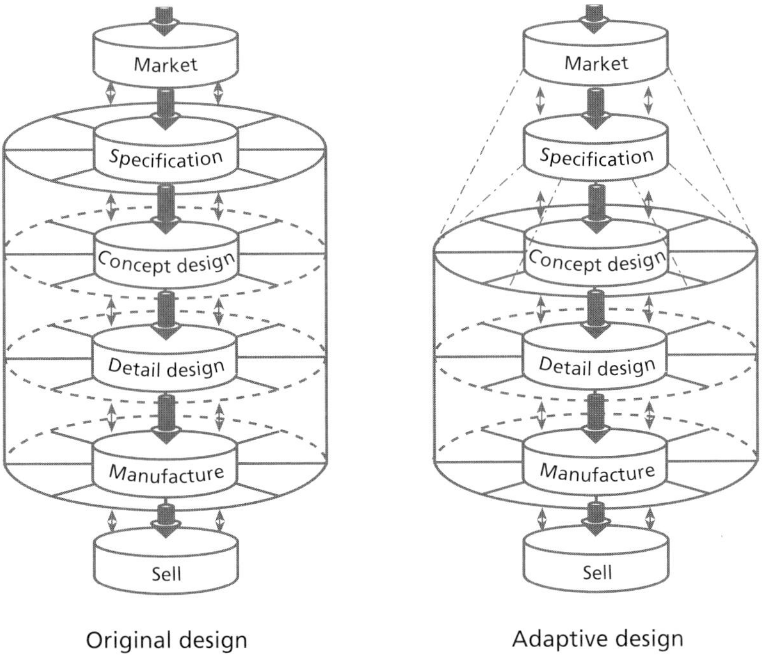

1.14 Original vs adaptive design process (adapted from Total design by Pugh, 1991)
Pearson Education Ltd.
© Addison-Wesley Publishers Ltd.
© Addison-Wesley Publishing Company Inc.

Project-focused literature

The project-focused literature discussed in the remainder of this chapter places emphasis on understanding the context of the design process, including such cost-related activities as product planning, marketing and risk management (e.g. Baxter, 1995). In other words, project-focused literature concentrates on product development as opposed to product design, defined by Roozenberg and Eekels (1995) as the development of a new business activity around a new product, where understanding the interaction between developing new products and new business is considered the key to success.

Integration of personnel and disciplines

Focusing on concept development, Ulrich and Eppinger (2003) propose methods intended to facilitate problem solving and decision making by integrating personnel from a variety of backgrounds and perspectives. They include the following as key challenges in new product development:

- Recognising, understanding and managing product related trade-offs, such as weight vs manufacturing cost.
- Working in an environment of constant change. As technologies and customer demands evolve and competitors introduce new products, there is a constant time pressure on all design and development activities.
- Understanding the economics of product development from marketing through to manufacture and sales, so that a return can be made on initial investments.

Taking a similar viewpoint, Pugh writes that industry is concerned with 'total design', i.e.:

> *"the systematic activity necessary, from the identification of the market/user need, to the selling of the successful product to satisfy that need—an activity that encompasses product, process, people and organisation."*
> (Pugh, 1991)

In other words, the development of any product requires the input of personnel familiar with many different disciplines, including those of technical engineering, engineering design, and many other non-technical fields. This integration of disciplines requires that all participants have a common view of the total design activity and can, therefore, subscribe to a common objective with a minimum of misconceptions. Pugh believes that visibility of operational structure is key to this common understanding, so that "everyone can find out what people are doing and why". He proposes that a disciplined and structured approach is necessary to achieve this.

Consideration of manufacturing constraints

Manufacturing issues need to be considered in early decision making.

A special emphasis is often placed on the importance of early analytical work to clarify design requirements – efficient design processes are of little use if the wrong product is being developed – but many early design models make little mention of manufacturing issues. Recognising that the manufacturability and the cost of the product are intimately tied to all design decisions, especially in the early phases where the concept is chosen, most modern methods highlight consideration of manufacturing concerns as an essential component of the successful design project. For example, Ullman (2003) proposes a stage-based method broadly similar to that of Pahl and Beitz, but includes design for manufacture and design for assembly as explicit steps at the end of the process.

Many other authors represent manufacturability as an influence which affects all stages of the project lifecycle. Andreasen and Hein (1987), for example, describe product development as the simultaneous development of market, product and production (Figure 1.15).

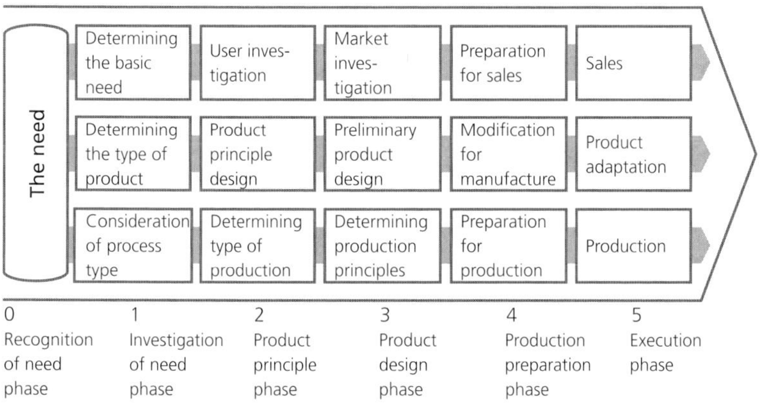

1.15 Integrated product development (reproduced from Andreasen and Hein, 1987)
IFS Publications Ltd.

Verification and validation

In a number of safety-critical industry sectors, such as aerospace, healthcare and power generation, formal demonstration of a product's 'fitness for purpose' is required before it can be released on the market. Good software development practice also places an emphasis on the provision of systematic proof of fitness of function. However, this aspect of design is not explicit in many of the models presented thus far in this chapter. Whilst 'evaluation' and 'iteration' are mentioned, their presence is more a reflection of observed practice than a formal description of the steps necessary to ensure fitness for purpose.

A particularly useful model of the design process, which highlights the role of evaluation, is the 'waterfall model' used by the US Food and Drug Administration's (FDA) Center for Devices and Radiological Health to promote good design practice (FDA, 1997). This model, whilst it says very little about design, illustrates the important and complementary roles of *verification*, *validation* and *review* in medical device development (Figure 1.16).

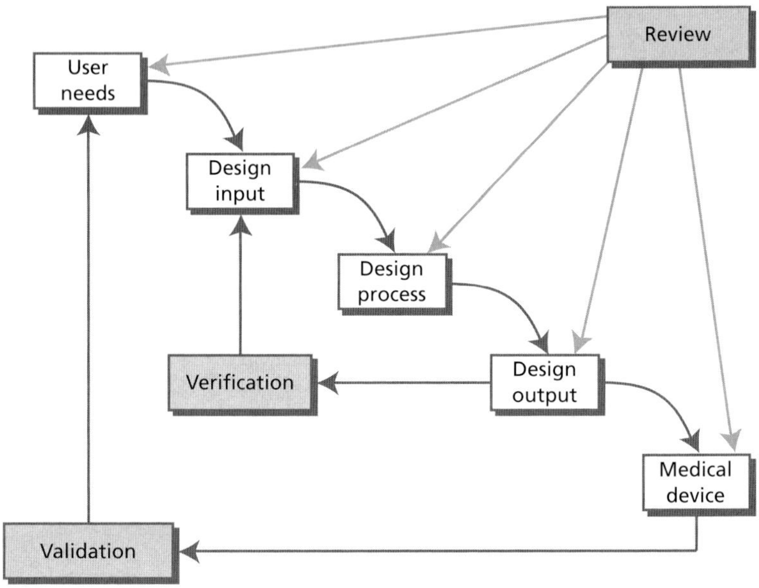

1.16 The waterfall model of medical device design
Medical Devices Bureau, Health Canada

The waterfall model has at its core a five-stage design process supported by three evaluation processes, namely:

- *verification*, which establishes whether the device design described by the design output conforms to the requirements described by the design input;
- *validation*, which establishes whether the medical device, produced in accordance to the design output, actually satisfies the users' needs;
- *review*, an activity undertaken regularly to ensure that good practice is followed at all times.

Validation is evidently a more involved process than verification and is usually understood to be the cumulative sum of all the verification efforts.

The overall philosophy of validation is the same whether it is applied to a device or its manufacturing process (Alexander *et al.*, 2001). For a device, validation is ultimately achieved by showing that the final device meets the original user needs and intended uses. For a process, validation is achieved by showing that the process equipment meets its original needs and intended uses, reviewing both the process equipment design and the corresponding production development (Figure 1.17).

The key to successful verification and validation lies in the early definition of the validation requirements. This helps to ensure that a design emerges that is not only fit for purpose, but may be proven to be so. Problems arising

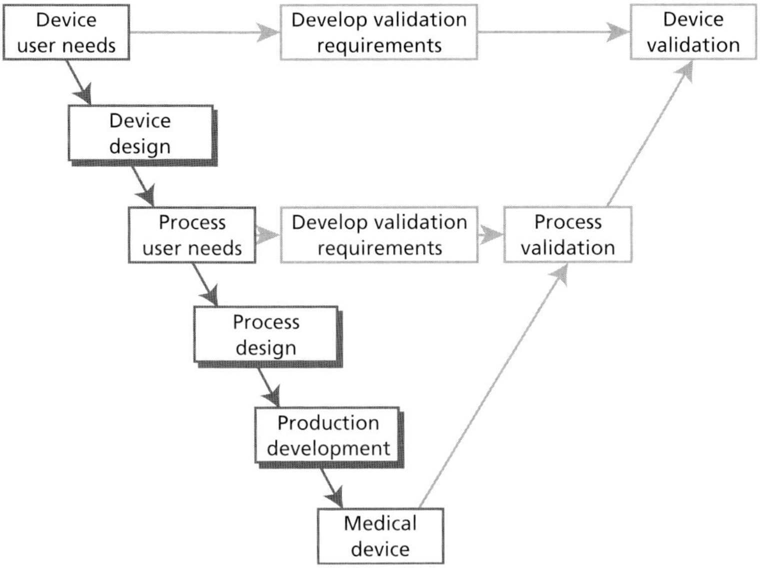

1.17 The medical device V-model (Alexander and Clarkson, 2000b)
Reproduced with permission of Taylor & Francis Ltd.

during validation inevitably result in some level of design iteration and further validation. Figure 1.17 may be further expanded to include device and process verification activities (Alexander and Clarkson, 2000a, b). Such models are applicable to all products and their associated manufacturing processes. However, the extent to which formal proof of validation must be presented will depend on the regulations governing a particular industry sector.

More recently, the FDA released a further model of the design process as part of their strategic plan (FDA, 2001). The total product life-cycle model (Figure 1.18) is a bold departure from the waterfall model (Figure 1.16) and is intended to highlight the iterative nature of device development and the connectivity between all stages of development. It alludes to issues such as Design for X, verification (preclinical), validation (clinical), quality, risk management, and so on. While it does not advise how to design a device, it does highlight many of the key issues.

Managing influences on the project

Projects are always influenced by a large number of factors which often have very little to do with the design process itself. Influences vary from project to project, ensuring that each is unique. This poses a management challenge, which Hales (2004) believes is best resolved by promoting awareness of

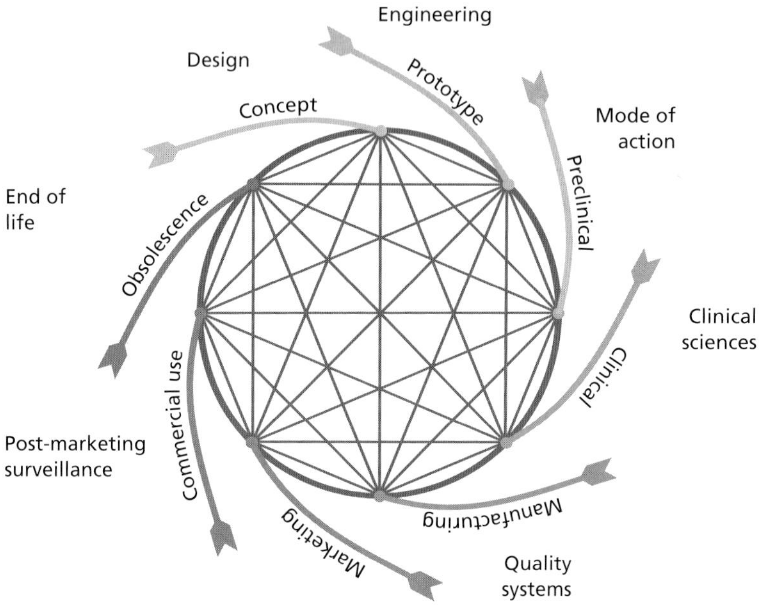

1.18 The total product life-cycle model for medical device design (adapted from FDA, 2001)

the influencing factors and their possible impacts on the project. He provides a comprehensive list of such influences at several different levels, including the macro-economic, micro-economic and corporate scales, summarised in diagram form in Figure 1.19. His method advocates the explicit consideration of each item on these checklists so that design managers can gain a broader perspective and make more informed decisions. The figure shows the now familiar stage-based view of the design process, placed into context within the project, company and market.

Individual projects must often compete for limited resources within the company. Furthermore, the development of complex products typically requires the coordination of many organisations, of which the individual companies may have responsibilities ranging from subsystem design to component manufacture; in either situation, successful integration of inter-organisational processes is critical to prevent delays to the project. For example, specification errors can be extremely costly for an externally designed, long lead-time component. Other influencing factors are further removed from the project and cannot be directly managed or influenced. For example, changes in organisational structure, government legislation or available manufacturing technology may cause a project to fail or to be cancelled.

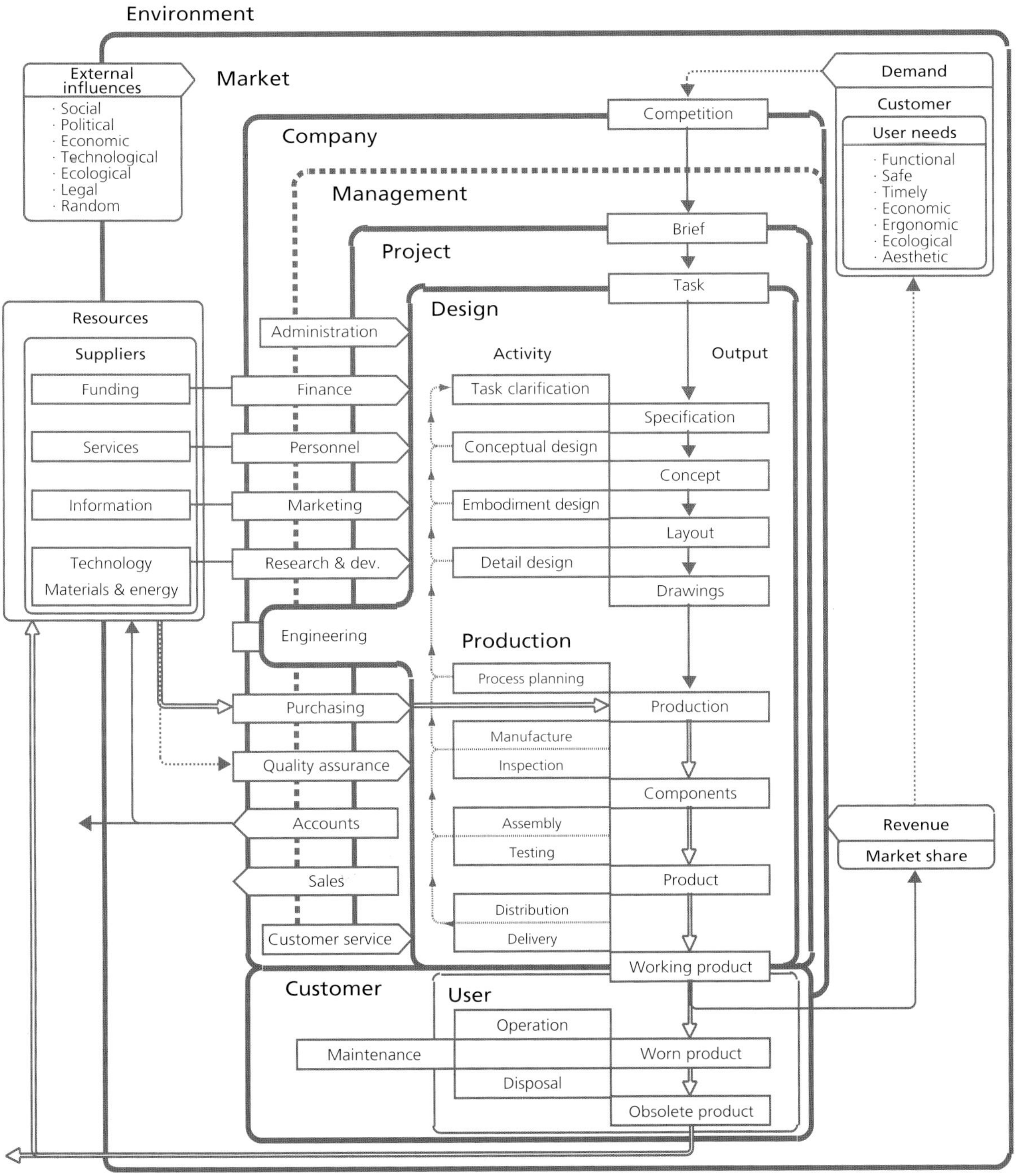

1.19 The project set in context (reproduced from Hales, 2004)

Process control and evaluation

The problem-oriented perspective of the design process as a linear progression through a series of stages is popular in industry and has been adopted in a variety of forms by many successful companies, including DuPont, 3M, Hewlett-Packard, Proctor and Gamble, ICI-UK, IBM, Polaroid, Black and Decker and Exxon Chemicals (Cooper, 1994). The 'gates' between stages, through which each project must pass to continue, are a dual-purpose structure used both for rationalising decisions and for planning. The well-defined deliverables from each stage are convenient documents with which to assess whether a project is likely to succeed, and the timing of these milestones anchors the schedule of the overall development project.

The artificial division of the process provides management with a quality control structure in which each gate represents an opportunity to recognise and halt a failing project; if the criteria for passing each gate are chosen wisely, following a prescribed process is one way of assuring the quality of the resulting product (Ulrich and Eppinger, 2003). Implementing such procedures allows a company to comply with quality standards such as ISO9000 (1994) or APQP (1995). This is obligatory for large engineering firms, as most European companies require their suppliers to gain such accreditation.

However, Cooper (1994) argues that there are many practical weaknesses to this form of gated process control. The system can be inefficient, in that projects must wait at a gate until all necessary activities have been completed. The overlapping of stages is impossible in most cases, although it is often desirable in the above situation. There can be high bureaucratic overheads at each gate, and the individual project perspective means there is little provision for managing the division of resources across a portfolio (Figure 1.20).

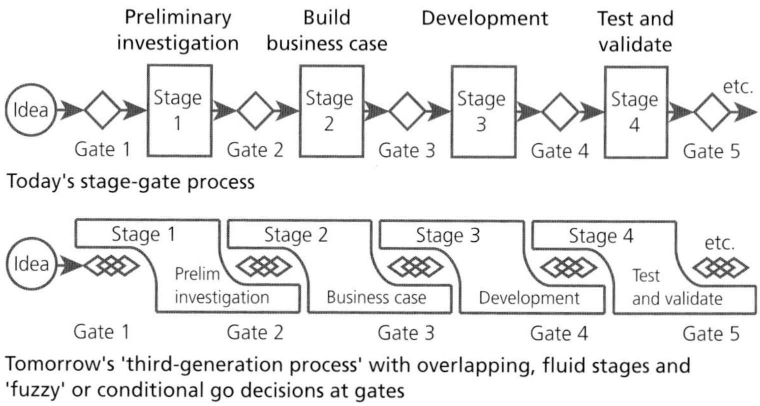

1.20 'The third-generation process' (Cooper, 1994)
Reproduce with permission of Blackwell Publishing Ltd.

Cooper proposes that these systems should be made more fluid and adaptable, should incorporate 'fuzzy gates' which are situational and conditional, should provide for sharper focus of resources and better management of the portfolio of products under development, and should be generally more flexible than the current stage-gate model.

Discussion

Many authors have proposed models and methods to support design project management. However, for each situation which may be improved by the application of such methods, there are many more which cannot; those problems highly dependent on human factors have proved particularly resistant. Another common difficulty lies in the balancing of activities and resources across a portfolio of projects under development; many methods are strongly focused on individual projects and offer little useful guidance in such a situation.

A key weakness of all the literature reviewed here is the difficulty of application to real design problems. Such methods range from the broad but abstract through to the concrete but limited in scope, and each design project represents a unique combination of a wide range of factors. Maffin *et al.* (1995) argue that although most process models are too general in scope and prescriptive in nature for easy application they can be interpreted for use in each design company (Table 1.21).

Company structure	Manufacturing process	Suppliers
Establishment size	Process complexity	Rationalisation
Independence	Process flexibility	Degree of control
Centralisation	Process constraint	Collaboration
	Production volume	Locality
	Internal span of process	

Market/customer	Product	Local environment
Market type	Product type	Local labour market
Market size and share	Product variety/range	Skills
Market complexity	Product complexity	Training
Market infrastructure	Product structure	Local support infrastructure
Exports	Product status	Financing and grants
Number of competitors	Innovation rate	
Competitive criteria	Design capability	

1.21 Company classification dimensions and key factors (Maffin, 1995)
Taylor & Francis Ltd. at http://www.tandf.co.uk/journals/

It is clearly important to make informed decisions about the nature of the models to adopt in the context of a particular situation. Maffin proposes that a set of critical factors which define the organisation and the product are influential upon the product development process, and that classifying companies according to this framework could form the basis for guiding the application of models in industry. Within a company, and for a particular line of products, many of these factors may be considered fairly constant.

Conclusions

To conclude, the theories, models and methods found in the literature span a diverse range of design problems and disciplines. To gain a balanced perspective it is necessary to study many such texts. Only by attempting to understand the design process from all points of view, from the individual designer's problem-solving process through to the need for continual business development, is it possible to begin to effect process improvement.

However, while these models all offer insight into the nature of the design project, they are far too general to help with project planning activities or to guide the daily decisions which must be made by design managers. We believe that recent advances in analytical techniques offer great potential in this area; such techniques are discussed in Chapter 2.

References

Alexander KL, Clarkson PJ (2000a) Good design practice for medical devices and equipment, part I: a review of current literature. Journal of Medical Engineering and Technology, 24(1): 5–13

Alexander KL, Clarkson PJ (2000b) Good design practice for medical devices and equipment, part II: design for verification. Journal of Medical Engineering and Technology, 24(2): 53–62. http://www.tandf.co.uk/journals

Alexander KL, Clarkson PJ, Bishop D, Fox S (2001) Good design practice for medical devices and equipment – a framework. Institute for Manufacturing, University of Cambridge, UK

Andreasen MM (1980) Machine design methods based on a systematic approach. PhD thesis, Lund University, Sweden

Andreasen MM (1992) The theory of domains. Workshop on understanding function and function-to-form evolution, University of Cambridge

Andreasen MM, Hansen CT, Mortensen NH (1997) On the identification of product structure laws. 3rd WDK Workshop on Product Structuring, Delft University of Technology, The Netherlands

Andreasen MM, Hein L (1987) Integrated product development. IFS (Publications)/Springer
APQP (1995) Advanced product quality planning and control plan (APQP). http://www.aiag/publications/quality/APQP2.asp
Archer LB (1965) Systematic method for designers. The Design Council
Asimow M (1962) Introduction to design. Prentice Hall
Bahrami A, Dagli CH (1993) Models of design processes. In: Concurrent engineering: contemporary issues and modern design tools. Chapman and Hall
Baxter MR (1995) Product design: practical methods for the systematic development of new products. Chapman and Hall
Birmingham R, Cleland G, Driver R, Maffin D (1997) Understanding engineering design: context, theory and practice. Prentice Hall
Blessing LTM (1994) A process-based approach to computer-supported engineering design. PhD thesis, University of Twente, The Netherlands
Browning TR (2001) Applying the design structure matrix system to decomposition and integration problems: a review and new directions. IEEE Transactions on Engineering Management, 48: 292–306
Bucciarelli LL (1996) Designing engineers. MIT Press
Cooper RG (1994) Third generation new product processes. Journal of Product Innovation Management, 11: 3–14. Blackwell
Cross N ed. (1984) Developments in design methodology. John Wiley. Figure 1.4 reproduced with permission of John Wiley and Sons Limited
Cross N (1994) Engineering design methods: strategies for product design. Figure 1.7 reproduced with permission of John Wiley and Sons Limited
Cross N, Roozenburg N (1992) Modelling the design process in engineering and in architecture. Journal of Engineering Design, 3(4): 325–337
Darke J (1979) The primary generator and the design process. Design Studies, 1(1): 36–44
Dym CL, Little P (2000) Engineering design: a project-based introduction. John Wiley
Ehrlenspiel K (1995) Integrierte Produktentwicklung: Methoden für die Prozessorganisation, Produkterstellung und Konstruktion. Hanser
Eppinger SD, Whitney DE, Smith RP, Gebala DA (1994) A model-based method for organizing tasks in product development. Research in Engineering Design, 6(1): 1–13

Evans JH (1959) Basic design concepts. American Society of Naval Engineers Journal, 71(4): 671-678

FDA (1997) Design control guidance for medical device manufacturers. Centre for Devices and Radiological Health, Rockville, MD, USA. http://www.fda.gov/cdrh

FDA (2001) CDRH strategic plan. Centre for Devices and Radiological Health, Rockville, MD, USA. http://www.fda.gov/cdrh

Finger S, Dixon JR (1989) A review of research in mechanical engineering design, part I: descriptive, prescriptive and computer-based models of design processes. Research in Engineering Design, 1(1): 51–68

French MJ (1999) Conceptual design for engineers. Springer

Frost RB (1992) A converging model of the design process: analysis and creativity, the ingredients of synthesis. Journal of Engineering Design, 3(2): 117–126

Hales C (2004) Managing engineering design. Springer

Hall AD (1962) A methodology for systems engineering. Van Nostrand Reinhold

Hillier B, Musgrove J, O'Sullivan P (1972) Knowledge and design. In: Environmental design: research and practice. University of California, USA

Hubka V (1982) Principles of engineering design. Butterworth

Hubka V, Eder E (1988) Theory of technical systems. Springer

ISO9000 (1994) BS EN ISO 9000-1: quality management and quality assurance standards – guidelines for selection and use. http://www.bsi-global.com

Jones JC (1963) A method of systematic design. In: Conference on design methods. Pergamon Press

Jones JC (1970) Design methods: seeds of human futures. Wiley

Kusiak A (1999) Engineering design: products, processes and systems. Academic Press

Lawson B (1980) How designers think. The Architectural Press

Maffin D, Alderman N, Braiden PM, Hills W, Thwaites A (1995) Company classification: a new perspective on modelling the engineering design and product development process. Journal of Engineering Design, 6(4): 275–289

March L (1984) The logic of design. In: Cross N ed. Developments in design methodology. Figure 2 on page 270 (copyright John Wiley and Sons limited) is reproduced with permission of John Wiley

O'Donovan BD (2004) Modelling and simulation of design processes. PhD thesis, University of Cambridge, UK
Pahl G, Beitz W (1996) Engineering design: a systematic approach. Page 66, Figure 3.3. Springer
Peirce CS (1923) Chance, love and logic. Kegan Paul
Pugh S (1991) Total design: integrated methods for successful product engineering. Addison-Wesley
Rawson KJ, Tupper EC (1994) Basic ship theory. Longman
Rittel H, Webber MM (1984) Planning problems are wicked problems. In: Developments in design methodology. John Wiley
Roozenburg NFM, Cross N (1991) Models of the design process: integrating across the disciplines. ICED'91, Zürich, Switzerland
Roozenberg NFM, Eekels J (1995) Product design: fundamentals and methods. John Wiley
Simon HA (1996) The sciences of the artificial. MIT Press
Steward DV (1981) The design structure system: a method for managing the design of complex systems. IEEE Transactions on Engineering Management, EM-28(3): 71–74
Ullman DG (2003) The mechanical design process. McGraw-Hill
Ulrich KT, Eppinger SD (2003) Product design and development. McGraw-Hill

Chapter 2

Design planning and modelling

Brendan O'Donovan, Claudia Eckert and John Clarkson
University of Cambridge

Tyson R Browning
Texas Christian University

The success or failure of any design process depends crucially on finishing on time and to budget. To achieve this, a successful design process is just as important as a high-quality product. The effective and efficient execution of a design project depends on the understanding of the design managers and the quality and utility of their project plans.

Design process planning where there is any degree of complexity is notoriously difficult (Figure 2.1). Companies with great confidence in their technical abilities often are very dismissive of their understanding of design process planning. They can be world leaders in their respective technologies, yet they may not understand the process through which they have generated them, or through which they will incorporate the technology into a product. In view of the crucial importance of planning and its frequent lack of effectiveness in practice, this chapter reviews planning and modelling approaches currently used in industry and developed at universities.

What is a design process? Intuitively, this is not a difficult question – if engineers were asked to give an example from their own work of a design process, few would find the question difficult. Furthermore, if different engineers within a company or industry were to compare their answers, then it is likely that there would be general consensus that all the processes proposed were indeed design processes. To some extent, therefore, the question is unnecessary – people are capable of defining 'design processes' that are of relevance or of interest to them.

A definition of a design process is required for two reasons. First, while there may be general consensus as to what is and is not a design process, such views will not be unanimous. Engineers may disagree whether computer generation of designs is a design process, and there may be even greater disagreement between engineers and other design professionals, such as architects and product designers. Second, a definition of what is meant by 'design process' is essential here if we are to talk about it in general as well as in specific terms.

Industry uses the term 'design process' to mean one of two things: the generic, high-level approach each design project would follow, or the set of activities that actually happen. For example, a company might have an official change process describing the formal steps designers need to go through to get a change approved, executed and validated. However, this says very little about how the designers actually carry out a change. In this chapter we are interested in the latter, the actual activities that are in reality required to design a product or meet new customer needs.

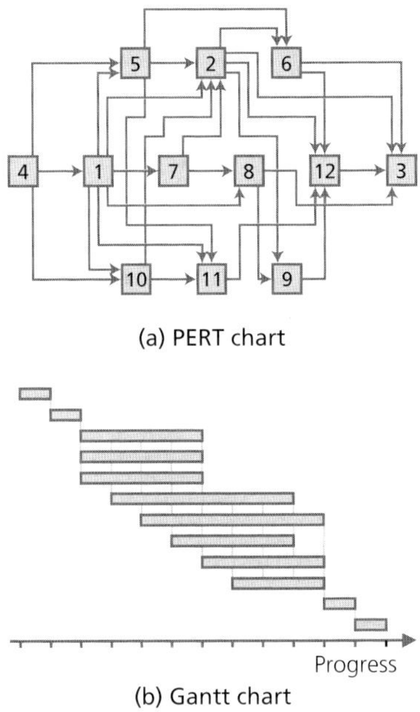

2.1 Design process planning where there is any degree of complexity is notoriously difficult (Adapted from Clarkson *et al.*, 2000)
© 2004 IEEE

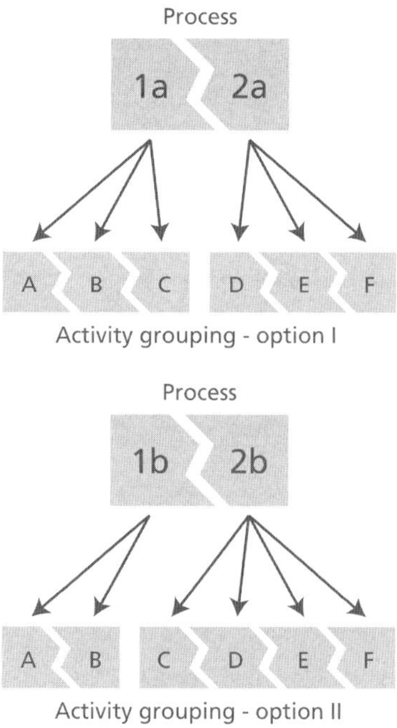

2.2 The set of activities chosen to represent a process is not unique – there may be many meaningful ways of grouping actions

As an initial definition, we might use "the network of activities performed with the goal of producing a design." However, there are a number of considerations beyond this basic definition that should be noted. When we talk about a series of tasks, this should not be taken to mean that tasks in a design process are performed in a purely serial fashion – the connectivity and interdependency of tasks in a design process is often far more complex than this; and, finally, what is meant by a 'design' cannot be taken as self-evident, particularly with regard to variations on or additions to an existing design (however, an intuitive understanding of the meaning of 'design' will suffice here). In everyday language the terms 'task' and 'activity' are often used interchangeably; however, in this chapter we adopt the definition that a task is goal directed, whereas an activity is not necessarily so. A task can be seen as grouping the low-level actions that make up a design process (the mouse-clicks, key-presses, words of dialogue, *etc.*).

It is important to distinguish between the 'design process' and 'design-process model'. A design process is a real, actual way in which design work is done and designs are produced. A design-process model is an attempt to describe a real design process in an abstract way. Models must make choices about how and to what extent to abstract from reality. Such decisions should align with the purposes or intended uses of the model (of which there may be many). Hence, different modelers may produce very different descriptions of the same design process (Figure 2.2). Activity models are groupings of activities, which are based solely on what is meaningful in the context of the overall process.

The set of activities chosen to represent a process is not unique – there may be many meaningful ways of grouping actions. This gives rise to the range of plans discussed below, and also means that different people producing the same style of plan for the same process might produce different results. It also indicates that activity models are inherently hierarchical, but this hierarchy is not unique, and it also might not fall into a neat tree structure (as complex problems rarely do). Thus, while a process is composed of related activities from one point of view, each of those activities may be a process in itself from another point of view – and from a 'higher' perspective, the whole process is enmeshed in a context of (*i.e.* may be seen as an activity within) a much larger process (such as a supply chain). All of these issues contribute to the challenge of satisfactorily representing a real process with a process model.

Another key distinction is between descriptive and prescriptive process models. Descriptive models seek to represent the *actual* design process for purposes such as understanding process behaviour, estimating process cost and duration, *etc.* They are built by asking questions such as "What work is done?" and "How are the results produced?" On the other hand, prescriptive models seek to specify what and how work *should* be done. Prescriptive models often exist in practice as standard processes or procedures. They seek to be canonical in nature. Unfortunately, many of the prescriptive models used in practice have not been adequately verified or validated as effective, efficient, or even feasible, which is a key reason why many designers do not follow them exactly.

This chapter will discuss engineering design in contrast to other processes to identify the specific requirements for design-process models and discuss the ways industry currently handles its planning needs. Before discussing particular modelling approaches we will revisit the relationships between process models and the process itself. All models are abstractions, which, by definition, means they are 'less' than the entity itself. In this sense George Box's famous quote holds true: "All models are wrong – but some are useful" (Box, 1976).

Engineering design

In order to discuss modelling approaches appropriate for the engineering design process, it is important to consider how engineering design differs from other similar activities, such as:

- other types of design process – typographic (Figure 2.3), architectural, services;
- other engineering processes – manufacture;
- other business processes – personnel, marketing, purchasing, order fulfilment.

In considering what makes engineering design unique, we will illustrate why existing modelling tools and approaches may have shortcomings when applied to engineering design. We will also begin to identify the specific issues that must be addressed in an engineering design-process model.

9 font family consisting *of* **5 weights** & *4 italics*

2.3 How is font design different from engineering design?

The distinct characteristics of engineering design processes can arise from the complex nature of the product itself, the complexity of the process and the difficulty of capturing this in any kind of model. These factors are highly interdependent, and it is often impossible to distinguish the cause and effect relationships between them.

Engineering projects are typically on a very different *scale* from other types of design project. The design of an aeroplane or production facility can involve thousands of person-years, requiring a high degree of *specialist* knowledge. Such a project requires active management of the development team. Whilst simple engineering projects do exist, any modelling and planning approach should be capable of scaling up to any feasible size required.

Design projects are by definition *unique*: if the product already existed, it would not need to be designed. The key difference between design and business processes can be summed up as: design processes seek to do something novel, once, whereas many other business processes seek to do the same thing repetitively. Plans and models of design processes can, therefore, only be developed on the basis of experience with similar projects. This in itself can introduce risk into a design process through mismappings and misunderstanding (Earl *et al.*, 2001).

Some engineering projects are also very *infrequent*. An extreme example would be the design of a helicopter, where a company might develop a new model every 10 or 20 years. Even in companies producing large numbers of customised designs for a range of customers, the total number of product designs will usually be in the tens or hundreds over a number of decades. This contributes to the difficulty of gathering data about the process, but it also means that design processes cannot be considered to be 'steady state' – unlike, for example, production. As a result, design-process models may have residual *ambiguity* regarding what activities and relationships to include.

There are many different sources of *uncertainty* in the engineering design process, such as the time required to develop solutions, the performance level of the proposed solutions and the time and money required to verify performance. While other types of process may exhibit some uncertainty (for example, tolerances in manufacture), the type of the uncertainty is often known, whereas in engineering design the nature of the uncertainty itself it often unknown. The range of uncertainty in design, the complexity arising from interactions, and the difficulty of identifying and quantifying such uncertainty creates unique difficulties.

Design processes seek to do something novel, once, whereas many other business processes seek to do the same thing repetitively.

Engineering design processes are highly *constrained* by external factors, such as requirements, resources and deadlines. They are also constrained by the nature of many technical engineering products, where components, functions and systems are strongly coupled (Suh, 1990). This makes finding an acceptable solution more difficult than in other problems where elements of the problem to be solved can be separated.

The range of acceptable solutions may also be smaller in engineering, where quantitative targets can be set, than in other fields of design, where the suitability of a given design is judged by more subjective criteria. Meeting conflicting constraints in coupled problems almost inevitably leads to iteration. This is a major feature of many engineering design processes. It may take the form of unplanned iteration, where redesign is necessary due to the failure of the initial design to meet given constraints and requirements, or planned iteration, where it is expected that several iterations will be needed to refine the performance of the product to a satisfactory level (Smith and Eppinger, 1997). Business processes tend, in general, to be more serial in nature.

Design processes are among the most difficult processes to understand, and thus modelling them is fraught with *ambiguity*, especially initially. Here, ambiguity comes from lack of knowledge regarding what variables to include in a model, whereas uncertainty comes from the inability to pinpoint the value of a variable. Over time, as further observation of design processes occurs, the ambiguity in a design-process model can be reduced.

Design processes are among the most difficult processes to understand, and thus modelling them is fraught with ambiguity.

Applications of a model

In order to understand the range of applications for a design-process modelling methodology, it is helpful to consider the market for such models. Some of these are summarised in Figure 2.4. Although in this chapter we primarily consider application of a model to issues of planning, both before and during an engineering design process, these modelling frameworks may also serve a much wider spectrum of needs.

Supporting all these applications is the greater understanding of the design process that is gained through the building of a process model. This in turn facilitates, directly or indirectly, better process planning and management. Process models support all participants by making their own and each other's tacit assumptions explicit, and allowing them to reflect on them and communicate about them using a common vocabulary.

Planning may occur at the beginning of a project (initial planning), and throughout the project (operational planning) in response to new events. *Initial planning* occurs before or at the start of a project. The likely cost and duration of the design process is identified, along with the resource requirements, at the strategic level. On the operational level, a plan of work is created detailing the activities that will be completed, expected timings, gateways in the process, *etc*. The major risks are identified and assessed, and activities are planned to mitigate these risks.

Dynamic planning is required when original plans need to be modified in response to unexpected outcomes or other events that have an impact on the project. Again, there is both a strategic dimension – as the expected

Visibility and Juxtaposability. Ability to view components easily. How easy is it to see or find the various parts of the notation while it is being created or changed? If the users need to compare or combine different parts, can they see them at the same time?

Strategic management. At the highest level in a company, the directors and CEO will make decisions about which products to develop, the allocation of resources and contingency funds for each project, and about responding to requests for tender . Strategically, companies are also interested in knowledge retention and management, which can be structured by process models.

Operational management. At the level of the project managers, functional managers and chief engineers, the primary concerns are in directing the course of a project, or a number of projects, as quickly, cheaply and effectively as possible. This may involve choosing tools and methods, scheduling, planning, allocating resources and facilitating communication of information. Process models aid in the standardisation of methods, tools, and training within and across projects.

Design engineers. These are the participants most directly involved with creating the product design. They require the critical information for their activities to be delivered, and guidance as to what the 'next step' in the process should be.

Process newcomers. New recruits to a company or division of a company will require training about the characteristics of the design process.

Supply chain. The increasingly close integration with the supply chain observed in the automotive and aerospace companies visited creates opposing pressures: on the one hand, to provide information and understanding of the design process to ensure that external work is compatible with internal design effort both technically and in terms of timing; but on the other hand to control the information released to the suppliers to attempt to slow the migration of expertise out of the core company and into the supplier base. Shared models of the design process may help to build trust, assisting in explaining the causes of delays or problems with the work.

Academia. Because many existing case studies are carried out with a fairly narrow focus, or specific purpose in mind, it is difficult to adapt existing fieldwork to test a new idea. A rich, standardised model format would facilitate transfer of data from one project to another, increasing the quantity of data available to researchers, and so assisting or even driving the hypothesis creation and testing process. If the model is sufficiently rich, it may be possible to test some hypotheses without the need to gather additional data, simply by analysing existing models in a new way.

cost and duration of the process change and decisions must be made about cancelling the project, or increasing the resource allocation – and an operational dimension, as resource allocations, activity orderings, *etc.* must be changed to ensure effective progress.

In planning design processes, managers need to make decisions about the division of activities, the allocation of resources and the likelihood of iteration. They need to play through *what-if* scenarios. Currently, these are not well supported, and design managers play through different scenarios mentally or with rudimentary tools. Process support systems should support, visualise and potentially automate those alternative hypotheses.

Individual designers use plans to understand what work they are required to do on a day-to-day basis. Currently, however, most plans only provide rough *process guidance*. Ideally, a design support system should be used to guide the selection of activities in such a way that the overall process is the most efficient possible. The system should also provide the design data or documents required by the activity, as well as suggesting items of information which are non-essential but potentially useful (Clarkson and Hamilton, 2000).

Information transfer from one activity to another, and from one person to another, is important for an effective design process. A model of the design process could assist design *communication* by identifying who needs to be informed about what, not just because they are directly involved in supplying or using an item of data, but because of indirect interests in the information. For example, people may be interested in decisions being made that will eventually reduce the tolerance margins available in their part of the design process. Visualisation of a plan also allows a broader spectrum of participants to critique its assumptions.

Design-process models can also support the overall organisation. In order to support the *training* of new staff in understanding the structure of the design process, the model should provide a readily comprehensible visualisation of the process. Interactive tools, allowing the trainee to interact with an example process, would also be of value.

Design-process models can also be employed as *knowledge management* tools by providing an organising structure and/or repository for design-process knowledge. In a simple form, a process model can provide links to other databases where various types of knowledge are actually stored. By organising knowledge around the work to be done, designers can have the best information, tools, and guidance at their fingertips when they carry out a particular activity. And as they do an activity, part of its completion can entail updating this knowledge based on lessons learned.

Design-process models can also be employed as knowledge management tools by providing an organising structure and/or repository for design-process knowledge.

Existing planning practice

Through case studies and consultation with industry, a number of types of planning activity were observed within engineering development projects (Eckert and Clarkson, 2003), as shown in Figure 2.5. Some of these plans are generated solely for planning purposes, while others serve as plans but have different primary functions. For example, a bill of material has many functions, such as costing components and describing assembly sequences, but one of them is also serving as a plan.

Strategic product portfolio plans. At the strategic level there are numerous plans, many of which are so distinct from the engineering design process that they are not encountered in this field of study (for example, financial plans),

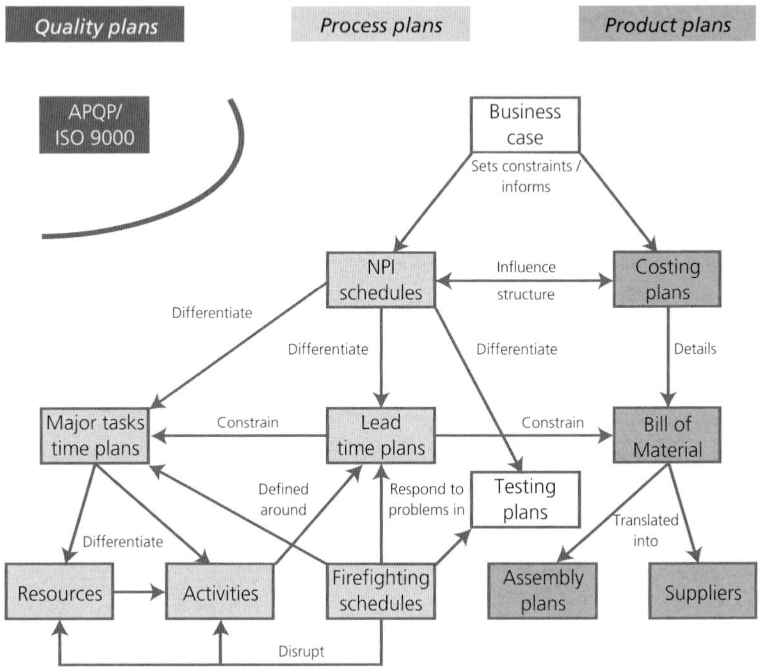

2.5 Different design process plans and their interaction (Eckert and Clarkson, 2003)
Reproduced with permission of the Design Society

but the overall product development strategy is of great direct importance. This plan both triggers activity reflected in other, more detailed plans for specific products, and also reacts to information supplied by the more detailed planning stages. The strategic product development plan is used to mediate between the internal factors of a development project, and also the external factors in the wider company and the market.

Bill of material (BoM). All engineering products have explicit BoMs indicating the components comprising the product. This is important for purchasing, manufacturing and other logistics functions in the company, but in some instances observed it also plays a significant role in directing design activity. Often the BoM of a previous, similar product is used as the basis at the start of a new project. For a relatively established product, such as a car, the names of components and structure of the product will change relatively little from one product generation to another. At the start of the project, expected lead-times for completing supplier negotiations, production development and purchasing, *etc*. are attached to each component. Subtracting these times from the planned introduction date of the new product gives latest dates for the completion of each design. These timings are used to prioritise specific design activities. Despite this planning function, the BoM is not explicitly recognised as a plan, perhaps because it is viewed by those responsible for planning as a description of the product, not the process.

Milestone plans. In long-duration development projects, it is common practice to decompose the overall goal into a set of sub-goals. These goals may relate to: the phase of design (concept, layout, detailed design, *etc.*); the areas of design completed (chassis design, electronics); or specific performance goals, risk reductions, or test passes (for example, road test of a 'mule' hybrid between the previous and new generation of a car, crash test passes). Often, a number of different types of goal are combined in a single milestone plan. The milestones state what should have been achieved at each point in the design process, but do not explicitly state how these goals should be reached. Experts can infer likely courses of action from the milestones, and these are then represented in the form of action plans. Milestones are also important for review of the development process; progress against time and cost schedules can be assessed by managers who may not be familiar with the detail of the process between milestones

Quality plans. Quality plans represent the actions necessary for compliance with ISO9000 or internal quality standards. These plans encompass varying amounts of the process: just start-up and close-down, milestone/gateway reviews, or, less frequently, a wider part of the development process. Unfortunately, perverse incentives are at work here. To maximise the probability of passing an audit or assessment, a company will tend to make a process model as general and ambiguous as possible, so that whatever a worker says they are doing will safely fall under the umbrella of the process description. Of course, this approach leads to process models that the workforce finds little value in consulting to accomplish their jobs – except in the time just before an audit or assessment!

The most common and diverse type of plan is the activity or task plan. It describes a set of actions to be performed by teams or individuals.

Activity plans. The most common and diverse type of plan is the activity or task plan. This describes a set of actions to be performed by teams or individuals working on the project. Not all action plans are intended to prescribe activities directly; in large or complex projects there may be hierarchical levels of activity plan with only the lowest-level plans actually driving the work and the higher-level plans providing structure and coordination. Activity plans may assign responsibilities, resources, timings and costs to specific actions, and are generally created by project and functional leaders within projects. There are many specialised instances of activity plans, such as those for testing or assembly.

Personal plans. Individuals working within a design process may create plans for their own actions in a number of forms, ranging from simple 'to do' lists to Microsoft Project plans. The need for planning at this level depends on the context in which they are working. It will tend to be greater in cases where an individual is working on multiple projects, has multiple concurrent responsibilities in a single project, needs to expand significantly on the details of actions requested of them or must adopt a particularly systematic approach to their actions (for example, testing).

These different forms of plan are only partially reconciled with each other, partly due to the volume of information and management overhead that full reconciliation would require. Differences of form, focus and accessibility all present barriers to the unification of the plans which exist within a company. In practise, coherence between different plans is achieved because different individuals work with more than one plan at any one time and map between these plans. However, as Figure 2.6 illustrates, the formal overlap between different types of plan is sparse.

Miscommunication between one plan and another can produce difficulties and even crises in a development project – for example, the failure to communicate a critical lead time from a BoM to an activity or milestone plan, or the failure to communicate increases in expected programme cost back to the strategic level. Conversely, diversity of plans provides for greater flexibility, with plans able to be changed locally to respond to events without the need to propagate every alteration back up the hierarchy of plans. Optimal planning practice clearly lies somewhere between the extremes of a single unified plan and a set of disconnected and inconsistent plan fragments.

A process model is usually generated at the beginning of a design project or when sufficient changes occur to previous plans to warrant a new plan.

Processes and process models

A process model is usually generated at the beginning of a design project or when sufficient changes occur to previous plans to warrant a new one. While people have a tacit understanding of when a process plan is no longer relevant, it is difficult to describe the relationship between the process plan and the process that actually occurs. This has practical and theoretical reasons. Process models are typically generated to plan, *i.e.* before the project, and hardly any company goes to the trouble of comparing the model with the process that actually exists. Process postmortems are rarely done, because everybody is busy moving onto the next project. While some main lesson might be learned, this is rarely about the process model itself.

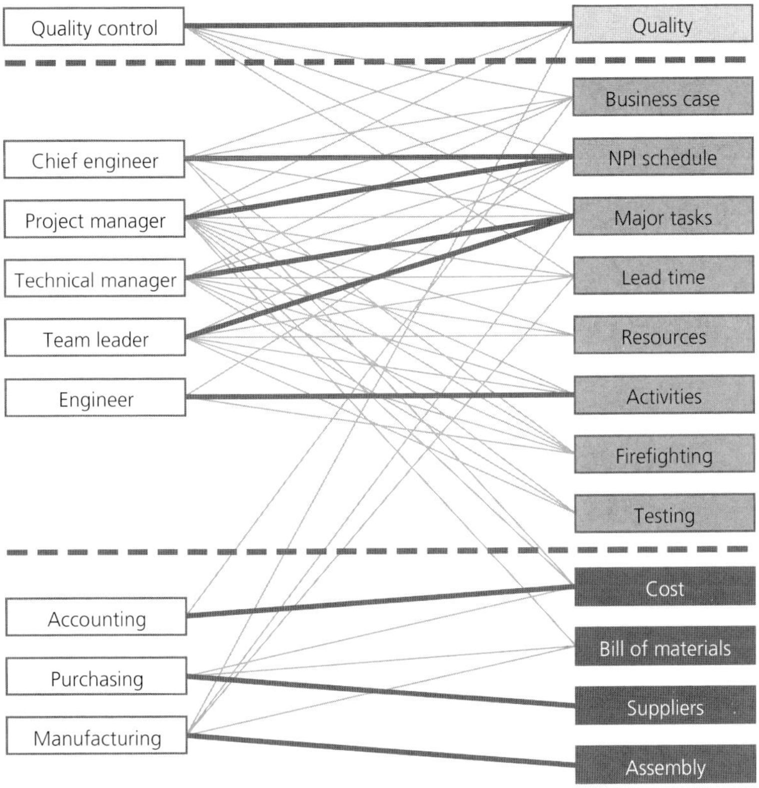

2.6 Plan and plan owners (adapted from Eckert and Clarkson, 2003)
Reproduced with permission of the Design Society

To generate a product, a process of designing is required; however, the actual process is impossible to capture completely. The soft system methodologists, such as Checkland (1981), point out that organisational models or processes do not have a real existence: they only exist in the perceptions of the individuals who take part in them. However, processes are treated by most people as real. Behind process planning often lies the assumption that there is a possible or even optimal process out there that can be followed. In some of the following models the processes are treated as deterministic – once task A has finished, task B will begin – whilst others are probabilistic.

The way a model is described influences the way people think and act in processes far beyond the scope of planning itself. If a process is modelled through input and output documents, as in many IDEF models, designers will strive to generate these documents. A DSM will focus the view on parameter dependencies and feedback loops, and a Signposting model might draw attention to the maturity of information.

As pointed out previously, the selection of factors included in a model is purpose-driven. This makes a process model very personal to the individual or team who generates it. It also influences the focus of attention in design processes. For example, the parameters modelled explicitly in a DSM might have more attention paid to them than those hidden from view.

As far as we are aware, to date no specific research has been done to look at how the structure of models influences the way designers think about processes and how they directly or indirectly influence the activities to be done. However, as Chapter 8 on psychology argues, the representation of the product profoundly influences the product that is created with it. Anecdotal evidence from the way design managers speak about processes indicates that the models of the processes and the way they are used for planning greatly influence the outcome.

Frameworks and models

The modelling approaches discussed here should be considered as two distinct elements: frameworks and process instances built on these frameworks. A modelling framework is a generic approach which may be applied to modelling any situation within its scope, but which in itself provides little specific guidance or insight. Process instances are the models created on a framework which provide specific guidance related to their content. Taking an analogy from a form of 3D modelling familiar to most children, the framework is the sandbox, while the instances are the sandcastles built within it. Just as the properties of sand limit the forms that may be created, so too a framework places constraints on the features of models that may be built within it.

The framework is the sandbox, while the instances are the sandcastles built within it. Just as the properties of sand limit the forms that may be created, so too a framework places constraints on the features of models that may be built.

The boundary between prescription and description is blurred in the case of instances built on a modelling framework. The initial construction of a model instance is essentially a description of a design process, whether observed directly, inferred or a statement of intended actions. Once an instance has been created in this way, it may be used for prescription, either to run a project that matches the original description, or the design-process model may be manipulated to obtain a process that is in some way 'better' or optimised. The effectiveness of this prescription based on description is dependent on the accuracy with which the model reflects the real process. Also important is the range of manipulation of the model instance which is possible: a more flexible framework will allow the exploration of a wider range of alternative design processes around the original configuration.

All of the frameworks that follow provide for the capture and representation of design processes in terms of component activities. Given the generality of some of these frameworks, it may be possible to express the same model instance in terms of a number of them, e.g. converting a DSM model into Signposting, or a Signposting model into IDEF. Nonetheless, these frameworks are by no means identical or equivalent, and significant differences exist in terms of the process phenomena that may be modelled, the tools and methods for analysis and representation that exist, and the balance between the cost of building one of these models and the detail that is captured. Each framework highlights certain aspects of the process, while hiding others. Therefore, each framework biases the process it describes.

Modelling approaches

This section reviews a number of modelling approaches that are used or proposed in industry, and which capture a specific design process rather than prescribe a procedure for carrying out all processes.

Modelling approaches may be used to capture a specific design process rather than prescribe a procedure for carrying out all processes.

PERT / CPM

The process evaluation and review technique (PERT) and critical-path method (CPM) are the two best known examples of the more general Precedence Diagramming Method (PDM) (PMI Standards Committee, 2000). In all these methods the activities are shown as nodes or boxes on a network and arrows joining the nodes signify the flow of information or material from one task to another. An alternative, less-used representation is the Arrow Diagramming Method (ADM, also known as Activity-On-Arrow, AOA) in which nodes denote system states and arrows represent activities.

Both techniques are primarily concerned with deriving the degree of 'float' (slack) or scheduling flexibility for each of the activities in a process, differing primarily in the value of activity duration which is used (CPM uses the modal duration, while PERT uses a weighted average of lowest, highest and most likely durations). A forward pass, propagating durations forward from a planned start date, is combined with a backward pass, propagated back from the target project completion date. These combine to give an earliest and latest start and finish date for each activity, indicating which activities are most important or carry the highest risk. An alternative view is that of the 'critical path' – the longest duration sequence of activities involved in the process. Any delay on the critical path will (without management intervention) cause a delay in delivery.

Code	IDEF method
IDEF0	Function modelling
IDEF1	Information modelling
IDEF1X	Data modelling
IDEF2	Simulation model design
IDEF3	Process description capture
IDEF4	Object-oriented design
IDEF5	Ontology description capture
IDEF6	Design rationale capture
IDEF7	Information system auditing
IDEF8	User interface modelling
IDEF9	Scenario-driven IS design
IDEF10	Implementation arch. modelling
IDEF11	Information artifact modelling
IDEF12	Organization modelling
IDEF13	Three schema mapping design
IDEF14	Network design

2.7 The IDEF family of models

PDM are routinely used by planners, and form the core of many commercially available project management software tools. They are particularly useful in situations where activities are more predictable, such as manufacturing. These methods are simple and in their basic form cannot represent process features such as iteration, although extensions to the methods (such as GERT and Q-GERT) have been proposed by a number of authors.

IDEF

IDEF is not a single model, but rather a family consisting currently of 16 different modelling structures (IDEF0 – IDEF14, including IDEF1X), although IDEF5 and above are still in development. The first IDEF models, IDEF0, IDEF1 and IDEF2, were intended to support systems engineering and analysis, but the scope of the IDEF family has widened, as has the application of the existing standards (see Table 2.7). Integration DEFinition or IDEF models have their origins in a 1981 US Air Force programme for integrated computer-aided manufacturing (ICAM).

Of the developed IDEF models, IDEF0 and IDEF3 have particular relevance for this work. IDEF1 is generally used in the structuring and relating of information without action or interaction. IDEF2 is a formal structure for representing scenario analysis within simulations, but is rarely used in practice. It is intended to provide a common language between simulation experts and domain experts, but in this research, as elsewhere, it seems likely that adding an extra layer of complexity would detract from, rather than improve, the usability of the model. IDEF4 is an object-oriented design method, most commonly used in software development.

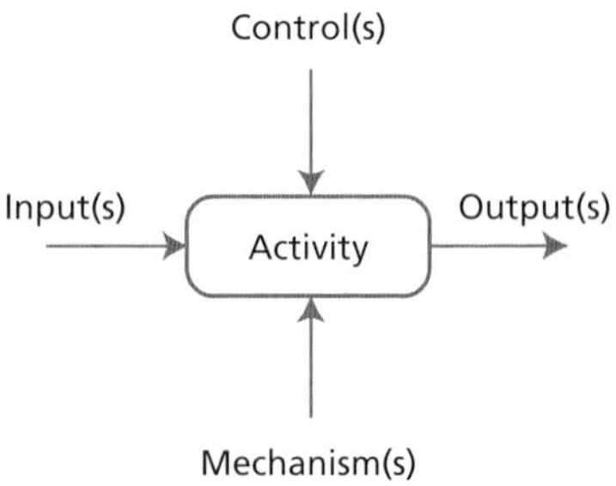

2.8 A task represented in IDEF0

IDEF0 (NIST, 1993) was the first standard to be introduced, and is well established and widely used. Originally intended to model the functional behaviour of engineering systems, it has since been applied to a broad spectrum of business processes, including design. (The process models observed in development at Airbus are based on IDEF0.) IDEF0 represents a process as being composed of a network of functions or activities, each having inputs, controls (for example, policies, standard working practices), outputs and mechanisms (for example, people, tools), referred to as ICOMs (Figure 2.8).

IDEF0 applied to design-process modelling (Godwin *et al.*, 1989; Kusiak and Wang, 1993) indicates flows of information and resources, both those consumed by activities (information, materials, money – represented as inputs) and unconsumed (people, tools – represented as mechanisms). It indicates precedences of activities (driven by information dependencies) but not the timing: adjacent activities in a model may occur at very different points in the process. In contrast, IDEF3 (Mayer *et al.*, 1995; Noran, 2000), the *process description capture method*, is intended to capture the dynamic behaviour of a process. There are two modelling approaches contained within IDEF3: the *object state transition description*, which is more appropriate for software and manufacturing applications, and the *process flow description* (Figure 2.9), which comprises a flow diagram of units of behaviour (UOBs). The latter indicates both temporal and information-based precedences of activities, and may incorporate such features as random outcomes and iteration.

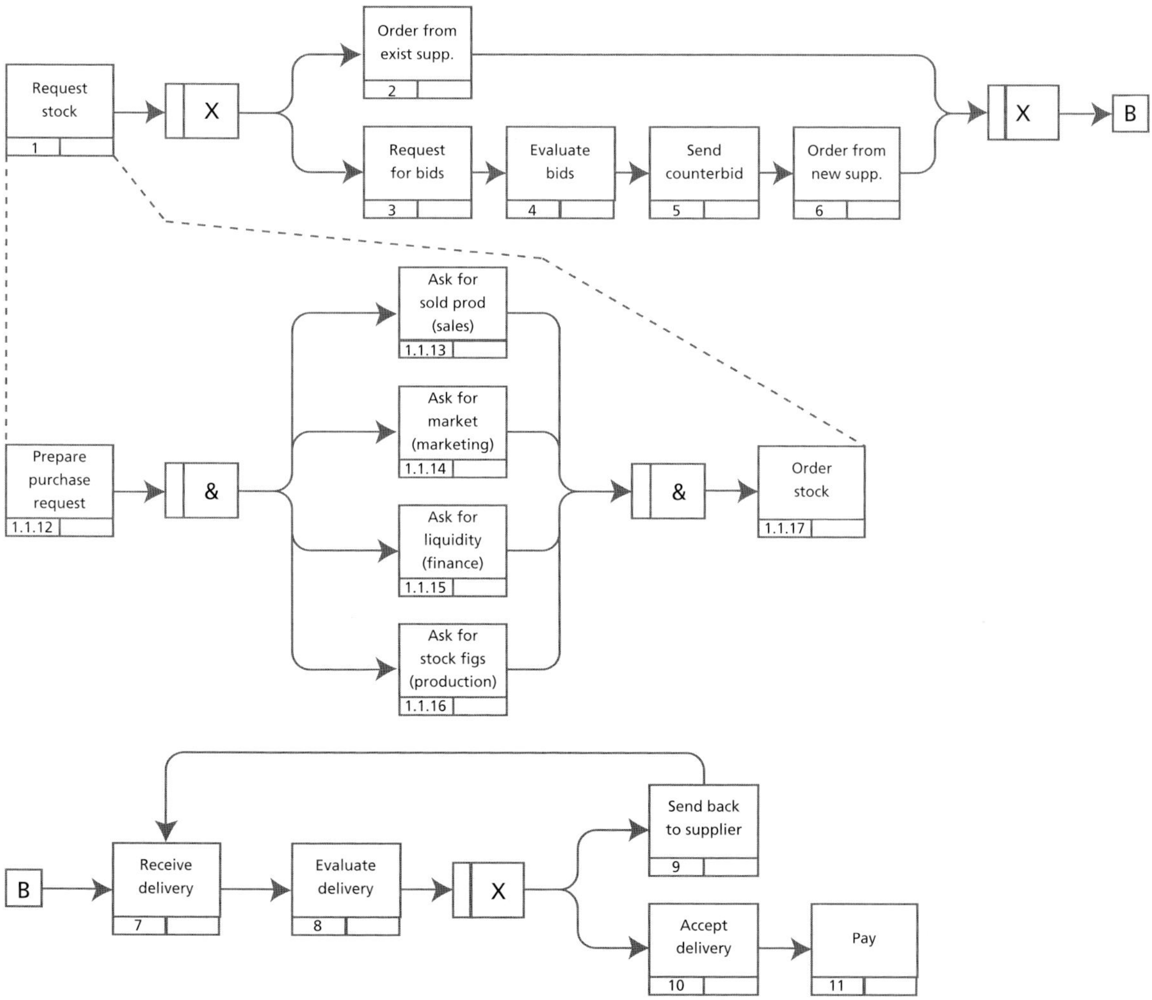

2.9 An IDEF3 process model (Reproduced with permission – Noran, 2000)

Both IDEF0 and IDEF3 have strong hierarchical structures through which activities or UOBs on one level of detail may be decomposed into component activities/UOBs at a lower level. In the short study at Airbus, it was observed that the IDEF0 model which had been created there had six levels of hierarchy, illustrating both the strength of the IDEF techniques in dealing with hierarchical structures and the complexity of the aerospace design process.

DSM

DSM (variously expanded as the design structure matrix, problem solving matrix (PSM), dependency structure matrix and design precedence matrix) is a modelling approach created by Steward (1981) based on matrix algebra and precedence diagram work in the 1960s and further developed by a number of researchers, most notably the DSM group at MIT (Eppinger *et al.*, 1994; Smith and Eppinger, 1997; Browning, 2001). DSMs are used to represent and analyse process models. Other applications of the DSM have included product modelling (links between systems or components) and team modelling (links between people or teams). Unlike product and team structure DSMs, the links in a process DSM are directional, with information flowing from one activity to another.

As shown in Figure 2.10, a DSM is a square matrix with corresponding rows and columns. The diagonal cells represent activities in a process, listed temporally. Off-diagonal cells indicate the dependency of one activity on

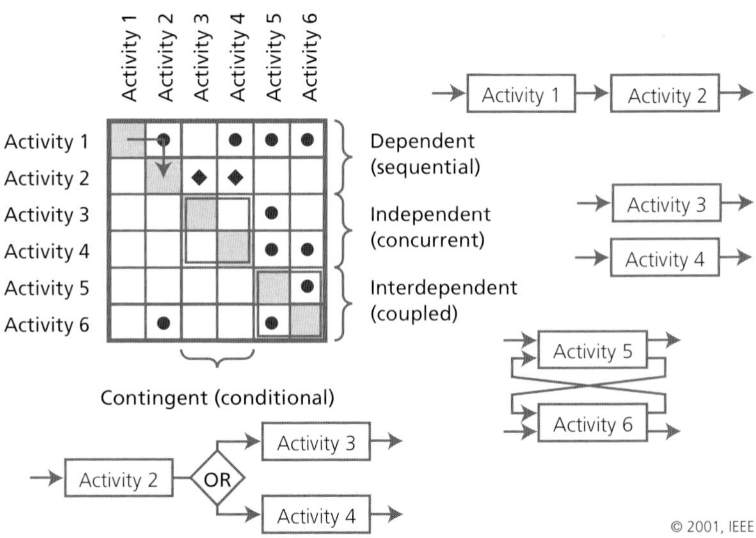

2.10 An example DSM (adapted from Browning, 2001. IEEE Transactions on Engineering Management)
© 2001 IEEE

another. Dependencies typically imply needs for work, products or information. Reading down a column shows sources; reading across a row shows sinks. For example, row 1 indicates that Activity 1 provides one or more deliverables to Activities 2, 4, 5, and 6. Column 2 shows that Activity 2 depends on something from Activities 1 and 6. Some DSM literature reverses this row column definition (*i.e.* transposing the matrix), but both conventions convey the same information.

Figure 2.10 also shows how the DSM displays dependent, independent, and interdependent activity relationships. Since Activity 2 depends on information from Activity 1, these two activities will probably be executed sequentially in the workflow. Activities 3 and 4 do not depend on each other for information, so they may safely proceed in parallel (barring resource constraints). Activities 5 and 6 both depend on each other's outputs. These activities are said to be interdependent or coupled. A decision is a kind of activity, one which produces information upon which other activities depend. Their sequence in a process will have a great bearing on its efficiency and effectiveness.

The DSM approach has been applied very widely in case studies, in a wide spectrum of industry sectors, notably aerospace, automotive and architecture, engineering and construction (AEC).

Of particular interest are the cases where marks appear in the lower triangular region of the DSM. Such marks indicate the dependence of an upstream activity on inputs created downstream. If project planners decide to execute the activities in this order, Activity 2 will have to make an assumption (an input proxy) about the input it needs from Activity 6. After Activity 6 finishes, Activity 2 may have rework if the assumption was incorrect. The DSM conveniently highlights iteration and rework, especially when it stems from activities working with potentially flawed inputs.

When we see a mark in the lower left corner of the DSM, we know that there is a chance of having to return to the beginning of the process, which could have a catastrophic impact on cost and schedule. The marks in the lower-left corner of the DSM may represent key drivers of cost and schedule risk.

Rearranging the activity sequence (by rearranging the rows and columns in the DSM) can bring some subdiagonal marks above or closer to the diagonal, thereby reducing their impact. Simple algorithms automate this exercise. Adding quantitative information to the DSM and using simulation can quantify the impacts of process architecture changes on cost and schedule risk (Browning and Eppinger, 2002) (Figure 2.11).

Sometimes a subdiagonal mark cannot be brought above the diagonal without pushing another mark below it. This is a case of interdependent

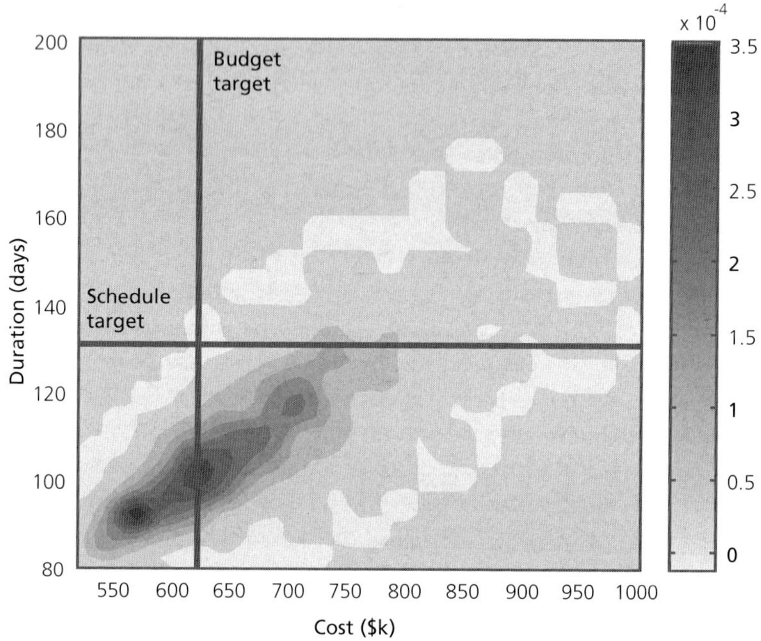

2.11 Simulation applied to DSM models (from Browning and Eppinger, 2002)

activities, such as Activities 5 and 6. Each activity depends on the other. They must work together to resolve a 'chicken-and-egg' problem. Typically, coupled activities work concurrently, exchanging preliminary information frequently. If a subset of coupled activities must begin before the rest, the more robust (less volatile and/or sensitive) deliverables should be the ones appearing below the diagonal in the DSM. If coupled activities are functionally based, an opportunity may exist to fold the activities into a single activity assigned to a cross-functional team.

The DSM provides a concise, visual format for representing processes.

Integration, test, and design review activities typically have marks in their rows to the left of the diagonal. These activities create outputs (including results of decisions) that may cause changes to (and rework for) previously executed activities. Unfortunately, most process planners 'plan to succeed' and their process models fail to account for these possibilities. Fortunately, the DSM provides an easy way to document potential 'process failure modes' and their effects on other activities. The simple marks in the DSM can be replaced by numbers indicating the relative probability of input change, iteration, *etc.* This enables an analysis of process failure modes and their effects on cost, schedule, and risk. Process improvement investments target mitigation of the biggest risk drivers.

By accounting for contingent activities and feedback loops, the DSM provides a basis for exploring adaptive processes. While the DSM itself is a static view of a process, it can be updated over time to reflect a current situation. The remaining activities in such a situation can then be quickly resequenced in an advantageous way, providing rapid project replanning.

The DSM provides a concise, visual format for representing processes. A process flowchart consuming an entire conference room wall can be reduced to a single-page DSM. After a quick orientation, everyone can see how his or her activity affects a large process. People can see where information comes from and where it goes. They can see why delaying the activities they depend on forces them to make assumptions, which may trigger rework later. It becomes apparent that certain changes tend to cause rework. Such situation visibility and awareness lead to process innovation and improved coordination. The DSM can provide a portal to a process knowledge base from which the foundations of process plans and risk assessments can be drawn. Moreover, the DSM is amenable to some simple yet powerful analyses.

The DSM approach has been applied very widely in case studies, in a wide spectrum of industry sectors (Browning, 2001), notably aerospace, automotive and AEC. The simplicity and accessibility of the core representation to industry appears to have contributed to the success of the method, particularly as a focus for very large-scale elicitation exercises involving groups of 50 or more (Guivarch, 2002). An adaptation of the DSM method, ADePT, created at Loughborough University (Austin *et al.*, 2000) has been successfully commercialised as the Planweaver software tool. This integrates a dependency structure matrix with a library of generic processes (originally targeting the AEC industry) and a set of tools for performing DSM and project scheduling operations (exportable to MS Project, *etc.*).

2.12 'Signposting', a response to the challenges associated with helicopter rotor blade design (Clarkson and Hamilton, 2000)
© AgustaWestland

Signposting

Signposting is an activity-based model of the design process which represents activities in terms of their information input/output characteristics. This framework was created in the Cambridge Engineering Design Centre as a response to the challenge of modelling helicopter rotor blade design (Figure 2.12) at Westland (Clarkson and Hamilton, 2000). This work introduced the concept of 'confidence' to describe the designer's belief in the suitability of parameter values during the process, providing a measure of the contextual meaning of the design data which is independent of the actual values of that data. This decoupling of value and contextual meaning of the design definition is of significant value in building generic classes of model that can be applied to a range of future projects.

Confidence levels (usually none/low/medium/high) are used to describe the state of all significant aspects of a design, giving rise to the 'design state', which may be visualised as a vector of the confidence values for all parameters. Not all design states are reachable, as the moves possible in 'design-state space' are restricted by the design activities. Activities take the form of transformations from one design state to another, being described in terms of input and output design states (O'Donovan *et al.*, 2004).

The original application of the Signposting model was to model the highly iterative rotor blade design process at Westland in a form that could be accessed by novice designers. It was observed in industry that many engineers had competence at individual activities within a design-process, but a lack of understanding of the overall strategy of the design work. Signposting aimed to address this 'what to do next?' problem by colour coding activities according to whether they were possible (required inputs available) and useful (the output would contribute to improving the design process, shown as an increase in confidence levels) – red for activities which were not possible, yellow for possible but not useful, and green for possible and useful.

One observation from the original project was that at certain points during the process there were too many 'green' activities to choose from, and that some further indication was needed as to which of the possible, useful activities should be performed next. The research carried out by

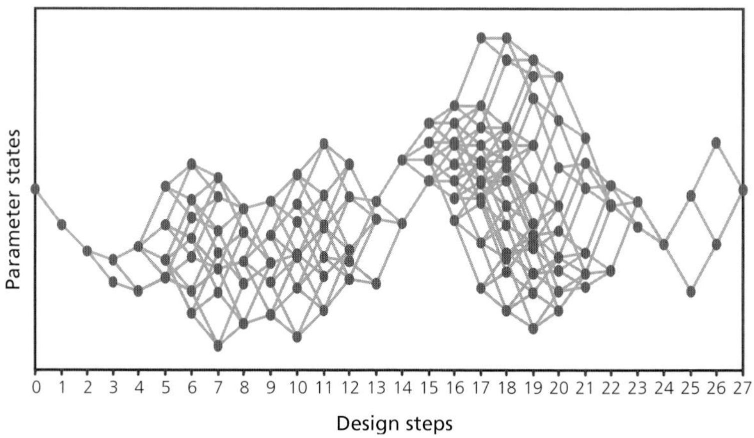

2.13 The gradient diagram showing routes through a process (Clarkson *et al.*, 2000)
© 2004 IEEE

Melo and Clarkson (2002) addressed the question of prioritisation of activities by looking at the reasons behind the 'soft' precedences identified by experts (activity precedences which were preferred, but could be broken). It was found that many of these precedences were linked to the risk of rework, e.g. performing testing before carrying out design work which could be invalidated if the basic design failed the test. In order to investigate these factors, the emphasis of the work shifted from identifying a 'next activity' to looking for complete routes through the design process.

The Signposting model was subsequently extended to include probabilistic data. Iteration in the design process was captured through 'failure' outcomes from activities and associated failure probabilities that determine which outcome occurs. Markov chain analysis was used to find routes through the design process which carried the lowest risk or expected cost, and the families of best routes were expressed in the form of a DSM. In this work the emphasis moved from support of engineers working on a design project to support for the project leaders managing the project, and a number of visualisations, notably the gradient diagram (Figure 2.13), were created to serve this audience (Clarkson *et al.*, 2000).

More recently, the Signposting framework has been updated by O'Donovan *et al.* (2003) to incorporate a number of significant aspects of the design process that could not be modelled with the previous versions. These include resource limitation, activity concurrency, learning/experience curve effects and the impact of trading off quality against process duration or cost. Models built on this extended framework may be extremely rich and detailed, but this must be weighed against the additional cost required to create them. For this reason, the extensions must be seen not as a fundamental change to the core of the model, but as a set of modular options that may be used or ignored to suit different needs.

As these models have become more complex, the information that may be extracted from them increases beyond what can be extracted simply by looking at a record of the activities. Hence, a simulation-based approach was adopted in order to reveal the distributions of process cost and duration (Figure 2.14), importance of activities and optimum sequencing of, and resource allocation to activities (O'Donovan *et al.*, 2004).

Signposting is now being developed to assess what-if scenarios, based on the simulation algorithms. This allows the project leaders to experiment with resources, times, cost factors and probabilities and assess the effect of the changes on the overall performance of the design process.

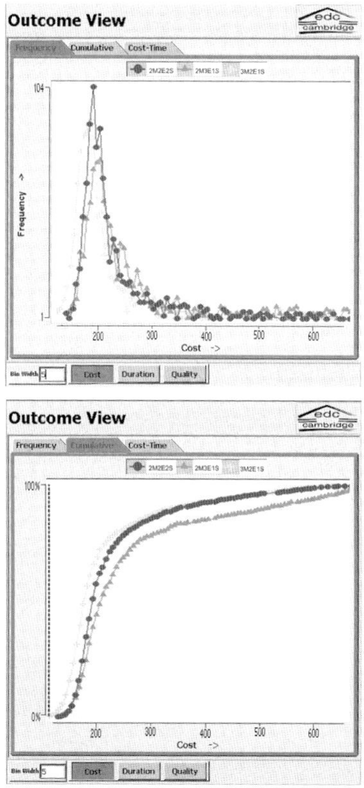

2.14 Typical simulation outputs including cost/frequency and distribution diagrams

Petri nets

Petri nets are a well established and developed tool for graphing and simulating discrete event systems (Peterson, 1981). In decomposing a process into activities, a discrete event structure is created which may be explored through Petri nets.

The central components of the method's representation are places, transitions, arcs and tokens. Places, shown as circles, are nodes which generally represent inputs to, or outputs from, activities in the process being modelled – in the case of engineering design processes these may be the implicit and explicit properties of the design object – auxiliary information (McMahon *et al.*, 1993), or resources (people and tools) needed to perform an activity (Dou and Cai, 2002; Puangpool and Damrianant, 2002). Transitions are shown as bars, and represent the actions possible in the system – design activities in the engineering design process. Arcs are arrows that point from places to transitions (indicating inputs to a transition/task) or from transitions to places (indicating outputs). Tokens are dots within the circle indicating a place, which show the presence/absence/quantity of the object represented by the place.

The basic scheme of simulating the actions of a process/system represented by a Petri net is that transitions are 'fired' when all input places have a token in them (Figure 2.15). The tokens are removed from the inputs, and a new token is placed on each of the output places. A large number of standard and non-standard extensions to this basic representation have been made, including OR inputs, stochastic outputs and extensions of the visual representation.

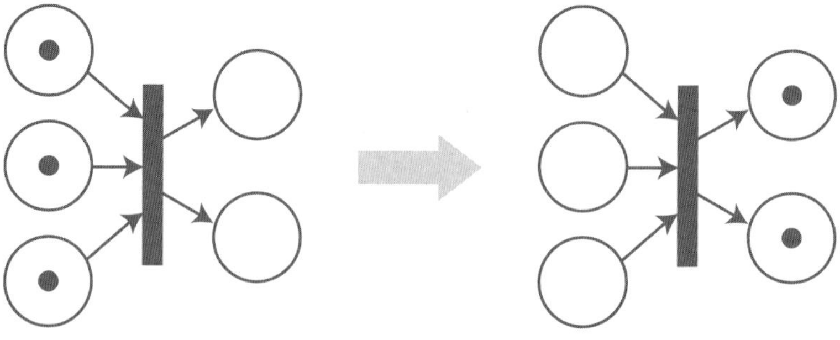

Once a token is present in each input 'place', the transition is triggered

After the transition has fired, tokens are placed in the output places

2.15 Transition in a simple Petri net

Petri nets have been used for modelling a very wide range of systems, from software and electronic systems to manufacturing. In the field of engineering design, models have been proposed by a number of authors. Of these, some (Dou and Cai, 2002; Puangpool and Damrianant, 2002) are highly specific applications of Petri nets to specific engineering design problems, while others (McMahon *et al.*, 1993) are more general, acting as design-process-specific frameworks within the broader Petri net framework.

Other frameworks

The Generic Design Model (GDM; van Langen, 2002) is an attempt to produce a comprehensive formal structure for describing the design process. In this work, van Langen extends predicate logic with a design-specific vocabulary, defining generic taxonomies of tasks and other entities (requirements, design data, *etc.*). This model is difficult to apply to the totality of the design process due to the volume of information that would be required and the need to represent formally all of the significant design concepts involved. However, a number of examples taken from parts of the design process suitable for the approach are provided.

Petri nets have been used for modelling a very wide range of systems, from software and electronic systems to manufacturing and engineering design.

The GDM is interesting, in that the activity taxonomy is not simply a set of labels, but instead a means of relating activities in a specific instance to highly detailed generic activity models such as 'Requirement Qualifier Set modification' and 'Design object data manipulation', which have comprehensive heuristics for interaction with each other and the coordinating goals of the design process. Once the generic model has been labelled with the correct elements of the specific instance, and extended with the domain-specific knowledge necessary, it is actually capable of executing the design without human intervention. As such it is of more interest for research into process automation and intelligent-agent applications in design than for designer support. The design process here is seen as a decision-making process, and the model captures this.

MILOS, or 'Minimally Invasive Long-term Organisational Support' is a methodology being developed in a joint project between the software process support group at the University of Calgary and the Artificial Intelligence Group at the University of Kaiserslauten (Dellen and Maurer, 1996; Maurer *et al.*, 2000). MILOS seeks to integrate project planning with workflow management systems for the software engineering industry. The process model within this tool is activity based, defining activities in terms of input and output parameters. These activities are dynamically assembled into a process by matching inputs and outputs. Although not intended for modelling general engineering design processes, this system provides an example of a process model being used to create a dynamic plan which adapts to the changing circumstances in a project. The choice between alternative activities is handled through basic scheduling algorithms. Iteration is not a major feature of the model.

The UML is a modelling framework which was developed for modelling software, but has been adapted for modelling business processes.

The Universal Modelling Language (UML) (Booch *et al.*, 1998; Jacobson *et al.*, 1999; Rumbaugh *et al.*, 1999) is a modelling framework which has evolved from earlier object-oriented software design methodologies such as the Booch Method (Booch, 1993) and OOSE (Jacobson *et al.*, 1992). Although still primarily applied in software engineering, the basic modelling framework is highly generic and has been adapted for modelling business processes. For example, in the Ericsson–Penker Business Extensions (Noran, 2000), the activity model in UML resembles the IDEF0 representation, but is supplemented by the other 'diagrams' which indicate the organisation, support systems, interactions, *etc*. Business process modelling under UML is still being developed and, where timing and resourcing issues are not critical, most of the concepts can be applied quite directly.

Conclusion

Planning their design processes is a great challenge for many companies as they strive to complete projects to time and to budget. Currently, companies often use multiple plans that meet the needs of individuals, but rarely have detailed high-level plans. Many plans are represented in Gantt and PERT charts, which do not show the nature of connectivity between activities and hide activity failure and iteration in its representation. As a result, many organisations do not understand or even appreciate the complexity of their design processes. Plans based on Gantt and PERT charts are inevitably more a work of fiction than a representation of reality.

No modelling framework is available at present to capture the entire richness of design process – although IDEF, DSM and Signposting make it possible to focus on particular features of design processes. However, these planning approaches still have limitations. It is inherent in the nature of design process execution that new activities and connections emerge. Provided these can be captured, the process can then be re-planned or re-represented.

Current modelling frameworks are based on the assumption of a lack of ambiguity, *i.e.* the models do not formally account for potentially missing activities at the chosen level of abstraction (they can account for them at the lower levels by allowing for uncertainty in the chosen activities). If the decisions to be made are assumed to be unknown *a priori*, then that is a highly ambiguous or chaotic project using Pich *et al.*'s (2002) terms and most modelling frameworks will be challenged to represent it. Additional research is needed to provide a framework to accommodate emergent and adaptive process structures.

Despite the variety of modelling frameworks and approaches, they tend to have a common kernel, a basis on an activity network. There is the potential to unify many of these modelling frameworks into an object-oriented one, in which the activities are among many potential objects, each with a number of potential attributes (Browning, 2002). Various objects and attributes may then be used, or not, depending on a particular model's purpose. Moreover, a rich model serving a variety of purposes could be presented partially to any specific user, or to support any specific use, by employing principles of information hiding. Hence, many of the model instances discussed above become partial 'views' of a simpler, more generic, yet richer modelling framework. Future research will carry forward this idea.

References

Austin S, Baldwin A, Li B, Waskett P (2000) Analytical design planning technique (ADePT): a dependency structure matrix tool to schedule the building design process. Construction Management and Economics, 18: 173–182

Booch G (1993) Object-oriented analysis and design with applications. Addison-Wesley

Booch G, Jacobson I, Rumbaugh J (1998) Unified modelling language user guide. Addison-Wesley

Box GEP (1976) Science and statistics. Journal of the American Statistical Association, 71: 791–799

Browning TR (2001) Applying the design structure matrix to system decomposition and integration problems: a review and new directions. IEEE Transactions on Engineering Management, 48(3): 292–306

Browning TR (2002) Process integration using the design structure matrix. Systems Engineering, 5(3): 180–193

Browning TR, Eppinger SD (2002) Modeling impacts of process architecture on cost and schedule risk in product development. IEEE Transactions on Engineering Management, 49(4): 428–442

Checkland P (1981) Systems thinking, systems practice. John Wiley

Clarkson PJ, Hamilton JR (2000) Signposting: a parameter-driven task-based model of the design process. Research in Engineering Design, 12(1): 18–38

Clarkson PJ, Melo A, Eckert CM (2000) Visualization of routes in design process planning. IV'00, London, UK

Dellen B, Maurer F (1996) Integrating planning and execution in software development processes. WET ICE'96, Stanford, CA, USA

Dou WC, Cai SJ (2002) Hybrid system for functional analysis on different stages of production system lifecycle. CE2002, Cranfield, UK

Earl CF, Eckert CM, Johnson JH (2001) Complexity of planning in design. ASME DETC'01, Pittsburgh, PA, USA

Eckert CM, Clarkson PJ (2003) The reality of design process planning. ICED'03, Stockholm, Sweden

Eppinger SD, Whitney DE, Smith RP, Gebala DA (1994) A model-based method for organizing tasks in product development. Research in Engineering Design, 6: 1–13

Godwin AN, Gleeson JW, Gwiflian D (1989) An assessment of the IDEF notations as descriptive tools. Information Systems, 14(1): 13–28

Guivarch AD (2002) Resolving timing issues using DSMs: application at Ford Motor company. DSM'02, Boston, MA, USA

Jacobson I, Christerson M, Jonsson P, Overgaard G (1992) Object-oriented software engineering: a use case driven approach. Addison-Wesley

Jacobson I, Booch G, Rumbaugh J (1999) Unified software development process. Addison-Wesley

Kusiak A, Wang J (1993) Efficient organizing of design activities. International Journal of Production Research, 31(4): 753–769

Van Langen PHG (2002) The anatomy of design: foundations, models and applications. PhD thesis, Vrije Universiteit Amsterdam, The Netherlands

Maurer F, Dellen B, Bendeck F, Goldmann S, Holz H, Kötting B, Schaaf M (2000) Merging project planning and Web-enabled dynamic workflow technologies. IEEE Internet Computing, 4(3): 65–74

Mayer R, Menzer C, Painter M, deWitte P, Blinn T, Perakath B (1995) IDEF3 process description capture method report. Knowledge Based Systems Incorporated, Texas, TX, USA http://www.idef.com/Downloads/pdf/Idef3_fn.pdf

McMahon CA, Sims Williams JH, Brown KN (1993) A transformation model for the integration of design computing. ICED'93, The Hague, The Netherlands
Melo AF, Clarkson PJ (2002) Planning and scheduling based on an explicit representation of the state of the design. ASME DETC'02, Montreal, Canada
Meyer C (1998) An introduction to complexity and business. Complexity, 3(4): 21–22
NIST (1993) Integration definition for function modeling (IDEF0). Federal Information Processing Standards (FIPS) Publication 183, National Institute of Standards and Technology, Gaithersburg, MD, USA
Noran OS (2000) Business modelling: UML vs. IDEF. Griffith University, Nathan, QLD, Australia
http://www.cit.gu.edu.au/~noran/Docs/UMLvsIDEF.pdf
O'Donovan B, Clarkson PJ, Eckert CM (2003) Signposting: modelling uncertainty in design processes. ICED03, Stockholm, Sweden
O'Donovan B, Clarkson PJ, Eckert CM (2004) Simulating design processes to assist design process planning. ASME DETC'04, Salt Lake City, UT, USA
Peterson JL (1981) Petri net theory and the modeling of systems. Prentice Hall
Pich M, Loch C, De Meyer A (2002) On uncertainty and complexity on project management. Management Science, 48(8): 1008–1023
PMI Standards Committee (2000) A project management body of knowledge (PMBOK). Project Management Institute
Puangpool S, Damrianant J (2002) An application of restricted colored petri nets for concurrent engineering in construction. CE2002, Cranfield, UK
Rumbaugh J, Booch G, Jacobson I (1999) Unified modelling language reference manual. Addison-Wesley
Smith RP, Eppinger SD (1997) Identifying controlling features of engineering design iteration. Management Science, 43(3): 276–293
Steward DV (1981) The design structure matrix: a method for managing the design of complex systems. IEEE Transactions on Engineering Management, EM-28(3): 71–74
Suh NP (1990) The principles of design. Oxford University Press

Chapter 3

Systems engineering

Peter Buckle
University of Surrey

John Clarkson
University of Cambridge

Despite advances in engineering knowledge and technology the everyday experience of the engineered world provides, all too often, evidence of failure as well as success. For example, as a literate and healthy human is it unreasonable to expect:

- to be readily able to set the alarm function on my digital watch and to be confident that it will work?
- to be able to read the instructions on food packaging?
- to correctly change batteries, first time, on an electric toothbrush?
- not to have to move every few minutes to prevent the office being plunged into darkness by a motion sensitive, power-saving system?
- to have my 'patient's notes' present at the same time as myself in an otherwise high-tech clinic?

All these problems, and more, have beset our group recently. The list is long, the explanation occasionally obvious (for example, the batteries were inserted incorrectly because it is almost impossible to see the polarity signs embossed on the internal base of the toothbrush battery casing) but the implications for engineering are enormous. Quite simply, they force us to ask whether the engineering process itself is correct.

All engineered environments and artefacts have human involvement. Even so-called 'fully automated processes' are anything but that. On analysis we find that they are specified and designed by humans, tested by humans, commissioned by humans, maintained by humans, and subsequently decommissioned and disposed of by humans. The need for a systematic approach to design that is inclusive of 'the human factor' is evident, but is it acted upon?

All engineered environments and artefacts have human involvement.

Even when the 'human factor' in the system is considered, it is often forgotten that whilst humans may come as individuals, they always work as groups, teams, organisations and, even, societies. Understanding the resultant needs, behaviours and attitudes is integral to systems engineering. Pheasant (1996) identified five fallacies of engineering design (Table 3.1). The common thread that runs through them all is the need to recognise that design, to be successful, must adopt a systems approach. How then to avoid such traps and develop systems that truly reflect modern thinking and knowledge?

The following sections present an introduction to systems engineering and ergonomics, focusing on the way in which they should influence the design process. Examples are presented to illustrate the key issues. Many are from the healthcare industry, where safety can only assured if a systems approach is adopted.

Fundamental fallacies regarding design
1. The design is satisfactory for me – it will therefore be satisfactory for everybody else
2. The design is satisfactory for the average person – it will therefore be satisfactory for everybody else
3. The variability of human beings is so great that they cannot possibly be catered for in any design – but since people are wonderfully adaptable it does not matter anyway
4. Ergonomics is expensive since products are actually purchased on appearance and styling, ergonomic considerations may conveniently be ignored
5. Ergonomics is an excellent idea. I always design things with ergonomics in mind - but I do it intuitively and rely on my common sense so I do not need tables of data or empirical studies

3.1 Fundamental fallacies (Pheasant, 1996)
Reproduced with permission of Taylor and Francis Ltd.

Systems engineering and ergonomics

Systems engineering is a process through which the analysis of existing systems and appropriate knowledge can be applied to new design problems. The emphasis is placed very clearly on the process and not the product. In reality, this will require addressing the needs of all stakeholders, including the end users.

In August 2000, the International Ergonomics Association Council adopted an official definition of the discipline of ergonomics. This states that:

> *ergonomics (or human factors) is the scientific discipline concerned with the understanding of interactions among humans and other elements of a system, and the profession that applies theory, principles, data and methods to design in order to optimise human well-being and overall system performance.*
>
> (IEA, 2000)

The very close relationship between systems engineering and ergonomics is readily apparent.

Human factor mismatches in work systems

For existing systems, a simple model has been presented to enable an appreciation of the need to consider how to avoid mismatches between users and work systems, in particular the managed and the engineered environments. Whilst the examples worked through below focus on 'mismatches' or problems that occur in systems, the same thinking may be applied to enhancing systems that are already deemed to be working 'satisfactorily'.

By way of explaining this model, we may start with the fact that in any work system work tasks are performed in order to meet specific goals. For example, cars are assembled, accounts are processed, customer enquiries are received and dealt with, software is installed, *etc*. On the right-hand side of Figure 3.2 we can observe that some of these tasks are allocated to machines (for example, production lines carry components around the workspace, tools exert high forces to secure components in place during assembly, computers store large quantities of detailed numerical data) whilst other tasks are allocated to the human operators (for example, saving and retrieving data, operating tools, fixing breakdowns, talking to customers). Task analysis is a specialised topic (see Annett and Stanton (2000)) and is an essential part of the process of understanding existing systems and subsequently developing new ones.

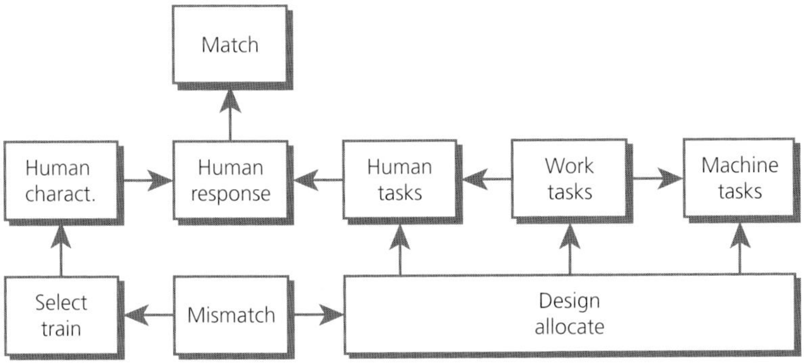

3.2 Human factor mismatches in work systems

Having undertaken such an analysis, the first critical question that often emerges is "on what basis are specific functions (and hence tasks) allocated to either people or machines?" Often the answer is "unclear!" Closer inspection frequently reveals a 'default' decision process, in that if there is a machine that can do it then, use the machine, and if not, get a human operator to do the task. Such an approach affords little attention to the relative advantages of people versus machines and is, in any event, unlikely to lead to coherent, meaningful jobs or sets of tasks for the worker(s).

On the left-hand side of Figure 3.2 is a box labelled 'human characteristics'. Most work systems employ, or engage with, a wide range of people. Usually, little attention has been paid to their capacities, needs or abilities (Coleman, 1999; Clarkson and Keates, 2001). Too often, much is assumed and little researched. The consequences of this are serious.

It would be inconceivable, for example, to imagine an engineer designing a control panel without careful consideration being given to, for example, the power required to illuminate a warning light and whether the circuit had power back-up. A legitimate question that follows from this is whether similar care and attention is paid to the component in the system that has to detect the signal, make decisions and act on it (i.e. the human operator).

At this point, many questions may be raised. For example, how conspicuous must the light be to be clearly visible under all operating conditions (see, for example, Figure 3.3), what other tasks is the operator required to perform that might interfere with his/her ability to detect or respond, will all operators behave in the same way, how might a history of earlier 'false alarms' affect the operator's performance in the event of a true alarm signal occurring and how might the culture of the organisation in dealing with false alarms affect the operator?

3.3 Warning lights

One framework for closer examination of these complex interactions is shown in Figure 3.4, where the interface between the operator and the machine at a given point in time is shown. Note, however, that such a model is best considered a state model, with inherent dangers if states are assumed to be steady and stable over time or if all operators are seen as homogeneous and identical. Corlett and Clark (1995) provide a thorough introduction to engineering/ergonomics design for workspaces and machines.

The reality of failing to take a systems approach is all too often evidenced as a failure or as an inefficient process. Indeed, much of the time it is the occurrence of mismatches (bottom centre of Figure 3.2) that triggers an awareness that not all is well with a given system. Thus, the accident, the injury, the poor output, or the uncompleted maintenance schedule all alert us to 'a problem'.

However, the response to this problem often shows further evidence of inappropriate systems thinking. The common practice of 'fixing' the problem by taking the route on the left-hand side of the model is best described as "changing the operator". This usually comprises either selection or training of the operator.

In the case of the visual alarm, taking this approach might lead to recruiting only those with a high degree of visual acuity or to train operatives to be 'more careful' when detecting or responding to alarms. However, it is well accepted, that reliance on both the selection and training strategies fails to recognise their inherent dangers. If the system contains latent design errors, e.g. a light that cannot easily be seen when the display has sunlight falling on it, then no amount of selection or training will make a substantial difference. On the contrary, the raised stress level of the operator (i.e. knowing that they "should be able to cope" when they cannot) might even exacerbate the situation and lead to a greater likelihood of error.

Those engaged in ergonomics and human engineering have long since recognised that the preferred route for preventing problems and enhancing systems performance in existing systems is to follow the right-hand pathway in Figure 3.2. This places the emphasis on design/re-design. This may require a consideration of a range of issues which include:

- the system goals;
- the task allocation;
- the equipment design;
- interactions between sets of equipment and groups of people;
- the work organisation;
- the job design.

Whilst methods (e.g. Wilson and Corlett, 1995) exist for the analysis of all these components of the system, the complexity of such an approach is, at first sight, daunting.

A recent model (Moray, 2000) attempts to draw together the components of systems that need to be considered if we are to take this systems design or systems engineering approach. This model enables the various levels of the system to be conceptualised for the purpose of understanding, interpreting, evaluating, information collection, and design purposes. Such an approach and understanding is required for successful systems analysis and design. Further understanding of the 'big systems' picture can be found in Hendrick and Kleimer (2002).

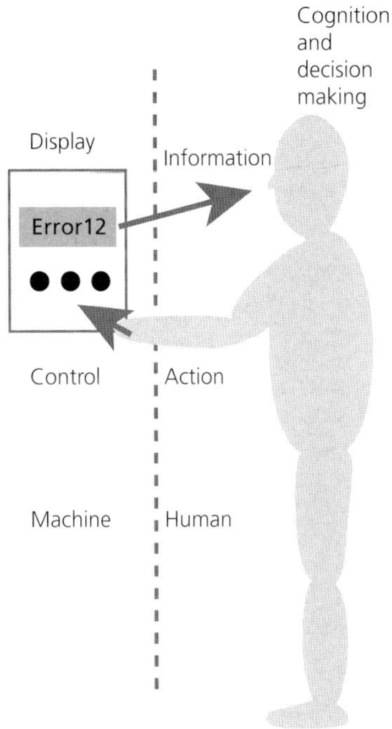

3.4 Human–machine interface

Error and systems engineering

In order to see how systems might be analysed it is perhaps helpful to consider specific examples. A recent study (Cambridge, RCA, Surrey, 2004) took a systems approach when reviewing the problem of medical error. Each year in the UK an estimated 850,000 people are involved in an adverse event caused by a medical error. The Medicines Control Agency received 18,196 reports of adverse drug reactions and the Medical Devices Agency received 6,610 reports of adverse incidents. The evidence of adverse incidents is almost entirely based on occurrences in secondary care (hospital) (Leape *et al.*, 1991, 1995; Wilson *et al.*, 1999).

In a study of adverse events by Wilson *et al.* (1999), Department of Health categories were identified as:

- a complication of or failure in technical performance of an indicated procedure or operation;
- the failure to synthesise, decide and/or act on available information;
- the failure to request or arrange an investigation, procedure or consultation;
- lack of care and attention or failure to attend to the patient.

A review of the current knowledge base showed that the problem is extensive, that there is little information about these problems outside of the secondary care setting (hospital), and that any engineered design solutions should, as a minimum, consider how they will address each of the four adverse events categories shown above. Case study 1 (below) considers an equipment interface and illustrates current problems.

The relevant information needed to reduce error in the design of equipment to be used by humans is readily available.
(Moray, 1994)

According to Moray (1994), the relevant information needed to reduce error in the design of equipment to be used by humans is readily available. However, even when all the ergonomic knowledge is applied to design of equipment the probability of error cannot be completely eliminated. The factors at work in a complex human–machine system have far greater potency for causing errors than do ergonomic factors. It is these factors that call for the notion of systems design.

Moray's model (Figure 3.5) is a representation of the causal structure of a complex hierarchical human–machine system. It is very general and is able to encompass bureaucratic organisations as well as the systems in which humans interact with complex machinery. By way of illustration, each level of the system is now briefly considered with respect to medical error.

Physical devices

At the centre of the system is the physical device or tool being used. There are many illustrations and examples of errors and difficulties associated with the use of equipment (see Obradovich and Woods (1996)).

One particular category of equipment, *i.e.* infusion devices, is often cited in adverse incident reports (Williams and Lefever, 2000). Setting infusion devices at the wrong rate is a frequent occurrence. Explanations for this type of error include the fact that confusion can exist between mg/hour and ml/hour when setting the infusion rate (Poster and Pelletier, 1988). This problem is exacerbated because users are often hindered by a lack of

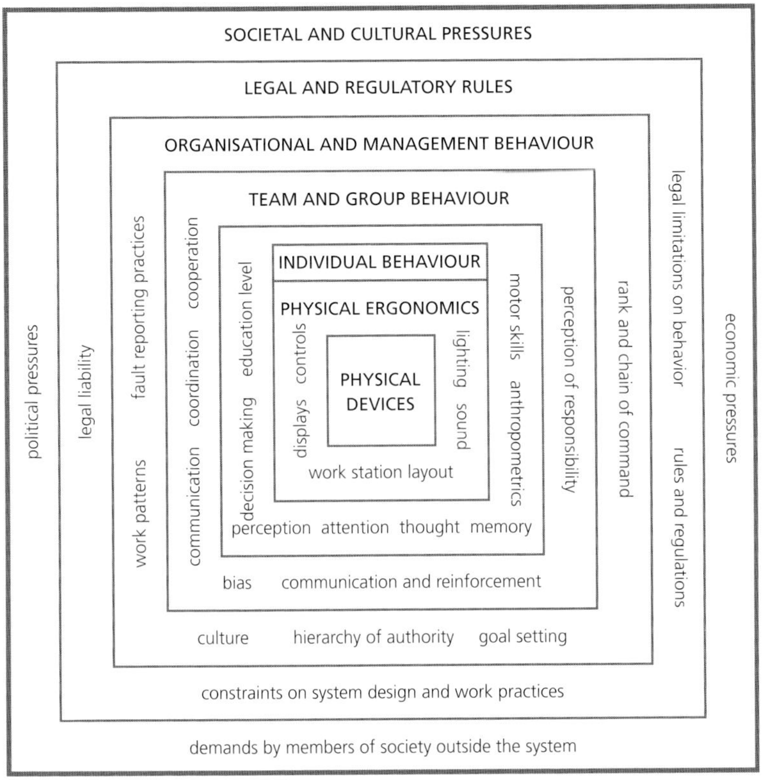

3.5 A model of ergonomics systems (Moray, 2000
Taylor and Francis Ltd. at http://www.tandf.co.uk/journals

feedback from the display and are frequently unable to detect which operational mode they are in (Garmer *et al.*, 2000). A fuller exposition of the user issues associated with the design of the interface of these devices has been included in case study 1.

Other aspects related to using physical devices include, for example, such issues as the legibility of labels on bottles and equipment and confusion over the identity of bottles with similar shapes and colours. Anaesthesiologists report that the colour of the ampoule containing a drug to be used and its label were both "extremely important" for ampoule recognition, as were the colour of the vial and cap. The text colour and external packaging were the most important features for pre-filled syringes, whilst for self-prepared syringes the drug label and syringe size were the most important features.

Knowledge of such factors is therefore critical for the systems engineering approach.

Individual behaviour

Omissions (*i.e.* the failure to carry out some of the actions required to achieve a desired goal (Reason, 1990)) were identified as the most common type of error (Poster and Pelletier, 1988). The role of such errors is evident when considering the giving of drugs to the wrong patient. This is frequently connected with failing to check the patient's identity bracelet and is often associated with distraction by other patients or interruptions because of the high level of ward activity. Administering the incorrect drug is most often associated with failing to read (or understand) the prescription chart or the drug label and the lack of knowledge of a particular drug (Gladstone, 1995).

Omissions are the most common type of error.
(Poster and Pelletier, 1988)

Physical ergonomics

Noise levels in working environments may cause messages to be misunderstood and can lead to interruptions. Chisholm *et al.* (2000) studied the number and type of interruptions occurring in emergency departments. Emergency physicians were frequently interrupted (about 31 times in 180 minutes). In primary care settings (general practice), nurses reported that interruptions were distracting, affected patient flow, and that the confidential nature of some consultations was irrevocably damaged by constant disturbances (Paxton *et al.*, 1996).

Team and group behaviour

Most people work within some kind of team, and so a consideration of factors such as communication, supervision and responsibility is required. Absence of, or poor, communication between and within teams is likely to contribute to errors (Dean *et al.*, 2002). For example, in a hospital setting the most junior medical officer is usually called upon to take a patient's medication history on admission. These doctors are often called upon to prescribe drugs and do so without asking questions under the assumption that this is the correct procedure. In some instances supervision is seen as inadequate, and other issues, e.g. overlapping responsibilities between teams, also contribute to errors (Dean *et al.*, 2002).

Absence of, or poor, communication between and within teams is likely to contribute to errors.
(Dean *et al.*, 2002)

Traditionally, information flows vertically through a hierarchy and orders are sent from the top down with the expectation that lower levels will implement them (West, 2000). Adverse events can occur because individuals of lower status experience difficulties challenging decisions of a person of higher status. Sexton *et al.* (2000), comparing medicine with aviation, suggest that poor communication is the equivalent of poor threat and error management. Effective cockpit crews use one-third of their communications to discuss threats and errors in their environment, whereas poorly performing teams spend about 5% of their time.

Organisational and management behaviour

Although factors affecting individuals have been highlighted, there is limited value in focusing on individual activity, as this tends to perpetuate a blame culture. The focus needs to widen to include systems issues underlying the problems that are present in any complex work environment (Anderson and Webster, 2001). Leape *et al.* (1995) carried out a study to identify and evaluate the areas of systems failure that underlie drug errors. They identified:

- drug knowledge;
- dissemination, dose and identity checking;
- availability of patient information;
- order transcription;
- allergy defence system;
- medication order tracking; and
- inter-service communication.

These failures were underpinned by impaired access to information and resulted from design faults. These included:

- defects in conceptualisation and planning;
- failure to recognise service needs; and
- failure to adapt systems to changing demands and changing technology.

Leape *et al.* identified other systems failures in such areas as:

- issues surrounding device use;
- standardisation of doses and frequencies;
- standardisation of drug distribution within the unit;
- standardisation of procedures;
- preparation of intravenous medications by nurses;
- transfers/transition procedures;
- conflict resolution;
- staffing and work assignments; and
- feedback about adverse drug events.

System failures are sometimes difficult for 'front line' staff to recognise because the decisions underpinning these systems may have been made in the past by those at a higher level of the organisation (Leape *et al.*, 1995). System changes suggested to reduce errors included adjusting work schedules to simplify work systems and enlisting the help of frontline personnel.

System failures are sometimes difficult for 'front line' staff to recognise because the decisions underpinning these systems may have been made in the past by those at a higher level of the organisation.
(Leape *et al.*, 1995)

The behavioural options available to those working in a system may be tightly constrained by regulatory rules.
(Moray, 1994)

Legal and regulatory rules

The behavioural options available to those working in a system may be tightly constrained by regulatory rules (Moray, 1994). For example, only certain drugs may be administered or procedures undertaken. As systems become more complex, the task of regulation becomes ever more difficult. For example, how do regulators cope with the issues that arise when multiple pieces of equipment are used conjointly or when 'intelligent' software is embedded within drug-delivery systems, thereby blurring the boundaries between equipment design and clinical decision-making?

Much has also been written on the role of standardisation in systems design. For example, West (2000) suggests standardisation and formalisation of tasks in an effort to reduce the complexity of work. The implications for systems design of such an approach again become apparent if specific contexts are considered. Equipment and environments would need to become standardised (for example, the aircraft cockpit) and the formalisation of tasks would require clarification of roles, rules and procedures.

Currently, many errors stem from the absence of controlled vocabulary for use in the medical setting (Senders, 1994). However, it is not inconceivable that all communication of medical orders and the names of medical preparations and devices could conform to the standards of a controlled vocabulary. This might help, for example, to reduce the number of prescription errors due to the use of non-standard abbreviations.

Societal and cultural pressures

The development of any large system is also likely to be subject to economic and political pressures, and demands by members of society outside of the system. Therefore, it is important to be aware of the potential impact of these pressures on the desired behaviours by those within the system when specifying, designing and implementing it.

Currently, many errors stem from the absence of controlled vocabulary for use in the medical setting.
(Senders, 1994)

A systems approach to patient safety

Design is the process by which something is created, whether it be a product, a protocol or a service. It is helpful to consider what design is in the context of systems development, since this will shed light on the role of design in improving patient safety.

There are many models of design that help to describe the nature of the process. One of the simplest may be found in British Standard 7000 Part 1. It describes the product life cycle as comprising of three key stages:

design, production and operation (these are illustrated in Figure 3.6). This model ignores the subtlety of design and paints a rather optimistic view of the process, and in reality there can be much iteration. Forecasting is necessary if the designer is to be able to design a product that can be made at the right price and used by the right people. Such forecasting is generally possible only if feedback is obtained about the performance of previous products or prototypes of the emerging product.

This model of design applies to products, services and systems. For example, if a new prescribing form is to be designed, a means must be defined to encourage the adoption of the form (production). In addition, the layout of the form must encourage its effective use (operation), both in terms of its ability to convey the required information accurately and its ability to be completed (and read) within an acceptable period of time.

Design is often then subdivided into a series of activities that enable the initial market need or idea to be converted into the manufacturing instructions that fully describe the product that is to be made (Figure 3.7). In reality, these stages are not strictly serial and may show significant overlap. The simple model also hides many significant influences that may affect the design process. These influences begin to show that product design is not simply an isolated activity, but is critically dependent upon and critically defines the business process. Indeed, the model presented by Moray (2000) (shown in Figure 3.5), derived from an ergonomic viewpoint, is remarkably similar to that presented by Hales (2004) (shown in Figure 3.8), derived from an engineering design viewpoint.

It is important to note that one person's product may be another person's component. For example, the Rolls-Royce Trent 700 jet engine becomes a component for an Airbus 340-500. Thus, a product may be made up of a complex mix of components and/or be one of a number of products required to contribute to a particular task or service. For example, the provision of a domestic electricity supply relies on a number of products configured in the generation, transmission and supply system.

As far as design is concerned, nothing is changed in dealing with a system, although there are usually more users, more requirements, and generally more demands and influences on the product, but the stages of design remain the same. However, in the case of systems the simple models of design do not help the design team and more rigorous design strategies are required. In addition, there is a need to develop methods better suited to ensuring the safety of the final product.

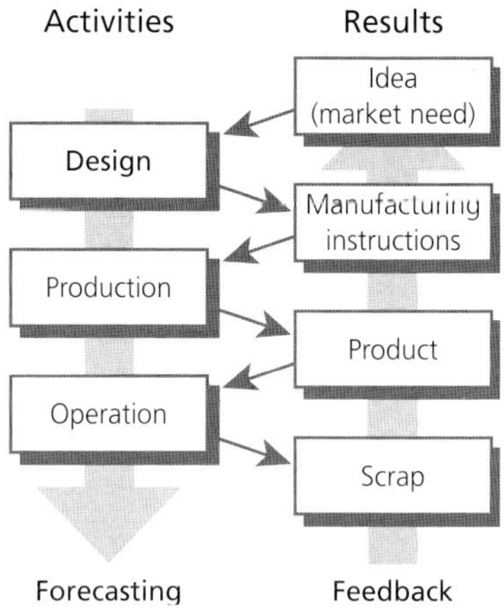

3.6 The product life cycle, adapted from the BS7000 product introduction process

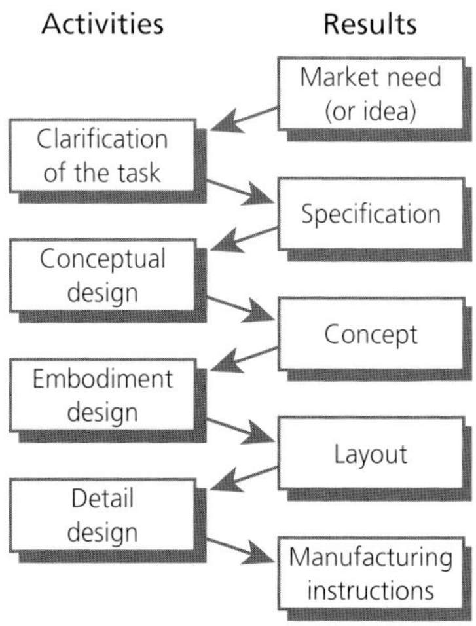

3.7 Elements of design

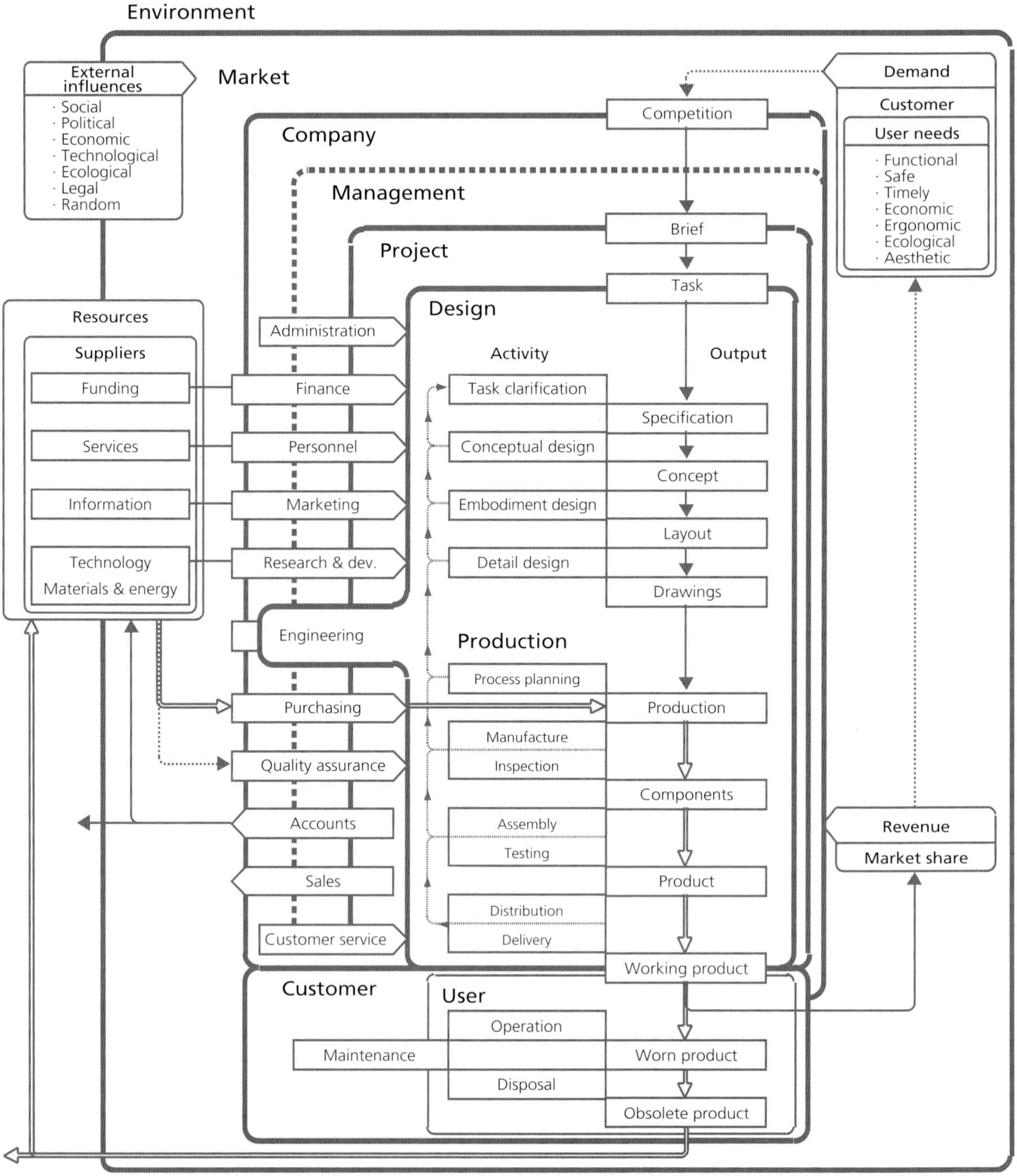

3.8 Engineering design in an industrial context (redrawn from Hales, 2004)

Better models of design

Thus far, all the discussion has been based on common descriptions of product design. However, they generally do not map well to the requirements of medical device or equipment design. More emphasis is required on the product safety requirements, whether the product be a medical device or medical procedure. In both cases, one way of ensuring safety is to evaluate the performance of the emerging product or system rigorously. Methods adapted from software engineering are useful for this purpose. One such adaptation is shown in Figures 3.9 and 3.10.

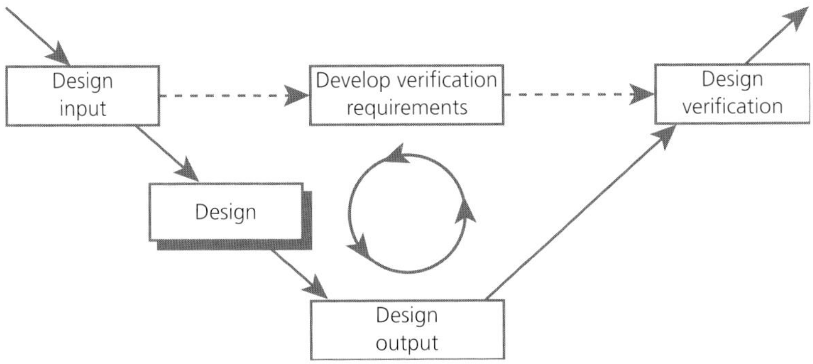

3.9 System verification – does it meet users' needs? (Alexander and Clarkson, 2000)
Taylor & Francis Ltd. at http://www.tandf.co.uk/journals/

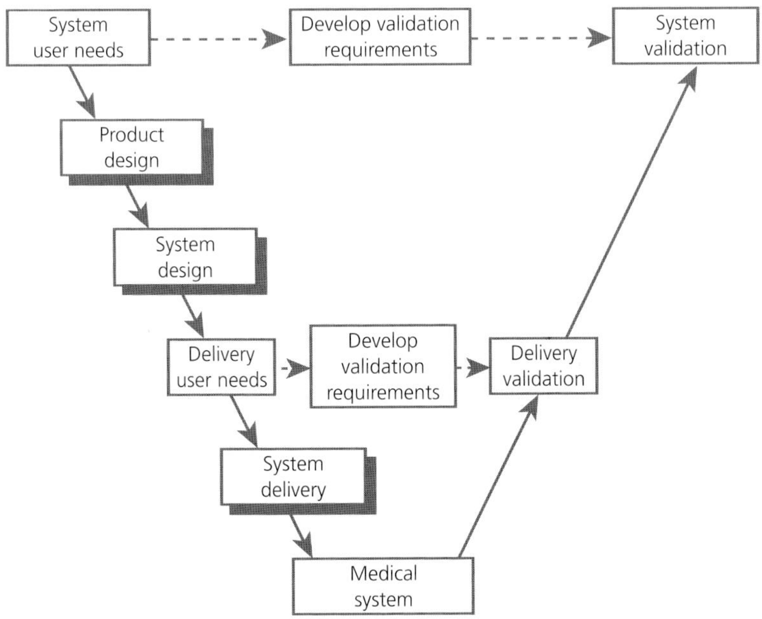

3.10 System validation – is it to specification? (Adapted from Alexander et al., 2001)

Evaluation, in the form of verification and validation, emerges as a critical component of all engineering design, in particular, medical device and equipment design.

Figure 3.9 shows the role of verification in the design of a system. Figure 3.10 shows the development of the system along with its delivery, highlighting the need for validation of the system and its delivery process. Put simply, verification and validation may be defined by:

> *"Verification: 'Are we building the thing right?'"*
> *"Validation: 'Have we built the right thing?'"*
>
> (Alexander et al., 2001)

Evaluation, in the form of verification and validation, emerges as a critical component of medical device and equipment design, ensuring that evidence of satisfactory performance is available. Of particular importance is the early definition of the evaluation requirements, which in turn may influence the design. The evaluation of medical devices or equipment must, in addition, be done in the context of their expected use.

Ideally, this involves a range of tests, including user trials, to provide representative performance data. Where a product is used as part of a system, the full system must be evaluated. The same is true for services, where every part of the service chain should be evaluated. For example, if a new treatment protocol is to be evaluated, all those activities required for the preparation, execution and monitoring of the protocol should be evaluated. Inevitably, this leads to the evaluation of human/equipment systems.

The systems engineering approach to design

The International Council on Systems Engineering (INCOSE) states that:

Systems Engineering is an interdisciplinary approach and means to enable the realisation of successful systems.
(INCOSE, 2004)

> *Systems Engineering is an interdisciplinary approach and means to enable the realisation of successful systems. Systems Engineering focuses on defining customer needs and required functionality early in the development cycle, documenting requirements, then proceeding with design synthesis and system validation while considering the complete problem: operations; performance; testing; manufacturing; cost and schedule; training and support; and disposal.*
>
> *Systems Engineering integrates all the disciplines and specialty groups into a team effort forming a structured development process that proceeds from concept to production to operation. Systems Engineering considers both the business and the technical needs of all customers with the goal of providing a quality product that meets the user needs.*
>
> (INCOSE, 2004)

It can be seen from these definitions that systems engineering is no different from design. However, its distinguishing feature is its complexity, brought about by its multi-disciplinary, multi-product or multi-user approach.

The validation model can be extended to provide the basis for a systems engineering approach to meet the needs of the NHS. The model, an extension of Figure 3.10, is based on the definitions and issues presented above (Figure 3.12). At the heart of this model is the innovation/procurement activity (within the inner box) which represents the design activity shown earlier. This process will be unique to a particular product or service, and should be informed by all the relevant stakeholders and agencies, and be actively managed to minimise technical and commercial risk.

3.11 An unusually complex design
© Airbus

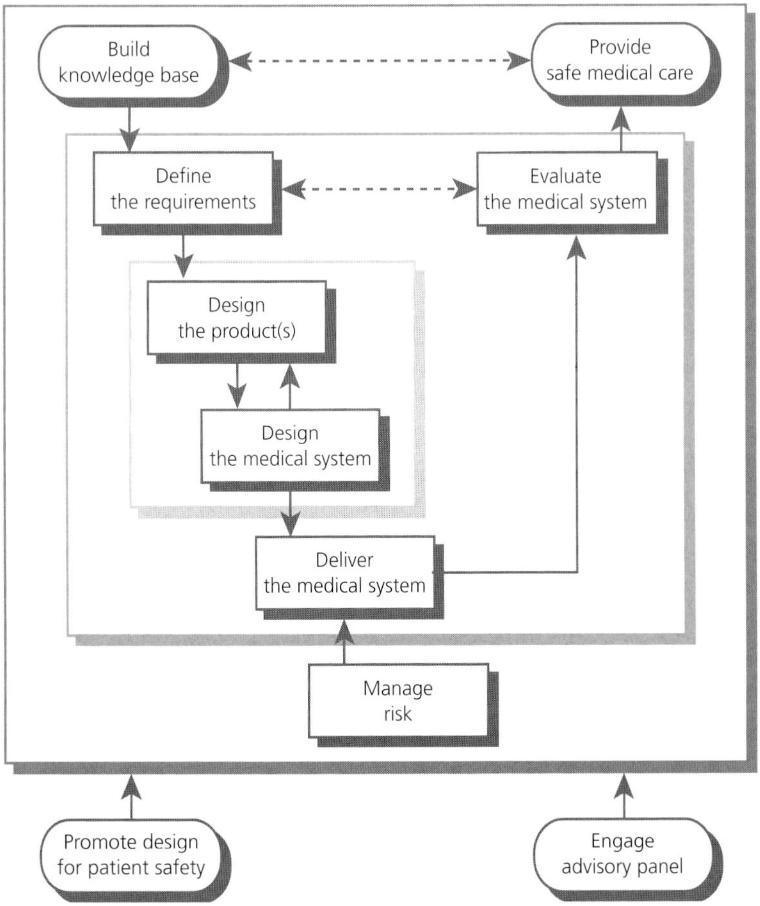

3.12 A systems-based user-centred approach to healthcare design

Successful product or service development cannot be done in isolation from the system or environment into which it will be introduced.

Successful product or service development cannot be done in isolation from the system or environment into which it will be introduced. Therefore, that system must be well understood, for instance by building an effective NHS knowledge base element. This improved understanding will in turn lead to the setting of more effective design requirements by the NHS, a prerequisite to improvements in procurement and innovation practice. This whole process could be informed and assisted by an advisory panel made up of industry and academic experts.

Figure 3.12 represents a convergence of views from the fields of ergonomics, engineering design and user-centred design. Thus, it presents a strong case for a systems-based user-centred approach to healthcare design.

Systems engineering and ergonomics as a process

Three case studies are presented to illustrate some of the processes and methods available to inform the systems design approach. The first is the assessment of the usability of a commonly used drug-delivery system known as an infusion device. This device enables fluid medication to be delivered to a patient at a regular rate, without the need for constant intervention by the healthcare deliverer.

The second illustrates the benefits of engaging with the end users of systems during the design phase. As part of the development of a new supermarket checkout system, the designers worked with checkout operatives to help select appropriate technology, design the physical layout of the workstations and evaluate and test the designs in an iterative fashion from concept to installation.

The final case study illustrates the breadth of methods that might be used in tackling complex systems where the existing knowledge base is weak. This approach, used to map healthcare delivery systems, helped to obtain a clear understanding of the systems and of where changes might be beneficial.

Case study 1: computer-based infusion devices

The design of computer-based infusion devices has been considered by Obradovich and Woods (1996). A study of devices adapted for terbutaline infusion showed how the device characteristics increased the potential for error. They also studied strategies that have been developed by users to protect themselves from failure. Amongst the conditions they identified as deficient were complex and arbitrary sequences of operations, mode errors due to poor differentiation of operating modes, ambiguous alarms and the problem of the user 'getting lost' in multiple displays. There was also poor feedback on the device state and behaviour.

Analysis of existing interface design

Garmer *et al.* (2002) have considered the development of a new user interface for an infusion pump using the human factors/ergonomics approach. Usability analysis was undertaken on existing designs based on observations, interviews, reported incidents and the theoretical basis for memory and human error. A new interface was developed based on a number of ergonomics principles (Table 3.13). An evaluation of the reduction in errors was undertaken. The number of errors was reduced but remained significant.

Equipment design improvements for the existing interface (Garmer *et al.*, 2002)
* Larger numbering in the display window
* Buttons for setting the numerical values to be placed on the display window
* Plainer messages to be left in the display window
* One button for volume to be infused and one for flow rate
* To replace symbols by words
* To avoid several functions on the same button
* To make it easier to see if the volume to be infused is activated

3.13 Infusions devices interface design improvements (Garmer et al., 2002)

Garmer *et al.* suggest that further tests are needed to improve the interface. They have identified, in particular, the need to provide more effective mode operation (for example, with the use of spring-loaded buttons). With regard to the process for finding solutions, they emphasise the importance of usability testing with a wide range of methods. They also emphasise the need to study both competent, experienced users and novice or learner users.

Currently, both the range of equipment and variety of interfaces have serious implications for the transfer of skills and the need for elaborate and complex training.

Examples from the Garmer *et al.* (2002) study illustrate how basic, but important, some of the design changes might be. For example, they identified that the pump should always have the same start-up mode and that this should be the mode most frequently used. Other modes should be user maintained. They also note that numerical information should be presented using only significant numbers, that if a decimal point is used, then it must be readable from all positions in the environment of use, and that all buttons should be marked with all of their functions. Many of these basic feedback and display topics are well understood and, through appropriate guidance, could lead to the development of far more effective/user-friendly interfaces.

It can be seen from Table 3.14 that many of the features imply simple design changes. However, these changes have hitherto not been reported in the literature, nor is there evidence that the medical device industry has researched these in any depth.

New interface design requirements (Garmer *et al.*, 2002)
* No decimal units, as these increase the risk of errors
* A different colour on the decimal unit in the display window
* It should be easier to see if an infusion is activated (with a movable line or movable drops in the display window)
* A sound that indicates set values
* When looking at the interface it should be easy to understand how to zero the device
* There should not be a requirement to press two buttons simultaneously when zeroing
* In the display window itself, it should be possible to get a description of how to set the volume to be infused
* When values have been set, the system should confirm when it has been done correctly

3.14 Infusion devices interface design requirements (Garmer et al., 2002)

The design of the alarm systems for such devices also illustrates the need for a systems approach to design. The journal *Health Devices* reports frequent system error messages disabling one particular model of infusion pumps. It appears to be well recognised that alarms are frequently triggered in situations of normal 'use'. The users in these situations often learn to ignore these alarms without considering the possible implications should the alarm reflect a truly abnormal operating situation.

Currently, there appear to be no formal or informal standards available for the design of interfaces for infusion devices (Garmer *et al.*, 2002). Thus, it is scarcely surprising that a multitude of interfaces exist and that many of these confuse the operators.

Implications

This case study is one of very few that has examined the user interface of equipment used in healthcare settings. The information base that such studies generate is essential as part of informing the systems engineering approach. However, the need to recognise the role that humans play within the system remains imperative if safe, reliable and efficient systems are to be developed.

For example, Kim *et al.* (1999) describe an ambulatory infusion device, which has been developed to provide perinatal drug delivery at a precisely controlled rate. The device uses the concept of electro-hydrolysis of a negatively charged hydro-gel. The system comprises a pump unit and an electronic control unit. Whilst the accuracy and precision of the device have been verified, there has been little discussion of the potential user-related issues. Technological advances that have failed to recognise the importance of usability are indicative of an industry that has yet to fully appreciate the concept of systems and the place of technology within such systems.

Case study 2: participatory design in a supermarket

A leading UK supermarket chain, employing up to 70,000 checkout operators, had concerns over the health and safety of checkout operators (especially musculoskeletal disorders of the back, neck and upper limbs). A new checkout carcass was drawn with the checkout operator area left completely blank. A participative approach was to be used to develop, test and agree the final design (see Figure 3.15). A series of earlier modifications to existing checkouts and a selection of individual new technological components had also used a participatory approach, but this project was the first to consider the complete design. The checkout design team was therefore mandated with a clear brief by the operational board to develop the new checkout to ensure the best possible operator environment, within specified cost and time restraints.

Description of the system

The checkout was to be installed in all new large supermarkets and to be retrofitted into the existing larger stores according to a strict time schedule. The work of a checkout operator involves highly repetitive handling of goods, often with significant time pressures imposed by customer demand. The checkout operator is also seen as crucial in establishing and maintaining good customer relations. For many customers this is their only point of contact with the organisation, and staff wellbeing is recognised as being important to enhance this interaction.

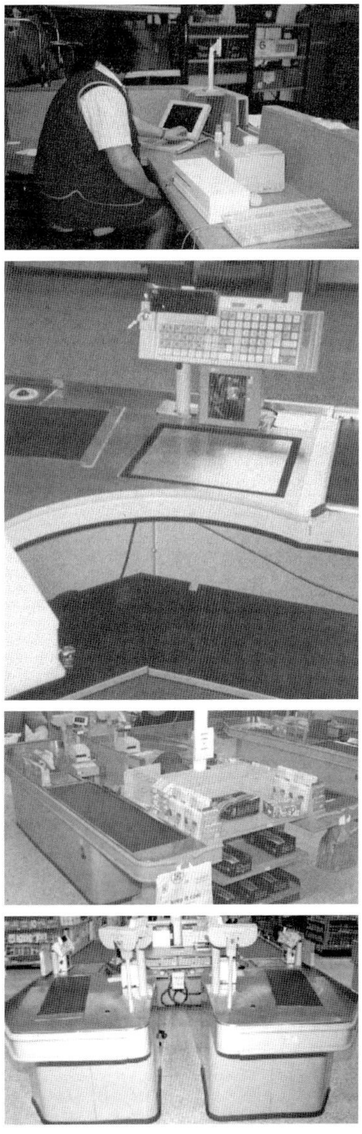

3.15 Checkout operator areas

The participation of users in the design process

Representatives of the checkout operators were selected from three stores. They included experienced and novice members of the workforce. Females and males were included. Representatives of each part of the engineering process were also part of the team, as were representatives of the organisation's health-and-safety team and customer relations department and an external ergonomist. The ergonomist acted as facilitator in the early stages. As the project progressed, other facilitators from the engineering project team were also able to adopt this role.

Regular meetings were held with the end users, the checkout operators. The response included comments that they "loved" the idea of only having bits of wood to look at and not a finished checkout to "comment on". They felt this really showed they could have some influence on the design. Mock-ups were built after each session and then commented on and tested through simulations at each subsequent meeting. Many changes were required. These were always agreed by all those present. This iterative process was used throughout.

User involvement can lead to high customer satisfaction and smooth project execution.

Final testing was carried out at a trial store over a period of several weeks. A number of minor modifications were made. It was noticeable that members of the team who were not checkout operators came to increasingly respect the views of those who actually used the equipment, as the project developed. Whilst the focus of the participation was the checkout operators, the requirements of customers (also end users) were also evaluated.

The project ran according to plan and to budget. The post-implementation report highlighted the role the checkout operators had played in the design and their preference for the new design, particularly for its space, layout of and design of equipment, choice of standing or sitting working posture and comfort. Customers also showed high satisfaction with the new design.

In this example, a wide range of stakeholders were involved throughout the design process. Much of the early work took place at the 'concept building' belonging to the organisation. This was important, as it was away from the shop floor and not located at the company's headquarters either. It was a 'neutral' location that encouraged each contributor to think in an open way and enabled all ideas to be received equally. As the project developed, the participatory process was moved to the checkout manufacturer's offices and the final meetings were held at the store where the in-store trials were being run.

Design problems identified by end users

The first focus group showed there to be some 50 significant problems identified with the existing design. These related to both customer and staff problems. The richness of this information enabled most of the problems to be identified very quickly. These were then classified as to how easy the problem was to overcome, if possible, in the new design. The types of problem reported for staff included lack of comfort, too great reach requirements, postural demands (especially the need to twist), cleaning and maintenance difficulties, snagging of clothing on protuberances, inefficient operation, and feelings of insecurity.

Improvements made

As a result of the participatory design approach, an ergonomically designed work space was designed including: the provision of sit or stand option, acceptable reach requirements, improved location of peripherals and technological devices (for example, scanner, scales, displays) through task analysis, improved customer interface, tested and improved scanner, better chair, a full footrest and a secure 'back-to-back' checkout design. Many improvements were also made for the customer, notably with regard to packing and ease of communication with checkout operators. In addition, the checkout operators felt they were co-owners of the new design. The post-implementation follow-up was reported. Some minor modifications were required and were to be addressed in subsequent checkouts.

Wider implications for the organisation

The checkout operators were co-owners of the new design, which was significantly better than could have been achieved by the design team without their input. The additional cost was insignificant. All parties adjudged the process successful.

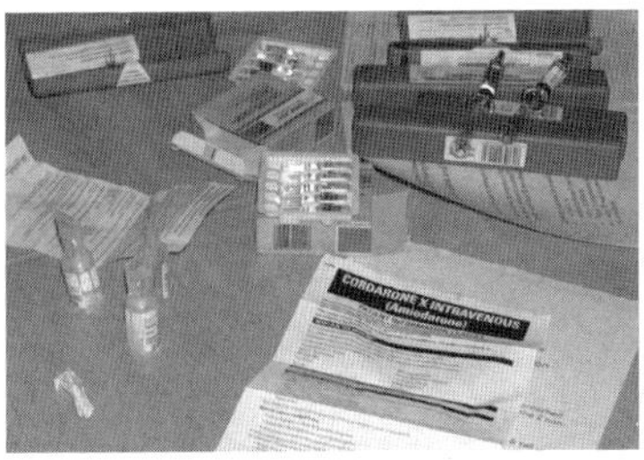

3.16 A medication delivery system

Case study 3: mapping healthcare delivery systems

Many systems comprise a complex system of interactions between diverse stakeholder groups, the environments in which they work, the associated information, equipment and changes over time. Experience of such systems has demonstrated that successful design interventions are unlikely to be made without the introduction of a systems approach to the design process, design analysis and, where appropriate, risk assessment and risk management. Mapping the system is an important element in any such intervention.

Methods

A recent study (Cambridge, RCA, Surrey, 2004) from the UK National Health Service illustrates the methods that were used to help achieve a suitable knowledge set on which to base design decisions. The methods used in this process are detailed in the Table 3.17 along with the objectives being sought. Inspection of the table reveals that all except one of the objectives has at least two methods associated with it. In this way, convergence between methods can be identified, thereby allowing greater confidence in the findings.

Method \ Objectives	1 Mapping the problem	2 Investigation of special cases	3 Identifying problems	4 Identifying best practice	5 Facilitating change	6 Solving the problems	7 Making recommendations	8 Communicating findings	9 Achieving action
Systematic literature review: journals	✓	✓	✓	✓		✓	✓		
Literature review: reports and 'grey' literature	✓	✓	✓	✓		✓			
Information exchange with international experts	✓		✓						
Prior experience of research team	✓		✓	✓	✓	✓	✓		✓
Workshop: other safety critical sectors				✓	✓	✓	✓		
Input from senior health service and agency personnel	✓			✓					✓
Interviews with healthcare practitioners	✓	✓	✓	✓					✓
Focus groups with healthcare practitioners	✓	✓	✓	✓					✓
Workshop with primary/secondary healthcare deliverers	✓	✓	✓	✓	✓	✓	✓		✓
Workshop with supply chain stakeholders	✓		✓	✓	✓	✓	✓		✓
Workshop with patient support group	✓		✓		✓	✓	✓		✓
Workshop with designers					✓	✓	✓		✓
Systematic consensus/priority setting by the research team					✓	✓	✓		✓
Iterative review of report with stakeholders								✓	

3.17 Methods and objectives for understanding and mapping healthcare

One extremely productive method was that of stakeholder workshops. In building the map, it became apparent that the intricacy of the systems they worked in surprised even the participants and pointed to key underlying problems related to fragmentation, parochialism and lack of communication and integration. As the interfaces between stakeholder groups became apparent, then so too did the potential for the emergence of error and hotspots. Such mapping exercises allowed key challenges to be identified and prioritised.

Summary

The use of a range of soft and hard methods enabled an understanding of the problems to be reached. For complex systems it is often not possible to include all stakeholders. Bias that might result from the selection of stakeholders or that arising from experts can be minimised by using multiple methods to address each objective and by prioritising data that are congruent.

Conclusions

This chapter has outlined the need for systems engineering and shown how the process can be achieved. It has also demonstrated that systems engineering and ergonomics are closely allied. Both are characterised by the interrelatedness of components relevant to the successful operation of the system in question. Developing an understanding of the human factor throughout the systems design process is essential, whether it be the implicit biases of those involved in the design process or an analysis of the use (and users) of existing systems.

The process of systems engineering demands rigorous use of appropriate methods and the objective evaluation of resultant information. To apply the approach successfully will almost certainly require multi-professional teams, engagement with relevant stakeholders and iterative stages in development. The benefits of applying this process are great, whereas the failures associated with any other design approach remain all too evident.

References

Alexander KL, Clarkson PJ (2000) Good design practice for medical devices and equipment, part II: design for verification. Journal of Medical Engineering and Technology, 24(2): 53–62. http://www.tandf.co.uk/journals

Alexander KL, Clarkson PJ, Bishop D, Fox S (2001) Good design practice for medical devices and equipment – a framework. Institute for Manufacturing, University of Cambridge, UK

Anderson DJ, Webster CS (2001) A systems approach to the reduction of medication error on the hospital ward. Journal of Advanced Nursing, 35: 35–41

Annett J, Stanton N (2000) Task analysis. Taylor and Francis

Cambridge, RCA, Surrey (2004) Designing for patient safety. A scoping study to identify how the effective use of design could help to reduce medical accidents. EDC, University of Cambridge

Chisholm CD, Collison EK, Nelson DR, Cordell WH (2000) Emergency department workplace interruptions: are emergency physicians interrupt-driven and multitasking? Academic Emergency Medicine, 7: 1239–1243

Clarkson PJ, Keates LS (2001) Towards an inclusive design approach. ICED'01, Glasgow, UK

Coleman R (1999) Inclusive design in human factors. In: Product design: current practice and future trends. Taylor and Francis

Corlett EN, Clark TS (1995) The ergonomics of workspaces and machines. Taylor and Francis

Dean B, Schachter M, Vincent C, Barber N (2002) Causes of prescribing errors in hospital inpatients: a prospective study. Lancet, 359: 1373–1378

Garmer K, Liljegren E, Osvalder AL, Dahlman S (2002) Application of usability testing to the development of medical equipment. Usability testing of a frequently used infusion pump and a new user interface for an infusion pump developed with a human factors approach. Industrial Ergonomics, 29: 145–159. Figures 3.13 and 3.14 reproduced with permission of Elsevier

Gladstone J (1995) Drug administration errors: a study into the factors underlying the occurrence and reporting of drug errors in a district general hospital. Journal of Advanced Nursing, 22: 628–637

Hales C (2004) Managing engineering Design. Springer

Hendrick H, Kleiner B (2002) Macroergonomics: theory, methods and applications. Lawrence Erlbaum

IEA (2000) The discipline of ergonomics. International Ergonomics Association. http://www.iea.cc/ergonomics/

INCOSE (2004) What is systems engineering? A Consensus of the INCOSE Fellows. http://www.incose.org/

Kim HC, Bae YH, Kim SW (1999) Innovative ambulatory drug delivery system using an electrolytic hydrogel infusion pump. IEEE Transactions on Biomedical Engineering, 46(6): 663–669

Leape LL, Brennan TA, Laird N *et al.* (1991) The nature of adverse events in hospitalized patients (results of the Harvard Medical Practice Study II). New England Journal of Medicine, 324(6): 377–384

Leape LL, Bates DW, Cullen DJ *et al.* (1995) Systems analysis of adverse drug events. Journal of the American Medical Association, 274: 35–43

Moray N (1994) Error reduction as a systems problem. In: Human error in medicine. Lawrence Erlbaum
Moray N (2000) Culture, politics and ergonomics. Ergonomics, 43(7): 858–868. Francis and Taylor Ltd., http://www.tandf.co.uk/journals
Obradovich JH, Woods DD (1996) Users as designers: how people cope with poor HCI design in computer-based medical devices. Human Factors, 38: 574–592
Paxton F, Heaney D, Howie J, Porter BA (1996) A study of interruption rates for practice nurses and GPs. Nursing Standard, 10(43): 33–36
Pheasant S (1996) Bodyspace: anthropometry, ergonomics and the design of work. Table 1.1 on p. 10. Taylor and Francis
Poster EC, Pelletier L (1988) Primary versus functional medication administration: monitoring and evaluating medication error rates. Journal of Nursing Quality Assurance, 2: 68–76
Reason J (1990) Human error. Cambridge University Press
Senders JW (1994) Medical devices, medical errors, and medical accidents. In: Human error in medicine. Lawrence Erlbaum
Sexton JB, Thomas EJ, Helmreich RL (2000) Error, stress, and teamwork in medicine and aviation cross sectional surveys. British Medical Journal, 320: 745–749
West E (2000) Organisational sources of safety and danger: sociological contributions to the study of adverse incidents. Quality in Healthcare, 9: 120–126
Williams C, Lefever J (2000) Reducing the risk of user error with infusion pumps. Professional Nurse, 15: 382–384
Wilson JR, Corlett EN eds (1995) Evaluation of human work. Taylor and Francis
Wilson RMcL, Harrison BT, Gibberd RW, Hamilton JD (1999) An analysis of the causes of adverse events from the quality in Australian health care study. Medical Journal of Australia, 170: 411–415

Perspectives on design

Chapter 4
Requirements engineering

Pericles Loucopoulos
UMIST

In many systems engineering activities the elicitation of requirements is regarded as a central activity for the efficient and effective functioning of the intended system. In recent years, requirements engineering (RE) has been established as a distinct field of investigation and practice. Its application has evolved from being concerned initially with software systems (IEEE-Std. 729, 1983; IEEE-Std. 830, 1984) to a broader perspective that extends to incorporate other aspects of systems and of the environment in which these systems function (Greenspan *et al.*, 1994; Loucopoulos and Karakostas, 1995; Pohl, 1996; Yu, 1997; Zave, 1997).

This broader view of RE is based on the premise that, in designing systems, requirements engineers aim to 'improve' organisational situations which are seen as problematic – or, at least, as needing some change. Hence, the problem of system design moves closer to addressing a wider set of problems found within organisational settings. Within this context, requirements are usually classified as *functional* requirements and *non-functional* (or quality) requirements. Whilst the former are concerned with the identification of intended system behaviour, the latter address issues relating to service provision for the intended usage of the system.

The success of the requirements engineering process often depends on the ability to proceed from informal, fuzzy individual statements of requirements to a formal specification that is understood and agreed by all stakeholders.

RE typically deals with a class of problems that have been termed "ill-structured problems" (Reitman, 1965; Rittel and Webber, 1984; Simon, 1984). The problem state is not *a priori* specified and there is no definitive formulation. To a great extent, formulating the problem amounts to solving it.

The success of the RE process often depends on the ability to proceed from informal, fuzzy individual statements of requirements to a formal specification that is understood and agreed by all stakeholders. However, the process is far from deterministic or straightforward.

The aim of this chapter is to outline the process of RE and to focus on two complementary techniques that facilitate this process, namely, *goal modelling* and *business rules modelling*.

Conventional methods of system development offer a *prescriptive* approach to RE. The traditional view of requirements definition is that this phase of systems development begins with an informal description of '*what*' the system is expected to do. However, recent research, supported by experiences in the industrial domain, recognises that successful system development relies upon the ability to understand and represent not only what the system should do, but also '*why*' (Loucopoulos and Kavakli, 1995; Yu, 1997; Yu and Mylopoulos, 1998).

Understanding the 'why' (teleological) dimension in systems development is necessary to ascertain and justify the presence of requirements components which may not be comprehensible to clients and users. Goals express intention and capture the reason for the system to be built. According to their degree of specificity, enterprise goals can be organised into goal hierarchies. Vague, high-level goals are refined into concrete, formal goals. This refinement is necessary because only simple primitive goals can be *operationalised*.

Operationalisation is the process of refining goals so that the resulting sub-goals have an operational definition (Anton *et al.*, 1994). The most common approach to goal operationalisation is that of *goal reduction*. Goal modelling techniques that are used within the RE process are presented later.

A process that is allied to goal operationalisation is *business rules modelling*. This activity concerns the definition of static and dynamic constraints (Loucopoulos *et al.*, 1991). Naturally, business rules, as part of requirements gathering and systems analysis, have not been ignored by structured analysis, information engineering or object-oriented analysis approaches (Moriarty, 1993), which, to varying degrees, subsume or represent business rules as part of notation schemes used to specify application requirements (Gottesdiener, 1999). The way that business rules relate to stakeholder goals and are modelled for the purpose of describing the key issues of the application domain is also presented later.

A requirements specification provides a focal point for the process of trying to understand correctly the needs of the customers and users of the intended system.

An overview of the requirements engineering process

A definition of requirements is given in IEEE-Std. 610 (1990) as:

1. A condition or capability needed by a user to solve a problem or achieve an objective.
2. A condition or capability that must be met or possessed by a system or system component to satisfy a contract, standard, specification, or other formally imposed document.
3. A documented representation of a condition or capability as in (1) or (2).

A requirements specification provides a focal point for the process of trying to understand correctly the needs of the customers and users of the intended system. It is the means by which a potentially large and diverse population of requirements stakeholders and requirements analysts communicate.

A specification may be part of the contractual arrangements, especially when an organisation wishes to procure a system from a vendor rather than develop it 'in house'. In such a situation, the specification is used for evaluating the final product and may play a leading role in any acceptance tests agreed between system consumer and system supplier. The requirements specification life-cycle is defined as:

> *the systematic process of developing requirements through an iterative co-operative process of analysing the problem, documenting the resulting observations in a variety of representation formats and checking the accuracy of the understanding gained.*
>
> (Pohl, 1993)

This reflects the view that the requirements specification involves an interplay of representation and social and cognitive concerns (Pohl, 1993). Issues of representation range from informal descriptions, such as natural language expressions and hypertext, to formal conceptual modelling languages. In the social domain, consideration is given to the complex social process by which the communication and co-operative interaction between the stakeholders of the requirements determines the quality of the final product. Issues in the cognitive domain concern different model orientations, in terms of understanding the process itself and validating the requirements.

A requirements specification is likely to change many times before proceeding to design and needs to be subjected to evaluation in order to gain confidence in its validity. The RE process generally consists of four tasks (Figure 4.1):

- requirements elicitation;
- requirements negotiation;
- requirements specification;
- requirements validation.

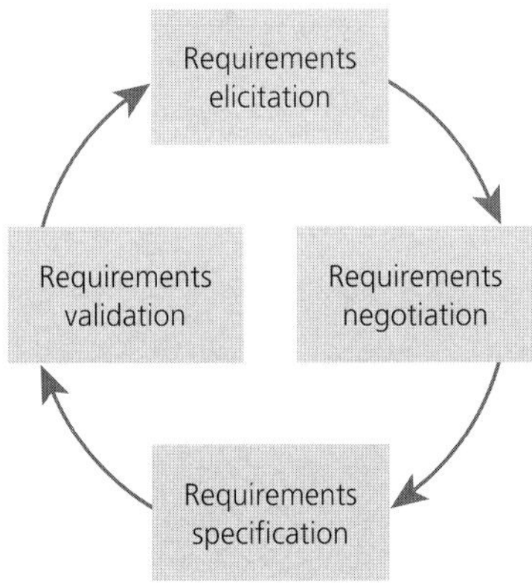

4.1 The four stages of RE

Requirements *elicitation* is about understanding the organisational situation that the system under consideration aims to improve and describing the needs and constraints concerning the system under development. The relevant knowledge about the problem (system) is distributed among many stakeholders.

The objective of *negotiation* is to establish agreement on the requirements of the system among the various stakeholders. The requirements *specification* involves a mapping of real-world needs onto a requirements model. Finally, the *validation* task intends to ensure that the derived specification corresponds to the original stakeholder needs and conforms to the internal and/or external constraints set by the enterprise and its environment.

To facilitate these four activities, a number of techniques and associated tools have been developed and the remainder of this chapter focuses on two techniques, *goal modelling* and *business rules modelling*.

Goal modelling

Stakeholder goals and their role in defining and solving design problems are topics of longstanding interest in the field of RE. Approaches to goal analysis emphasise the use of the 'notion of a goal' in order to understand or describe aspects of the real world. This, in turn, supports attempts to find better ways of coping *rationally* with the complexity of human affairs. Hence, RE addresses the problems associated with, for example, *business goals*, plans, processes, and *systems* to be developed or evolved in order to achieve organisational objectives (Loucopoulos and Karakostas, 1995; van Lamsweerde, 2001).

RE projects require the involvement of multiple stakeholders (e.g. the sponsor organisation, the system developers and users, and external regulators). The question here is: how do stakeholders co-ordinate their actions in order to provide a common result? Research in the areas of process modelling, workflow analysis and computer-supported collaborative working (CSCW) (Ellis and Wainer, 1994; Schedin, 1995; Nurcan and Rolland, 1997) endorses this goal-directed view.

This view is based on the premise that, in collaborative work situations, people do not follow rules or procedures strictly, rather they are aware of the personal and group goals and act accordingly (Smith and Boldyreff, 1995). This is especially true when people are not faced with well-structured, repetitive processes, but are tackling ill-structured problems where both the intended outcome and the possible routes to reach this outcome need to be specified, which is usually the case in RE (Bubenko, 1995; Loucopoulos and Kavakli, 1997). The role of goal-oriented approaches in relation to the four RE activities is summarised in Table 4.2.

The modelling of goals has been proposed during requirements elicitation in order to describe current organisational behaviour (e.g. goal-based workflow, i*, enterprise knowledge development (EKD) and goals, operations, methods and selection (GOMS)) and to set the objectives for change (e.g. information systems work and analysis of changes (ISAC) and F^3). Equally, goal analysis techniques have been used in the context of requirements negotiation in order to assist reasoning about the need for organisational change and to provide the context for deliberation during RE (e.g. SIBYL, the *reasoning loop model* and REMAP).

Modelling of goals has also been used in requirements specification to describe how organisational change can be implemented in terms of the new system's components by relating business goals to functional and non-functional system specifications (e.g. KAOS, goal-based requirements analysis

RE activity	Goal analysis contribution	Goal-oriented approach
Requirements elicitation	1. Understanding the current organisational situation	GOMS, goal-based workflow, i*, EKD
	2. Understanding the need for change	ISAC, F^3, EKD-CMM
Requirements negotiation	3. Providing the context within which deliberation occurs during the RE process	SIBYL, REMAP, the reasoning loop model
Requirements specification	4. Relating business goals to functional and non-functional system components	KAOS, GBRAM, the NFR framework, the goal-scenario coupling framework
Requirements validation	5. Validating system specifications against stakeholders' goals	GSN, GQM

4.2 The role of goal analysis in relation to RE activities

method (GBRAM), the non-functional requirements (NFR) framework and the goal-scenario coupling framework).

Finally, in the context of requirements validation, goal analysis techniques have been used to define the stakeholders' criteria against which the fitness of system components is assessed (for example, goal-structuring notation (GSN) and goal-questions–metrics (GQM)).

A description of goal modelling approaches

This section gives a brief description of the main contributions in each of the five classes of goal analysis given in Table 4.2.

Understanding the current organisational situation

Work in this area focuses on conceptual techniques and tools for explicitly capturing and representing, in a structured way, the domain knowledge subsequently used to drive the system development phases. This falls into two broad categories: *enterprise modelling* and *cognitive task analysis*.

Techniques in *enterprise modelling* describe the business environment as co-operation among different organisational actors, (for example, human individuals, IT systems and workgroups) based on the assumption that these actors share common goals and act towards their fulfilment. Enterprise models, implicitly or explicitly, represent the goals of individuals, groups, or organisations, whereby a goal is a desired condition potentially attained at the end of an action (or process). Goals are considered a potential motivator to action, and are distinct from plans, procedures or other means of attaining the goal.

The **i*** approach (Yu, 1997; Yu *et al.*, 1995, 2001; Castro *et al.*, 2002) provides a description of work organisation in terms of dependency relationships among actors. This approach acknowledges the fact that actors have freedom of action, within the social (inter-actor) constraints, called *strategic dependencies*. An actor is an active entity that carries out actions to achieve goals. Intentional components, *i.e.* *goals* to be achieved, *tasks* to be accomplished, *resources* to be produced and *softgoals* (non-functional requirements) to be satisficed are made specific, embedded in the dependencies between actors.

In the goal-based workflow approach proposed by Ellis and Wainer (1994), an organisation is seen as a tuple (G, A, R) where G is a set of goals, A is a set of actors, and R is a set of resources. Actors act collaboratively using resources in order to attain their goals. In goal-based workflow the focus is on people and goals rather than on procedures and activities.

Finally, in the EKD approach (Loucopoulos and Kavakli, 1997; Loucopoulos *et al.*, 1997; Kavakli and Loucopoulos, 1999), a business enterprise is described as a network of related *business processes* which collaboratively realise *business goals*. The EKD approach uses a 'network' of goals to express the causal relationships between enterprise objects. This it does in terms of the *goals–means* relations from the 'intentional' objectives, that control and govern the system operation, to the actual 'physical' enterprise processes and activities available for achieving these objectives.

Techniques in *cognitive task analysis*, e.g. GOMS (Card *et al.*, 1983), are focused on human tasks. In this context, a goal (also called an external task) is defined as a state of a system that the human wishes to achieve. A goal is achieved using some instrument, method, agent, tool, technique, skill or, generally, some *device* which is able to change the system to the desired state. A *task* (or internal task) is defined as the activities required, used or believed to be necessary to achieve a goal using a particular device. A task is a structured set of activities. An *action* is defined as a task that involves no problem solving or control-structure components.

Understanding the need for change

Work in this area focuses on methodologies for planning, organisation and control of enterprises. Discussion of goals in this context is considered not at an individual level (as for the ones discussed earlier) but at a broader organisational level.

In the ISAC approach (Lundeberg, 1982), goal analysis is considered during the early stages of RE, namely during the business change analysis phase. The purpose of the analysis is to ensure that the business problems to be solved are identified and that these problems are diagnosed correctly. The relationship between problems and goals can be represented by means of a problem/goal matrix. Use of this matrix assists the identification of clusters of similar problems that relate to similar goals. Each cluster defines a 'change need' that will be a goal of the development process.

A richer formalism for expressing goals and goal relationships for change is described in the Objectives Model (OM) of the F^3 framework (Bubenko, 1994; Loucopoulos, 1995). The OM is used for describing the intentional and motivational perspective of the enterprise, *i.e.* the enterprise goals along with the problems obstructing achievement of the goals. It is used to encourage communication between enterprise stakeholders in order to understand current problems and explicitly identify future goals and opportunities.

EKD-CMM (Rolland *et al.*, 1999) is a systematic approach for developing and documenting enterprise knowledge. It helps organisations consciously to develop schemes for implementing changes, which do not necessarily concern the development of computerised systems. Indeed, the decision to develop a software system forms part of the *derived* solution that meets stakeholder needs.

Organisational change must be considered at the same time as system requirements.

Organisational change concerns the transition from an initial 'as is' organisation situation, which is unsatisfactory in some aspect, to a desired 'to be' situation where the problem is resolved. Both the future state and the possible change routes that can be followed to reach this state have to be specified. To this end, organisational stakeholders develop hypotheses (termed *scenarios*) as to the nature of the desired solution.

Scenario formulation is based on the systematic specification of *change goals* and their causal relationships on the basis of:

- current-state goals;
- stakeholder intentions;
- contextual forces.

The confluence of these three components results in a set of change goals, which are presented in a *change goal model*. This model is subsequently utilised for the definition of alternative scenarios. The appropriateness of a proposed scenario may depend on a number of criteria (termed *evaluation goals*), such as implementation costs, efficacy of proposed transformation. Such criteria cannot be known in advance but need to be defined within the context of the particular change application.

Providing the context within which deliberation occurs

Work in this area aims at providing conceptualisations of the RE process, as well as supplying methods for improving activities such as problem solving and decision making (Loucopoulos *et al.*, 1996; Louridas and Loucopoulos, 2000). In this context, goals have been used in order to document, and subsequently trace, the history of the rationale of decisions concerned either with the system that is being designed or with the design process itself.

SIBYL (Lee and Lai, 1991) is a system designed to help users represent and manage the qualitative elements of the decision making process. SIBYL is organised around decision graphs, which record the pros and cons of choosing from a set of alternatives to satisfy a goal. In the *reasoning loop model* (Loucopoulos *et al.*, 1996; Louridas and Loucopoulos, 2000) a generic non-prescriptive approach is presented that combines informational with operational primitives in order to define, reason about and resolve design problems. In particular, it employs the notion of a goal to denote the designer's intentions (e.g. objectives to be reached, demands to be satisfied or problems to be solved). Achieving these goals is based on the generation of hypotheses as to the design actions to be taken.

A similar approach, but with a stronger focus on capturing the design rationale during RE in a structured manner, is REMAP (Ramesh and Dhar, 1992). The REMAP model is based on the issue based information systems (IBIS) design rationale model (Rittel and Weber, 1973) and uses goals to provide the context in which design deliberations occur in RE. *Goals* express requirements that the system should fulfil, derived requirements that emerge because of higher-level design decisions, or constraints.

Goal-oriented approaches explicitly link business needs and objectives to non-functional or functional system components.

Relating business goals to system components

Work in this area is based on the premise that system components satisfy some higher goal in the larger environment (Loucopoulos and Kavakli, 1995). By putting emphasis on goal analysis, goal-oriented approaches explicitly link business needs and objectives to non-functional or functional system components. The relationship between business goals and the intended functionality of a system and its quality is addressed in terms of three broad categories: *goal elaboration*, *scenario definition* and *non-functional requirements definition*.

In goal elaboration, KAOS (Dardenne *et al.*, 1993; Darimont, 1995; van Lamsweerde *et al.*, 1995; Letier and van Lamsweerde, 2002) highlights the importance of explicitly representing and modelling organisational goals and their relations to operational system components. The KAOS methodology is aimed at supporting the process of requirements elaboration – from the high-level goals that should be achieved by the composite system to the operations, objects and constraints to be implemented by the software.

GBRAM (Anton, 1996; Anton *et al.*, 2001) offers prescriptive guidelines on how to extract goals from different sources into one ordered goal set. The operationalised goals, responsible agents, stakeholders, scenarios and obstacles are ultimately consolidated into a set of goal schemas.

In many *scenario approaches* (e.g. Leite and Haumer, 1997) goals are considered as a contextual property of a scenario (*i.e.* a property that relates the scenario to its organisational context). Cockburn (1995) goes further and suggests the use of goals to structure scenarios by connecting every action in a scenario to a goal assigned to an actor. In a similar way Ben Achour *et al.* (1998) proposes the organisation of scenarios using goal hierarchies.

A goal is defined as something a stakeholder hopes to achieve in the future, whilst a scenario expresses a possible way in which the goal can be achieved. By assigning goals to scenarios and organising the goals using 3 types of relationship (refine, AND, OR) a structure for managing scenarios is also established. An interesting aspect of this approach is that it advocates a bi-directional goal–scenario coupling: just as goals can help in structuring scenarios, scenarios are also used to discover new goals.

In the NFR definition, Chung *et al.* (2000) and Mylopoulos *et al.* (1992) define a framework which provides for the representation of non-functional requirements in terms of interrelated goals. Such goals can be refined through refinement methods and can be evaluated to determine how far a set of non-functional requirements is supported by a particular design. The NFR model consists of goals that represent non-functional requirements (*NFR goals*), design decisions (*satisficing goals*), arguments for or against other goals (*argumentation goals*), and goal relationships for relating goals to other goals.

Validating system specifications against stakeholders' goals

System validation aims at certifying that the system specification produced is in accordance with the users' needs. The objective is to ensure a solution that is right for the user needs rather than a correct (i.e. consistent and unambiguous) specification.

The objective of validation is to ensure a solution that is right for the user needs.

System validation is of major importance, especially when dealing with the design of safety-critical applications. For these, validation is performed through the construction of a safety case, a collection of documents and data which, together, present clear, comprehensive and defensible arguments

that an application will be acceptably safe throughout its life. In this context, the modelling of goals has been suggested in order to:
1. Give safety cases a better structure.
2. Explicitly link safety goals to alysis results and evidence.
3. Make the rationale, assumptions and justifications explicit.
For example, in the safety case approach described by Kelly and McDermid (2001) and Wilson *et al.* (1995), a GSN was developed in order to express safety requirements as goals. It is also able to capture assumptions, justifications, proof in the general sense, and rationale.

An alternative approach to system validation is to define a set of metrics (qualitative or quantitative) against which system properties can be measured. Again, the use of goals has proven useful in this context. In particular, the GQM approach (Basili and Rombach, 1988; Basili, 1993) supports the identification of metrics from goals through the use of appropriate questions.

"Statements about how the business is done, *i.e.*, about guidelines and restrictions with respect to states and processes in an organisation".
(Herbst, 1996)

The construction of a GQM model starts with the formulation of the measurement goals. Each goal is refined into a set of questions which collectively represent an operational definition of the goal at hand. Each question in turn defines a number of metrics. The GQM process of setting goals and refining them into quantifiable questions is supported by a template for defining goals and a set of guidelines for deriving questions and metrics. Aspects of GQM are supported by software tools such as GQMaspect (Hoffmann *et al.*, 1996).

Business rules modelling

In recent years there has been increasing interest in the information system (IS) community regarding business rules. This has resulted in a number of dedicated rule-centric modelling frameworks and methodologies (Zaniolo *et al.*, 1997). The term "business rule" has been used by different methodologists in different ways. For example, Rosca *et al.* (1997) describe business rules as "statements of goals, policies, or constraints on an enterprise's way of doing business". Herbst (1996) defines them as

> *statements about how the business is done, i.e., about guidelines and restrictions with respect to states and processes in an organisation.*
> (Herbst, 1996)

Kramer (1997) considers them as "programmatic implementations of the policies and practices of a business organisation", whilst von Halle (1994) states that

> *depending on whom you ask, business rules may encompass some or all relationship verbs, mathematical calculations, inference rules, step-by-step instructions, database constraints, business goals and policies, and business definitions.*
>
> (Halle, 1994)

In general, we can also distinguish between three types of business rule (Kardasis and Loucopoulos, 2003): *intentional rules*, *operational rules* and *IS implementation rules*.

Intentional rules are expressions of business rules seen from a *business context perspective*. They express laws, external regulations, or principles and good practices which constrain the way an organisation conducts business. Laws are imposed by the legal system of the environment in which the organisation operates (for example, the State enforces taxation laws). Regulations are not legally binding but are imposed by other organisations as a prerequisite for interacting with them (for example, an organisation may have regulations about the content, structure and appearance of service offerings submitted to them). Principals and good practices are recommended ways of working, leading to the acceptance of an organisation by its environment (for example, a company may adopt the principle of equal opportunities for all).

Operational rules are expressions of business rules, approached from a *business process perspective*. They prescribe action on the occurrence of some business event, or describe valid states of an organisation's informational entities.

IS implementation rules are expressions of business rules examined from an *IS architecture perspective*. They describe valid states of data entities, or prescribe action on the occurrence of some systems event.

A framework for relating business rules to goals

A framework for the specification of business rules involves four main activities, as shown in Figure 4.3.

- *Project boundaries identification* deals with the identification of specific project goals and boundaries of the application area to be addressed by the project.
- *Business charting* includes the study of the main business processes and information entities that are of interest within the particular project.
- In *stakeholder viewpoints analysis*, representatives of different enterprise departments (for example, legal and financial) are asked to explain the objectives of the organisation from their own viewpoint, taking into consideration the particular project boundaries, which leads to the definition of a set of unstructured rule expressions.

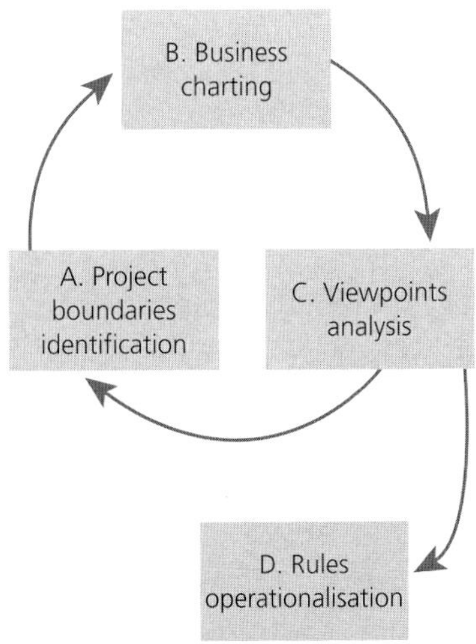

4.3 Activities for the specification of business rules

- In *rules operationalisation*, previously collected rules are further decomposed to more detailed rules, and are finally transformed to purely operational rule expressions (*i.e.* statements that prescribe action on the occurrence of specific business events), prior to the identification of rule conflicts.

Project boundaries identification

The boundaries of a project can be explored through the use of a goal-based approach.

The boundaries of a project can be explored through the use of a goal-based approach. Goals represent the purpose, rationale and motivations behind enterprise structures and operations, as well as the intentions, objectives and visions of stakeholders regarding future states. A high-level goal graph specifying the boundaries of an example project is presented in Figure 4.4.

The objective of this project is to design an electronic procurement system, which will assist a medium-sized construction company in purchasing raw materials from their suppliers and in sub-contracting services to other companies. According to the goal model of Figure 4.4, the main aims of the project are "to manage requests for proposals (RFPs) electronically" and "to perform electronic catalogue procurement".

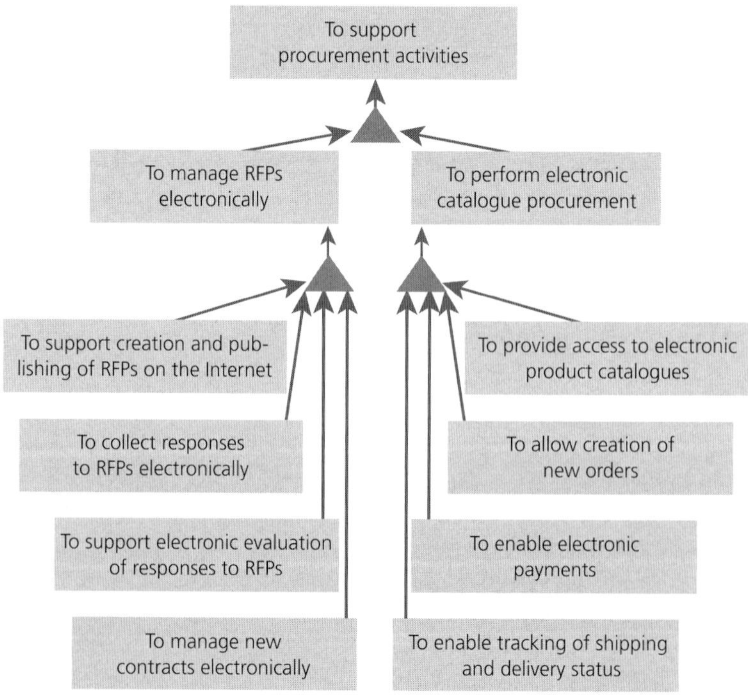

4.4 An example goal graph towards the development of an electronic procurement system

RFPs are the requests for proposals from sub-contractors and involve the publishing of specifications for sub-contracted services, the evaluation of proposals and the management of contracts. Catalogue procurement concerns the purchasing of raw materials, and includes creation of purchase orders, invoicing, payments and monitoring of order execution.

Business charting

Goal graphs created during *project boundaries Identification* generate sufficient input for identifying the main business activities to be affected by the project. *Business charting* deals with the production of comprehensive models, describing the business activities along with the business events that trigger them and the informational entities produced or used by them.

Formalisms that allow description of business activities at a high level (or escalating levels) of detail are most suitable. For example, the process modelling component of the IEEE IDEF (IDEF0, 1993) formalisms provides the necessary semantics and notations for depicting major groups of business activities and the ways they interact with each other, along with the relevant informational entities and business events.

Data flow diagrams (Robinson and Berrisford, 1994), from the structured systems analysis and design method (SSADM), are also an option for representing business activities and produced or used informational entities.

Informational entities can be presented in two types of model: business object diagrams, which are a variation of the unified modelling language class diagrams, and object lifecycle diagrams, which are able to "describe the states of business objects, from the time they are created until they are destroyed" (Whitten and Bentley, 1998).

Business object diagrams are concerned with the way in which informational entities are associated with their attributes and with each other. Object lifecycle diagrams approach informational entities from a different perspective, evolving as a result of different business events.

Fulfilment of strategic goals (objectives) depends on the satisfaction of operational goals (tactics), which are eventually implemented in business rules (policies).

Viewpoint analysis

Ross and Lam (1998) present the relationship between enterprise *missions* and *objectives*, *business tactics* and *policies* in the following two definitions:

- A tactic is defined as a course of action that can be followed to meet objectives.

- A policy expresses that some specific or quantified constraint is needed to meet objectives.

They also state that "... a policy must always be sufficiently detailed and precise to give direct guidance to workers such that they know what to do or how to make some decision in relevant circumstances". Thus, the fulfilment of strategic goals (objectives) depends on the satisfaction of operational goals (tactics), which are eventually implemented in business rules (policies).

The difference between goals and rules is illustrated by the following examples. "To support evaluation of responses to RFPs" is an enterprise goal at the strategic level. "To identify an RFP response that represents an acceptable technical solution at the lowest possible price" is an enterprise goal at the operational level, and represents a tactic. The corresponding business rule is "To accept an RFP response, if the financial offer is the lowest of all and the technical offer score is above 80%".

Stakeholders with different views are usually aware of different business constraints (in other words, different business rules). These stakeholders need to participate in the requirements analysis phase of the project, by stating their own goals and by discussing how these goals are translated to business rules.

The selection of stakeholders is a crucial issue, and needs to be tackled by the project management with appropriate respect for the importance of the project, its complexity and its size. In the procurement example, there are five viewpoints which are of interest: the viewpoint of the construction company, the supplier, the sub-contractor, the client and the State. Given that many of the relevant stakeholders (e.g. client and State) may not be accessible, their role must be played by their 'representatives' in the organisation (e.g. corresponding project managers and legal department).

The outcome of different stakeholders' involvement in the requirements analysis process is, initially, a set of complementary goal graphs (one or more per stakeholder). The goal graphs of Figures 4.5 and 4.6 represent the view-points of the construction company and of the sub-contractor respectively.

The reader may notice that the same project goals are refined differently for each of the viewpoints. For example, as far as the electronic evaluation of proposals is concerned, the construction company has three goals:

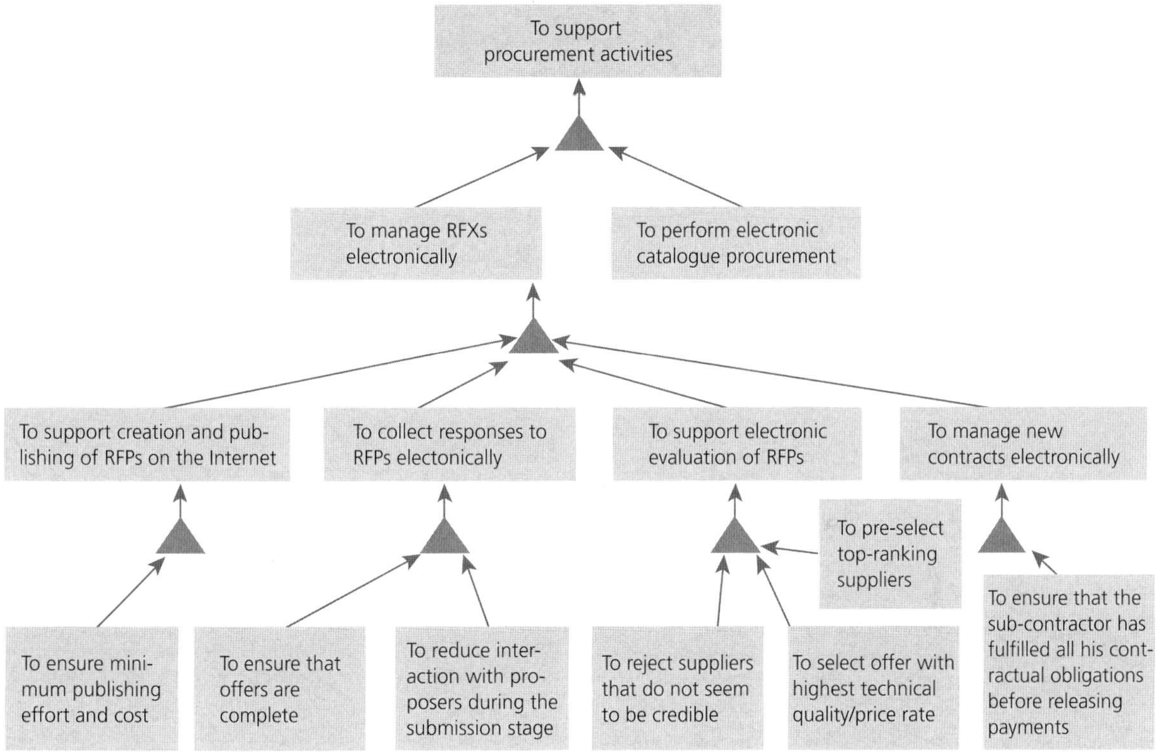

4.5 The construction company's perspective

- to pre-select top ranking suppliers;
- to reject suppliers that do not seem to be credible;
- to select proposals with highest technical quality for price rate.

On the other hand, the objectives of the sub-contractors are:

- to ensure transparency of the evaluation process;
- to ensure opportunities to new entrants in the field;
- to ensure that there can be further negotiations after the proposal stage.

Rule operationalisation

During rule operationalisation the following actions are to be taken:

- Unstructured rules need to be identified as *descriptive* and *prescriptive*, where descriptive rules explain the meaning of a business term, whilst prescriptive rules explain what has to be done on the occurrence of a certain event, given that various conditions hold.

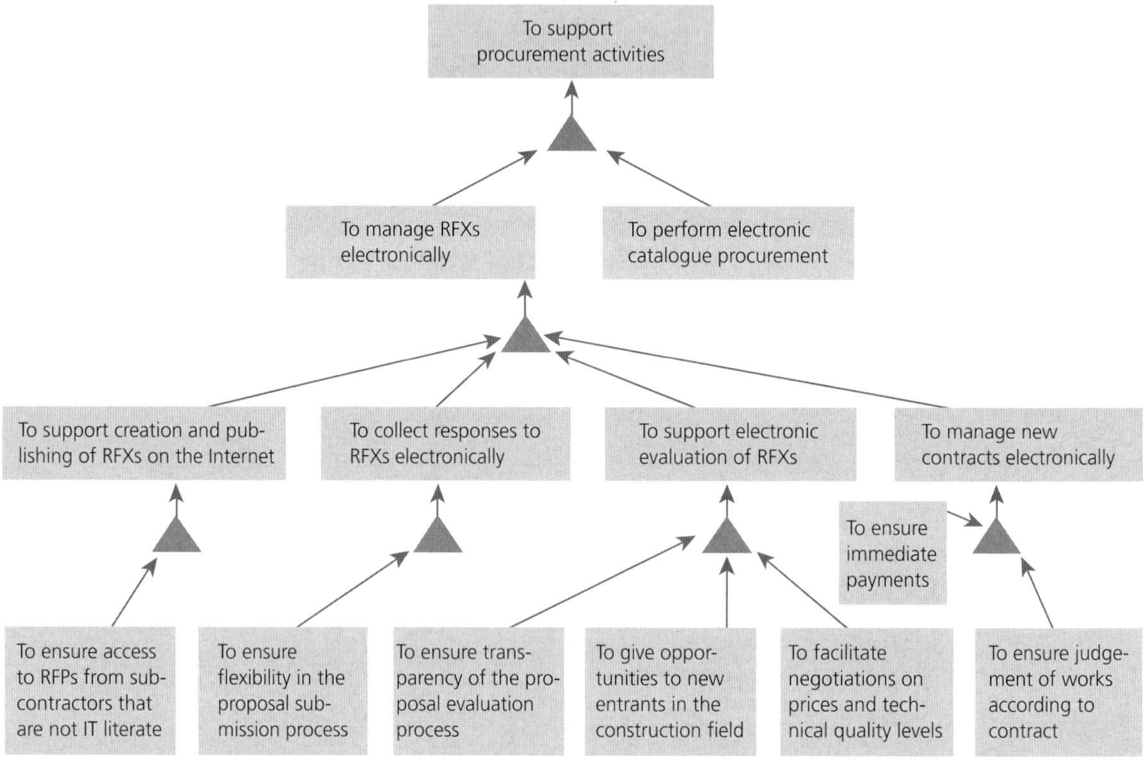

4.6 The sub-contractor's perspective

- All rules need to be refined to the lowest possible level of operational detail, deriving a number of prescriptive rules from the existing prescriptive rules and, where necessary, deriving prescriptive rules from descriptive rules.
- All rule expressions need to be rephrased in a way consistent with the models utilised for business charting, adopting an appropriate rule language.
- Prescriptive rules need to be rephrased in such a way that it is clear: (a) what the event is that triggers their execution; (b) what the conditions are that need to be checked; and (c) what the action is that needs to be taken.

The result of rule operationalisation is a set of structured rule expressions following the event-condition-action (ECA) paradigm (van Assche *et al.*, 1988). These rules include:

- *Derivations* which aim to derive a certain object attribute from other attributes of the same or different business objects (see for example Figure 4.7).

WHEN part	–
IF part	–
THEN part	*Proposal.overall_rank= Proposal.financial_rank+Proposal.technical_rank*

4.7 A sample derivation

- *Information assertion* rules that examine a set of conditions on the occurrence of some event, and perform information updates accordingly (see for example Figure 4.8).

WHEN part	*Proposal_evaluation_initiated*
IF part	*Subcontractor.market_presence_rank>=15* *Subcontractor.experience_rank>=10* *Subcontractor.capacity_rank>=50*
THEN part	*Subcontractor.overall_rank=5*

4.8 A sample information assertions rule

- *Action assertion* rules that examine a set of conditions on the occurrence of some event, and determine what workflow actions need to be taken (see for example Figure 4.9).

WHEN part	*Proposal_evaluation_initiated*
IF part	*Proposal.technical_offer_status="submitted"* *Proposal.financial_offer_status="submitted"*
THEN part	*Perform_proposal-technical_evaluation*

4.9 A sample action assertions rule

Conclusions

RE represents an influential activity in the lifecycle of system development. Correct, unambiguous and complete requirements play a key role in the development of systems that are effective, efficient and maintainable. Whilst there are many techniques and methods that are used in RE, the two complementary approaches of goal modelling and business rules modelling discussed in this chapter are thought to be the two key techniques required to achieve high-quality requirements specifications.

An intentional view of requirements can simultaneously reflect an enterprise or stakeholder requirement and the goal of the designer in attempting to meet these requirements. The task of someone using goal modelling is to determine the means by which an ultimate goal will be realised. In attempting to achieve this, the process is governed by causal relationships within a network of goals. At every step, the process is controlled or driven by the goal at hand. Every node represents successively refined design goals. The actions chosen for attaining the goals represent *working hypotheses*. As goals and subgoals are established, these are tentative at least until they are tested for their fitness for purpose.

A practical implication of hypothesis formulation is the presence of design constraints. Requirements constrain the design solution and the exact balance of satisfying these requirements cannot be known in advance of producing a design. Consider, for example, an air traffic control system that is being commissioned for design. The client (airport authority) may have a goal of 'increasing the throughput of aircraft', the users (air traffic controllers) may have as a goal 'reducing stress at work' and the legislators, 'conforming to safety regulations'. These three goals, three requirements, are at least to some extent in conflict. How should they be satisfied, if at all? Should they be considered at the outset and arbitrarily assigned some priority? How do they interact with other goals of other stakeholders (or even of the same stakeholders)? The designer will need to explore various possibilities for balancing satisfaction of these three requirements.

Design constraints arise from required or desired relationships between two or more elements. These relationships may refer either:

1. Entirely to elements of the object being designed, in the above example the air traffic control system for an airport.
2. To elements of interaction between the object being designed and its environment. In the above example this might be a compatibility relationship between the designed system and other air traffic control systems.

These constraints can be modelled in terms of business rules techniques. In principle, business rules are considered to be either the effect of enterprise goal operationalisation, or the reason behind the way goals are operationalised. Based on this premise, rules are 'harvested' by examining the views of different stakeholders with regard to their organisation's objectives, and on the relationship between these objectives and existing business constraints. In this way, it is guaranteed that the collected rules represent the policies and tactics that enterprise management has determined. Moreover, it is guaranteed that parameters that employees may not be aware of (for example, complex legal issues and delicate ethical concerns) will also be taken into consideration.

References

Anton A (1996) Goal-based requirements analysis. ICRE '96, Colorado Springs, CO, USA

Anton A, McCracken WM, Potts C (1994) Goal decomposition and scenario analysis in business process reengineering. CAiSE'94, Utrecht, The Netherlands

Anton A, Carter A, Dempster J, Siege D (2001) Deriving goals from a use-case based requirements specification for an electronic commerce system. Requirements Engineering, 6: 63–73

Basili VR (1993) Applying the goal/question/metric paradigm in the experience factory. In: Software quality assurance and measurement: a worldwide perspective. Chapman and Hall

Basili VR, Rombach DD (1988) The TAME project: towards improvement-oriented software environments. IEEE Transactions on Software Engineering, 14(6): 758–773

Ben Achour C, Rolland C, Souveyet C (1998) A proposal for improving the quality of the organisation of scenario collections. REFSQ'98, Pisa, Italy

Bubenko J (1994) Enterprise modelling. Ingenierie des Systems d'Information, 2(6): 657–678

Bubenko J (1995) Challenges in requirements engineering. RE'95, York, UK

Card S, Moran T, Newell A (1983) The psychology of human–computer interaction. Lawrence Erlbaum

Castro J, Kolp M, Mylopoulos J (2002) Towards requirements-driven information systems engineering: the Tropos project. Information Systems, 27(2002): 365–389

Chung L, Nixon BA, Yu E, Mylopoulos J (2000) Non-functional requirements in software engineering. Kluwer Academic

Cockburn A (1995) Structuring use cases with goals, human and technology. http://members.aol.com/acockburn/papers/usecases.htm

Dardenne A, van Lamsweerde A, Fickas S (1993) Goal-directed requirements acquisition. Science of Computer Programming, 20: 3–50

Darimont R (1995) Process support for requirements elaboration. PhD thesis, Universite Catholique de Louvain, Belgium

Ellis CA, Wainer J (1994) Goal-based models of collaboration. Collaborative Computing, 1: 61–86

Gottesdiener E (1999) Discovering an organisation's knowledge: facilitating business rules workshops. IAF'99, Williamsburg, VA, USA

Greenspan S, Mylopoulos J, Borgida A (1994) On formal requirements modeling languages: RML revisited. ICSE'94, Sorrento, Italy

Herbst H (1996) Business rule oriented conceptual modelling. Physica

Hoffmann M, Birk A, van Els F, Kempkens R (1996) GQMaspect v1.0. User manual

IDEF0 (1993) Integration definition for function modeling. FIPS 183

IEEE-Std. 610 (1990) IEEE standard glossary of software engineering terminology

IEEE-Std. 729 (1983) IEEE standard glossary of software engineering terminology

IEEE-Std. 830 (1984) IEEE guide to software requirements specifications

Kardasis P, Loucopoulos P (2003) Managing business rules during the requirements engineering process in rule-intensive IT projects. BIS'03, Colorado Springs, CO, USA

Kavakli V, Loucopoulos P (1999) Goal-driven business process analysis - application in electricity deregulation. Information Systems, 24(3): 187–207

Kelly TP, McDermid JA (2001) A systematic approach to safety case maintenance. Reliability Engineering and System Safety, 71: 271–284

Kramer MI (1997) Business rules: automating business policies and practices. Distributed Computing Monitor, (May)

Lee J, Lai K-Y (1991) What's in design rationale? Human–Computer Interaction, 6(3/4): 251–280

Leite J, Haumer P (1997) Enhancing a requirements baseline with scenarios. RE'97, Annapolis, MD, USA

Letier E, van Lamsweerde A (2002) Agent-based tactics for goal-oriented requirements elaboration. ICSE'02, Orlando, FL, USA

Loucopoulos P (1995) The F^3 (from fuzzy to formal) view on requirements engineering. Ingénierie des Systèmes d'Information, 2(6): 639–655

Loucopoulos P, Karakostas V (1995) System requirements engineering. McGraw-Hill

Loucopoulos P, Kavakli E (1995) Enterprise modelling and the teleological approach to requirements engineering. Intelligent and Cooperative Information Systems, 4(1): 45–79

Loucopoulos P, Kavakli V (1997) Enterprise knowledge management and conceptual modelling. International Symposium on Past, Present and Future of Conceptual Modeling. Springer

Loucopoulos P, McBrien P, Schumacker F, Theodoulidis B, Kopanas V, Wangler B (1991) Integrating database technology. Rule-based systems and temporal reasoning for effective information systems: the TEMPORA paradigm. Information Systems, 1(2): 129–152

Loucopoulos P, Louridas P, Kavakli V, Filippidou D, Tzanaki A (1996) The enterprise requirements analysis approach. Technical Report, UMIST

Loucopoulos P, Kavakli V, Prekas N, Rolland C, Grosz G, Nurcan S (1997) Using the EKD approach: the modelling component. ELEKTRA/WP2, UMIST

Louridas P, Loucopoulos P (2000) A generic model for reflective design. ACM TOSEM, 9(2): 199–237

Lundeberg M (1982) The ISAC approach to specification of information systems and its application to the organisation of an IFIP working conference. In: Information systems design methodologies: a comparative review. IEEE Computer Society Press

Moriarty T (1993) The next paradigm. Database Programming and Design

Mylopoulos J, Chung L, Nixon B (1992) Representing and using non-functional requirements: a process-oriented approach. IEEE Transactions on Software Engineering, 18(6): 483–497

Nurcan S, Rolland C (1997) A meta-model for goal-driven cooperative work processes. MEPE'97, Gammarth, Tunis

Pohl K (1993) The three dimensions of requirements engineering. CAiSE'93, Paris, France

Pohl K (1996) Process-centered requirements engineering. Research Studies Press

Ramesh B, Dhar V (1992) Supporting systems development by capturing deliberations during requirements engineering. IEEE Transactions on Software Engineering, 18(6): 498–510

Reitman WR (1965) Cognition and thought: an information processing approach. John Wiley

Rittel H, Webber M (1973) Dilemmas in a general theory of planning. Policy Science, 4: 155–169

Rittel HW, Webber MM (1984) Planning problems are wicked problems. In: Developments in design methodology. John Wiley

Robinson K, Berrisford G (1994) Object-oriented SSADM. Prentice Hall

Rolland C, Loucopoulos P, Kavakli V, Nurcan S (1999) Intention based modelling of organisational change: an experience report. CAiSE/IFIP 8.1, EMMSAD '99, Heidelberg, Germany

Rolland C, Souveyet C, Ben Achour C (1998) Guiding goal modeling using scenarios. IEEE Transactions on Software Engineering, 24(12): 1055–1071

Rosca D, Greenspan S, Feblowitz M, Wild C (1997) A decision support methodology in support of the business rules lifecycle. ISRE'97, Annapolis, MD, USA

Ross RG, Lam G (1998) Putting business rules to work: a tutorial and workshop on business rules, business tactics and policies. Business Rule Forum'98, Technology Transfer Institute, Chicago, IL, USA

Schedin K (1995) Goals in business modelling. MSc thesis, Kungl Tekniska Hogskolan, Sweden

Simon HA (1984) The structure of ill-structured problems. In: Developments in design methodology. John Wiley

Smith S, Boldyreff A (1995) Towards an enterprise method for CSCW systems. ISADS '95, Phoenix, AZ, USA

Van Assche F, Layzell PL, Loucopoulos P, Speltinx G (1988) Information systems development: a rule based approach. Journal of Knowledge-Based Systems, 1(4): 227–234

Van Lamsweerde A, Darimont R, Massonet P (1995) Goal-directed elaboration of requirements for a meeting scheduler: problems and lessons learnt. RE'95, York, UK

Van Lamsweerde A (2001) Goal-oriented requirements engineering: a guided tour. RE'01, Toronto, Canada

Von Halle B (1994) Back to business rule basics. Database programming and Design, 7(10): 15–18

Whitten JL, Bentley DL (1998) Systems analysis and design methods. McGraw-Hill

Wilson SP, Kelly TP, McDermid JA (1995) Safety case development: current practice, future prospects. ENCRESS'95, Bruges, Belgium

Yu E (1997) Towards modelling and reasoning support for early phase requirements engineering. RE'97, Annapolis, MD, USA

Yu E, Mylopoulos J (1998) Why goal-oriented requirements engineering. REFSQ'98, Pisa, Italy

Yu E, Du Bois P, Dubois E, Mylopoulos J (1995) From organization models to system requirements – a "cooperating agents" approach. CoopIS'95, Vienna, Austria

Yu E, Liu L, Li Y (2001) Modelling strategic actor relationships to support intellectual property management. ER-2001, Yokohama, Japan

Zaniolo C, Ceri S, Snodgrass R (1997) Advanced database systems. Morgan Kaufmann

Zave P (1997) Classification of research efforts in requirements engineering. ACM Computing Surveys, 29(4): 315–321

Chapter 5
Human resources

Martina Dürndorfer
Management Perspective

Global economic reality is currently undergoing a period of transformation. The so-called "New Economy" has been declared dead; most of the related companies and their products have disappeared from the market or have been merged with other organisations. What remain are organisations faced with intensified competition and increased pressure on costs and profitability. At the same time they are going to be judged by how successfully they innovate and attract new customers, and how they open up new distribution channels in global markets in ever shorter times.

Every time managers are convinced that they have implemented the appropriate strategy to guarantee long-term organisational success they are confronted by new technology, new products, new competitors, new customers or new market requirements, calling again for new management solutions. This dynamic of a knowledge-driven society and the specific structural and interactive nature of network-innovative markets characterise our society and act on companies from within. It is against this background that innovation regularly occurs.

As the business context changes, new management solutions are continuously required.

It is vitally important to grasp the fundamental perspectives and changing demands of the competitive business landscape in order to identify specific challenges and risks for technological processes. This chapter focuses on:

- the interface between management and technological challenges;
- predominant developments and management approaches in the context of design;
- the implementation and deployment of processes;
- process '*enablers*';
- the role of human resources as a key dimension among other managerial implications.

Process-based management approaches

At the beginning of the 1990s dramatic shifts in the business environment loomed, affecting both strategic management thinking and the introduction or renewal of management approaches and methods. Companies started concentrating on value-added processes, on their resources, capabilities and core competencies. This was in response to growing customer demands and to the acceptance of the fact that customer fulfilment and the development of core competencies, as much as quality and price, are criteria which can signpost unique positions of competitive advantage. Companies began to reconsider their business processes and internal routines with regard to service level, terms of delivery and customer fulfilment. They started initiatives to optimise these by using their own inherent technological and human potential, and to transform them into innovative and beneficial, value-adding solutions.

The functional partition of, thus far, valid organisational structures and the hierarchic alignment of companies, however, made flexibility difficult and hampered action in the marketplace. Companies urgently needed appropriate organisational forms and leadership instruments, adapted to their altered circumstances. Alternative organisational structures in the form of process models were developed. The essence of those management approaches, known as a resource-based view, consists of an undeviating focus on the customer and on value-creation and the organisation of internal and external business processes. Traditional functional structures and organisational boundaries lost their importance and were replaced by process forms.

Companies started to identify core processes and to define core competencies so they could configure products and services according to strategic relevance and added value. Core processes were defined as those characterised by highly interdependent and essential tasks, decisions, information and allocation of resources, which decisively promote the company's added value and competitiveness. They derive from the company's core competencies.

Support processes remain necessary to make sure that core processes function smoothly. Surveys identify product development, product supply to customers and the maintenance of customer relationships as the most decisive core processes. Those tasks and activities not crucial for the company's added value were subsequently, as far as possible, reduced or outsourced.

Support processes remain necessary to ensure that core processes function smoothly.

Classification of business processes

Within the discussion of processes as a means of corporate management, many problems and misunderstandings derive from differing perspectives. Management often takes a different view of the naming and content of business processes from that of a technologically oriented department.

From a management perspective a process is seen as comprising different business tasks with defined outcomes and objectives to be realised in given and standardised sequences of operations. In a process, internal and external customers have to be provided with services and information.

The following business processes can be distinguished:

- *achievement* processes which, in turn, are comprised of *service* processes and *production* processes;
- *support* processes;
- *leadership* processes.

Hence, a company may be interpreted as a bundle of processes striving for consistency, i.e. avoiding interfaces and friction in the resulting process organisation. This results in a paradigm shift from the fundamental inertia of the previous corporate organisation and management systems.

Those changes of corporate management can be carried out either radically (business re-engineering) or in moderation (for example, lean management and supply-chain management). Nonetheless, all process-oriented management approaches pursue, in principle, identical objectives (see Figure 5.1). Concentration on processes rather than on functions is the most common. In fact, process-based and resource-based management approaches and management systems did incorporate numerous well-known concepts of organisational management, such as results-oriented organisation or project-oriented organisation. Yet a careful look at them reveals a crucial difference from those of process re-engineering: it is the simultaneous advancement of technology deployment and the resulting opportunities of collaboration and automation.

To release the power of new technological possibilities, using software systems in the restructuring and renewal of business processes turned out to be inevitable. Workflow management is an example of a systematic process-based application and transfer of knowledge across heterogeneous applications. The potential for technological and administrative savings was the primary focus of networking within process management approaches.

Networking, as an integral element of a company's processes, was established as a key factor for added value and corporate success. The importance of networking dynamics, the variations and opportunities as well as the subsequent results, needs to be better understood and appreciated, both inside and outside the company.

A further development of resource-based strategy can be seen in the concept of core competencies (Prahalad and Hamel, 1990). According to this approach, a corporate strategy should concentrate on the development of a core-competencies portfolio (as a set of capabilities resolving distinct tasks and problems) rather than on actions tied to corporate positions vis-à-vis the product/market matrix. The original concept of core competencies relied strongly on production technique and the technology steering it, and was subject to ongoing enlargement. Subsequently, knowledge and the underlying processes for knowledge creation were identified as crucial resources for competitive strengths and advantages in corporate management (Kim and Mauborgne, 1999).

Resources and capabilities:

- Resource-based view for competitive advantage

Main concerns:

- Sources of competitive advantage within the firm

Principal concepts and tools:

- Resource analyses
- Core competency analyses
- BPR (business process re-engineering)
- BSC (balanced scorecard)
- TQM (total quality management)

Organisational and implementation issues:

- Restructuring around key resource competence
- Outsourcing
- Focus on building core competencies
- Alliances

5.1 Strategic management approaches in the 1990s (adapted from Leibold *et al.*, 2002)

Current challenges for management

The existing strategic management system, based on the definition and examination of business objectives, is focused on the identification and delivery of core competencies and process efficiency. However, it is vulnerable to the fundamental, rapidly changing business environment. Those far-reaching economic changes at the beginning of the 1990s had a profound effect on corporate management and management approaches. Only a few years later companies are once again faced with huge challenges and renewal requirements. Those approaches are no longer sufficient for a highly dynamic knowledge-driven and network-oriented economy.

"The existing strategic management system ... ... is more likely to be a source of organisational inertia than a proactive force for dynamic change."
(Leibold et al., 2002)

> *The existing strategic management system – including defined purpose (vision, mission, objectives, etc.), organisational structure, planning processes, measurement practices, core competency focus, human resource management, cultural norms, and evaluation and reward systems – is more likely to be a source of organisational inertia than a proactive force for dynamic change. Prior experiences, business process re-engineering, balancing and 'mapping' strategic processes, and historic 'formulas' for success increasingly become impediments to the innovative strategic management required for dealing with a turbulent knowledge economy*
>
> (Leibold et al., 2002)

Close ties between technological, organisational and social progress are already part of process-oriented management approaches, as can be seen in the process innovation approach taken by Davenport (1993) who describes it as:

> *the envisioning of new work strategies, the actual process design activity, and the implementation of the change in all its complex technological, human and organisational dimensions*
>
> (Davenport, 1993)

Nevertheless, it took a long time for resource-based process-management approaches to shift their focus and strategic scope: the employees and their networks, their customer relationships and knowledge are the most decisive value drivers in a modern, innovation-promoting company, not the corporate structure by itself.

Current economic challenges can be summarised under four headings according to Leibold et al. (2002):

- the dramatic shift from visible assets and invisible customers to invisible assets and visible customers;
- the reality that vertical and horizontal organisations are being displaced by networks of intra-company, extra-company and intercompany relationships;
- displacement of the focus on competition (and competitive 'outperformance') to a focus on collaboration (and 'unique performance' and sustainability);
- descriptive and reactive traditional strategic management mind-sets are being forced to shift to creative, proactive strategic mindsets.

In a resource-focused and process-focused company the 'dynamic capabilities' are expressed in both processes and competencies and in its capability of triggering reconfiguration and learning processes (Teece *et al.*, 1997).

Smooth running reconfiguration and learning processes for communication, documentation, skill development are required to assure technical competence.

In order to develop competitive advantages, the reconfiguration and learning processes have to be encouraged to respond swiftly to changing business environmental parameters. Processes enabling and ensuring

- knowledge transfer and communication (social processes and networks)
- generation, transfer and documentation of knowledge and
- support, encouragement and development of knowledge-workers and knowledge-explorers (people)

are the most important value-drivers and catalysts for resource-based and capability-based companies to respond promptly and flexibly to a fast-changing business environment.

In summary, the focus of analysis shifts from products and companies to people, organisational networks, and the social processes that bind them together in ongoing relationships.

Management in (business) networks

The past years have shown economy and society reflecting a world that is increasingly interconnected and in which the pace of technological change has been accelerating. In the course of the rapid progress of information and communication technologies, process management marked the beginning of a deployment of far-reaching network structures within and between companies. (Examples can be seen in Workflow Management Systems, computer supported co-operative work and in further virtual corporate structures.) Organisations "are changing more and more from well-structured and manageable systems into interwoven network systems with blurred boundaries" (Seufert *et al.*, 1999), where the underlying corporate architecture approaches the "ideal of the 'boundaryless' organisation" (Cross *et al.*, 2002). This implies fundamental and heterogeneous challenges in the management of corporate processes: boundaries and rules for information, communication and decision-making processes are going to be transformed, corporate boundaries no longer rigidly separate information and communication processes between customers, suppliers and co-operating communities of practice (CoP). The permeability of boundaries exists both within corporate structures (*i.e.* between departments and teams, between core and support processes) and at the interface of companies and external stakeholders.

Interorganisational networks

In recent years, collaboration has become more and more an integral part of doing business with suppliers and customers. Business dynamics demand an even more determined integration of customers in order to build efficient innovation processes. So, what is new compared with the current practice of customer relationship management (CRM), which already claims that the customer is the centre of company interest? The concept of customer value propositions (product and services configuration) changes drastically.

In recent years, collaboration has become more and more an integral part of doing business with suppliers and customers.

Customers function as product designers, as catalysts of innovation processes, not only in R&D, but in the broad spectrum of searching, generating and selecting ideas and market expectations, as agents for market response and as critical end users. As a logical consequence, CRM further develops to customer knowledge management (CKM; Leibold *et al.*, 2002) and weaves stronger ties between customers with their market, product and competitor knowledge, and the company concerned.

Network communities

Whereas innovations in the 'traditional' economy are often triggered by independently acting research institutes, R&D departments and other institutions, in a knowledge-based economy innovations are more likely to be influenced by networks and collaborating communities. Temporary as well as persistent learning and innovation networks arise. Companies strive for closer ties with supplier and customer networks to stimulate new ideas, create technology and improve value–chain management and business processes beyond corporate boundaries.

These network communities often use networking opportunities supported by Internet technology: approaches such as collaborative commerce (c-commerce) or E-collaboration have gained ground in the last couple of years. What emerges is an intensive networking of different companies collaborating in the fields of product development, design and resource management, with the vision of profiting from mutual transfer of knowledge in a knowledge community. Objectives and criteria for success are, for example, the use of external ideas for innovation appropriate for a shared market, trend-scouting and active dialogue with customers and knowledge multipliers (research institutes, R&D institutions). Collaborative innovation, collaborative design, collaborative marketing, collaborative selling, collaborative support, and collaborative communication are terms symptomatic of a development integrating end users as co-creators in business processes.

With the intensified integration of customers in production processes and sales, product and service development, obligations and responsibilities have increased and continue to rise, for both companies and customers. In return for tailor-made products and services customers are expected to reveal specific information about their business and strategy. That means an investment of trust in the collaborating company, and offers, in return, a constant, reliable but flexible fulfilment of the customer's needs. This requires far-sighted and prudent deployment and development of corporate quality management strategies, going far beyond the mere product and service-focused customer–supplier relationship. In that sense, the existing mental models of the company are fundamentally challenged. It is forced to transform the, thus far, valid "'command-and-control' mentality, that characterised the age of information inequity, to the 'connect-and-collaborate' mentality needed in the age of information democracy" (Sawhney, 2002).

In return for tailor-made products and services customers are expected to reveal specific information about their business and strategy.

Experience proves that there is no guarantee for successful implementation of a connect-and-collaborate strategy in operative business. There are various risks associated with the complexity of customer-oriented strategies, which can fail because of deficiencies in corporate structure and culture. The creation of transparency, of trust and sustainability is even more valuable in intercompany networks and structures than it was in former business environments. In particular, times of massive downsizing show clearly that handling customer – employee relationships with care is a crucial factor for success, but, conversely, it can become a bottleneck.

Continual damage to customer networks by labour force reduction can endanger strategically important access to the customer and to the market and subsequently threaten corporate success.

Implications for management

From a management perspective, the following conclusions can be drawn. First, social capital lays the foundation for the networks which are decisive for innovations. Costs of mutual synchronisation (within and between companies and institutions) can be reduced by shared norms and beliefs rather than by hierarchic norms (Giddens, 2000).

Second, there is a trend to organic organisational forms. Irrespective of the specific character of the form (modular or cellular organisation), all are autonomous, self-organising systems that lack hierarchic or even lasting structures, but are bound together through trust relationships. Methods and guidelines on how to manage knowledge creation and transfer in this flexible context of networking and knowledge network life-cycles have to be developed and analysed by scientific research.

Third, cost consciousness and competition will intensify in a lasting economic downturn, simultaneously with high quality standards and differentiated customer needs. Profitability, efficiency and flexibility requirements of network-actors ('Networkers') and their importance as a competitive factor are consequently on the rise. The shorter 'half-life' of knowledge and the sharp increase of implicit knowledge force organisations to alter their behaviour. Efforts have to be undertaken to retain employees in the corporate network in order to gain stability and reliability and, more importantly, to build and make use of new network ties. The importance of personal factors, such as commitment, job satisfaction, job and work place design, motivation, corporate culture and corporate values, is on the rise and should kick off a renewal of organisational arrangements.

Efforts have to be undertaken to retain employees in the corporate network in order to gain stability and reliability and, more importantly, to build and make use of new network ties.

Knowledge, networks and processes

In the preceding section we saw that there are various reasons for companies to join networks. The advantage of network structures initialising and supporting knowledge creation and knowledge-sharing processes is one of the most crucial points. It emphasises that knowledge is the most important resource of a company in the current economic situation. This is symptomatic of a trend in the current information and knowledge society where tangible means of production lose relevance as indicators of corporate performance ability compared with intangible assets, such as customer relationships, image, innovation skills, and human and intellectual capital (Stewart, 1997). With knowledge rapidly evolving, the basis of economic growth undergoes a remarkable alteration where the value of a company is derived not solely from tangible assets.

Many firms now have intellectual property as their major asset. Intellectual capital is the value that companies are able to extract from intellectual property – product innovation, patents, copyrights, know-how and corporate knowledge. To realise its value, companies must understand what intellectual capital is, where it resides, how to invest it, use it and determine the strategic value of all of the company's intellectual assets, as well as how to express the value to the marketplace in order to turn it into a competitive advantage. Generally, those companies that can be expected to achieve higher value growth by means of innovative products and services have a higher market value. Competitive advantage can often be equated with knowledge and capability, so companies cannot allow knowledge as a value driver to become a bottleneck.

To realise value, companies must understand what intellectual capital is, where it resides, how to invest it, use it and determine the strategic value of all the company's intellectual assets ...

Intangible resources are difficult to acquire on external markets. Therefore, they have to be developed and harvested within companies themselves with a view to the future. Knowledge management solutions consequently have to safeguard the stability of the corporate knowledge pool. Besides being integrated into production, problem-solving and decision-making processes, employees have to be given opportunities to develop their skills further and to acquire specific knowledge regarding products and services, customers and networks. For this, organisational forms have to be instituted to facilitate creation and transfer of knowledge between individuals, groups and networks, both in organisations and across processes.

Critical elements and measures for success can best be described by the following elements according to a *dynamic knowledge management model* proposed by Seufert *et al.* (1999):

1. Interconnect the different levels and areas of knowledge:
   - enable networking between individual knowledge types (explicit and implicit);
   - enable networking between different levels (for example, individual, group, organisation);
   - enable networking between different areas of knowledge (for example, customer knowledge, R&D knowledge);
2. Interconnect knowledge work processes and knowledge network architecture:
   - knowledge creation and transfer can occur at different real, virtual or mental 'places';
   - knowledge creation and transfer can establish themselves in formal or informal networks;
3. Interconnect knowledge work processes and facilitating conditions.

Implications for management

When discussing organisational capabilities and dynamics, the approaches of process management and knowledge management are closely bound. Knowledge is seen as an indicator both for today's and tomorrow's innovative power of a company. Knowledge as a resource is created by the interaction of individuals within a business network, in "micro-communities of knowledge" (von Krogh *et al.*, 2000) and in business processes with different types and contents of knowledge.

Corporate knowledge management has to take into account the specific characteristics of knowledge creation and transfer and the characteristics of formal and informal networks in order to understand how new, relevant knowledge can be created and multiplied.

A method is needed that not only allows the gathering of knowledge but also encourages the development of knowledge competence. The kind and quality of network relations, their underlying mental models, and the respective embedding in cultural, political, socio-economic frameworks are increasingly important.

Knowledge management becomes a corporate strategic resource and core competence if the company succeeds in shaping idiosyncratic knowledge creation and transfer processes not easy for competitors to imitate.

People make the process

Process management is about the structured co-ordination of people and information.

Often process management and knowledge management are discussed as two interconnected dimensions of a prevalent business approach. However, these are two basically different perspectives. Process management is about the structured co-ordination of people and information. It is organised in a top-down manner, based on the assumption that it is easy to codify value creation, supposing that corporate processes can successfully be controlled by means of rules, routines and control mechanisms largely irrespective of individuals.

On the other hand, knowledge and innovation management is organised in a bottom-up manner and assumes that managers can best encourage knowledge creation by responding to the inventive, improvisational ways by which people actually get things done (Brown and Duguid, 2001).

As a consequence, companies face a dilemma that is difficult to handle. This demonstrates the conceptual weaknesses of previous process-based management approaches: on the one hand, processes are organised in such a manner that working methods and employees' performance are embedded in standardised core processes with strict boundaries and performance indicators; on the other hand, companies profit from giving employees scope and a free hand to create innovative solutions. This gets more difficult the more widespread it becomes in business – acting in highly innovative informal networks and communities means, as a rule, acting beyond core process boundaries. Therefore, companies have to keep their balance between implementation and control of process frameworks and creation and systematic support of individual liberties.

Consequently, the leverage for success for process-based and knowledge-network-based companies comes from their human capital. Core competencies are difficult to imitate; knowledge pools fit for future challenges and innovations become strategic factors for success and move employees into the centre of company interest. If companies manage to enhance knowledge-worker productivity then the payoffs will be enormous. Recruiting and retaining the most highly skilled workers are vital elements of a company's success. The identification and promotion of relevant enablers become major strategic and operational concerns of companies.

Work is performed as a collective act, expedited through the investment of capabilities, skills and motivation by the employees. How work and working conditions are experienced differs from person to person and is based on various mental, individual and organisational attributes. The following core dimensions can be differentiated:

- personal (attitudes, motivation, commitment);
- inter-individual (standards, values shaped by social interactions and socialisation);
- organisational (rules, routines, instruments of working environment).

To lead a company requires particular qualities and capabilities in the knowledge economy. Leadership should have the ability to provide context and meaning to the organisation and its networks. The structural supports for a knowledge-network-based economic reality are defined in the following subsections.

Work is performed as a collective act, expedited through the investment of capabilities, skills and motivation by the employees.

Enabler: collaborative organisational context

Assessing and supporting informal networks and social networks is regarded as a key enabler. Informal networks are especially important in knowledge-intensive sectors, where people use personal relationships to find information and do their job (Cross *et al.*, 2002).

Within companies new ideas arise in highly flexible networks. The manner by which they manage to bridge 'structural holes' (Burt, 1992) and closed circles is crucial to their success. The way members of different networks (e.g. business economists, engineers, researchers, experts in human resources or finance) constructively work and communicate together is decisive for the efficiency of joint innovation and knowledge creation. Not only the circulation, but also the production of valuable information and innovative ideas derive from employees with 'bridging' and 'gatekeeper' functions and in large part from their individual and social capital as well as the given opportunities of modification (Coleman, 1988; Burt, 1992).

Knowledge management is likely to work more efficiently when knowledge workers have unhindered access to various knowledge sources and partners.

Specialists work on a basis of expert knowledge acquired in special training sessions. However, knowledge-workers work on a basis of implicit knowledge of their own and of transferred, explicit knowledge of others (Scarbrough, 1999). Therefore, they are dependent on a collaboration across subjects, organisations and functions. Consequently, knowledge management is likely to work more efficiently when knowledge workers have unhindered access to various knowledge sources and partners. But a subject-based division of labour institutionalised in firms obstructs this path. It triggers both visible and invisible boundaries between departments and working communities as well as hinders knowledge sharing and knowledge generation.

Enabler: autonomy and personal liberty

A further step in creating a framework to encourage innovation consists of the enhancement of commitment and self-organising power within a company: innovative firms let the employees participate in decision-making processes and give them responsibility. This is to increase the intrinsic motivation of employees, essential for the development of creativity and innovation (Axtell *et al.*, 2000).

Enabler: social capital

As mentioned previously, social capital is a valuable asset for those networks involved in fostering innovation. Costs of mutual synchronisation can be reduced by shared experiences, values and standards much better than through hierarchies or bureaucratic rules (Giddens, 2000). Social capital is influential and flexible: it can permeate boundaries, as can be seen in the successful co-operation of companies, research institutes and related associations.

As a consequence, firms have to make efforts and take appropriate measures to connect individual, inter-individual and organisational processes in an idiosyncratic way, i.e. difficult for competitors to imitate. Knowledge and people as resources become strategic factors for success and competitive advantages when management recognises the potential of the combined economic, mental and motivational aspects of process performance and takes measures to transform them into added value. How far human resource management can help enable this transformation is to be shown in the following section.

Implications for management

In the course of process and core-competencies orientation and continuous innovation in technology, the role and focus of human resources (HR) has changed strategically and operationally. HR management (HRM) has been forced, more and more, to improve its own customer orientation both within the company and as a service provider for external customers. Above all, an adaptation to the new economic challenges means that innovative HRM has to support the above-mentioned 'enablers' with compatible HR measures and instruments.

It is not sufficient for HR to perform the role of important but increasingly narrow technical specialists. It is more about serving as reactive advisers and hand-maidens to line management (Clark, 1993). Additionally, HRM is to be regarded as an effective enabler for social processes in the context of innovation processes, as a guide and coach for networks, as a supporter of empowerment, and as a supplier of appropriate HR tools (e.g. recruiting and personnel development tools) and data of available human resources (HR metrics). Thus, HR professionals become fully recognised members of the management team, strategic change-makers, specialist advisers and monitoring 'auditors' in the planning, implementation and operation of change (Clark, 1993).

It is not sufficient for HR to perform the role of important, but increasingly narrow, technical specialists. It is more about serving as reactive advisers and hand-maidens to line management, acting as:

- enablers;
- business partners;
- designers; and
- supporters.

(Clark, 1993)

Personnel development and organisational development, knowledge management and organisational management have to work closely together to support the strategic development of corporate management (see Table 5.2).

The *core tasks* identified as essential for management in innovative organisations and the subsequent *HR roles* supporting and developing the company's human capital in order to ensure sustainable competitive advantages are:

1. Creation and adaptation of organisational routines and forms:
   - *Enabler* – design, coaching and monitoring of HR tools and measures forming a framework supporting motivation in value-chain processes (e.g. HR incentives, workplace design, design of working time models).

	Competition for products and markets	Competition for resources and competencies	Competencies for creativity and staff development
Perspective on employees	People viewed as factors of production	People viewed as valuable resources	People viewed as "talent investors"
HR's role in strategy	Implementation, support	Contributory	Central
Key HR activity	Administering of recruitment, training and benefits	Aligning resources and capabilities to achieve strategic intent	Building human capital as a core source of competitive advantage

5.2 The evolving focus of strategy (Bartlett and Ghoshal, 2002)

- *Business partner* – integration of HR professionals within business line, working with executive teams to create value-driven people strategies and consistent knowledge creation and innovation.
- *Designer* – design and implementation of flexible organisational structures supporting learning and innovation processes (e.g. project teams, formal and informal communities of practice).
- *Supporter* – HRM simplifies or takes over administrative work of employees and business managers in order to achieve a customer-oriented efficiency beyond value-chain processes.

2. Co-operation and communication:
   - *Mediator* – social processes and networks need experts capable of mediating and moderating in conflict situations.
   - *Strategic partner of management* – design and implementation of a personnel strategy as an integral part of a predominant corporate strategy.
   - *Communicant* – HR must not only account for an organisation's human capital, but also channel and communicate this capital wisely. A lack of managerial attention will harden individual and organisational barriers which already exist.
3. Corporate culture:
   - *Cultural agent* – HR plays an important role in shaping a corporate culture, communicating corporate values and designing instruments to promote a culture of innovation and learning.
   - *Commitment agent* – in order to boost commitment, HR has to establish a matching organisational framework (e.g. measures such as partial transfer of responsibility, integration of employees in corporate strategy

and values, transparency of company's goals and processes, intrinsically and extrinsically attractive rewards for individual performance readiness) that is supported at all levels of management.

4. Leadership and development:
   - *Trainer* – coach employees and superiors in working out and accepting joint and binding leadership principles.
   - *Consultant* – selection, placement and development of capabilities and skills of employees and teams appropriate to handle upcoming economic challenges.
   - *Value driver* – transforming HR departments in companies from cost-producers to value-drivers of human capital.

The importance of human capital as an indicator and integral part of production capacity and corporate competitiveness is on the rise. Personnel management activities and human engineering as a means to manage competitive business dynamics and the subsequent challenges have to be further enhanced, since today's managers

> *must compete not just for product markets or technical expertise, but for the 'hearts and minds' of talented and capable people and (...) ensure that those valuable individuals become engaged in the organisation's ongoing learning processes and stay committed to the company's aspirations*
>
> (Bartlett and Ghoshal, 2002)

"[managers] must compete not just for product markets or technical expertise, but for the 'hearts and minds' of talented and capable people ..."
(Bartlett and Ghoshal, 2002)

Conclusion

This chapter has shown that knowledge creation and innovation are always a social as well as an individual process. Moreover, every company has various organisational, social and individual barriers that obstruct knowledge creation and process performance. Hence, the most important organisational challenges in a knowledge-network economy are to lead and guide knowledge workers, to make effective use of technology for improving productivity and communication, to guide heterogeneous teams, and to adapt organisational structures to fit the needs of different businesses. Especially important is how organisations create links through business processes and project-based assignments to address the need for innovation, speed and effective execution. Following a serious period of fundamental down-sizing and restructuring, most corporations need to revitalise their human side in order to make better use of their inherent potentials and rise to challenges.

References

Axtell CM, Holman DJ, Unsworth KL, Wall TD, Waterson PE (2000) Shopfloor innovation: facilitating the suggestion and implementation of ideas. Occupational and Organizational Psychology, 73: 265–285

Bartlett CA, Ghoshal S (2002) Building competitive advantage through people. MIT Sloan Management Review, 44(2): 34–41. Figure 5.2 reprinted with permission of Massachusetts Institute of Technology

Brown JS, Duguid P (2001) Creativity versus structure: a useful tension. MIT Sloan Management Review, 42: 93–94

Burt R (1992) Structural holes: the social structure of competitions. Harvard University Press

Clark J (1993) Human resource management and technical change. Sage

Coleman J (1988) Social capital in the creation of human capital. American Journal of Sociology, 94: 95–120

Cross R, Nohria N, Parker A (2002) Six myths about informal networks – and how to overcome them. MIT Sloan Management Review, 43(3): 67–75

Davenport TH (1993) Process innovation – re-engineering work through information technology. Harvard Business School Press

Giddens A (2000) The third way and its critics. Polity Press

Kim W, Mauborgne R (1999) Strategy, value innovation, and the knowledge economy. MIT Sloan Management Review, 40(3): 41–54

Leibold M, Gilbert JB, Probst MG (2002) Strategic management in the knowledge economy: new approaches and business applications. Wiley

Prahalad CK, Hamel G (1990) Core competences of the corporation. Harvard Business Review, 3: 79–93

Sawhney M (2002) Don't just relate – collaborate. MIT Sloan Management Review, 43(3): 96

Scarbrough H (1999) Knowledge as work: conflicts in the management of knowledge workers. Technology Analysis and Strategic Management, 11: 5–16

Seufert A, von Krogh G, Bach A (1999) Towards knowledge networking. Knowledge Management, 3(3): 180–190

Stewart TA (1997) Intellectual capital – the new wealth of organizations. Doubleday Dell

Teece DJ, Pisano G, Shuen A (1997) Dynamic capabilities and strategic management. Strategic Management, 18: 509–533

von Krogh G, Ichijo K, Nonaka I (2000) Enabling knowledge creation – how to unlock the mystery of tacit knowledge and release the power of innovation. Oxford University Press

Chapter 6

Artificial intelligence for design process improvement

David C Brown
Worcester Polytechnic Institute

This chapter presents some ways in which artificial intelligence (AI) research can be used to improve the way that agents (people or machines) design things (*i.e.* design process improvement). There are a variety of definitions of AI, influenced by the goals of the researchers involved (Russell and Norvig, 2003). The best known is Marvin Minsky's statement that it is the science of making machines do things that would require intelligence if done by humans. This highlights the common AI paradigm of producing some theory about how a task might be done: in terms of specifying the knowledge and reasoning, and possibly also details of sensing, action and communication. The theory is then implemented in some computational form (typically a computer program) to see whether it can exhibit the appropriate intelligent behaviour. The tasks studied are usually those for which no efficient solution is known, and usually (but not always) those which intelligent beings can solve. Some researchers focus on a more cognitive point of view:

> *By 'artificial intelligence' I therefore mean the use of computer programs and programming techniques to cast light on the principles of intelligence in general and human thought in particular.*
>
> (Boden, 1977)

while some seek to study AI in more absolute terms:

> *...studying the structure of information and the structure of problem solving processes independently of applications and independently of its realisation in animals or humans.*
>
> (McCarthy, 1974)

"By 'artificial intelligence' I therefore mean the use of computer programs and programming techniques to cast light on the principles of intelligence in general and human thought in particular."
(Boden, 1977)

The area of 'AI in design' (AI-in-D) has flourished since the early 1980s. It attempts to use the techniques and approaches of AI to study design processes, most often engineering or architectural design. As it is so closely tied to AI, its researchers have also focused on different outcomes. The field has produced:

- software systems that design artifacts;
- software systems that provide assistance to designers (for example, by critiquing design choices);
- theories about how designers reason;
- studies and analyses of actual designer activities;
- models and descriptions of natural categories of design activity (for example, routine parametric design, or configuration);
- guidance about how to apply existing AI techniques to design problems.

Most AI-in-D researchers believe that engineering design is not a mysterious art and that there are core reasoning 'skills', and specific types of knowledge that apply to the same type of design task (e.g. component selection), even across domains.

An overview of the history of the AI-in-D field can be found by looking at the following sources: the collective *Proceedings of the AI in Design* conferences; the *AI EDAM* journal (Cambridge University Press); the *IEEE Expert AI in Design* special issues (Brown and Birmingham, 1997); Stahovich's (2001) survey ; and the Encyclopedia of Artificial Intelligence article on Design (Brown, 1992).

The field has progressed over time by attempting to understand and replicate increasingly less-well understood design activities. Early work focused on parametric, routine and case-based design, moving gradually via configuration to functional reasoning and creative design, and from solo designers to teams.

There is a vibrant group of researchers active in the AI-in-D area world-wide. It has its own major conference, the *International Conference on Design Computing and Cognition* (until recently the *International Conference on AI in Design*), lots of related specialised workshops, and its own Webliography and list of AI-in-D books which can be found at

http://www.cs.wpi.edu/Research/aidg/AIinD-hotlist.html

http://www.cs.wpi.edu/Research/aidg/AIinD-books.html

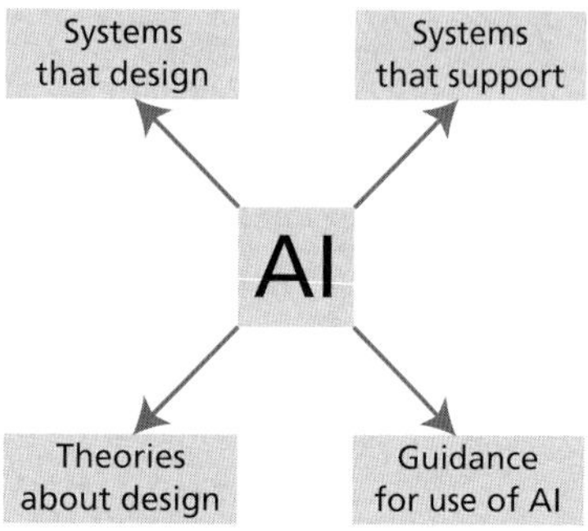

6.1 AI in design

In this chapter, distinction will be made between:

- how AI has contributed to producing better theories about design processes;
- how AI can be involved in the process itself to help improve it;
- how AI can be used to produce better processes.

Artificial intelligence producing better theories

When expert systems (Jackson, 1999) were first introduced, it was quite quickly noticed that an immediate benefit of studying an expert's reasoning and knowledge, in enough detail that a software system could be built to replace him or her, was that a previously private process became public and understandable. Sometimes that yielded enough knowledge to improve the process without developing a system.

The field of AI-in-D has had a similar effect. Theories and models of design activities have been produced that make conjectures about exactly what kinds of knowledge and what kind of reasoning are necessary in different design situations. Once this is well understood, then this information can be taught, and design or design assistance systems built, all of which can improve design processes.

The rest of this section provides a brief introduction to some of this work, and points to where further descriptions might be found.

The basis

The initial AI-in-D research on parametric design, and routine (or near-routine) design resulted in several models of design processes. Two that shaped the way that researchers looked at design were Chandrasekaran's (1990) and McDermott's (1988).

In Chandrasekaran's paper, and in others subsequently, he proposes a task-method analysis where designing consists of picking an appropriate method from a set that will help address a task. Each method suggests its own sub-tasks, and they too have alternative methods. This process repeats. Methods can be selected on the basis of available knowledge, the likelihood of success, or a variety of other reasons. He points out that many design tasks can be characterised as having the sub-tasks 'propose', 'critique', and 'modify' (Figure 6.2). Proposing solutions might be done by case retrieval, constraint reasoning, or some other AI technique, or the problem might be reduced by some decomposition method into lower level sub-tasks.

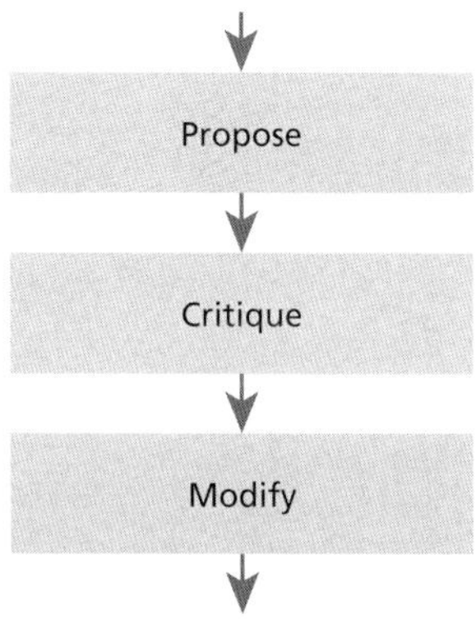

6.2 Many design tasks can be characterised as having sub-tasks (Chandrasekaran, 1990)

McDermott, in a similar proposal, focuses more on suggesting that we try to tease out what kinds of knowledge are needed, what roles they play, and how they can be represented. For example, he identifies design extension knowledge (propose), constraints, and fixes (to help correct constraint violations).

Methods for parametric design have been extensively studied by the AI-in-D community (Motta, 1999; Fensel, 2000). For example, 'propose and backtrack' starts by extending incomplete but consistent designs, and then restores consistency if constraints fail by backtracking to prior design extension choice points; 'propose and revise' starts by extending incomplete but consistent designs, and then revises them to restore consistency when necessary using special-purpose fixes; 'propose and improve' starts with a complete solution and then attempts to improve it. Unfortunately, these terms are not always used consistently.

Note that many parametric design problems can be viewed as constraint satisfaction problems. For example, see the discussion in Russell and Norvig (2003) of the min-conflicts heuristic for constraint satisfaction problems as an example of Propose and Improve.

The DSPL language (Brown and Chandrasekaran, 1989) was developed to allow the expression of design knowledge. It recognises distinct pieces of knowledge that represent what a designer did for major sub-problems, individual design decisions, groups of decisions, constraints, suggestions about what to do if failure occurs during designing, plans, plan selection, and several other aspects. Once these actions are captured and expressed in DSPL, it forms a design expert system.

The basis for this language was that during routine reasoning (Brown, 1996) the knowledge needed at every step is known, so that decisions can be made with essentially no searching or planning. For some problems the kinds of knowledge in DSPL will suffice. It can be used to do routine parametric design problems and routine configuration problems in a domain-independent manner. However, for many problems, designers move in and out of situations that are routine for them. They need to search and plan for a while until they get to a sub-task that they recognise and can treat as routine (Brown, 1996).

extend-design
find-constraints
test-constraints
suggest-fixes
select-fix
modify-design
find-constraint
test-constraints
propagate-changes
test-if-done

6.3 Basic methods with known knowledge types (Balkany *et al.*, 1991)

Balkany *et al.* (1991) studied several design systems, including the DSPL-based AIR-CYL system, attempting to compare and contrast them in terms of basic methods that have known knowledge types, such as extend-design, find-constraints, test-constraints, suggest-fixes, select-fix, modify-design, find-constraint, test-constraints, propagate-changes, and test-if-done (Figure 6.3). As the systems all contained a significant amount of parametric design, and a lot of routine reasoning, they found a large amount in common.

All the research mentioned above points out that identification of these basic reasoning 'skills' (the ingredients of designing) and the knowledge they need allows focused 'knowledge acquisition'. For example, pointed questions can be asked, such as "when the constraint on the size of part A fails, what is the best way to alter the design so that the constraint no longer fails?" Hence, formerly impenetrable processes can be understood in terms of these basic skills.

Once researchers in a field think something is well enough understood there is the inevitable move towards some kind of toolkit. In the area of problem-solving methods (PSMs; Motta, 1999; Fensel, 2000) there have been attempts to build catalogues of templates or abstracted modules for different tasks. The reasoning is characterised in terms of patterns of inferences and general methods for carrying out the task. The goal is to provide reusable modules that, when completed with the right knowledge, can be used to build systems.

CommonKADS (e.g. Bernaras, 1994; Schreiber *et al.*, 1994, 1999) recognises many types of tasks, including synthesis, configuration design, assignment, and planning. The 'generic tasks' effort (Chandrasekaran and Johnson, 1993) is focused on diagnosis and design.

Smithers (1998) has a detailed formal 'knowledge-level' theory of designing in which he identifies the types of knowledge "involved in designing and the different roles they play in the overall process". He shows how these different types of knowledge are connected by processes, but does not specify the processes themselves, unlike in the PSM approaches.

Configuration and learning

Most of the work that focuses on routine and parametric design makes the assumption that the requirements are given and remain essentially the same throughout the design process. This is not always true. The AI-in-D work sharing this more general view has become known as exploration-based models of design (Smithers and Troxell, 1990). Both the candidate sets of requirements and the candidate sets of designs are specified, explored and refined during designing (Brazier *et al.*, 1994). A similar view of this process can be found in research on co-evolutionary design (Maher, 2000).

Configuration is the process of taking a set of predefined components and attempting to find a set of relationships between them that satisfies a set of constraints and meets some requirements.

Another more recent area of research has been on configuration (Mittal and Frayman, 1989; Brown, 1998; Wielinga and Schreiber, 1997). If you take a set of predefined components and attempt to find a set of relationships between them (for example, an assembly or arrangement) that satisfies a set of constraints and meets some requirements then you've done a configuration task. Simple depth-first (propose and backtrack) methods for configuration are very inefficient and require significant knowledge-based guidance.

As there are extensive commercial applications of configuration (Faltings and Freuder, 1998), this work has received a lot of attention, and powerful constraint-based and logic-based approaches have been developed (Soininen and Stumptner, 2003). Knowledge can help with the process, as, for example, if you know something about the physical or functional organisation of the desired configuration, perhaps at an abstract level, then this can be used to guide a refinement or instantiation process.

Another more recent area of study has been how to model the learning that takes place while designing. Designers accumulate all sorts of knowledge as a consequence of the process of designing, and that knowledge affects their present and future design activity.

Influenced by an analysis of how the learning that takes place in design might be classified (Grecu and Brown, 1998), Sim and Duffy (1998, 2000) took a simple model of generic design activity, in terms of the flow of knowledge, and coupled it to a model of learning. The models include eight kinds of knowledge and two kinds of process. By coupling the models in different ways they were able to characterise three kinds of design-related learning. 'Retrospective' learning occurs after the designing is completed; '*in-situ*' learning occurs during the designing activity; and 'provisional' learning takes place prior to design activity that will require the knowledge being learned.

Creativity

There is a very large body of literature on creativity, but less about creative design (Christiaans, 1992; Gero and Maher, 1993, 2001; Dasgupta, 1994, 1996; Goel, 1997). Creativity is seen as being relative to a standard of some kind, such as the designer's own previous designs, or the designs of some community (Boden, 1994). As a consequence, it is hard to come up with a model of this activity.

Analogical reasoning involves the retrieval of the solution to a similar design problem and the transfer of its relevant aspects for use in the solution to the original problem.

It is possible, however, to point to some situations and approaches that tend to produce results that are judged to be creative. In a situation where the design variables need to be determined and can change, and the ranges of their possible values can change, then the design process (and probably the result) will be creative as the designer will be taken outside his or her normal experience during much of the process. The approaches that tend to produce creative results include the use of analogy (Goel, 1997), functional reasoning (Umeda and Tomiyama, 1997), mutation (de Silva Garza and Maher, 2000) and reasoning from first principles (Williams, 1992).

Analogical reasoning involves the retrieval of the solution to a similar design problem and the transfer of its relevant aspects for use in the solution to the original problem. Analogy can be used to provide designs, methods, plans or other useful knowledge.

Analogical design research usually focuses on conceptual design, as more abstract descriptions are easier to work with. Representations of function are also at this level and can be used in analogical design, in conjunction with representations of structure and behaviour (Balazs and Brown, 2001).

Functional representations can also be used to guide the refinement of a design, as once the necessary functionality has been determined it can be refined into types of component that can provide it (Umeda and Tomiyama, 1997). They can also be used to provide high-level simulations of a proposed design. AI provides a richer set of representations than that currently found in engineering design views of function (Wood and Greer, 2001).

Mutation is used to explore the space of possible designs by making changes to a proposed (partial) design. This might then lead the designer into new sub-problems not faced before, or allow him to make unusual associations that lead to interesting analogies. Reasoning from first principles allows the designer to work from scratch and to be less biased by his existing knowledge.

Artificial intelligence involved in the process

There are plenty of places in the design process where a designer can use a little more assistance. Intelligent tools, for example, can provide assistance with generating design ideas, can critique design proposals, and can actually do some of the designing to relieve the designer of the more mundane tasks.

A finer grained analysis of roles that knowledge-based techniques and systems can play during design can be found in Brown (1992). These design sub-tasks include: abstraction; analysis; conflict resolution; criticism; decomposition; estimation; evaluation; interpretation; learning; negotiation; patching; prediction; redesign; refinement; retraction; suggestion making; and selection.

There are a wide variety of AI techniques that can be used to assist with these sub-tasks to help improve the process (Brown, 1992; Stahovich, 2001). These include expert systems, genetic algorithms (GAs), case-based reasoning (GBR) and formal grammars. Other examples include neural networks, qualitative reasoning, heuristic search, planning and multi-agent systems.

Expert systems (David *et al.*, 1993; Jackson, 1999) are computer programs that solve problems or give advice about some specialised subject by reasoning using representations of knowledge. They are normally associated with tasks that would be done by an expert, such as diagnosis or design. They usually involve heuristic reasoning, structuring the reasoning the way the expert would, and using knowledge based on the experiences of an expert. The majority of such systems have been built using rules. Each rule describes some key aspect of a situation (for example, the partial diagnosis and the symptoms available) and suggests what action to take in that situation. They can be used 'forwards', recognising and acting, or 'backwards', where the

Expert systems are computer programs that solve problems or give advice about some specialised subject by heuristic reasoning using representations of knowledge.

system hypothesises that the result exists (for example, a disease, for diagnosis) and reasons back through the rules to see whether all the necessary conditions for its existence are actually present. Rule-based systems are appropriate for a very wide variety of applications.

GAs (Bentley, 1999; Bentley and Corne, 2001; Lee *et al.*, 2001) attempt to mimic evolution. It is a very flexible and efficient technique for searching a large space. In this approach, descriptions (of designs or design processes, for example) evolve and improve. A whole 'population' of descriptions is considered at once.

In order to make progress towards better and better solutions, the descriptions are evaluated in every generation to determine their fitness, as fit solutions are worth keeping. Fitness might be, for example, how well the design represented by the description meets the requirements. You'd also like to keep the qualities of the best items and propagate them to the next generation. To enable this, randomly selected fit parents are used to generate offspring descriptions by doing a 'crossover' between their two descriptions (Figure 6.4). Typically the two parts of description A (A1, A2) are mixed with the two parts of description B (B1, B2), to generate two new descriptions (A1, B2) and (B1, A2). Sometimes new descriptions are also generated by random mutations of a few descriptions. The fittest are kept and passed to the new generation.

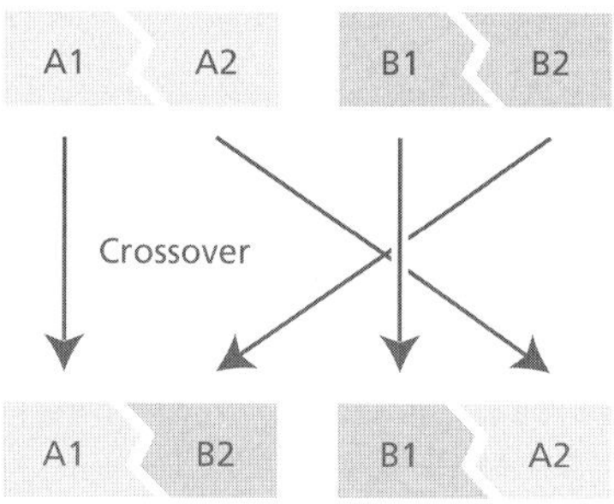

6.4 GAs, an attempt to mimic evolution

This process of generating offspring, evaluation, and passing fit members of the population to the next generation is repeated until some stopping criteria are reached – for example, the best description does not improve over several generations.

CBR (Maher *et al.*, 1995; Maher and de Silva Garza, 1997; Maher and Pu, 1997) is a method for reusing experience. Cases, past design solutions for example, are collected and 'indexed' by key features. For design cases, the features used to index them might include the requirements of the problem for which that design was the solution. Hence, when a new design is needed the index is used to 'recall' the stored cases that appear to be the most likely to be solutions (because the new requirements are similar to the old ones). Once those candidates are retrieved and evaluated, either one case is the perfect answer, or one or more recalled cases need to be 'adapted' to improve them. Typically, CBR systems rely on many cases and small adaptations. But if a human designer is doing the adaptation then the cases can be presented to him or her as starting points for the design process, encouraging him or her to build on well-tried results, and also to try things that would not necessarily have been thought of.

Formal grammars (Cagan, 2001; Brown, 1997) are a way of representing the structure of things precisely. Grammars for languages include rules such as "a sentence (S) is a noun phrase (NP) followed by a verb phrase (VP)", written "S → NP VP". Rules such as these can be used to recognise (by seeing an NP next to a VP and reporting that they form an S), or to generate (by seeing an S and replacing it by NP followed by VP). Rules describing an NP, a VP and their components allow recognition and generation of sentences.

This example operates in one dimension (the row of words), but grammars for design need to be for two or three dimensions. Shape grammars allow representations of shapes to be on both the left- and right-hand sides of a rule. Hence, more complex shapes (*i.e.* designs) can be generated by replacing examples of the shape from the left-hand side of a rule with the (usually more complex) shape from the right-hand side of the rule. Even more flexibility can be gained by adding parameters to the shapes in the rules. Semantic checking can also be added by allowing attributes to be attached to portions of the shape being generated, and constraints on them to be associated with the rules.

Artificial intelligence producing better processes

AI has had a lot of indirect impact on design processes. Models have been developed that have affected the way we describe design processes; types of knowledge and reasoning have been identified, and AI has led to new tools and new processes (for example, the use of GAs). In general, AI has affected our ways of thinking about designing, so that we now see it as a potentially understandable and rational information-processing task that requires lots of knowledge, and lots of different reasoning skills.

AI has affected our ways of thinking about designing, so that we now see it as a potentially understandable and rational information-processing task that requires lots of knowledge, and lots of different reasoning skills.

The expert systems, PSM and modelling efforts have raised the awareness of the importance of knowledge acquisition (not only of expert knowledge), and have led to studies of knowledge sharing (including the use of ontologies).

But there are opportunities for AI to affect the design process directly. The following sections present three examples: agent learning; methodology generation; and planning.

Agent learning

First, some work that shows ways in which learning can occur during designing will be presented (Grecu and Brown, 1996, 2000). Our experiences greatly affect how we design, as past successes and failures, for example, change the way we do things and change our evaluations of potential design decisions.

Sometimes during designing an attempt is made to decompose the problem so that decisions can be made more independently. Even for an individual designer, one decision can affect another in unanticipated ways. Separation of decision makers, in design teams for example, makes it even harder for one decision maker to know what impact his decisions might have on another. Usually, however, sub-tasks and decisions really are not completely independent, and the composition of partial solutions that follows from decomposition of the problem reveals conflicts.

In Grecu and Brown's single function agent-based design system small knowledge-based programs, known as agents, interact to solve a spring design problem. Each agent has a particular 'role' to play, dictated mainly by its function (what kinds of input it needs and output it produces).

For example, one agent might select a value for a parameter, while another acts as a critic, offering an explanation of why the chosen value isn't good. The selector agent builds a model (*i.e.* learns) of which selections to avoid in which design situations, based on feedback from critics. This reduces the number of interactions between agents by about half, thus reducing communication overhead and speeding up the whole process.

An experienced designer can use expectations to predict from available information what will happen later in the design process.

Another form of knowledge that can be learned is expectations (Grecu and Brown, 2000). An experienced designer can use expectations to predict from available information what will happen later in the design process: for example, that a constraint may fail. This can be done despite having only part of the information required to know this definitely.

In Grecu and Brown's LEAD system, "causal attribution" is used to collect together all the factors that might be an input to the expectation in the situation for which an expectation is desired. This might include relevant constraints, design decisions already made, or the state of other agents. Once those features are collected, an inductive learning phase, known as covariational analysis, uses data collected from past design processes to filter out those features that do not contribute. It builds a description of the expectation, which is validated prior to use in the system. The expectations change the design process, allowing better design choices, and reducing failures.

Methodology generation

The robot designer (RD) system (Shakeri and Brown, 2004) is concerned with methodology generation for multi-disciplinary design (for example, robot arm design). A design methodology is a scheme for organising reasoning steps and domain knowledge in order to construct a solution effectively and efficiently. A good methodology guides a designer towards a successful design.

The RD system was built as a collection of agents, where each did a very small portion of the design from one disciplinary point of view. By breaking disciplinary knowledge up into many small pieces it becomes possible to interleave decisions from different disciplines, and even make them in parallel. Agents were opportunistic and were able to contribute when their preconditions were met.

The agent-based system was given many sets of systematically slightly varying requirements. The resulting design traces were analysed. In the 960 experiments run, only 4% of the 2304 possible traces generated successful designs. In addition, some sets of traces were similar. By clustering them, it was possible to identify families of traces and describe to what situations they applied. By generalising these clusters it was possible to produce methodologies.

Planning

A large sub-area of AI is planning (Allen *et al.*, 1994; Russell and Norvig, 2003), which involves producing a set of actions, in advance of their execution, that is expected to achieve a goal. Clearly, planning is used in designing, as the design process itself needs to be ordered and structured.

Planning techniques involving searching for a plan in a space of possible plans, or proving a plan using logic, are not efficient. Actions are typically represented using a description of what must be true before the action can be used (preconditions), a list of what is no longer true after the action, and a list of what new conditions are true after the action.

A good approach to planning is to commit to where actions should go in the plan as little as possible until it is absolutely necessary. Hence a "partial order plan" might determine that the action sequences (I then J) and (P then Q) are between actions A and B, but makes no further commitment. This results in six possible sequential plans; for example, (A, I, P, J, Q, B) (see Figure 6.5). The process of building a partial order plan typically works back from the goal, finding actions that can satisfy outstanding preconditions for actions already selected. In addition, actions that invalidate other preconditions are moved or replaced. More complex planning techniques are required for realistic planning situations involving resources (scheduling), uncertainty, and multiple agents planning together.

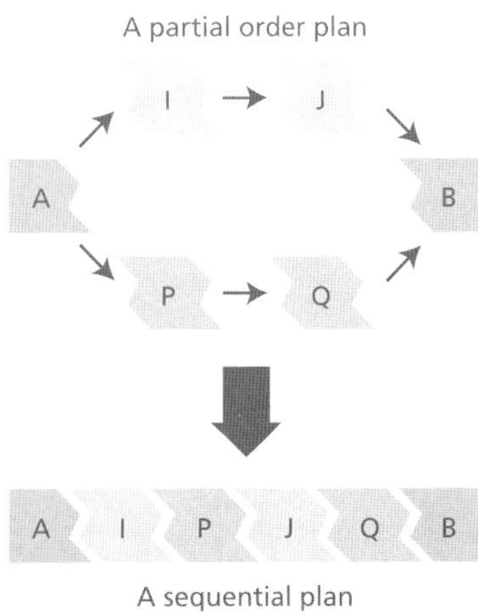

6.5 AI in planning

Conclusions

In this chapter, some of the areas in which AI can impact the design processes have been revealed. The study of AI, and subsequently AI-in-D, has led to a number of theories and models of the design process, and of subareas such as configuration, learning and creativity. There are many uses during designing for AI techniques such as expert Ssystems, GAs, CBR and formal grammars. They can be used in tools to aid the designer. Finally, AI can be used to produce better processes themselves, e.g. via agent learning, methodology generation, and planning.

It's evident that AI has contributed to producing better theories about design processes, that AI can be involved in the process itself to help improve it, and that AI can be used to produce better processes.

References

Allen J, Tate A, Hendler J eds (1994) Readings in planning. Morgan Kaufmann

Balazs ME, Brown DC (2001) Design simplification by analogical reasoning. In: Knowledge intensive computer aided design. Kluwer Academic

Balkany A, Birmingham WP, Tommelein ID (1991) A knowledge-level analysis of several design tools. AID'91, Edinburgh, UK

Bentley PJ ed. (1999) Evolutionary design by computers. Morgan Kaufmann

Bentley PJ, Corne DW (2001) Creative evolutionary systems. Morgan Kaufmann

Bernaras A (1994) Problem-oriented and task-oriented models of design. AID'94, Lausanne, Switzerland

Boden M (1977) Artificial intelligence and natural man. Basic Books

Boden M (1994) What is creativity? In: Dimensions of creativity. MIT Press

Brazier F, van Langen P, van Ruttkay P, Truer J (1994) On formal specification of design tasks. AID'94, Lausanne, Switzerland

Brown DC (1992) Design. In: Encyclopedia of artificial intelligence. John Wiley

Brown DC (1996) Routineness revisited. In: Mechanical design: theory and methodology. Springer

Brown DC (1998) Defining configuring. AI EDAM, special issue on Configuration, 12(4): 301–305

Brown DC, Birmingham WP (1997) IEEE Expert, special issues on AI in Design, 12(2) and 12(3)

Brown DC, Chandrasekaran B (1989) Design problem solving: knowledge structures and control strategies. In: Research notes in artificial intelligence series. Pitman

Brown KN (1997) Grammatical design. IEEE Expert, special issue on AI in Design, 12(2): 27–33

Cagan J (2001) Engineering shape grammars. In: Formal engineering design synthesis. Cambridge University Press

Chandrasekaran B (1990) Design problem solving: a task analysis. AI Magazine, 11(4): 59–71

Chandrasekaran B, Johnson TR (1993) Generic tasks and task structures: history, critique and new directions. In: Second generation expert systems. Springer

Christiaans HHCM (1992) Creativity in design: the role of domain knowledge in designing. Lemma

Dasgupta S (1994) Creativity in invention and design. Cambridge University Press

Dasgupta S (1996) Technology and creativity. Oxford University Press

David JM, Krivine JP, Simmons R eds (1993) Second generation expert systems. Springer

De Silva Garza AG, Maher ML (2000) A process model for evolutionary design case adaptation. AID'00, Worcester Polytechnic Institute, MA, USA

Faltings B, Freuder E eds (1998) IEEE Intelligent Systems Engineering, special issue on Configuration, 13(4): 32–33

Fensel D (2000) Problem-solving methods: understanding, description, development, and reuse. Lecture Notes in AI, LNAI 1791. Springer

Gero JS, Maher ML eds (1993) Modeling creativity and knowledge-based creative design. Lawrence Erlbaum

Gero JS, Maher ML eds (2001) Computational and cognitive models of creative design. HI'01, Heron Island, Queensland, Australia

Goel AK (1997) Design, analogy, and creativity. IEEE Expert, 12(3): 62–70

Grecu DL, Brown DC (1996) Learning by single function agents during spring design. AID'96, Stanford, CA, USA

Grecu DL, Brown DC (1998) Dimensions of machine learning in design. AI EDAM, special issue on Machine Learning in Design, 12(2): 117–121

Grecu DL, Brown DC (2000) Expectation formation in multi-agent design systems. AID'00, Worcester Polytechnic Institute, MA, USA

Jackson P (1999) Introduction to expert systems. Addison Wesley Longman

Lee C-Y, Ma L, Antonsson EK (2001) Evolutionary and adaptive synthesis methods. In: Formal engineering design synthesis. Cambridge University Press

Maher ML (2000) A model of co-evolutionary design. Engineering with Computers, 16(3/4): 195–208

Maher ML, de Silva Garza AG (1997) Case-based reasoning in design. IEEE Expert, 12(2): 34–41

Maher ML, Pu P eds (1997) Issues and applications of case-based reasoning in design. Lawrence Erlbaum

Maher ML, Balachandran M, Zhang DM (1995) Case-based reasoning in design. Lawrence Erlbaum

McCarthy J (1974) Review of "Artificial intelligence: a general survey". Artificial Intelligence, 5: 317–322

McDermott J (1988) Preliminary steps towards a taxonomy of problem-solving methods. In: Automating knowledge acquisition for expert systems. Kluwer Academic

Mittal S, Frayman F (1989) Towards a generic model of configuration tasks. IJCAI'89, Detroit, MI, USA

Motta E (1999) Reusable components for knowledge modelling. IOS Press

Russell S, Norvig P (2003) Artificial intelligence: a modern approach. Prentice Hall

Schreiber G, Wielinga B, de Hoog R, Akkermans H, Van de Velde W (1994) CommonKADS: a comprehensive methodology for KBS development. IEEE Expert, 9(6): 28–37

Schreiber G, Akkermans H, Anjewierden A, de Hoog R, Shadbolt N, Van de Velde W, Wielinga B (1999) Knowledge engineering and management: the CommonKADS methodology. MIT Press

Shakeri C, Brown DC (2004) Constructing design methodologies using multi-agent systems. AI EDAM, 18(3)

Sim SK, Duffy AHB (1998) A foundation for machine learning in design. AI EDAM, 12(2): 193–209

Sim SK, Duffy AHB (2000) Evaluating a model of learning in design using protocol analysis. AID'00, Worcester Polytechnic Institute, MA, USA

Smithers T (1998) Towards a knowledge level theory of design process. AID'98, Lisbon, Portugal

Smithers T, Troxell W (1990) Design is intelligent behaviour, but what's the formalism. AI EDAM, 4: 89–98

Soininen T, Stumptner M (2003) AI EDAM, special issue on Configuration, 17(1): 1–2

Stahovich TF (2001) Artificial intelligence for design. In: Formal engineering design synthesis. Cambridge University Press
Umeda Y, Tomiyama T (1997) Functional reasoning in design. IEEE Expert, 12(2): 42–48
Wielinga R, Schreiber G (1997) Configuration-design problem solving. IEEE Expert, 12(2): 49–56
Williams BC (1992) Interaction-based design: constructing novel devices from first principles. In: Intelligent computer aided design. Elsevier Science
Wood KL, Greer JL (2001) Function-based synthesis methods in engineering design. In: Formal engineering design synthesis. Cambridge University Press

Chapter 7
Complexity

Chris Earl and Jeffrey Johnson
The Open University

Claudia Eckert
University of Cambridge

Complexity is a widely used term; it has many formal and informal meanings. The aim of this chapter is to examine the relation between complexity and design. Several formal models of complexity can be applied to designs and design processes. This argument runs in two ways.

First, designing provides insights into how to respond to complex systems – how to manage, plan and control them. Second, the overwhelming complexity of many design projects leads us to examine how better understanding of complexity theory can lead to improved designs and processes. This second direction is the focus of this chapter. We start with observations on where complexity arises in design, followed by an overview of the scientific background to complexity to introduce the wider context in which the concepts and methods of complexity theory have arisen.

Many involved with design recognise some area of their work as complex. Figure 7.1 shows the main sites for complexity in design and designing. First, the product/service/system under construction may be complex in its own right, in structure as well as behaviour in use. Second, the process of designing may contain many interrelated tasks, each having many subtasks. Third, the designer and their part in the organisation of project teams integrating complex sets of capabilities and experience. Fourth, users, and those more widely affected through life-cycle effects such as environmental impacts, provide a complex context for designs.

The relationships between designs (products, services or systems), processes, designers and users create yet another level of complexity. For example, the relation between design and users includes the difficult and complex problems of sustainability – the widespread impacts of a design across populations and into the future. Figure 7.1 also indicates the wider context of designing which forms another level of complexity.

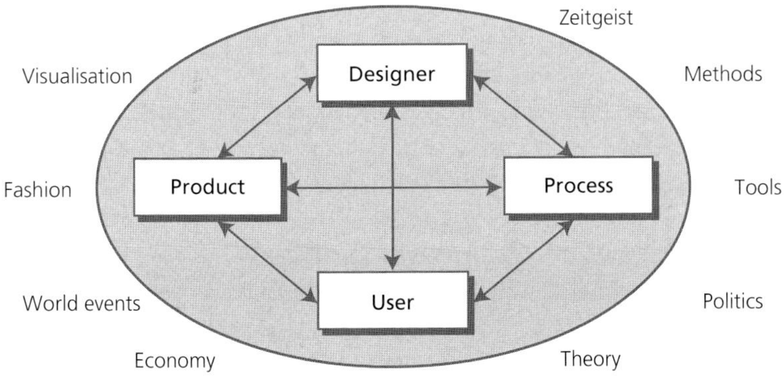

7.1 Designing in context

The relation between product and process is critical and is frequently the source of complexity. For example, scheduling the product across available design resources and capabilities which make up the process is a difficult task, not least because individual design activities in the process have uncertain durations.

The way that a product 'flows' across the resources and capabilities in the design process, with associated interactions between parts of the process, is complex. Managing these flows is a challenging task. As a design develops (through process) it is represented in several different ways. These representations and models may be complex in their own right. They may also be used in complex ways. Representations change in type and content as design proceeds from concepts and sketches to computational models and prototypes.

Designing can certainly be complex in the informal senses in which it has been described above. These observed characteristics are mirrored by established formal models and ideas in the science of complexity. In Figure 7.3 we summarise briefly the main points of complexity theory (see Suh (1999) for a brief summary in relation to design). These models have evolved to describe particular systems and their properties, which accounts for some of their differences. Many complex systems display aspects of several of these views simultaneously.

There is one additional point we would like to make. The way that designing develops intention, through concept to final design, appears to be an exemplar of how to model a complex system by increasing detail in representations through a process of iterative evaluation. Indeed, there may be lessons for complexity science itself from analysis of the way that design is undertaken (Cross, 2000), especially recent work on comparing processes across different domains (Eckert *et al.*, 2004). We talk intuitively about complexity in design and know that it can cause problems. But can we understand and manage complexity in the different areas and levels of design? To answer this we do three things. First, we distinguish different kinds of complexity that are present in design. Second, we discuss the methods and techniques from complexity theory. Third, we seek to apply these to designing.

7.2 The EH101, complete with five composite rotor blades
© AgustaWestland

Complexity in an engineering context

A helicopter rotor blade is complex not only in its form and manufacture, but also in its functions. Its design process is complex to the extent that it eludes conventional process modelling, with a large number of closely

Differential and difference equation models represent

- Relations among variables describing the state of a system
- How state variables change with time
- Parameters which identify specific relations among variables
- Behaviour as described by solution trajectories → *system order* in equations and *behaviour uncertainty* in trajectories

Types of system

- Conservative systems (respond to perturbations with permanently altered behaviour)
- Dissipative systems *(absorb* perturbations, returning to a *steady state* behaviour) (Nicolis and Prigogine, 1989)

Types of behaviour

- Lyapunov stable – behaviour changes proportional to perturbation, e.g. planetary orbits
- Asymptotically stable – behaviour returns to steady state (an attractor) after a perturbation
- Unstable behaviour departs radically from the initial state
- Locally stable (below a threshold in perturbations)

Non-linear equations

- Combined effect on behaviour of perturbations (with small effects individually), is non-linear (superposition does not apply)
- Behaviour may be unstable and difficult to predict

Chaotic systems

- Different types of trajectory which are *very close* to one another at certain parameter values
- Small unmeasurable disturbances alter system parameters knocking the system from a stable to an unstable trajectory, or from an unstable to an asymptotically stable trajectory (a chaotic attractor)
- Behaviour cannot be predicted because of inherent measurement uncertainty
- Chaotic behaviour in one element can propagate across the entire design
- Designed systems may potentially chaotic, e.g. aerodynamic and road systems often perform best with parameters *on the edge of chaos*

Information measures of complexity

- Expected information (Jaynes, 1957) or algorithmic complexity (Chaitin, 1987)
- Balance system order and behaviour uncertainty

Synthetic systems models

- Rules and goals indicate order
- Simulation reveals uncertainty in behaviour

Nearly decomposable systems

- Strong relations within parts and weak relations between parts (Simon, 1969)
- Techniques for identifying near decomposability are widely used in models of design process (Eppinger *et al.*, 1994; Suh, 2001)

Fractals and cellular automata

- Simple rules generate complexity, e.g. fractals (Mandelbrot, 1983) and cellular automata (Wolfram, 2002)
- Applications, e.g. urban development (Batty and Longley, 1994; Wilson, 2000)

Adaptation and coevolution

- Adaptation – change behaviour in respose to environment
- Coevolution – mutual adaptation, e.g. simulation of both transportation infrastructure and land use (Barrett *et al.*, 2001)

7.3 Overviews of the theoretical models of complexity

interdependent and related shape and material parameters which are determined iteratively (Clarkson and Hamilton,2000). Off-road diesel engine designs are customised for users and subject to environmental impact legislation. Their complexity lies in the interactions between product and users (and the logistical effort involved in designing and producing thousands of slightly different products). Power generation switchgears are customisations of standard products completed on a contract basis. Managing several different products through the design and manufacture process produces complex scheduling problems under constraints of uncertainty and finite capacity resources (Earl *et al.*, 2001).

Product structure

A design may be structurally complex – an engine has many parts and specific functional relations between parts. Parts and relations between parts form a hierarchical structure which is not necessarily tree-like but may display more connected lattice properties (Figure 7.4). A bill of materials (BOM) for manufacture describes the structure of a product in terms of which parts are included in aggregate units. A BOM can go to the finest detail of components and is in the form of a tree-like 'explosion' of the product.

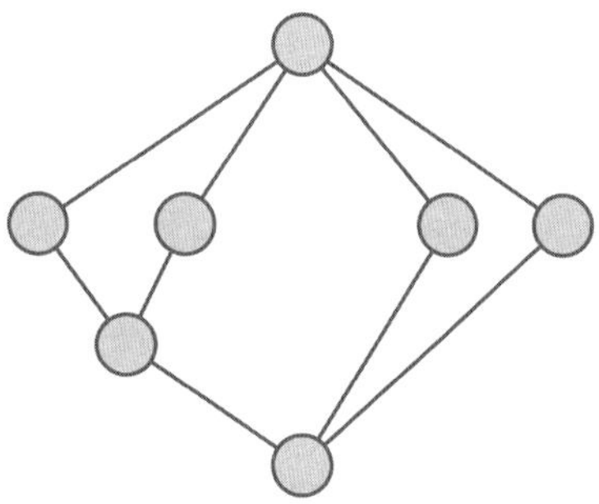

Lattice structure

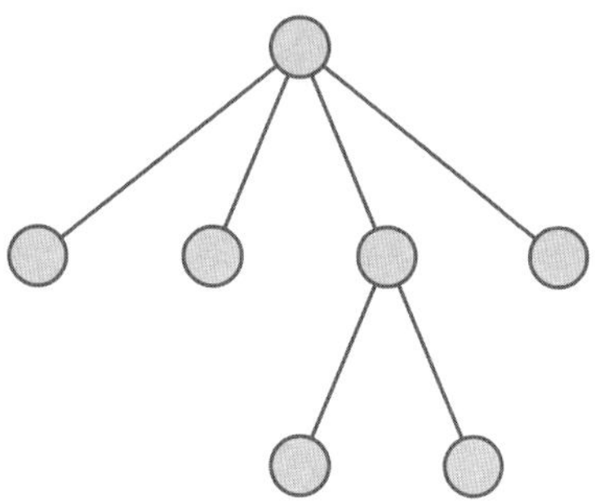

Tree structure

7.4 Tree and lattice structures

For a product with many components, the BOM may be a broad and deep structure with main parts having many subparts (breadth) and these in turn being decomposed repeatedly until the final manufactured components are reached. Companies can reduce the breadth and depth of BOMs trees by taking delivery of whole subsystems from suppliers. However, the BOM structure, although complicated, is not really complex. It has been handled by materials and manufacturing planning software which has proved an invaluable basis for manufacturing planning generally. A product has other structures associated with it during its development and it is the interaction among these structures which presents the complexity designers experience in product development.

Product structure is a decomposition which corresponds to functional parts of a design. Parts at one level of the decomposition may 'belong' to several larger functional parts. Thus, a rotor shaft in a jet engine 'belongs' to both the turbine and the compressor. The shaft itself has two parts, one for the turbine rotor and another for the compressor rotor. This kind of relationship among parts is not captured by a tree-like hierarchy, but requires a lattice hierarchy.

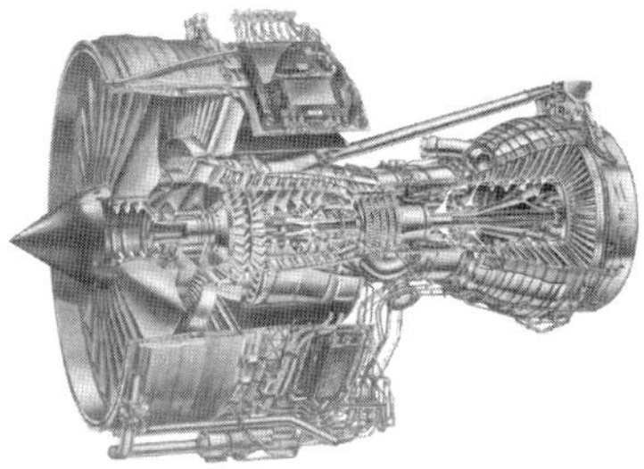

7.5 Rolls-Royce Trent series jet engine
Reproduced with the kind permission of Rolls-Royce plc

In these lattice structures, for any two parts there is a unique ('smallest') part at a higher level of the decomposition to which the two parts 'belong'. In the case of the rotor shaft, the two parts belong to the part 'rotor shaft' as well as to the turbine and compressor. These are functional parts of the whole jet engine, as is the rotor shaft. The 'smallest' higher level part to which both parts of the shaft belong is the rotor shaft. Notice that the decomposition of the engine we are describing here is not one which divides the design into distinct parts; there is considerable overlapping.

There are further descriptions. The manufacture and assembly of a product is described in terms of the precedences and sequences of operations. This structure may be quite different to the product structure as the simple example of the rotor shaft shows.

These different hierarchical descriptions may not be particularly complex in themselves. However, with several different descriptions used by different teams during product development, the result can be very complex. This is compounded by the nature of the design process in which descriptions are constantly changing as details of concept are completed, suppliers contracted and manufacturing planned. These structures are central to understanding complexity in design and are reviewed in greater detail later in the chapter.

The BOM illustrates many of the problems of describing a complex product. For manufacturing and assembly a BOM is fairly unproblematic. It indicates which parts are assembled together and it is used to track parts. Every part has a unique place in the BOM – it is a tree structure. However, in many companies a BOM is a problematic concept in design. Designers are interested in systems and their parts. The BOM is used to track progress in design, in terms of what percentage of the BOM has already been designed.

From a design point of view, conflicts can arise when several people work on the same part independently, or when nobody does. Important subsystems can easily be buried in a BOM, either because the parts are distributed or are defined by other parts. For example, the fuel tank of an Airbus emerges as the space between the parts of the wing. Similarly, clearances between separate parts (in this case functional subsystems) may be inadequately tracked by the BOM. Some companies advocate a single tree structure BOM, and suffer the consequences of severing the links between parts. Others have multiple BOMs and struggle with the translation between them. Often, individuals, computer programs or formal processes are blamed for problems that most fundamentally arise from trying to map a complex lattice structure to a tree.

The structure of relations among parts in a design takes many forms. These structures are dynamic, changing through the process of design.

Mismatches

The structure of relations between parts in a design thus takes many forms. These structures are also dynamic, changing through the process of design as details are specified and performance analysed. However, during the design process it is not only structures of relations between parts which change, but also performance and behaviour of successive design proposals at the various stages of the process.

Analysis at each stage in product development assesses performance or potential performance against specification. Mismatch can occur either in detail or type of behaviour. The former includes mismatch in performance parameters, e.g. fuel consumption or torque characteristics of an engine, whilst the latter includes unexpected behaviour, e.g. vibration resonance from new combinations of design features.

Mismatches in details are handled interactively, whilst mismatches in type resulting from new behaviours emerging in the product during the design process are more difficult to control. Exceptionally, these new behaviours may be desirable – the delightful serendipity of design – but, for the most part, engineering designers try to eliminate these unwanted characteristics.

The later stages of many complex design processes are dedicated to eradicating unwanted behaviour, such as vibration, noise, electrostatic interference (ESI), rumble, heat, *etc.* The design process converges in both these ways to a final design in which behaviour (within the context of use) is predictable and desirable. Surprising and emergent behaviours are evidence of complexity.

An effective process seeks to uncover these behaviours by analysis and test, removing them if possible or restricting the possibility of occurrence by limiting the conditions under which the product is used. In this sense the process seeks to lower complexity of design, especially in the relation between product and user.

Surprising and emergent behaviours are evidence of complexity.

Emergence

Processes for designs like the helicopter rotor blade are also complex because of the structure of many iterative cycles, each with inherent uncertainty, whilst together apparently convergent. The design process may have discernable overall emergent characteristics (such as convergence to satisfactory design) which may not be entirely predictable from the characteristics of its elements. Similarly, designs with internal structural complexity are often intended to behave robustly in a wide variety of contexts. For example, the helicopter rotor blade operates in a specified temperature range and a wide range of altitudes.

Simon (1969), in his *Sciences of the Artificial*, suggested that an essential aspect of designing takes place at the interface between design and context. However, this interface can be tricky because emergent types of behaviour – that is, surprising and unpredicted behaviour – may arise from interactions of elements or propagation of effects from one area of a design to another.

In the engineering context the aim is usually to reduce this complexity, restricting emergent behaviour of a design to intended function (and of corresponding process to intended outcomes). Designers generally try to avoid emergent behaviour that is random and chaotic by locating designs within margins – for example, compressor stall in a jet engine is avoided using design margins to keep pressure surge within limits.

A major source of complexity arises in the interaction of design and process. Recall the functional and modular groupings in a jet engine considered above. The compressor and turbine are commonly designed by separate teams and there are institutional company barriers to the flow of information, especially change information. Reaction blading changes in the turbine alter axial loads along the rotor, including requirements for compressor bearings and seals. The combination of the effects of design decisions made rationally by individual domain experts may only emerge at prototype test.

On the one hand, decoupling of processes for jet engine design has reduced complexity in designing but increased the complexity in the product and its behaviour, introducing unexpected 'emergent' behaviour. In this case the emergence may be failure of bearings or seals. This example emphasises again the importance of complexity in the relations between major elements in design – in this case product and process.

Although product and process elements are complex in their own right, with many subelements and relations, the major complexity arises from the way that the product lies across the process or, in the language of complexity, forms a 'traffic' through the network of activities and tasks in the design process (Johnson, 1995). As the 'traffic' of product moves through processes the 'product' changes (or strictly speaking its description changes) and new behaviour emerges. So, the three-way relation on the lower part of Figure 7.1 among 'product–user–process' is significant in the complexity of design.

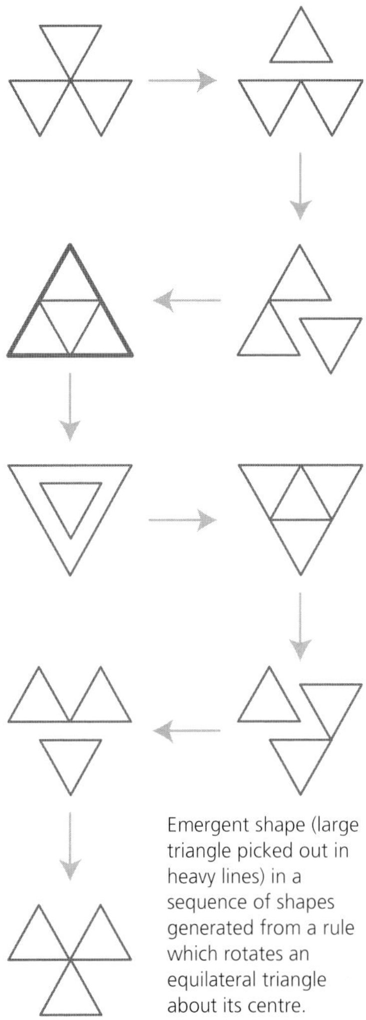

Emergent shape (large triangle picked out in heavy lines) in a sequence of shapes generated from a rule which rotates an equilateral triangle about its centre.

7.6 Exploiting emergence radically changes the shape development (Stiny, 2004)

Uncertainty

The discussion of examples of complexity in design reveals that complexity arises at several levels in the relations within and among products, processes, users and organisations. We turn to the idea that complexity arises from the balance of uncertainty and order which was introduced in the overview of complexity theory (Figure 7.3 – information measures of complexity).

This information complexity corresponds to entropy, which in its information sense (Jaynes 1957) is a measure of uncertainty relative to constraints (order). Maximising this entropy describes what balance can be expected between uncertainty and order. Complex physical systems seem to balance order and uncertainty at different levels. They might present patterns in overall behaviour but with extensive uncertainty at the microlevel. Conversely, microlevel order may be balanced by surprising overall uncertainties in aggregate behaviour. We would expect a complex design process, although containing a great number of uncertain events, to yield, overall, a satisfactory design. We would expect a complex product with many parts, possibly with uncertain performance early in the design process, to function and meet specification in ordered and predictable ways. Alternatively, a complex product may balance order and uncertainty differently. Extensive uncertainties in operating conditions may be balanced by an ordered behaviour, such as for example in intelligent systems. This section will describe several types of uncertainty which occur in design processes and the counteracting types of order.

Uncertainty is present in all areas of design and designing (products, processes, users, and organisations).

The balance between uncertainty and order can manifest itself in several ways. For example, at the beginning of a design process designers are uncertain about the details of configuration and parameters, but may have a detailed functional specification. So, despite some uncertainty, the specification implicitly restricts the selection of configuration and parameters. Company standards and policies will also direct design choices, thus imposing order, as do past designs and the experiences gathered through them. Without these features of order there would not be complexity. Designers can run out of ideas if there are no constraints. When the product provides few constraints, designers derive them from the wider context. For example, textile designers with few constraints will invoke contexts in prevailing fashion.

Uncertainty is present in all areas of design and designing (products, processes, users, and organisations). New designs have parameters and behaviours which are not known completely beforehand, processes have uncertain durations and uncertain effects, users and conditions of use can change, organisations change and, more widely, contexts, environments and long-term conditions of use are unpredictable. All these uncertainties make planning design processes harder by increasing the numbers and combinations of possible outcomes. Some have argued that uncertainty is at the core of design complexity (Suh, 1999). We will discriminate two basic types of uncertainty: 'unknown' and 'known' uncertainty. These basic types are present in two areas: (i) descriptions and (ii) data (which includes uncertainty in measurement). Similarly, we discriminate between several types of order: structural order of relations between parts, dynamic order of patterns of behaviour and the order imposed by constraints. Generally, complexity seems to occur when there are high levels of uncertainty combined with high levels of order. We focus here on the types of uncertainty (Figure 7.7).

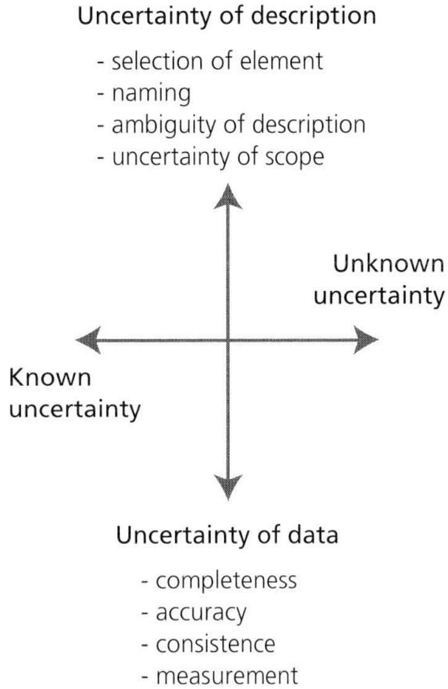

7.7 Types uncertainty in of design

'Known' uncertainty is based on variability in past cases. It can be characterised by probability distributions, e.g. of process task durations or the probabilities of a process (such as a computational analysis or prototype test) improving design performance. A key problem in design is the estimation of these known uncertainties in unique products and processes. Known uncertainties put limits on possibilities and describe them through probability distributions. In other cases, uncertainties may be known but their effects are unknown uncertainties in behaviour.

The uncertainty of surprise is an 'unknown' uncertainty in the sense that there is no particular expectation of such an event. Internal unknown uncertainties arise in the product, the process, the user or the organisation itself. These could range from unexpected material fatigue, to problems with software packages or employees getting pregnant. External unknown uncertainties come from the context in which the product or process operates, such as political events. 9/11 is an extreme example of unknown external uncertainty. Uncertainty in products is one of the sources of uncertainty in process. For example, uncertainty about vibration problems leads to uncertainty in process planning and scheduling. When managers do not know that vibration occurs, they won't plan in resources. When they do not know in which part of the product it will occur, they do not know to which team they need to allocate resources. For example, in helicopter design it is very difficult to predict where and to what extent vibration will occur. It is difficult to know up front what remedial actions will have to be taken and, therefore, what resources will be required. Design managers cope with this by analysing the design as much as possible, but scheduling time at the end of the process for sorting out these, as yet unknown, vibrations.

A key problem in design is the estimation of known uncertainties in unique products and processes.

Both types of uncertainty are present in the uncertainties of descriptions. Designers make choices on those aspects of a design which are to be included in a model and exactly how the model is constructed. For example, the selection of meshing points affects the results of vibration analysis. In modelling, subsystems are grouped together, making analysis within the grouping easier than outside it. The vibration models, for example, look at particular sets of components, but it might be the subtle interaction of these sets of components which causes a problem. These are uncertainties in what is included in the description. During the design process there are also uncertainties in the design itself: in its configuration and parameters and in its behaviour. These are also classified here as uncertainties of description. Besides the uncertainty in the selection of elements there is an ambiguity in how elements are grouped into meaningful concepts. Naming these elements or groups carries its own uncertainties. Each description implies a range of possible meanings, and often the boundaries of the interpretation are uncertain. For example, when a car is called a 'sports car' this may have significantly different meanings for different people. Further, the use of a particular label ('sports car') changes our perception of the design.

Many complex systems are characterised by voluminous heterogeneous data of variable quality and completeness.

Many complex systems are characterised by voluminous heterogeneous data of variable quality and completeness. Uncertainty in data lies not just in its accuracy but also its completeness and consistency. In design processes and product development, as designs are developed from concept to layout, and then to manufacture, many types of data are generated. Incompleteness is a characteristic of data during design, especially with speculative proposals. In some complex human systems it is impossible to have data that are complete or consistent, and the science of these systems has to accept this as one of its axioms. It is not simply a case of collecting better data to eliminate inconsistency – the issue is to provide robust predictions even though the data are incomplete and inconsistent.

There are underlying 'unknown' uncertainties in all measurements. In chaotic systems, the response to 'unmeasurable' differences in initial conditions is an unknown uncertainty. This randomness is an essential part of how a complex system behaves. But it is not necessarily due to internal uncertainties on the parameters or variables.

Continuous models which are entirely deterministic differential equations can, nevertheless, exhibit wildly random behaviour. In discrete models, state transition probabilities specify the known uncertainties. At a higher level of behaviour, the patterns of these transition probabilities make some types of transition more likely than others. This is the background of known uncertainty against which surprising events and behaviour occur.

Uncertainty is only one of several significant sources of complexity in design. We will now outline some of these in the context of general developments in complexity theory.

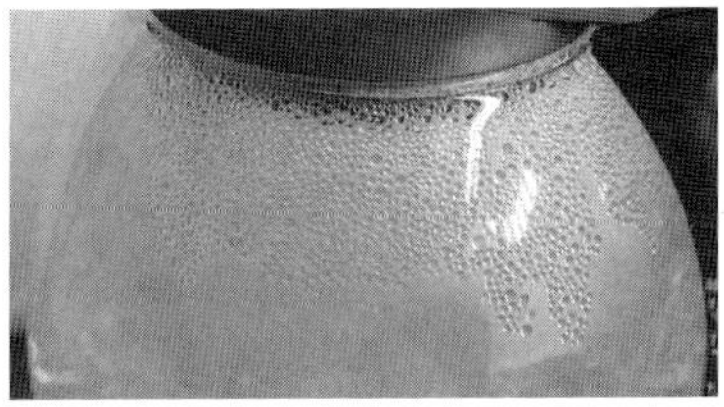

7.8 Water. Some systems, such as convection in water, show uncertain behaviour in detail. However, there is emergent, structured and bounded behaviour of the overall system (Nicolis and Prigogine, 1989).

Complexity theory and design

In this section we examine some specific characteristics of complexity which are pertinent to design. While these factors can be managed, they cannot be eliminated, because they are inherent in any complex system.

Dynamics

The main ingredients of deterministic chaos (see Figure 7.3 overview of complexity theory) are (i) sensitivity to initial conditions and (ii) boundedness. The first means that the slightest errors in measuring the initial conditions cause the behaviour to 'explode', but stay within bounds of 'normal' behaviour. Examples include many human and socio-technical systems. Designing and its processes are an example of such hard-to-predict systems. And many products themselves display these characteristics of uncertain behaviour, especially in the context of the wide spectrum of 'users' from the immediate customer to those affected during the design lifecycle and beyond.

Processes of engineering design cope, in practice, with cumulative small effects by redescribing the system at the different stages of the process. Through gateway processes companies force products and processes to reach certain well-defined points. This is a cyclic process of description and prediction. Suh (1999) advocates using this as a design principle for time-dependent systems, such as design processes and schedules. He advocates attempting to transform time-dependent combinatorial complexity (with increasing uncertainties into the future and their 'knock-on' effects) into periodic complexity (with uncertainties being reset at regular intervals). This is achieved by introducing 'gateways' or reducing the dependencies between parts of the design process.

Understanding the dynamics of many complex systems requires an appropriate notion of time. There is an interplay between the 'calendar' or 'clock' time of physics, and 'system time' defined by the structural 'events' of the system. For example, a product may be planned to be launched on a given day in a given year (calendar time), but the emergent system event "the

7.9 Clouds. Models of deterministic chaos were initially developed by Lorenz (1963) to model weather patterns.

7.10 Magic roundabout. Disentangling the paths of connections can improve overall performance, even though the whole system appears more complicated (Johnson, 1976).

product is ready" may not have happened. Such mismatches between system time and calendar time are well known, especially in the software industry. Understanding the complex interplay between events and time is fundamental in design, planning and management.

Connectivity in dynamic systems

A significant aspect of complex system dynamics is the transmission of energy, information and matter, e.g. vehicles and people in transportation systems, information in design teams and goods across the supply chain. These flows require appropriate channels connecting parts of the system. There is a conflict between facilitating essential communication and de-coupling parts of the system to eliminate undesirable interference and noise (as for example in reducing the options offered on a car). Designing an infrastructure 'backcloth' to carry the system 'traffic' is an essential part of applied complexity theory in planning and management (Johnson, 1995).

Flows take place on networks of connectivities. In design, several types of network may be present:

- product components are connected by function, geometry, manufacture and assembly;
- people, such as engineers, analysts and designers, are connected in team structures, hierarchies and even friendship;
- activities and tasks in the design process are connected by information and design representations, with process interfaces which may operate with checks or as gateways;
- a range of products in a company are connected by shared components, methods of manufacture, designers or design capabilities;
- supply networks include designing, manufacturing and service outsourcing.

The key factor in all these is the flow on the background structure of the connections. Complexity arises from the structure and connectivities of the network, but most importantly from the dynamics of the patterns of flows. However, dynamics can also manifest itself in another way. Some networks change rapidly over the course of a design project. One of these is the network of connectivities among the relevant knowledge of the participants. As the project proceeds, the connectivities will change as knowledge is acquired, analysed and embodied in a design. Other networks, such as the structure of teams, change more slowly during a project. Although connectivities may be present they may not be continuously active but rather are activated by events such as a competitor's new product or a scheduled project meeting.

One of the main challenges of design management is to keep an overview of these multiple connections through which information needs to flow, change and propagate (Jarratt *et al.*, 2004). A designed backcloth of connectivity, rather than an evolved one, would make it easier to analyse connectivity and the consequences of design decisions. However, this may not be possible, especially where structures are continually reinvented.

One of the main challenges of design management is to keep an overview of these multiple connections through which information needs to flow, change and propagate.
(Jarratt *et al.*, 2004)

Structure

Systems are described at different levels of aggregation by structures. To represent and reason about systems it is necessary that the corresponding structured objects have names (Figure 7.11). An important part of the design and management of complex systems involves constructing structured vocabularies. The BOM and other product structures discussed earlier illustrated the problems associated with using a tree structure to map a complex system.

A diagrammatic example with tree and lattice structures. It shows an object given the name **E**. It is made up from a set of 16 atomic objects named **s**. In (a) the squares are assembled into two structures named **V2** and **H4**, and these are assembled into **E**. In (b) the squares are assembled into two structures named **V7** and **H3**, and these are assembled into **E**. Note that superimposing the two structures gives a lattice structure (c) with the squares aggregating in different ways at the middle level. This illustrates that, in general, the intermediate structures (in this case (**V2**, **H4**), and (**V7**, **H3**)) and the names they are given are not unique. There are combinatorially many ways that hierarchies of named components can be constructed to represent a particular object. The selection of a particular hierarchical vocabulary rests with the designer, subject to constraints of how useful it is, and compatibility with pre-existing vocabulary, custom, and culture. It is not uncommon for the vocabulary to have inconsistencies, with the same object having more than one name, or more than one object having the same name. No matter how simple or complex a design, anomalies in vocabulary will increase the complexity, and act as a barrier to effective communication.

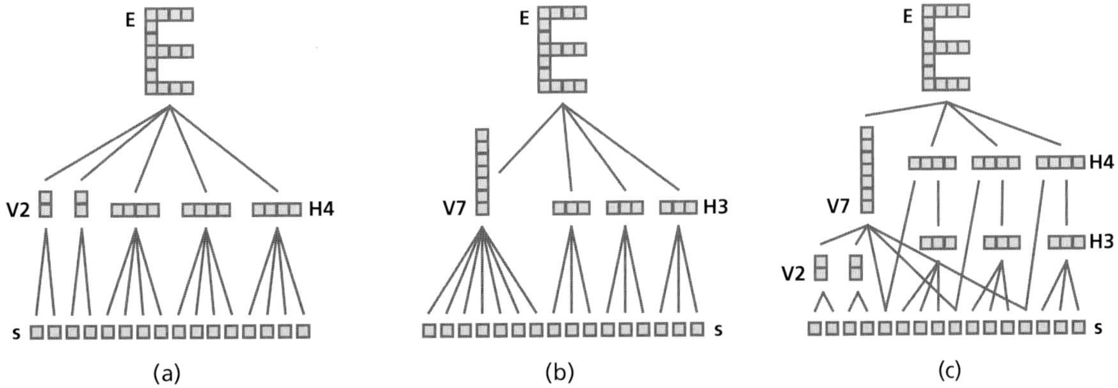

7.11 Hierarchies of assembly between a designed object and its atomic components

Hierarchical structures often have many intersections at all levels, leading to a more connected structure called a 'lattice'. Structural descriptions of parts and assemblies in products, or people in teams, fall into such a lattice structure, allowing many different possible groupings. Describing groupings in tree or lattice structures can be problematic in two ways:

- There are many possible descriptions leading to ambiguity, in the sense that each grouping describes the same object (Stiny, 2004).
- The group elements require names and labels whose construction involves a degree of negotiation. In mature products, naming of parts and groupings is often given by past products. The specific referents in past products can bias the design process significantly.

Search

Many complex systems have large numbers of interacting heterogeneous elements. Looking back at the example of the BOM, there are an enormous number of theoretically possible groupings of the elements in subsystems, which could affect the perception that people have of the BOM.

Figure 7.12 is an example of combinatorial explosion in which adding just one more element under relatively simple relational constraints can generate an order of magnitude more parts and relationships. Much of this emergent structure is not explicitly represented. Many complex systems have large numbers (hundreds, thousands, millions) of heterogeneous elements interacting. These systems do not have simple macroscopic properties, and system behaviour will be driven by micro-agent interactions. For example, an organisation may have hundreds of employees with distinct capabilities from whom to pick teams for projects. There are billions of ways to select six

Combinatorial explosion is illustrated by the graphical example below, where shapes are generated by adding a new rectangle at each stage. Even three rectangles generate dozens more, so that counting them all is a demanding task. It also generates new shapes, such as the inverted U at the top. Adding a fourth rectangle generates even more structure, while by the time a fifth is added, the resultant figure has hundreds of emergent shapes with hundreds of relations between them.

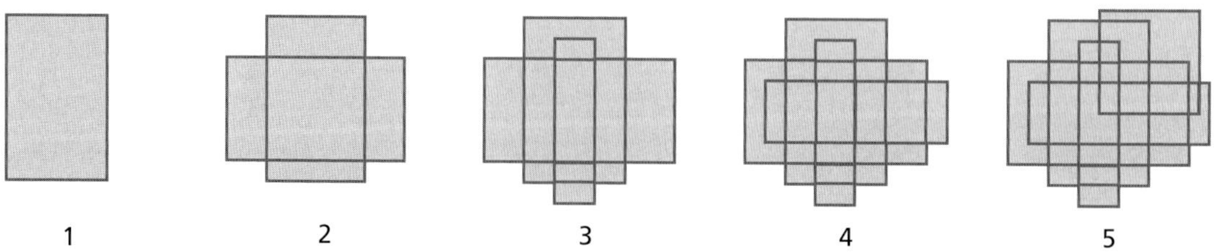

7.12 Small numbers of rectangles generate complex objects with many parts and relations

people from a hundred. Computational search of very large spaces has become an important tool in design and has highlighted the importance of the way problems are represented.

Well-formulated problems have a space of candidate solutions within the representation – the search space – with a subset that are actually solutions – solution space. This simple idea leads to techniques for problem solving based on searching for solutions. When a search space is small, examining every candidate can be a good approach. However, most search spaces are large and may have structures of connections like the lattices above. Exhaustive search is not feasible, so heuristics or random search techniques, such as simulated annealing or evolutionary algorithms, are applied.

The idea of searching for any solution soon leads to the idea of searching for the best or the optimum solution in the search space. Generally, it is impossible or impractical to be sure of finding the best solution to a problem, and optimisation becomes a process in which one seeks relatively good solutions. Design solutions often have to satisfy multiple criteria, so that a robust solution satisfies multiple goals as well as possible. Search is used both to find 'optimal' or satisfactory designs and then to search the possible modes of behaviour for each candidate by varying patterns of inputs and disturbance in a simulation of behaviour. Simulations are an important tool in managing complexity.

Managing complexity

Complexity is often inherent in systems and cannot be eradicated. However, it is possible to take active steps to reduce complexity in the hope of reducing the risk of problems occurring in the design process.

Simulation

The chaotic dynamics of many systems mean that it is impossible to make a point prediction that a certain event will occur at a certain time. Although the behaviour of most complex systems cannot be predicted in detail, there are many things that can be predicted. One answer is the generation of distributions of possible system states emerging from local dynamic interactions (Figure 7.13). Thus, simulations do not give 'point predictions' saying precisely what will happen when, but they give understanding of the spaces of 'possible worlds' in which things may happen, and they give information as to which of all the possible worlds are the most likely to be experienced.

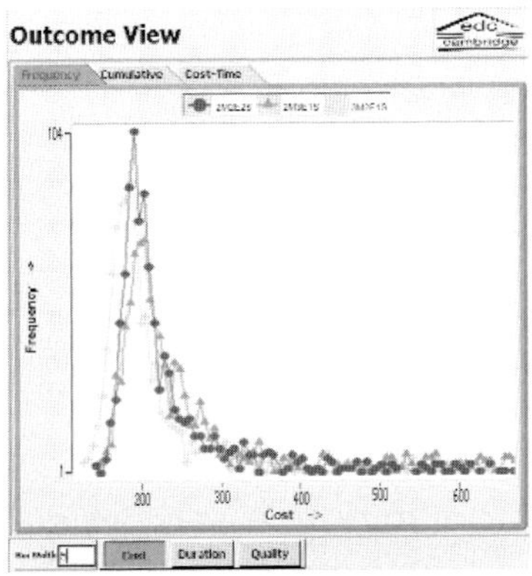

Figure 7.13 Distributions can be obtained by running a simulation many times

While designers and design managers are well aware of uncertainties in design processes, these are not necessarily accommodated in design-process planning tools.

Simulation is a major tool in design and design process evaluation.

As an example of this, consider simulating the behaviour of traffic on a crowded road, including calculating the number of vehicles passing a given point. The actual behaviour of the traffic depends on many things, including the kind of vehicles and the kinds of driver. For these reasons, road traffic dynamics are chaotic, and it is impossible to predict precisely what will happen. Sometimes there will be shock waves as a nervous driver touches the brakes, and sometimes there won't. By simulating the system many times a distribution can be obtained. Although it is impossible to give a point prediction of the flow, such distributions give useful information for the designers of road systems. However, it should be noted that the extent to which simulation results can be trusted (Johnson, 2001) remains a critical issue.

While designers and design managers are well aware of uncertainties in design processes, these are not necessarily accommodated in design-process planning tools (see Chapter 1 on process models). In reality, the duration and success of tasks are probabilistic. For example, the duration of a particular task or activity in a design process may not be known accurately in advance.

A computational fluid flow analysis may take anywhere between 10 and 100 hours. However, if a similar analysis has been done several times before, then we can consult the distribution of duration times and estimate an expected time for the new task. In design this is always problematic, since the historical data from which the distribution is constructed has not usually been acquired in a controlled way. The activity can change from occurrence to occurrence.

Lessons can be transferred between tasks so that great similarity to previous tasks will reduce both total time and the variation or spread of times expected for a new task. The 'observed' distributions for activity durations can be used in simulations of the whole process or important parts of it, perhaps a set of design tasks undertaken by a smaller team. In turn, a simulation then allows distributions for these sets of tasks to be created.

The modes of simulation for complex systems modelling are changing radically. Simulations of large socio-technical systems in areas such as transportation or sustainable development generate models starting from partial and incomplete data and progressively build models guided by convergence (and divergence) between model and practice (Barrett *et al.*, 2001). Simulation is a major tool in design evaluation, and there is considerable potential in using simulation for modelling design processes, and particularly the interactions between designs and their processes (Earl *et al.*, 2001; O'Donovan *et al.*, 2003).

Managing information complexity

Information about synthetic or designed systems is provided by descriptions and representations. One measure of the complexity of existing systems is how extensive their descriptions need to be to capture the features of the design or its behaviour. Algorithmic information theory (Chaitin, 1987) provides the basis for comparing such descriptions. The idea is that designs with compact descriptions, in terms of shorter procedures or fewer rules to generate them, have lower complexity. Designs exhibiting order and regularity in their behaviour may have short descriptions, whilst uncertain and unpredictable behaviour may require longer descriptions. However, taking this to an extreme, if behaviour is random then descriptions again become short as there is little information in the description. An intermediate representation or design proposal, created during the design process, also has an information complexity, although there are additional uncertainties in the design and its parameters. Provided there are statistically reliable estimations of uncertainties or variability, information measures of complexity can identify areas of a design where complexity might be reduced. Information complexity describes the balance between system order and behaviour uncertainty (Figure 7.14)

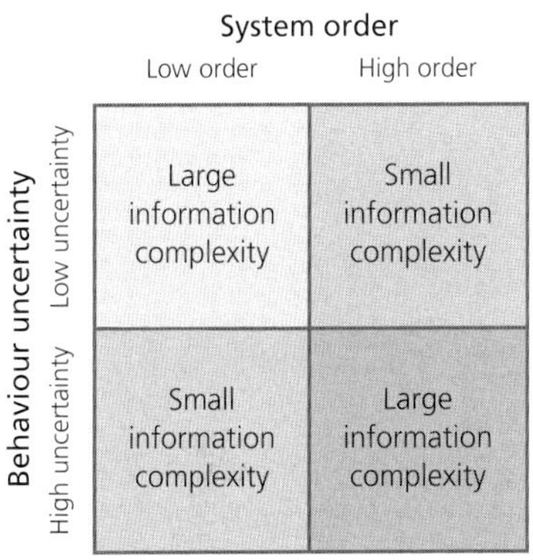

Figure 7.14 System order and behaviour uncertainty

Applying information complexity to the design process is problematic unless uncertainties can be estimated reliably. Many tasks within the process depend on the particular product being designed, the resources available and the 'memory' of similar products. Suh (1999, 2001) takes the view that complexity in design is mainly about uncertainty in parametric assignments. This approach may appear at odds with the idea of information complexity as 'balancing' order and uncertainty. However, with the order of the design process given by functional specification of final design it is feasible to measure complexity of design by uncertainties. In this view, complexity will change continually throughout design as uncertainties change for defined parameters and new parameters are defined and included in the design description.

Examples of complexity and design

The previous sections discussed aspects of complexity that are relevant to different areas of design. Here we describe briefly examples from design which exhibit some of these aspects.

Intelligent systems and control

The compressors of jet aircraft engines use combinations of static and rotating blades to drive air into the combustion chamber. As the blades attached to the rotor pass the fixed stator blades, there is a pressure gain. As with an aircraft wing, the pressure depends on the shape of the blades and their angles. By changing the geometry of the blade the pressure can be increased, but eventually the compressor becomes unstable, with small changes in the control variables causing large and sometimes undesirable changes in pressure. Engineers currently design engines to run 'on the edge of chaos', pushing the parameters to increase performance while (safely) keeping the system out of the dangerous chaotic region. Traditionally the blades were fixed, but some engines have mechanisms to set the angle of the blades optimally for take-off as well as cruising. Currently, a consortium of engineers is investigating the possibility of designing the blades to self-organise, with each blade acting as an agent, selecting its own optimal settings throughout operation of the engine (Johnson *et al.*, 2002).

Machines are becoming more intelligent, in terms of being able to sense their environment and respond to it. Cars have navigation systems that know their positions and can compute routes; aeroplanes fly on autopilot. We can expect new types of behaviour as machines communicate with each other, and with remote sources of information (Johnson and Iravani, 2004).

New types of system design have teams or swarms of intelligent machines working together. This approach to the design of engineering systems has many advantages. The members of swarms may self-organise to reconfigure themselves autonomously in order to perform new functions. In almost every area, from toys to domestic products, from industrial machines to transportation systems, designers are building in more autonomous intelligent behaviour.

Manufacturing as a complex system

Uncertainties and variability in manufacturing processes can, to some extent, be controlled.

It is instructive to look at how complexity is modelled in manufacturing systems. Descriptions of processes, relations between processes and dynamic flows through the structure of processes all contribute to understanding the behaviour of manufacturing as a complex system. Uncertainties and variability in manufacturing processes can to some extent be controlled – indeed, the focus on quality in manufacturing processes is about controlling variability in order to deliver a high-quality product to customers at low cost (in the broad sense of resources) to designer and manufacturer. With the possibility of measuring features of manufacturing system behaviour quantitatively (in terms of flows, lead times, inventories and queue size) information-theoretic complexity can be assessed.

Highly predictable processes will have low complexity, as do very variable processes. For these we know little about the overall process – or rather descriptions of what we know are limited. Either the patterns of behaviour are limited or they are so variable that no overall order or regularity can be discerned. We might say that these systems are unlikely to display emergent patterns of behaviour. However, the interesting cases from a complexity point of view are those with a balance of variability and order. Emergent behaviours will occur but the manufacturing system designer will want to limit these to desirable ones.

Static complexity is the "expected amount of information necessary to describe the state of the system" whilst dynamic complexity also includes the "expected amount of information required to report whether a facility is under control".
(Efstathiou *et al.*, 1999)

Complexity reduction by control of processes can increase the effectiveness of manufacturing. The excellent literature on manufacturing system complexity (Deshmukh *et al.*, 1998; Efstathiou *et al.*, 1999; Frizelle and Suhov, 2001) using information-theoretic models is a valuable resource for examining design processes. These models are based on entropy models which measure overall order in systems with high levels of local variability. However, the nature of the local variability is 'known' because processes are repeated and probability distributions can be constructed. However, we note that design is a rather different process, as the local variability is hard to quantify, processes can change and are susceptible to a wide range of external disturbances from customer, suppliers and, last but not least, competing design projects.

Finally, the distinction between static and dynamic complexity may be useful in design. Static complexity is the "expected amount of information necessary to describe the state of the system" whilst dynamic complexity also includes the "expected amount of information required to report whether a facility is under control" (Efstathiou *et al.*, 1999). Although these are useful concepts in understanding complexity, as we have already noted, quantitative information on design process would be required to apply these methods. This prompts a question as to whether this information can be acquired for design or whether design processes are inherently different.

Aerospace engineering design

Aerospace engineering provides illustrative examples of different types of complexity. For complexity arising from the interaction of design and process, the functional and modular groupings in a jet engine have already been considered.

Unexpected interactions between separately designed parts or between new parts and reused parts can also lead to unacceptable overall behaviour. In these cases, although it is in theory possible to analyse the whole design, this is often not done until test prototype. Because of complex multilevel structure and transmission through chains of connection, complexity effects are not picked up until the latter stages of design. Undesirable emergent behaviour is then, if possible, removed.

It is interesting to observe that emergent behaviour arises continuously throughout the process of taking a design from concept to embodiment and manufacture. In some cases this emergence represents new discovery and inspiration for design innovations (as in 'artistic' domains), whilst in other domains, such as engineering, the process of design is to remove undesirable emergent behaviours iteratively. The final design has behaviour which has 'minimal' complexity. This fits nicely with the information view of complexity, since a description of the possible behaviours of a 'well-behaved' design is relatively simple.

Complexity may exist in products, processes, users and management or organisation.

As an example, recall the design of the helicopter rotor blade discussed earlier. The process of design attempts to reduce complexity in behaviour so that it remains predictable. However, at the same time the search for 'optimal' or high-performing designs can lead to parameter values which are in the margins close to where behaviour becomes very unpredictable or chaotic.

Several complexity problems occur here. First, unexpected interactions between parts may cause behaviour to pass over the edges of the margin. Second, it may be that reductions in design process complexity through modularity give this higher design complexity in behaviour. Third, a design has a parameter envelope in which the design performs predictably, but optimal performance often occurs in the margins of this envelope.

Operating in the margins means that behaviour is complex and users require assistance to reduce complexity. An historical example is the comparison of the turning performance of Spitfire and Messerschmitt Me109 aircraft. Theoretically, the Spitfire had better performance in a wider envelope, but Me109s could be flown in narrower margins of their narrower envelope because they incorporated a passive moving element in the wings' leading edges. Although giving only a small aerodynamic improvement, these elements signalled to the pilot that the margin was being encountered. Inexperienced pilots could, therefore, avoid unstable behaviour, reducing complexity and improving performance (Morgan and Morris, 1940).

Conclusions

In this chapter we have shown that design can possess complexity in (a) products, (b) processes, (c) users and (d) designers (their organisation and capabilities). Although each of these elements can be complex, it is their combination that can cause the high levels of complexity that make the design process hard to understand and control.

To design successfully requires that this complexity be recognised and understood. Understanding complex behaviour allows designers and design managers to identify complexity as a root cause of some of their problems and take steps to reduce or manage it. This complexity can be conceptualised and described through a number of formal approaches that give insight into the behaviour of designs and design processes. However, there is no unified theory of complexity and no single theory captures all aspects of a complex system. Despite this limitation, we have shown that light can be usefully shed from differing angles on the problems of design complexity.

References

Barrett C *et al.* (2001) TRANSIMS: transportation analysis simulation system. Technical Report, LA-UR-00-1725, Los Alamos National Laboratory

Batty M, Longley P (1994) Fractal cities: a geometry of form and function. Academic Press

Chaitin GJ (1987) Algorithmic information theory. Wiley

Clarkson PJ, Hamilton JR (2000) Signposting: a parameter-driven task-based model of the design process. Research in Engineering Design, 12(1): 18–38

Cross N (2000) Engineering design methods: strategies for product design. Wiley

Deshmukh AV, Talavage JJ, Barash MM (1998) Complexity in manufacturing systems: part 1 – analysis of static complexity. IIE Transactions, 30(7): 645–655

Earl CF, Eckert CM, Johnson JH (2001) Complexity of planning in design. ASME DETC'01, Pittsburgh, PA, USA

Eckert CM, Blackwell A, Bucciarelli L, Clarkson PJ, Earl C, Knight T, MacMillan S, Stacey M, Whitney D (2004) What designers think we need to know about their processes: early results from a comparative study. Design 2004, Dubrovnik, Croatia

Efstathiou HJ, Tassano F, Sivadasan S, Shirazi R, Alves J, Frizelle G, Calinescu A (1999) Information complexity as a driver of emergent phenomena in the business community. IWES'99, Kobe, Japan

Eppinger SE, Whitney DE, Smith RP, Gebala DA (1994) A model-based method for organizing tasks in product development. Research in Engineering Design, 6: 1–13

Frizelle and Suhov (2001) An entropic measurement of queuing behaviour in a class of manufacturing operations. Proceedings of the Royal Society Series A, 457: 1579–1601

Jarratt TAW, Eckert CM, Clarkson PJ, Stacey MK (2004) Providing an overview during the design of complex products. DCC'04, MIT, Cambridge, MA, USA

Jaynes E (1957) Information theory and statistical mechanics. Physical Review, 106: 620–630

Johnson JH (1976) The q-analysis of road intersections. Journal of Man-Machine Studies, 8: 531–548

Johnson JH (1995) The multidimensional networks of complex systems. In: Networks in action. Springer

Johnson JH (2001) The "Can you trust it? " problem of simulation science in the design of socio-technical systems. Complexity, 6(2): 34–40

Johnson JH, Iravani P (2004) Robotics in the emergence of complexity science. AROB'04, Oita, Japan

Johnson JH, Lucas Smith A, Wiese P (2002) Complexity science for the design of intelligent geometry compressors for jet aircraft engines. AROB'02, Oita, Japan

Lorenz EN (1963) Deterministic non-periodic flow. Journal of the Atmospheric Sciences, 20: 130–141

Mandelbrot BB (1983) Fractal geometry of nature. Freeman

Morgan MB, Morris DE (1940) Messerschmitt Me 109 handling and manoeuvrability tests, Reports and memoranda no. 2361. In: Technical report of the Aeronautical Research Council Special Volume 1, 1955

Nicolis G, Prigogine I (1989) Exploring complexity. Freeman

O'Donovan BD, Clarkson PJ, Eckert CM (2003) The simulated design process. ICED'03, Stockholm, Sweden

Simon H (1969) Sciences of the artificial. MIT Press

Stiny G (in press) Shape: an essay on rules, ambiguity and design. MIT Press

Suh NP (1999) A theory of complexity, periodicity and the design axioms. Research in Engineering Design, 11(2): 116–132

Suh NP (2001) Axiomatic design – advances and applications. Oxford University Press

Wilson A (2000) Complex spatial systems: the modelling foundations of urban and regional analysis. Prentice Hall

Wolfram SA (2002) New kind of science. Wolfram Media

Chapter 8

Thinking and representing in design

Martin Stacey
De Montfort University

Kristina Lauche
University of Aberdeen

What are the mental processes of design thinking? How do different designers vary in their ability and skills? How do the powers and limitations of human thinking interact with the nature of design problems to shape the processes of designing? Some people claim design should proceed very analytically and rationally. Others emphasise the intuitive aspects of designing. This depends largely on the design task undertaken and the product that is designed. But it is also a question of how designers think and what they are thinking about. Experts can be more effective because they have different strategies from novices. Understanding how designing works as a human activity can be useful in understanding the causal connections in design processes, and for changing design processes in ways that exploit and enhance designers' abilities and take account of human limitations.

There is a large and increasing body of research on how designers think. This chapter does not attempt to survey it. Instead we concentrate on the relationship between design thinking and how designers interact with the representations that they generate in creating and reasoning about designs, such as sketches, diagrams and CAD models. While this is only one facet of the psychological factors that influence design processes, it is directly affected by changes to design processes, and influences the success of those changes.

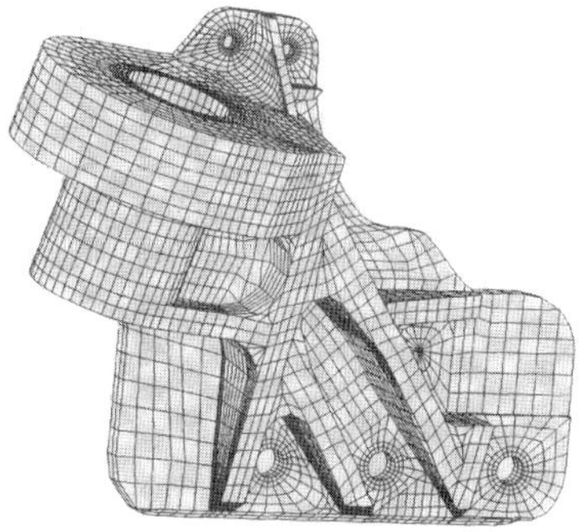

8.1 Generating a mesh for a finite element stress analysis. Experienced engineers estimate the value the analysis should produce and set the meshing points correctly. Less experienced engineers set the points almost at random and often don't recognise when the analysis result is out by an order of magnitude.

A psychological perspective

In this chapter we discuss design from our perspective as cognitive and organisational psychologists. Design activities have been analysed and studied in great detail, both in laboratory experiments and in field observations. The experimental approach means that a phenomenon can be studied in detail and any confounding factors can be eliminated, thereby allowing causal conclusions. The difficulty is that it often remains an open question whether the results can be generalised to designing in the 'real world' of a company. Also, a lot of research has focused on a small number of phenomena which can be studied comparatively easily, such as observations of sketching, but are not necessarily important to all types of design. Therefore, some researchers have argued that it is equally important to carry out field studies in the workplace. The danger with this approach is that all the results ever do is provide a description without any theoretically founded explanation or intervention.

We have included both types of research in the evidence we report. Psychological researchers have often shied away from studying complex

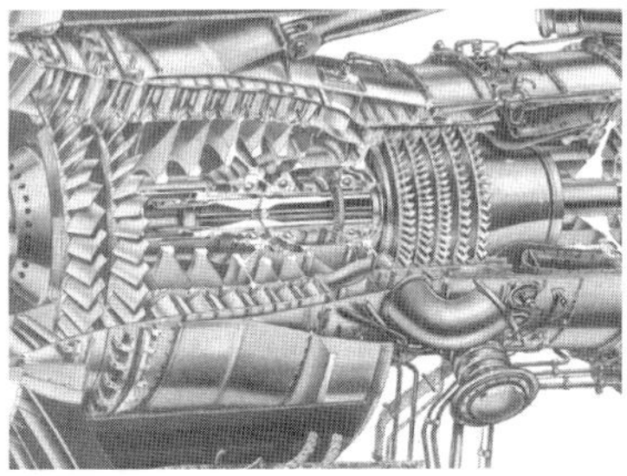

8.2 Reducing the rumble in a jet engine combustor. These low-frequency noises arise from subtle details in the shape of the combustor and the details of the combustion process. A few rumble experts know what changes to make, while other jet engine designers see this as a black art. Reproduced with the kind permission of Rolls-Royce plc

8.3 Reasoning about change propagation. Some engineers think through the sequences of connections between components, struggling to incorporate multiple propagation paths, while others reason by analogy to problem they have encountered with similar designs in the past.

activities in natural contexts, because they are too open-ended, and the volume of data overwhelms methods that give insight into finer-grained thinking processes in small-scale experiments. This can make research results difficult to relate to complex industrial situations, but observations are a valuable source of insights.

Researchers view design from a variety of theoretical perspectives. Information processing psychologists aim to understand how the mind works in terms of the mechanisms of mental processing, and how relatively simple mental operations combine to create complex behaviour. This includes seeking to understand the structure and content of mental representations – our internal descriptions of things outside ourselves (real, or possible, or impossible). In information processing psychology, other people are sources of sensory inputs. In social interactions, the content of our mental representations of the environment, goals and actions is different from what it is in solitary problem solving, but the mechanisms are the same.

Activity theory aims to understand human action in its cultural and historical context. The use of external representations for communication is a general psychological process: the symbols of language and the artefacts we create mediate between the individuals' minds and the task they are trying to achieve (Vygotsky, 1962) (Figure 8.4). Through these representations we can all access and contribute to our shared culture: language, received wisdom, historical artefacts and classical designs all form part of our common understanding of the task and influence design (Leont'ev, 1978). Design thinking means internalising what we see and externalising to others what we think. Only if mental representations are externalised can they become accessible to others. These external representations may in turn foster a shared mental model of the design object (Figure 8.5).

Sociological design researchers, such as Minneman (1991), Bucciarelli (1994), Glock (1998) and Henderson (1999), come from a research tradition that views the development of shared understanding as fundamentally problematic. They focus on how meaning gets communicated, in terms of the content of the talking, sketching, gesturing, exchange of documents and so on that comprise communication. The difficulties inherent in achieving a shared understanding of design problems and design solutions shape the interactions and working practices of designers, and understanding how and how far it does happen is a profound challenge for sociology, psychology and philosophy.

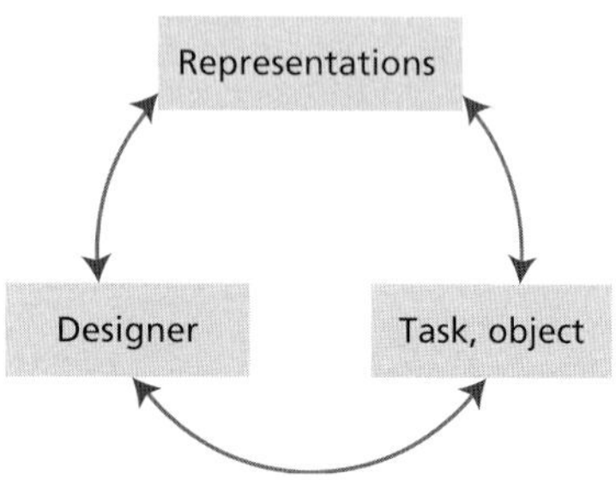

8.4 Representations mediate between the designer's intentions and the object they design

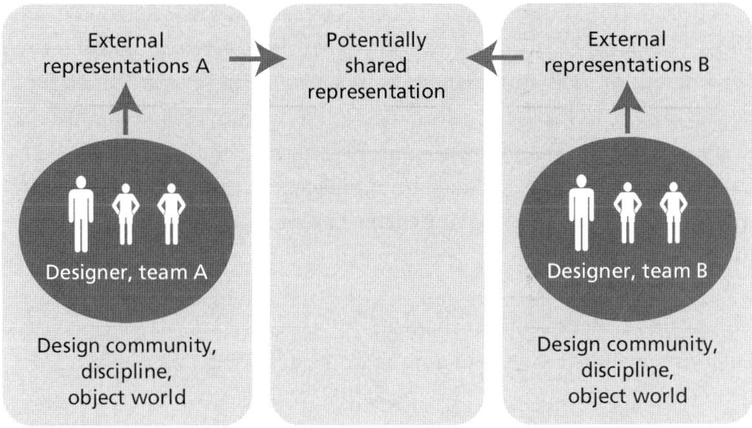

8.5 Building shared representations

Overview

The next section discusses how designers use graphic representations and models during their work and how this affects individual thinking and interactions with others. The following sections provide a more detailed explanation of the underlying mental processes, including: discussion of how designers form and use mental models of their designs; a description of some types of mental action that are characteristic of designing and a characterisation of expertise and skilled behaviour in design; discussion of some of the special characteristics of creative thinking; and some issues to think about in considering representations in new processes and methods, where the properties of representations can be usefully analysed using Green's (1989) cognitive dimensions framework.

The role of mental and physical representations in design

Understanding human thinking involves understanding how we create mental representations though the interaction of what we perceive and what we know, and how we use mental representations, both in tandem with direct perception of external things and independently (Figure 8.6). How designers work depends crucially on the interaction between their mental abilities and the representations in which they conceive, describe and communicate design ideas. New methods, procedures and computer tools require designers to represent design information differently and think about old problems in new ways. Effective choices of representations enable designers to use new methods and tools rather than struggle to work around their limitations.

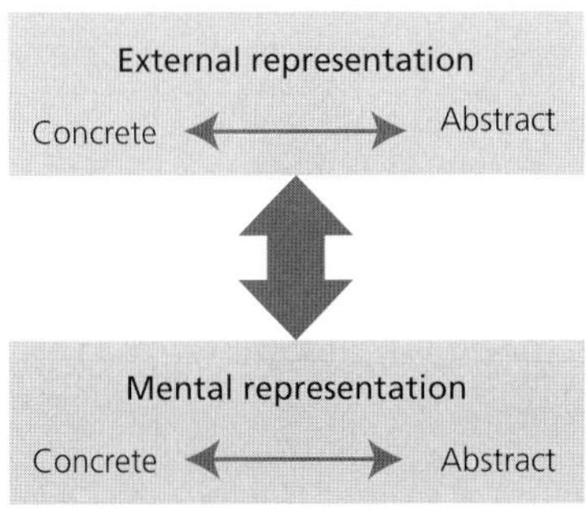

8.6 External representations vs mental representations

Designing involves both abstract and concrete thinking.

Effective combinations of representations can facilitate thinking fluently about design problems in a wider variety of alternative conceptual terms. (The design of tools and methods also demands careful consideration of task demands, to minimise effort and, where appropriate, to stimulate creative thinking). Designing involves both abstract and concrete thinking. Depending on the tasks and individual preferences, designers think about underlying physical principles, functional features or the concrete form of their design object, often in rapid alternation.

Variety of use of representations

Engineers use different types of representation depending on the task. Besides geometry they have to consider functional requirements and structural constraints, as well as information about the characteristics of components, materials, performance, construction processes and so on.

Eckert *et al.* (2004) observed and interviewed 20 helicopter design engineers about the representations that they used. Sketches were often used to generate or communicate ideas in early design phases. Many engineering tasks were not concerned with the creation of geometry directly but with function or performance. Engineers involved in these types of tasks typically did not use pictorial sketches. Avionics engineers sketched blobs and lines to outline the components and connectivity of systems. Software packages played a larger role in numerical modelling of stress or heat, and designers used the colour coding of the resulting diagrams tacitly to reason about shape. Rapid prototyping and testing complemented the computational analysis. More abstract representations, such as performance diagrams or matrices, were used to analyse the functionality or the relationship between parts or describe the connectivity.

Understanding the organisation of complex systems involves abstracting away from the form and detailed operation of individual components to focus on skeletal representations of what they do and how they are connected. A variety of notations have been developed for showing functional and causal relationships between abstract representations of design elements, as graphs (such as Petri nets and Bond graphs) and as tables.

Sketches are often used to generate or communicate ideas in early design phases.

Design research has focused on sketches of geometric form as a medium for generating ideas and communicating them. The following discussion will therefore use sketching to illustrate more general features of design cognition. Similar issues also apply to the representation of design processes, which are discussed in Chapter 2 on design process planning.

Representations support individual design problem solving

External media function as a way to unburden working memory for complex operations so that the designers can see their design ideas and thus have less need for accurate memories of their earlier thoughts. As we describe in the next section, the external representations function as cues for recalling and reconstructing elements of mental representations. Being able to transform mental representations into physical shape by sketching is for many designers a way of 'thinking with their hands'. For many designers, especially in early conceptual design or creative problem solving, design thinking is inseparable from physical action to create external representations; so, for them, sketching is generating ideas not describing ideas. Sketching, therefore, has an immediacy that other media do not have. Sachse and Hacker (1997) found that, when instructed to use sketching while designing on a CAD system, engineering students produce better solutions in the same time as the control group who only used the CAD system.

Schön (1983) influentially described designing as a dialogue between the designer and the sketch, in which the evolving sketch suggests interpretations of the design beyond what the designer intended to draw. Perception of sketches alternates between *seeing as* and *seeing that* (Schön and Wiggins, 1992; Goldschmidt, 1991); Goldschmidt (1991) observed designing progressing as an alternation between pictorial and non-pictorial reasoning. A common thread in research on sketching in design is that designers reinterpret ambiguous sketch elements to suggest new possibilities (see Purcell and Gero (1998) for a review). Engineers use sketches as representations of their mental concepts but sometimes also attribute a different meaning to the lines and see them as something else (Pache *et al.*, 2001). This reinterpretation of ambiguous sketches can serve as a means for stimulating creative, non-intended ideas, often triggered by dissatisfaction with the current design (McFadzean *et al*, 1999). New concepts or requirements can enable designers to re-examine sketch features from a new perspective (Suwa *et al*, 1999).

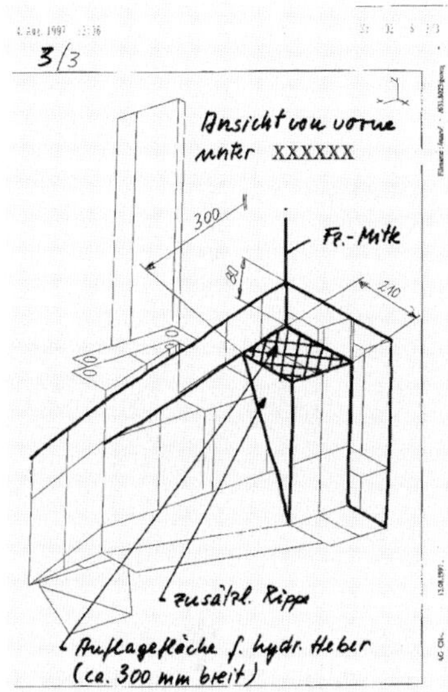

8.7 An example of an engineering sketch. Engineers like to sketch when they are solving problems. These sketches are often generated in meetings or brought to meetings as illustrations. Different interpretations can lead to different views of the problem. Engineers need to reach a shared understanding or risk costly changes later (reproduced with permission – Lauche *et al.*, 1999).

Joint designing and development of shared understanding

Engineering design is a collaborative activity: not only do designers engaged in different tasks need to exchange information and coordinate their activities, but, also, a lot of important decisions are made in meetings or informal discussions (Bucciarelli, 1994; Badke-Schaub *et al.*, 2001), and designs are sometimes created jointly. Design teams use a variety of shared artefacts. Rather than writing lengthy verbal descriptions, some designers generate largely non-annotated graphical representations (Weber *et al.*, 1999). During video-conferencing with limited bandwidth they give preference to seeing drawings rather than their colleagues' faces (Weber *et al.*, 1999). Minneman (1991) describes designers negotiating through proposal and counter-proposal for mutual understanding as much as agreement, and has argued that ambiguity in sketches has a beneficial effect in suggesting new ideas in design meetings. However, it is provisionality in design representations rather than ambiguity that matters: how strongly someone is committed to a proposal, the degree of precision intended, whether details are meant seriously or represent qualitative categories. The challenge lies in signalling provisionality while avoiding misinterpretation. Designers can signal provisionality in sketches relatively easily by degree of apparent roughness, but sketches are easy to misunderstand, especially when the creator cannot be consulted (Stacey and Eckert, 2003).

8.8 A much-praised computer tool for mass customisation. The Web interface allows customers to pick the material from a range of tarpaulins and make their own cutting patterns for their choice of bag. Image www.freitag.ch

Designers often fail to recognise that the resulting problems are communication problems (Eckert, 2001). Finished-looking graphic representations are often interpreted as more fixed than is intended or appropriate. For instance, designers are more ready to modify and change a joint representation when it is drawn by hand on a flipchart or electronic white board, rather than a spreadsheet or professional presentation (Kunz *et al.*, 2001). In meetings, designers use speech, sketches and gestures in combination to disambiguate each other, and signal how far decisions are open or negotiable, through subtleties of phrasing and tone of voice (Minneman, 1991; Neilson and Lee, 1994; Brereton *et al.*, 1996; Glock, 1998). These signals are missed by those who communicate across distances.

Communicating across object worlds

Communication between members of design teams can involve subtle problems when different specialists have mental representations of designs and design problems that comprise differing concepts, objects, features, properties and relationships – what Bucciarelli (1994) terms their *object worlds*. Members of design teams with different fields of interest and responsibility share and exchange sketches, diagrams, specifications, CAD models and so on, but interpret them differently.

Reading representations is a learned skill, and the mappings between the elements of sketches or diagrams and what they stand for depend on the conventions of a community as well as any geometric resemblance (Henderson, 1999). Similarly, terms for concepts can mean different things to different people (for instance Bucciarelli (1994, ch. 6) discusses the different meanings placed on the term 'module voltage' by the members of a team designing a photovoltaic generator).

Design researchers term the objects such as sketches and CAD models that are shared by different participants in a design process, and which convey information between them, *boundary objects* (Star, 1989; Bucciarelli, 1994). That is, objects that enable communication between object worlds, so that the inhabitants of the different object worlds have compatible understandings of the state of the design.

Mental representations

Although thinking usually involves direct interaction with one's environment, it happens in one's head: sensory perceptions create mental descriptions of what is out there – mental representations – which depend on one's memories. These mental representations trigger the direction of attention, the recall of memories, conscious reasoning and goal setting, the imaginative synthesis of mental representations of possible situations, and physical actions. In this section we outline how designers' mental representations of designs and design problems work, and how this governs the ways they use external representations such as sketches to cope with the size and complexity of design tasks.

Types of information

Humans have consciously accessible memories for three distinct types of information (Figure 8.9): visuospatial information, in which shape and extent, and sometimes movement, is inherent; propositional information that can be described in statements, and episodic information that is inherently experiential and time dependent. Designers' mental representations of designs combine visuo-spatial and propositional information (Goldschmidt, 1991). Episodic memory can play a role in envisioning how a design is used (Schön, 1988).

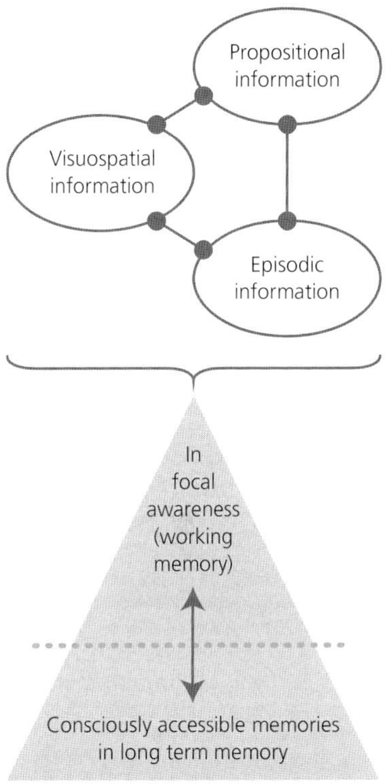

8.9 Types of information

Mental models

Mental models are representations of the form and properties of physical objects (or other kinds of systems with causally connected components), with which people envision their behaviour, to understand what the objects or systems do or predict what they will do (Johnson-Laird, 1983). The users of interactive computer systems and other consumer products form mental models of how they work, which often differ markedly from their designers' mental models of how they work (Norman, 1988).

People think about how physical systems behave through a combination of reasoning with consciously articulated propositional beliefs, and imagining changes in visuospatial forms and relationships. The relationships between the structure of physical systems and how they act and change are partly learned tacitly through experience of the world and encoded in visuospatial or kinaesthetic form. Tacit beliefs about how objects move can be wrong and misleading, and can often persist through quite a lot of education (Figure 8.10; McCloskey, 1983).

Relationships between form and behaviour are also partly learned consciously, from verbal descriptions of physical principles. DiSessa (1983, 1993) describes people reasoning about how physical systems behave by recalling and applying what he terms p-prims (for phenomenological primitives): elemental causal or explanatory relationships that appear to fit particular situations. As people learn more about physics, they learn new p-prims and alter the priority with which they recall them and apply them. Mental models of all but the simplest systems are invariably incomplete and unstable (Norman, 1983), and people may have multiple mental models of an object or system, which are not necessarily compatible. The users of interactive devices may use both models of the structure and causal processes within a device, and models of how it behaves in response to inputs – what Young (1983) called a task-action mapping model (Norman, 1988). Engineers reason about the characteristics of a variety of different abstractions of designs, for which they have different mental models, and construct models of these abstractions in explicit external form to perform different kinds of analyses (Hoover *et al.*, 1991).

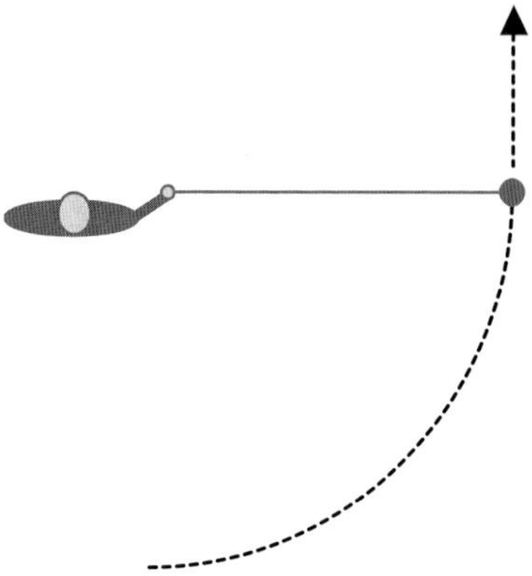

Imagine whirling a weight on the end of a rope in a horizontal circle around your head. How does it move when you suddenly let go?

People who have paid attention in physics, and internalised Newtonian mechanics, will imagine and describe the path of the flying weight as a straight line in the horizontal plane, perpendicular to the line of the rope at the moment of release; while the object drops with increasing speed in the vertical plane (due to gravity). People who have not learnt mechanics often imagine that the weight will move in a curved path, and give similarly wrong answers to other problems to do with force, motion and gravity (McCloskey, 1983). Their naïve intuitive physical beliefs correspond remarkably closely to theories of motion held by scholars in the centuries before Newton (McCloskey, 1983; Clement; 1983).

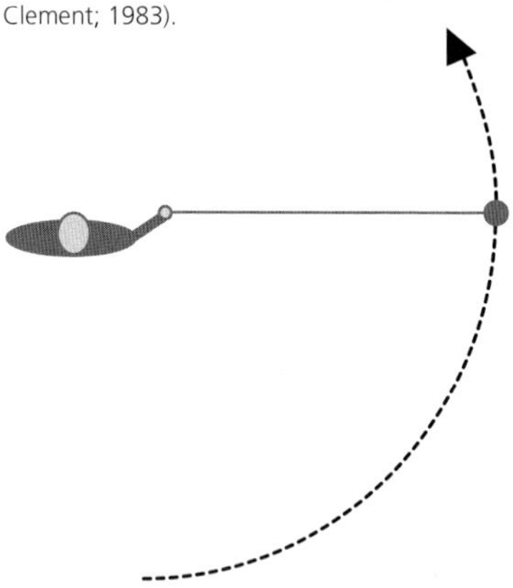

8.10 Naïve physics

Working memory

Reasoning about the behaviour of complex physical systems is limited by the capacity of working memory – what we currently hold in conscious attention. Humans can switch the focus of conscious attention extremely quickly, but it is impossible to hold all the components of a complex design and their relationships in mind at once. Miller (1956) famously estimated the capacity of working memory at seven plus or minus two chunks of information. The richness of the mental representations in conscious awareness depends on the size of the chunks – the combinations of elements of information that people have grouped into units that they retrieve from memory as a whole. As well as the size of chunks, the richness and strength of the associations between different chunks influence one's ability to retrieve related information reliably. For instance, Akin (1978) found that architects' memories for architectural drawings depend on schematic encoding of drawing chunks. The ease and accuracy with which items of information can be remembered depend on the number and meaningfulness of the associations that are formed to other items in memory; studies of expert–novice differences in various fields show that experts do not just know more, but their knowledge is structured so that situations trigger recall of both appropriate general principles and appropriate specific information (Bédard and Chi, 1992).

Memory recall is reconstruction

Research on what people recall from memory, and how, indicates that this is best viewed as an active process of constructing coherent mental representations from comparatively sparse and incoherent components, rather than as faithful and passive retrieval (Bartlett, 1932; Koriat *et al.*, 2000). Recognising an object or situation as a member of a category (such as being in a restaurant) activates a learned schema for constructing a mental representation of a situation of that type, creating expectations that it will include components with particular characteristics, roles and behaviour (Schank and Abelson, 1977; Schank, 1982).

The ease and accuracy with which items of information can be remembered depends on the number and meaningfulness of the associations that are formed to other items in memory.

When these expectations are violated the situation is perceived as being different or surprising. Studies of memory for drawings of faces (Wulf, 1922) and for stories (Bartlett, 1932) have shown that unusual features that are perceived as significant are highlighted and exaggerated, while other unusual features are smoothed towards what is standard for the category (Koriat *et al.*, 2000). Perceptual recognition of an object or scene as a member of a category (which involves the use of the category representation to construct a representation of the individual) can distort what people perceive, highlighting salient unusual features and minimising others, as well as enabling them to perceive the object or scene as a configuration of particular components (Goldstone, 1998).

Although designers' memories include details of both exact form and context, research on mental imagery, perceptual learning (Goldstone, 1998), and expertise in electronics (Egan and Schwartz, 1979), as well as radiology (Myles-Worsley *et al.*, 1988), and chess (Gobet and Simon, 1998), indicates that visuospatial representations are highly structured, incorporating categorisations of both structural features and emergent visual features. It is difficult to assess how much of the mental representation of an individual design, or a sketch, is unique to it, and how much is reconstructed from representations of more general categories. The structure and redundancy in visuospatial representations enable details to be reconstructed from sparse mental descriptions.

Research on memory has shown that the mental representations that guide thinking cannot be divided neatly into the content of conscious awareness controlling behaviour and memories passively awaiting recall. How readily memories for objects, concepts or events are recalled, or serve to influence other cognitive processing, depends on how recently and forcefully someone has perceived or been reminded of them. This effect works not only for the items themselves, but also for other related items (Baddeley, 1996). The priming of memories for recall by the activation of related memories depends on the organisation of memory; thus, it depends on both associations and generalisation across cases to create categories and abstract types (Anderson, 1983).

Categories and exemplars as reference points

8.11 Objects often serve as mental references. Famous buildings, such as Lloyds of London, are shared mental reference points amongst a community of professionals.

Design researchers have found that designers in a variety of fields make use of memories of both individual designs and design elements, and generalisations into categories. Schön (1988) describes functional types and references as forms of architectural design knowledge. Drawing on the cognitive theory of dynamic memory proposed by Schank and Abelson (1977; Schank, 1982), Oxman (1990) argues that precedents are used in design as prototypes, through a process of typification in which individual designs, problems *etc.* are used to create and refine more abstract generalisations, and are indexed in memory by these generalisations. Eckert and Stacey (2003b) argue (from observations of fashion and knitwear designers' working practices and how they describe designs to their colleagues) that remembered designs often serve as exemplars and indices for subtly differentiated categories. Eckert's later observations of engineers indicate that commonly known objects play an even larger role in engineering, where there are fewer potential reference designs than in textiles and they are shared by the members of multidisciplinary design teams.

Multiple mental representations

Designers can think about the same problems using very different mental models and reasoning strategies, according to how well they create concrete visuospatial representations of the structure of the design, and abstract propositional representations of its functions. Engineers who are highly skilled in applying analysis techniques reason about the consequences of making changes to a design by constructing lattices of causal connections, which are imagined in relatively abstract propositional terms. This approach yields a deep understanding of why changes have particular consequences, but reasoning mistakes can lead to completely wrong conclusions. Detailed and concrete visuospatial representations of structure and behaviour support retrieving related designs from memory and reasoning about similarities and differences. Other engineers reason about the same sorts of change by making predictions from how similar designs behaved in the past. While some very experienced engineers can make good predictions by similarity reference, the effects of small changes can easily be overlooked. By providing these predictions, they can enable their analytical colleagues to construct correct causal models for computing more precise results.

Visuospatial thinking is very important in most types of engineering designing, and many designers are good at it. Most engineering design creation involves relating visuospatially imagined structure to functions and constraints that are reasoned about in propositional terms. Moreover, many mechanical engineers think in terms of visually imagined concrete instances of mechanisms or machines when there is no actual need to do this for the problems they are solving. Some find it very hard to think in the abstract functional terms envisaged by top-down design methodologies (Andreasen, 1980) and by computer tools for designing in terms of networks of functions. For example, Nam Suh of the Massachusetts Institute of Technology has said that many engineers take naturally to his axiomatic design method (Suh, 1990, 2001), while others find it very difficult and unnatural. One reason for this is that mental representations of designs in terms of abstract statements of functions, and transformations and transmissions of matter, energy and information, form sparse networks of items of information with relatively arbitrary connections between them, with little redundancy. So they do not form large and strongly connected chunks as readily as spatial information from visual perception of sketches, diagrams and the artefacts themselves, which contain a lot of redundant and mutually reinforcing connections between elements.

8.12 Designers find it difficult to think about functions abstractly. They often make reference to known shapes to describe functions, finding it difficult to break away from such visual descriptions.

Another reason why many engineers find it difficult to reason abstractly without reference to particular physical embodiments is that functions and behaviour are hard to imagine except as the actions of concrete spatial things, and functions are usually associated in memory with examples of machines that embody those functions. Recalling a concept – a category of designs or design elements – cues the recall of features typically present in a design embodying the category, either as elements of a composite archetype, or because the representation of some aspect of the concept cues the recall of exemplars of the concept. Thus, thinking about designs in functional terms often imports structural and behavioural information into the designers' mental representations of the design situation. Conversely, a visuospatial representation of the form of an object is tied to its identity as a type of thing with functions and behaviours, and cues recall of its functions and behaviours. Ignoring these associations can prove impossible, even when one is actively trying (Jansson and Smith, 1991; Purcell and Gero, 1996).

Individual differences

The differences between individuals are larger than most people imagine. Reasoning abilities and styles differ according to how well people form different kinds of mental representations (Figure 8.13). At one extreme, some people have very little subjective mental imagery, and that is fleeting and fragmentary; while at the other end some people have images they subjectively experience as stable, detailed pictures of scenes and situations, and recall or generate them easily – sometimes too easily. However, there is no strong relationship between subjective mental imagery and the ability to solve a lot of visuospatial problems (Neisser, 1970). Some people who have rich static images find imagining movement difficult, as the rich detail and spontaneous retrieval of other images gets in the way of making changes to them. Strong associations to large coherent visual memories are likely to be an advantage in finding visuospatial analogies, and may be a handicap in reasoning about movement and causal processes.

8.13 Flight simulator. Some people find it difficult to generate mental images, while others find it difficult to picture movement.
© Airbus

The relationship between subjective mental imagery and the 'real' objects they refer to is also subtle and not fully understood. Psychological theorists still argue about whether mental imagery is essentially pictorial, most famously Kosslyn (1980, 1994), or essentially comprises symbolic descriptions, most famously Pylyshyn (2002, 2003). There is evidence that even when a rich mental image is subjectively experienced as complete, details within it do not exist until attention focuses on a part of the image (Kosslyn, 1980, 1994). But visuospatial representations that are subjectively experienced as images may not just be missing details, they may also be missing entire categories of information; and relationships or resemblances that are not an explicit part of the structure of the scene imagined may be invisible when they would be perceptually obvious in a picture.

Seeing objects, photographs, sketches, schematic diagrams and, so on triggers the creation of mental representations of designs through perceptual recognition. People can actively control the focus of attention to obtain the elements of an external representation that they need, so having an external memory enables them to use much more complex information than they can hold in a coherent mental representation otherwise. As we have seen, people can perceive features and relationships that were not previously part of their mental representations of the designs, though this usually requires active search. But the external representation functions as a set of cues for constructing mental representations. For this, accurate depiction is only required when fine details differ from category-normal in significant ways, and the appearance of roughness cues the inclusion of uncertainty or provisionality in the mental representation.

Mental actions

Designing comprises various sorts of mental and physical action, using and creating mental and external representations. In this section we view designing as mental action, at the level of individual moves through the spaces of possible designs, to examine how external representations contribute directly and indirectly to the actions that create new designs. We also look at design thinking as skill – learned capacities for constructing representations of design problems, making particular kinds of moves in design spaces, evaluating design proposals, and for structuring the design process.

The designer's understanding of the problem co-evolves with the solution.

Basic elements of design cognition

Analyses of design processes at different levels of detail converge to a view of designing, originally formulated by Asimow (1962), as comprising a cyclic process, of formulating the problem, making a change to the proposed design, evaluating the new state of the design, reformulating the problem, making another change to the design, and so on. The designer's understanding of the problem co-evolves with the solution (Dorst and Cross, 2001). What gives design thinking its characteristic form is that the design cycle is fractal down to the level of mental actions, with cyclic design processes for subproblems nested within a single stage of a larger task. Complex engineering design processes employ specialists to perform particular evaluations in the outer loops, while the smallest cycles of evaluating and changing happen entirely mentally in a few seconds. Like other problem-solving activities, designing involves means–ends analysis and a hierarchical structure of goals and subgoals (Simon, 1996). At each level the subtasks include decision making, retrieving information, recording information, and planning, as well as generating design ideas. Design activities include well-defined sub-problems, many requiring deductive reasoning and procedure following, rather than propose-and-evaluate idea construction. Successful creative thinking requires both fluency in idea generation – divergent thinking – and in linear problem solving – convergent thinking; these abilities are not highly correlated. Of course, designing also involves the many activities involved in managing processes and human relationships, which are beyond the scope of this chapter.

A lot of what we know how to do is tightly bound to particular situations.

Situated cognition

A lot of what we know how to do (what psychologists term procedural knowledge) is tightly bound to particular situations; and much of human thought is inseparable from perception of one's environment and action in direct response to it, guided by conscious and latent goals (Suchman, 1987; Clancey, 1997). A lot of problem solving, including design, proceeds by applying characteristic sequences of mental actions to situations of particular types, triggered by goals and elements of perceptions and mental representations that belong to particular categories – though the exact form of the actions depends on the subtle details of the mental representations of the situations. A lot of these actions have the character of heuristics: reasoning or decision-making steps that are potentially useful but not guaranteed to be right (Duncker, 1935; Newell and Simon, 1972; see Akin (1986) for a detailed theory of architectural design as problem solving). Conscious decisionmaking about what to do next (as opposed to larger-scale goal setting) is relatively rare. Conscious real-world goal-directed behaviour typically has the character 'remember or decide what ought to be done next, or think of something to do, and do it if the estimated benefits exceed the estimated costs' (Anderson, 1990). Consciously chosen actions (which are goals to be achieved by finer-grained actions) are only planned or decided about at the level of detail that is needed. Finer details are dealt with as they arise by unplanned situated actions. Plans and goals do not rigidly dictate behaviour but form part of the mental context for situated actions, functioning as resources to guide behaviour (Suchman, 1987; Clancey, 1997). Designers are guided by plans but act opportunistically to correct mistakes, respond to unexpected events and fulfil latent goals (Visser, 1990, 1994).

Facets / Trajectories	An artefact	A process	A relation
State of			
Making sense of			
Framing futures of			

8.14 Designers need to reason about all nine aspects of the Minneman (1991) matrix

Creating an understanding

Actively creating an understanding of the problem is a vitally important part of problem solving, especially in design. This involves both perception and reasoning. Designers face problems that are inherently ill-defined, that are underspecified and in which important constraints are implicit (Simon, 1973, 1996). Designers often reformulate the design problem, to add structure and to recast it in terms more useful for guiding its solution: categorising it, thus activating additional constraints, and implicitly selecting solution strategies and eliminating alternatives.

Finding the right view of a problem is often the key to solving it (Duncker, 1935). Such reformulations can be guided by established principles and guidelines, individual preferences, the recognition of a similarity to another problem, or be more-or-less arbitrary. But patterns of thinking actions are largely determined by the requirements of the task, and hence by the form of the product. Well-defined problems can predominate in the design of tightly specified products. Hoover *et al.* (1991) point out that designers generate different abstractions of their designs for particular practical purposes, such as modelling their performance, and develop their designs further by refining these abstractions by adding more concrete detail; refinements made from different abstractions may not be compatible.

Darke (1979) argued that the designs of the architects she studied were shaped by the aspects of the design problems that were explicit and salient in the architects' minds when they generated the essential features of their conceptual designs; and that the most prominent aspect of the problem situation for an architect is typically the physical characteristics of the site a building is being designed for.

Manipulating past designs

Designers' pattern synthesis actions that create or modify new designs, combine, manipulate and transform the objects, features and properties they have available in memory, often derived from past designs (Lawson, 1997). The most strongly available design elements are those in conscious awareness or available in the designer's visual field. This depends on what the notational conventions of sketches and other external representations make salient (Zhang, 1997). Knowledge of previous designs biases designing towards similar designs even when designers know they are actively trying to create something different – a phenomenon known to psychologists as fixation (Figure 8.15; Jansson and Smith, 1991; Purcell and Gero, 1996).

8.15 Child's beaker. In a study on fixation, Jansson and Smith (1991) showed design students a mug with a mouthpiece and told them to create a non-spill mug without a mouthpiece: despite this instruction, the majority of designs incorporated a mouthpiece.

8.16 Passenger jet. Engineers are often able to assess the feasibility of designs and recognise what analyses are required. Skilled specialists can, for example, predict aerodynamic properties and the transmission of forces and stresses.
© Airbus

Designers assess the quality of changes to designs (envisioned mentally or using external representations) perceptually, as well as by explicit reasoning. In some design fields, perceptual evaluations are very tightly coupled with design synthesis actions, and play a crucial role in the development of conceptual designs. Humans are extraordinarily good at perceiving the important features of their environment, including categories, symbols and meanings, as well as subtle similarities and differences. This ability is precisely tuned to the demands of the current task. Experienced designers know about and can recognise more perceptual features (Egan and Schwartz, 1979), and this is a highly trained skill in many design professions. Thus, designers create designs conforming to their perceptually recognised visuospatial constraints and requirements (within the limitations of the power of their pattern synthesis actions); and recognise the degree to which they conform to visuospatial constraints and requirements. In aesthetic design, perceptual visuospatial knowledge of the context and of what is required is an essential part of formulating the problem (Eckert and Stacey, 2001).

Designers rely on perceptual evaluations either when the problem is simple enough to see or too complex to analyse. For example, knitwear technicians can spot whether two curves of different shape have the same lengths. At the other extreme are complex emergent phenomena in jet engine design, such as combustor rumble, where only a few experts have a detailed tacit understanding of the relationship between combustor shape and rumble, and everybody perceives it as a black art. The experts have learned complex associations between features of combustors and levels of rumble. The interplay of perceptual and explicit reasoning can be seen when engineers build up analysis meshes. Experienced engineers know the order of magnitude of a result and conduct computer analysis to fine tune the value. They perceive the features of the object that are significant for the analysis and the relationships between them, and recognise correct meshing points or use situation-specific knowledge to reason about them, and get analyses close to the real value. Novices might put the points in the wrong places, and not even recognise when their solutions are out by several orders of magnitude.

Design as skill

Experienced designers usually know more than novices. Not only do they know more facts, rules, principles, guidelines and examples, but their knowledge is more highly organised so that it is more accessible and applicable when needed. But expertise, especially in design, is primarily skilled action, for perceiving, formulating and solving problems (see Bédard and Chi (1992) and Bolger (1995) for introductions; see Chi *et al.* (1988) and Ericsson and Smith (1991) for seminal research on expertise).

While most studies of expertise distinguish between experts, intermediates and novices, Raufaste *et al.* (1998) make a further distinction between experts and super-experts, leading authorities who spend a lot of time reflecting on very difficult cases; they point out that much of the research on expertise has contrasted super-experts with novices and neglected ordinary experts, who are competent but mostly deal quickly with routine cases.

8.17 Aircraft cockpit. Mechanical engineers and avionics engineers often know little about each other's tasks. Even experts are effectively novices in the other field.
© Airbus

Experienced engineers working outside the scope of their expertise may have more general strategic knowledge to call on but will suffer the same difficulties as novices in recognising significant features and formulating problems, and will need to reason backwards from their goals to how to achieve them (Figure 8.17).

Expert problem solving in any field requires a rich and powerful set of associations between different situations and appropriate actions. Experts (performing routine tasks) work forward from the present situation: they know how to recognise the pertinent features of the problem situation, they know what to do, and do it, without needing to formulate a plan. For experts in many fields, their task-specific problem-solving procedures include recalling and adapting solutions to previous problems; for designers, these are elements of previous designs.

Experts are subject to fixation on previous designs in a different way from novices. Because they possess memories of a greater stock of relevant designs, they will be better able to find an appropriate model, and escape a particular recent exemplar, but will find it harder to escape closer matches to the present situation and stronger situation–action associations. People with expert knowledge have both richer and stronger associations between elements of their factual knowledge, and more specialised mental procedures. Thus they can focus recall from memory and mental actions more narrowly. This can be an advantage, but mental actions can embody tacit constraints inherited from previous similar problem situations that are no longer relevant, leading to incorrect or unsuccessful problem solving (Wiley, 1998).

Novices, who lack task-specific situation–action associations, explore and learn from their mistakes. They reason backwards from what they want to how they can get it, applying general problem-solving strategies to the facts that they know. Task-specific procedures are created as the starting points

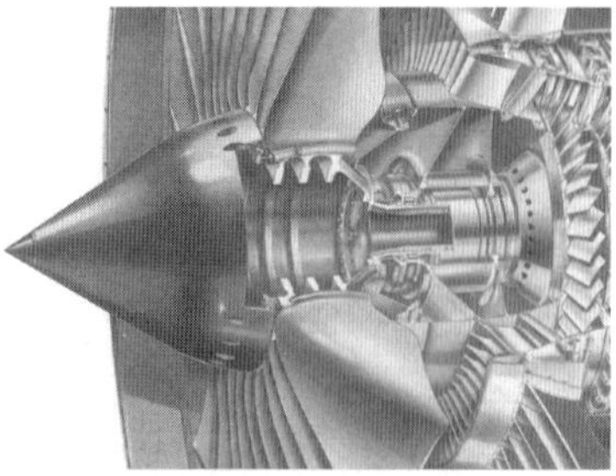

8.18 Jet engine. Ahmed and Wallace (2004) found that the novices were aware of their information needs in only one-third of their queries. Reproduced with the kind permission of Rolls-Royce plc

and outcomes of such reflective problem-solving processes are associated in memory, to create situation–action pairs. Now no reasoning is needed to go from recognising the situation to performing the action. Situation–action associations that are repeatedly successful are strengthened and generalised; when they fail, situations are differentiated so that more tightly specialised situation–action associations are formed (Anderson, 1983). People learn to avoid actions that are related to the appearance of failure, interpersonal conflict or other negative rewards. In non-routine situations, experts do means–ends reasoning just like novices, but their conscious, reflective problem-solving strategies are also a learned skill. By learning from the success and failure of their reasoning they develop more elaborate and powerful specialised strategies for the problems they meet in their field.

Expert designers put considerable effort into articulating their problems (typically more than novices). By collecting all the available constraints on the design, they minimise the range of designs they need to think about. As designers gain experience, they develop skills in recognising, formulating, prioritising requirements and constraints, and employing them in their design thinking. Skilled actions learned by expert engineers include identifying the different issues they need to consider and what information they need to solve a task (Ahmed *et al.*, 2003). Of course, skills that contribute to high performance include process management and cooperation with others (Sonnentag, 1998).

In many fields, the skills developed by experts include reading the notations and graphic conventions used in their field. Increasing skill in reading graphic conventions reduces the time and effort involved in generating appropriate mental representations from external design representations, as a greater variety of symbol combinations become perceptually recognisable. As Henderson (1999) notes, this is an important aspect of professional group membership and possession of a shared object world. In some industries designers employ active strategies for creating the mental representations they will require later for creating designs. This is more prevalent in fashion -driven industries, where designers learn categories that implicitly define the spaces of acceptable designs within current fashion (which the designers use to formulate design goals) and that provide components of the designer's own new designs (Eckert and Stacey, 2001, 2003a). While engineers study competitors' products and look for applicable solution principles when required, constant opportunistic gathering of sources of inspiration is seldom part of their work culture.

8.19 Sports car. Only a few areas of engineering designers employ active strategies for creating mental representations. For example, a car stylist employs a process very similar to a fashion designer.

Mental actions: creative thinking

In this section we consider some of the skills and mental actions required when standard solutions will not work, and expertise is not enough.

A great deal of engineering design is routine design, in the sense that it involves either modifications or transformations of existing design elements that do similar jobs – design by adaptation – or the application of well-understood procedures for creating concrete embodiments of standard solution principles – design by refinement (Oxman and Oxman, 1992).

In these situations the product architecture is understood – so designers can create mental representations of what the design should be in the form of skeletally imagined components, because they know the mappings from functions to structural elements to fulfil those functions. But sometimes more innovative designing is required (Figure 8.20), when straightforward adaptations of previous designs are insufficient. Not only are more radical transformations required, but finding a suitable design or solution principle to adapt may not be easy.

8.20 Helicopter. Often innovative design problems turn up within larger routine design problems, in enabling the use of existing components and approaches, and stopping changes propagating through a design, as Eckert *et al.* (2004) discovered in a study of the customisation of helicopters. Photo © AgustaWestland

Designing with constraints

Designing is guided by the constraints on the product. Hard constraints, to which the product must conform, act differently from guidelines, targets, and soft constraints, to which the product should conform (Figure 8.21). All these features of the problem formulation serve to activate learned problem solving procedures, including the recall of prefabricated solution chunks. Thus, they channel designers into repeating and adapting designs they have produced before.

When designers are unable to create designs conforming to all the soft constraints, they weaken or discard the less important constraints, to make their designs produced by their standard methods meet the task demands as well as possible. But when hard constraints are in conflict, they can prevent standard solutions from working. This situation forces designers to try to innovate, by exploring and using reflective problem-solving strategies, and progressively refining their understanding of the problem. From repeated failures and partial successes they refine their strategies for reformulating problems and generating novel ideas. The role of difficult combinations of hard constraints as a spur to creativity has been observed by many outstandingly creative people, for instance Gordon Murray, the racing car designer, who constantly needed to work around and exploit complex technical regulations (Cross and Clayburn Cross, 1996).

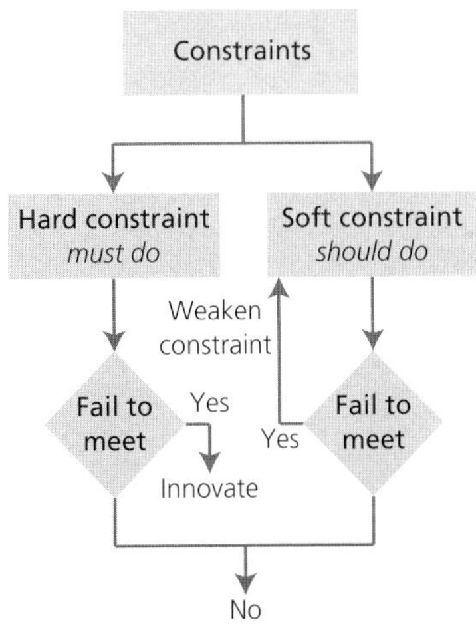

8.21 Different types of constraint

Analogical reasoning as a mechanism of creativity

We view the key creative step as the recognition of an analogy between the requirements of the current situation and some other machine or process or form. This can be a specific exemplar or an abstractly imagined category. The structure mapping theory views analogical reasoning as identifying a correspondence between the structures of the relationships between the components of two different composite entities (Gentner, 1983). The more different the characteristics of the components themselves, the more abstract and radical the analogy.

The difference between innovative and routine design is one of how far the formulation of the design problem needs to be abstracted away from the salient features of the design elements that perform similar tasks in similar designs, to guide the generation of solutions that do not share those features. This usually requires reframing the problem, by describing needs and constraints in different terms, as functions, or as different kinds of functions, so that different salient features of the problem guide the retrieval of different analogies from memory. In cognitive terms this is a difference of degree, as similarity between entities with similar components is recognised through the same mechanisms as analogy between entities with dissimilar components (Gentner and Markman, 1997). Nevertheless, finding abstract analogies is hard, because, first, the problem situation has associations in memory to more concretely analogous designs, on which designers fixate; second, there are no prior associations in memory between the problem situation and any abstract analogies to it; and third, reframing the problem is often difficult.

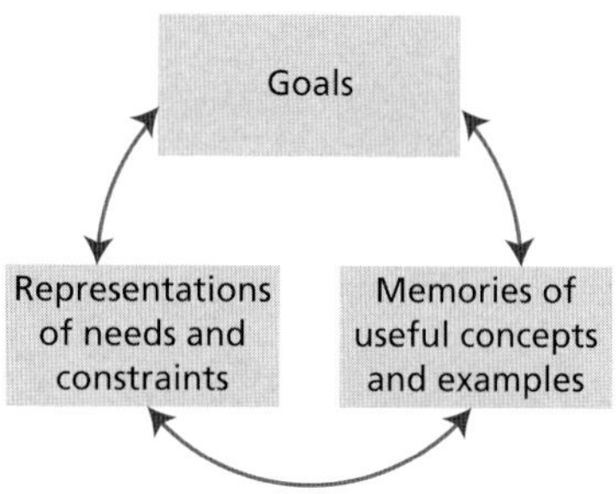

8.22 Elements of design cognition that are required to generate creative ideas

Constraints enhance creativity

The challenge in applying methods and processes for innovative design is to turn the narrowness and tight focus of most people's analogy recognition and design synthesis actions to advantage. This is achieved through enabling designers to formulate their design problems in ways that facilitate the generation of appropriate ideas (Figure 8.22).

Designers often elaborate the first promising idea they think of, investing time and effort in it and becoming emotionally committed to it, when instead they should look further for more and better initial ideas. A major purpose of some design methodologies is encouraging designers to look for a range of possible alternative designs in conceptual outline before committing to any one (either by conscious selection, or by investing too much effort in elaborating it). Some established methods for generating innovative design ideas, such as TRIZ (Figure 8.23), work by encouraging designers to formulate problems that have tight and novel constraints, or that make particular solution principles the most salient sources of analogical mappings. Brainstorming functions, in a loose and unsystematic way, to get designers to consider the relationships between the problem and the arbitrary constraints and potential analogical mappings suggested by the ideas put forward in the brainstorming session.

In many fields of endeavour, creative breakthroughs often come from finding a different problem to solve. In engineering this can be done by adding constraints to underconstrained problems, or by removing constraints from overconstrained problems. Engineers are taught to reformulate problems by stripping away assumptions about how a machine should work to obtain a more abstract, functional view of what their designs should achieve. As we have explained, excluding assumptions about physical embodiments from functional formulations of problems is not necessarily easy.

The phenomenon of fixation in design stems from the design synthesis actions being tacitly overconstrained by the association of functional requirements with particular physical embodiments. Simon (1996) explained the well-known phenomenon that insight in problem solving frequently occurs after a break (incubation) as due to the forgetting of unhelpful associations with the problem in memory. Designers in many fields routinely add constraints to underconstrained problems to define them clearly enough to solve, by choosing standard solutions, or, where there are none, by making major decisions about the form of the design arbitrarily or according to personal preference.

Finke (1990) got people to imagine combinations of arbitrary shapes (which he terms preinventive forms) and then use them to solve problems requiring creative thinking, thus giving them a much tighter set of constraints; he found that his subjects did better in the constrained condition than when allowed to think freely. Using chance forms to meet design goals is often a fruitful idea-generation strategy in artistic design fields.

In innovative designing, external representations are needed for the structure of the product architecture. Graphic representations of functions and behaviour can make designers' mental representations of functional aspects of design problems more salient and coherent, facilitating the search for radical analogies and novel embodiments of principles. Finke's (1990) results suggest that arbitrarily selected preinventive forms might also facilitate this (Benami and Jin, 2002). Finke *et al.* (1992) conceive of

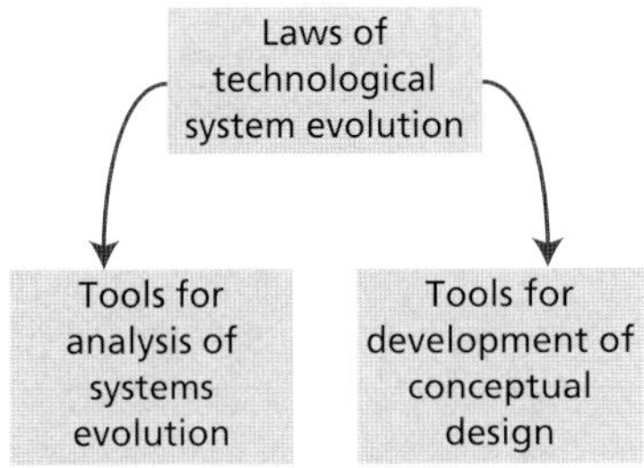

8.23 TRIZ is a systematic technique for generating innovation (Altshuller, 1994). It requires designers to formulate their problems in an abstract way using a matrix of 39 parameters, where each cell points to patented solutions. Altshuller also developed a set of 40 principles, such as replacing mechanical systems by optical, acoustic or thermal ones, or eliminating failure-prone processes altogether.

In many fields of endeavour, creative breakthroughs often come from finding a different problem to solve.

preinventive structures – novel visual patterns, object forms, mental blends, mental models, verbal combinations – as being initially formed without full anticipation of their resulting interpretation. Benami and Jin (2002), presenting a model of creative conceptual design in engineering, argue that the stimulating properties of preinventive entities in external representation are meaningfulness, relevance to the matter at hand, divergence (the capacity for finding multiple uses for the same entity), incongruity (conflict or contrast between elements) and emergence (the extent to which unexpected features appear).

Improving representations in design processes

In this section we discuss ways to improve design processes by improving the representations designers use. Negotiating a common understanding of shared representations is a first step to improving design processes. We will not attempt to survey the large body of academic research on developing better CAD systems or computer sketching systems (see Do (2002) for one indication of what is possible). Rather, we will discuss ways to think about the issues involved in choosing and using effective representations.

There are two challenges in improving engineering design processes at the level of designers' thinking:
- enabling designers to find the information they need;
- finding ways to display information graphically.

There are two challenges in improving engineering design processes at the level of designers' thinking. The first is enabling designers to find the information they need. Searching for information takes up a lot of their time; for designers, knowledge and procedures for analysing their information needs, and strategies for searching for information, are an important part of expertise (Kuffner and Ullman, 1991; Ahmed *et al.*, 2003; Ahmed and Wallace, 2004). The second is the concern of this chapter: finding ways to display information graphically that facilitate reasoning with it and manipulating it.

Visualisation

Most importantly, this involves ways of making significant features and relationships directly visible in the display, eliminating the need to reason about what they are. Analyses in terms of mental representations and operations are not needed for this. What is required are techniques for translating both geometric and abstract structures into graphic forms that make certain features and relationships salient. Tufte (1983, 1990, 1997) provides valuable guidance on how to do this in a wide variety of situations, though focusing primarily on data displays and maps. In changing procedures to use different representations, or to migrate manual activities onto computers, it is essential to understand the functions served by the representations used in the current process, and how they are used, to ensure that the new procedures support the same kinds of thinking.

In many situations designers need to consider and modify different aspects of their designs, and perform different kinds of actions – comparisons, identification of relationships, ordering into sequences, parameter changes, synthesis of new forms, discovering the consequences of changes, and so on. These may require different representations that make different features and relationships perceptually visible.

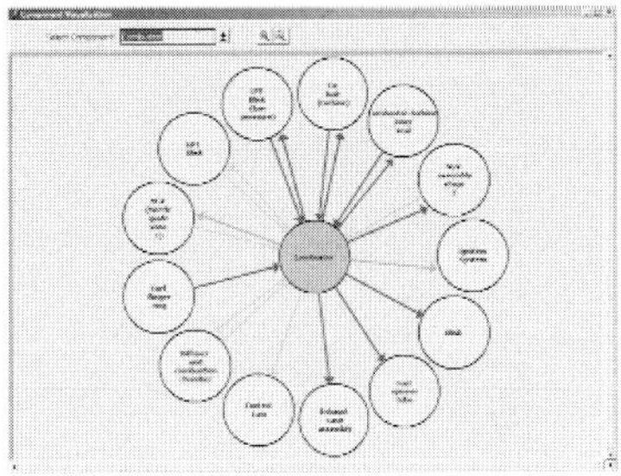

8.24 Connectivity. Graphic representations of the different dependencies between the components of a design are developed to aid change prediction (Jarratt *et al.*, 2004).

Designers of complex products, whose structure, function and performance need to be considered in different ways in parallel, need to switch between different mental models supported by different external representations and information sources. So a potentially fruitful avenue for academic research into design process improvement is the provision of alternative graphic notations for design information, both for computer tools and for designers to sketch with. An example is the research into graphic representations of the dependencies between the components of a design, and between the tasks in a design process. Another example is the research into graphic notations and computer tools for tracking design rationales (Buckingham Shum *et al.*, 1997; Heliades and Edmonds, 2000; Bracewell and Wallace, 2003).

Cognitive dimensions

Green (1989) argues that representations of complex information structures, such as the programming environments used by software developers (Green and Petre, 1996) and the graphical notations used in electronics (Petre and Green, 1992), can be considered as good or bad on a number of cognitive dimensions. The cognitive dimensions of information artefacts determine how easy or hard they are to use or modify in particular ways. Designing computer tools for displaying and manipulating complex information structures (like designs) involves making trade-offs (consciously or unconsciously) between different cognitive dimensions. Pencil and paper is a medium for representing information structures, like a CAD system, but with very different positions on the cognitive dimensions. Using a pencil frees designers to be inconsistent, go beyond standard notational conventions, and give symbols different meanings, but they are still bound to notational conventions (however idiosyncratic) that make some types of information salient rather than others, may fail to show significant dependencies, and may make certain kinds of comparisons and evaluations difficult. And as soon as a description needs to be both detailed and consistent, hand-drawn diagrams or drawings become a very viscous medium.

Visibility and juxtaposability. Ability to view components easily. How easy is it to see or find the various parts of the notation while it is being created or changed? If the users need to compare or combine different parts, can they see them at the same time?

Viscosity. Resistance to change. When the users need to make changes to previous work, how easy is it to make the change?

Hard mental operations. High demand on cognitive resources. What kinds of things require most mental effort with this notation? Do some things seem especially complex or difficult for the users to work out in their heads (for example, when combining several things)?

Closeness of mapping. Closeness of representation to domain. How closely related is the notation to the result that the users are describing? What parts seem to be a particularly strange way of doing or describing something?

Hidden dependencies. Important links between entities are visible. If the structure of the product means some parts are closely related to other parts, and changes to one may affect the other, are those dependencies visible?

Progressive evaluation. Work completed can be checked at any time. How easy is it for the users to stop in the middle of creating some notation, and check their work so far? Can they do this any time they like? Can the users find out how much progress they have made, or check what stage in their work they are up to? Can the users try out partially-completed versions of the product?

Provisionality. Degree of commitment to actions or marks. Is it possible for the users to sketch things out when they are playing with ideas, or when they are not sure which way to proceed? What features of the notation help them to do this? What sort of things can the users do when they do not want to be too precise about the exact result they are trying to get?

Premature commitment. Constraints on the order of doing things. When the users are working with the notation, can they go about the job in any order they like, or does the system force them to think ahead and make certain decisions first?

Secondary notation. Extra information in means other than formal syntax. Is it possible for the users to make notes to themselves, or express information that is not really recognised as part of the notation? If the notation was printed on a piece of paper that the users could annotate or scribble on, what would they write or draw? Do the users ever add extra marks (or colours or format choices) to clarify, emphasise or repeat what is there already?

Detail in context. Ability to see both complete descriptions of local information and their relation to a wider picture. Is it possible to see how elements relate to others within the same notational layer? Is it possible to move between them with sensible transitions?

Synopsis. Support for holistic views. Does the system provide an understanding of the whole structure when the user 'stands back and looks'?

Free rides. New information is generated as a result of following the notational rules. Can users read new information off, as a result of making measurements and observations of the things they put there previously?

Unevenness. Bias towards specific solutions or actions. Does the system push users' ideas in a certain direction because certain things are easier to do?

8.25 Some cognitive dimensions (selected from a fuller list in Blackwell *et al.*, 2003)

Conclusions

Various kinds of graphic representations and models are an important part of most aspects of engineering, but many engineers fail to recognise their influence on individual thinking, communication between designers, and the organisation of design processes. In some activities the entities and relationships the representations make explicit become the concepts designers think with. Many design processes could be improved if their participants understood each other's information needs and how information can be most effectively conveyed.

Changing processes, methods and tools changes designers' tasks and information needs; this changes the functions of existing representations of design information, and may create a need for new representations. The development of new methods and procedures should include a careful consideration of what designers' information needs are and what graphic representations of design ideas can best meet those needs. While this chapter has concentrated on representations of designs, most of the points it makes apply equally well to representations of processes, which are frequently important in guiding design processes but which have attracted little research.

References

Ahmed S, Wallace KM (2004) Understanding the knowledge needs of novice designers in the aerospace industry. Design Studies, 25: 155–173

Ahmed S, Wallace KM, Blessing LTM (2003) Understanding the differences between how novice and experienced designers approach design tasks. Research in Engineering Design, 14: 1–11

Akin Ö (1978) How do architects design? In: Artificial intelligence and pattern recognition in computer-aided design. North-Holland

Akin Ö (1986) Psychology of architectural design. Pion

Altshuller G (1994) And suddenly the inventor appeared. TRIZ, the theory of inventive problem solving. Technical Innovation Center, Worcester, MA, USA

Anderson JR (1983) The architecture of cognition. Harvard University Press

Anderson JR (1990) The adaptive character of thought. Lawrence Erlbaum

Andreasen MM (1980) Syntesemetoder på systemgrundlag. PhD thesis, Lunds Tekniska Högskola, Denmark

Asimow M (1962) Introduction to design. Prentice Hall

Baddeley AD (1996) Human memory. Psychology Press

Badke-Schaub P, Stempfle J, Wallmeier S (2001) Transfer of experience in critical design situations. ICED'01, Glasgow, UK

Bartlett FC (1932) Remembering. Cambridge University Press

Bédard J, Chi MTH (1992) Expertise. Current Directions in Psychological Science, 1: 135–139

Benami O, Jin Y (2002) Creative stimulation in conceptual design. ASME DETC'02, Montreal, Canada

Blackwell AF, Hewson RL, Green TRG (2003) Product design to support user abstractions. In: Handbook of cognitive task design. Lawrence Erlbaum

Bolger F (1995) Cognitive expertise research and knowledge engineering. Knowledge Engineering Review, 10: 3–19

Bracewell RH, Wallace KM (2003) A tool for capturing design rational. ICED'03, Stockholm, Sweden

Brereton MF, Cannon DM, Mabogunje A, Leifer LJ (1996) Collaboration in design teams: mediating design progress through social interaction. In: Analysing design activity. John Wiley

Bucciarelli LL (1994) Designing engineers. MIT Press

Buckingham Shum SJ, MacLean A, Bellotti VME, Hammond NV (1997) Graphical argumentation and design cognition. Human–Computer Interaction, 12: 267–300

Chi MTH, Glaser R, Farr MJ eds (1988) The nature of expertise. Lawrence Erlbaum

Clancey WJ (1997) Situated cognition. Cambridge University Press

Clement J (1983) A Conceptual Model Discussed by Galileo and Used Intuitively by Physics Students. In: Mental Models. Lawrence Erlbaum

Cross NG, Christiaans H, Dorst K eds (1996) Analysing design activity. John Wiley

Cross NG, Clayburn Cross A (1996) Winning by design: the methods of Gordon Murray racing car designer. Design Studies, 17: 91–107

Darke J (1979) The primary generator and the design process. Design Studies, 1: 36–44

DiSessa AA (1983) Phenomenology and the evolution of intuition. In: Mental models. Lawrence Erlbaum

DiSessa AA (1993) Toward an epistemology of physics. Cognition and Instruction, 10: 105–225

Do EY-L (2002) Drawing marks acts and reacts: toward a computational sketching interface for architectural design. Artificial Intelligence in Engineering Design Analysis and Manufacturing, 16: 149–171

Dorst K, Cross NG (2001) Creativity in the design process: co-evolution of problem-solution. Design Studies, 22: 425–438

Duncker K (1935) Zur Psychologie des produktiven Denkens. Springer

Eckert CM (2001) The communication bottleneck in knitwear design: analysis and computing solutions. Computer Supported Cooperative Work, 10: 29–74

Eckert CM, Stacey MK (2001) Designing in the context of fashion – designing the fashion context. Design Thinking Research Symposium, Delft, The Netherlands

Eckert CM, Stacey MK (2003a) Sources of inspiration in industrial practice: the case of knitwear design. Journal of Design Research, 3(2)

Eckert CM, Stacey MK (2003b) Adaptation of sources of inspiration in knitwear design. Creativity Research Journal, 15(4): 355–384

Eckert CM, Clarkson PJ, Zanker W (2004) Change and customisation in complex engineering domains. Research in Engineering Design, 15(1): 1–21

Egan DE, Schwartz BJ (1979) Chunking in recall of symbolic drawings. Memory and Cognition, 7: 149–158

Ericsson KA, Smith J eds (1991) Toward a general theory of expertise: prospects and limits. Cambridge University Press

Finke RA (1990) Creative imagery: discoveries and inventions in visualization. Lawrence Erlbaum

Finke RA, Ward TB, Smith SM (1992) Creative cognition: theory research and applications. MIT Press

Gentner D (1983) Structure-mapping: a theoretical framework for analogy. Cognitive Science, 7(2): 155–170

Gentner D, Markman AB (1997) Structure mapping in analogy and similarity. American Psychologist, 52(1): 45–56

Glock F (1998) Konstruieren als sozialer Prozess. Deutscher Universitätsverlag

Gobet F, Simon HA (1998) Expert chess memory: revisiting the chunking hypothesis. Memory, 6: 225–255

Goldschmidt G (1991) The dialectics of sketching. Creativity Research Journal, 4(2): 123–143

Goldstone RL (1998) Perceptual learning. Annual Review of Psychology, 49: 585–612

Green TRG (1989) Cognitive dimensions of notations. In: People and computers V. Cambridge University Press

Green TRG, Petre M (1996) Usability analysis of visual programming environments: a 'cognitive dimensions' framework. Journal of Visual Languages and Computing, 7: 131–174

Heliades GP, Edmonds EA (2000) Notation and nature of task in comprehending design rationale. Knowledge-Based Systems, 13: 215–224

Henderson K (1999) On line and on paper. MIT Press

Hoover SP, Rinderle JR, Finger S (1991) Models and abstractions in design. Design Studies, 12: 237–245

Jansson DG, Smith SM (1991) Design fixation. Design Studies, 12: 3–11

Jarratt TAW, Eckert CM, Clarkson PJ, Stacey MK (2004) Providing an overview during the design of complex products. DCC'04. Boston, MA, USA

Johnson-Laird PN (1983) Mental models. Harvard University Press

Koriat A, Goldsmith M, Pansky A (2000) Towards a psychology of memory accuracy. Annual Review of Psychology, 51: 481–537

Kosslyn SM (1980) Image and mind. Harvard University Press

Kosslyn SM (1994) Image and brain. MIT Press

Kuffner TA, Ullman DG (1991) The information requests of mechanical design engineers. Design Studies, 12: 42–50

Kunz A, Müller S, Kennel T, Lauche K, Mbiti K (2001) Supporting teamwork by new means of information technology. ICED'01, Glasgow, UK

Lauche K, Verbeck A, Weber WG (1999) Multifunktionale Teams in der Produkt- und Prozessentwicklung. In: Optimierung der Produkt- und Prozessentwicklung. Vdf Hochschulverlag

Lawson B (1997) How designers think – the design process demystified. Architectural Press

Leont'ev A (1978) Activity consciousness and personality. Prentice-Hall

McCloskey M (1983) Naïve theories of motion. In: Mental models. Lawrence Erlbaum

McFadzean J, Cross NG, Johnson JH (1999) Notation and cognition in conceptual sketching. In: Visual and spatial reasoning in design. MIT Press

Miller GA (1956) The magical number seven plus or minus two: some limits on our capacity for processing information. Psychological Review, 63: 81–97

Minneman SL (1991) The social construction of a technical reality: empirical studies of group engineering design practice. PhD thesis, Stanford University, CA, USA

Myles-Worsley M, Johnson WA, Simons MA (1988) The influence of expertise on X-ray image processing. Journal of Experimental Psychology: Learning Memory and Cognition, 14: 553–557

Neilson I, Lee J (1994) Conversations with graphics: implications for the design of natural language/graphics interfaces. Journal of Human–Computer Studies, 40: 509–541

Neisser U (1970) Visual imagery as process and as experience. In: Cognition and affect. Little Brown Co.

Newell A, Simon HA (1972) Human problem solving. Prentice Hall

Norman DA (1983) Some observations on mental models. In: Mental models. Lawrence Erlbaum

Norman DA (1988) The psychology of everyday things. Basic Books (later printings published as The design of everyday things)

Oxman RE (1990) Prior knowledge in design: a dynamic knowledge-based model of design and creativity. Design Studies, 11: 17–28

Oxman RE, Oxman RD (1992) Refinement and adaptation in design cognition. Design Studies, 13: 117–134

Pache M, Römer A, Lindemann U, Hacker W (2001) Re-interpretation of conceptual design sketches in mechanical engineering. ASME DETC'01, Pittsburgh, PA, USA

Petre M, Green TRG (1992) Requirements of graphical notations for professional users: electronics CAD systems as a case study. Le Travail Humain, 55: 47–70

Purcell AT, Gero JS (1996) Design and other types of fixation. Design Studies, 17: 363–383

Purcell AT, Gero JS (1998) Drawings and the design process. Design Studies, 19: 389–430

Pylyshyn ZW (2002) Return of the mental image: are there really pictures in the brain? Trends in Cognitive Science, 7: 113–118

Pylyshyn ZW (2003) Mental imagery: in search of a theory. Behavioural and Brain Sciences, 25: 157–237

Raufaste E, Eyrolle H, Mariné C (1998) Pertinence generation in radiological diagnosis: spreading activation and the nature of expertise. Cognitive Science, 22: 517–546

Sachse P, Hacker W (1997) Unterstützung des Denkens und Handelns bei Konstruieren durch Prototyping. Konstruktion, 49: 12–16

Schank RC (1982) Dynamic memory: a theory of reminding and learning in computers and people. Cambridge University Press

Schank RC, Abelson R (1977) Scripts, plans, goals and understanding. Lawrence Erlbaum
Schön DA (1983) The reflective practitioner: how professionals think in action. Basic Books
Schön DA (1988) Designing: rules types and worlds. Design Studies, 9:181–190
Schön DA, Wiggins G (1992) Kinds of seeing and their functions in designing. Design Studies, 13: 135–156
Simon HA (1973) The structure of ill-structured problems. Artificial Intelligence, 4: 181–202
Simon HA (1996) The sciences of the artificial. MIT Press
Sonnentag S (1998) Expertise in professional software design: a process study. Journal of Applied Psychology, 83: 702–715
Stacey MK, Eckert CM (2003) Against ambiguity. Computer Supported Cooperative Work, 12: 153–183
Star SL (1989) The structure of ill-structured solutions: heterogeneous problem-solving, boundary objects, and distributed artificial intelligence. In: Distributed artificial intelligence 2. Morgan Kaufmann
Suchman LA (1987) Plans and situated actions. Cambridge University Press
Suh NP (1990) Principles of design. Oxford University Press
Suh NP (2001) Axiomatic design: advances and applications. Oxford University Press
Suwa M, Gero JS, Purcell AT (1999) Unexpected discoveries: how designers discover hidden features in sketches. In: Visual and spatial reasoning in design. MIT Press
Tufte ER (1983) The visual display of quantitative information. Graphics Press
Tufte ER (1990) Envisioning information. Graphics Press
Tufte ER (1997) Visual explanations: images and quantities, evidence and narrative. Graphics Press
Visser W (1994) The organisation of design activities: opportunistic with hierarchical episodes. Interacting with Computers, 6: 235–274
Visser W (1990) More or less following a plan during design: opportunistic deviations in specification. Journal of Man–Machine Studies, 33: 247–278
Vygotsky LS (1962) Thought and language. MIT Press
Weber WG, Lauche K, Verbeck A (1999) Gemeinsame Vergegenständlichungen und kooperatives Handeln in multifunktionalen Teams. In: Design thinking. vdf

Wiley J (1998) Expertise as mental set: the effects of domain knowledge on creative problem solving. Memory and Cognition, 26: 716–730

Wulf F (1922) Beitrage zur Psychologie der Gestalt: VI. Über die Veränderung von Vorstellungen (Gedächtniss und Gestalt) Psychologische Forschung, 1: 333–375

Young RM (1983) Surrogates and mappings: two kinds of conceptual models for interactive devices. In: Mental models. Lawrence Erlbaum

Zhang J (1997) The nature of external representations in problem solving. Cognitive Science, 21: 179–217

Design practice

Chapter 9

Communication in design

Claudia Eckert and Anja Maier
University of Cambridge

Chris McMahon
University of Bath

Design is never a solitary activity. It is a social interactive process. Complex products are designed by teams of people, perhaps in single companies but more often distributed through a supply chain. An extreme example is the design of a new aircraft, where thousands of engineers may work together in collaborating companies. Hundreds of other engineers work on the design of a new aircraft engine in a first-tier supplier company. In addition, dozens of engineers work on the fuel pumps as second-tier suppliers, and this company will in turn have its own suppliers.

At the other extreme, a specialist engineer or a graphic designer might work alone on a design. Even in this example, communication takes place in many different forms, e.g. between the graphic designer and the customer. Communication is an essential part of any design process, and problems in design communication can lead to delays, mistakes and even the ultimate failure of this process.

Communication can happen between many different people or groups, such as different engineers, project teams, different departments within one company, or between the company, the supplier and the end customer. It has different directions, such as top-down from manager to design engineer, bottom-up or in-between. Communication can be formal or informal. It can happen at the same time – synchronously – or at different times – asynchronously. Transmitted information can take many different forms, e.g. verbal, written or pictorial.

Design is never a solitary process. It is a social interactive process.

For designers and design managers, it is important to understand how communication works in design, where it breaks down and how it can be managed and supported. This chapter will start off with a short theoretical background on the conceptualisation of communication. The central focus will then be on characterising communication in design and on the ways it can break down. Understanding how communication works and where it breaks down is an important step towards improving it. Finally, active management and support of communication through electronic media will be discussed.

Theoretical background

The following sections will set the scene by starting with a short discussion on data, information and knowledge and then introduce some general characteristics of communication.

Communication is about exchanging data and information, as well as creating knowledge. These are broad, abstract, complex and multi-faceted concepts and thus difficult to define. The question of their definition has occupied the minds of philosophers since the classical Greek era and has led to many epistemological debates. A commonly held view is that data consist of raw numbers and facts whereas information is defined as knowledge only once it has been understood and authenticated (Ahmed *et al.*, 1999).

In contrast, Alavi and Leidner (2001) suggest that the presumption of hierarchy from data to information to knowledge is inaccurate. Similarly, Dahlbom and Mathiassen (1995) argue that data, information and knowledge correspond to different forms of human activity. They contend that data are a formalised representation of information, and that information is essentially a charting of knowledge within a shared practice – where the reliance on shared practice and experience of situation is the key.

Tuomi (1999) provides an alternative view, arguing that the assumed hierarchy from data to knowledge is actually inverted. Knowledge must exist before information can be formulated and data can be measured to form information. Furthermore, 'raw data' does not exist *a priori*. Knowledge processes are always employed in identifying and collecting even the most basic data.

However, in the everyday language of design the terms data and information, as well as information and knowledge, are used fairly interchangeably for items of communication. This is because the same item can be data, information and knowledge at different times and to different people. This chapter will therefore refer to all items of communication as information.

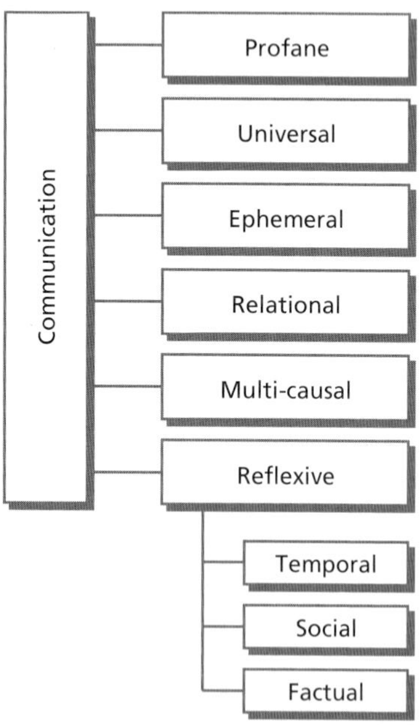

9.1 General characteristics of communication

General characteristics of communication

Sociologists have long tried to define communication, but find it difficult to come up with simple models. Merten (1999) points out that communication is profane, universal, ephemeral, relational, multi-causal and reflexive. His characterisation of communication is summarised in Figure 9.1. Everybody communicates all the time in many different ways, so we can never measure the quality of communication in general terms. Most of what we express only has meaning for a very short time. Even if we analyse a particular aspect or time span, much of the relevant communication cannot be captured.

Furthermore, communication is relational. It is a process that can never be attributed solely to the communicators (sender and receiver), nor to the message, but occurs as the specific relation between these units. Communication, therefore, cannot fully be modelled as an object but only as a relational category. Communication is also never static ; it is dynamic. It is multi-causal and inherently reflexive, *i.e.* influenced by other communication acts in three fundamental dimensions (Merten, 1976):

- *Temporal* – communication processes directly impact themselves.
- *Factual* – all communication processes require factual statements, but also require meta-statements to make communication understandable and targeted.
- *Social* – communication processes are orientated towards others.

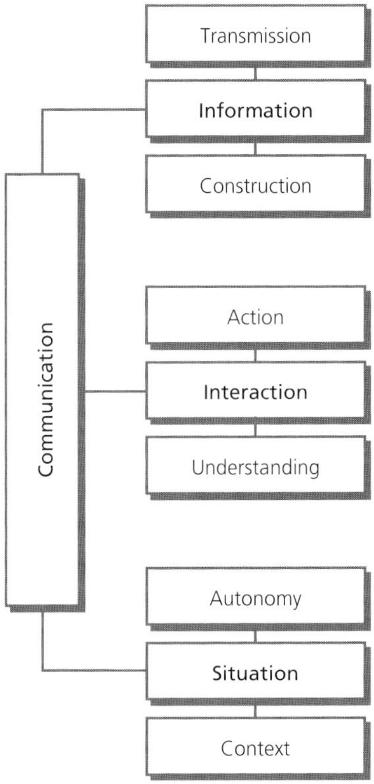

9.2 Theories of communication

Theoretical conceptualisations of communication

After having introduced general characteristics of communication, the following sections group existing theories according to their respective foci on information, on interaction or on the situation in which communication takes place, as illustrated in Figure 9.2. All these viewpoints are combined in a systemic view of communication.

Information-centred theories

Information-centred theories focus on the transmission and processing of information. There are basically two ways of looking at a communication process from an information-centred viewpoint. One either focuses on transmission of information, adapting reductionist models from electronic data transfer in electronics to human communication, or one focuses on the creation of sense by the communicators, following social constructivist viewpoints.

Models of information transmission, which were originally developed for noise reduction in electric data transmission, are simple sender–receiver–models that assume that the sender sends a message through a channel that is received by the receiver like an object. The Shannon–Weaver model (Shannon and Weaver, 1949) proposes that a basic system of communication may be thought of as being composed of five elements: source, transmitter, channel, receiver and destination as schematically represented in Figure 9.3.

The information source produces a message, which is encoded into a signal, which is transmitted across a channel. The receiver decodes the signal and the message arrives at the destination (Shannon and Weaver, 1949). A similar model is the 'conduit metaphor' (Reddy, 1979), where the sender is seen as putting a message and its meaning in a tube and out it comes at the other end.

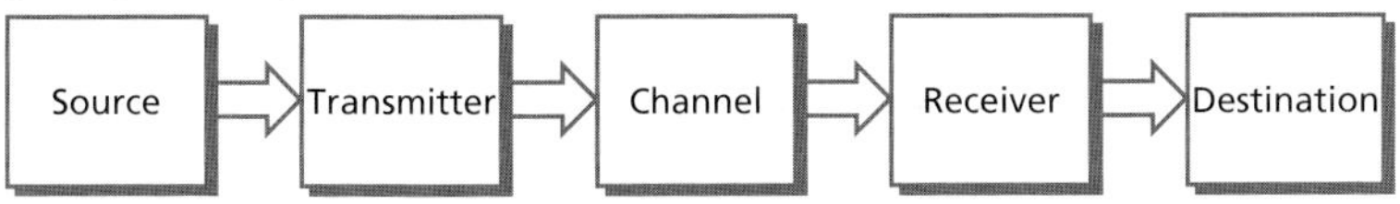

9.3 Shannon and Weaver's model of communication

Underlying both models is a mechanical, linear view of communication, which is seen as a one-way process. While these models are intuitive and shed light on many difficulties in communication, they ignore the individual cognitive processes of the communication partners and their interaction and do not consider the factors outlined in Figure 9.1.

Transmission theorists model communication as the passing on of information.

Whereas the transmission theories model communication in essence as the passing on of information, the constructivists model communication as the construction of new knowledge via the communication partners. What looks like a pure information flow on the surface is actually a process where the information flow is taken as a trigger to create social reality.

Delia (1977) contends that an essential determining factor of the communication process is the cognitive complexity of the communication partners, which becomes visible in interaction processes. The communicators interpret the received information. Underlying this interpretation – or the act of making sense – are cognitive schemes and categories, which develop as a result of interactions between the individuals and the challenges in their environment.

Interaction-centred theories

In addition to information-centred theories, which are mostly interested in internal processes within a communicator, interaction-centred theories focus on the relationship between communicators. For the latter theories, again, one can basically differentiate between two approaches. On the one hand, there is the view that interaction forms one unit insofar as the goals of the individual actions of the parties concerned will be reached. On the other hand, there is the view that communicative interactions represent an understanding between the partners on the basis of joint conventions.

Constructionists model communication as the construction of new knowledge via the communication partners.

In general, communication involves some kind of interrelation between several participants. Advocates of the speech–act theory (Austin, 1962; Searle, 1969) focus their analysis of the communicative interaction on the individual contribution of the participants. Communicative interaction is seen as a sequence of individual actions of the participating communication partners.

Goodwin (2000) sees the interrelation as co-operative, where the partners strive to pursue a common goal to reach consensus. This requires common understanding of conventions and the rules and obligations that can be inferred from them. Conventions form a specific basis for communication that is applicable to all members of one culture.

Habermas (1981) envisages a universal basis for communicative interaction by calling each participant of the interaction to agree on the validity of his or her respective expressions. However, the willingness to support a common understanding is just one among many other dispositions.

Psycholinguistic researchers, such as Herbert Clark, assume that speakers and listeners understand each other because of a common ground. This common ground is generated through cooperation between speakers and listeners (Clark and Carlson, 1982; Clark and Murphy, 1982; Clark *et al.*, 1983; Clark and Brennan, 1991) or, as later proposed, through coordination (Clark, 1992, 1996).

Situation-centred theories

The third basic dimension of communication is the communication situation, which influences the information transmission and the interaction. The situation does not just take the directly perceivable environment into account, but also the wider context, such as the nature of the team within which communication takes place, the organisation and the social background. There is as much debate as to how this happens as there is consensus over the fact the environmental factors influence the communication process. The classical linguistic tradition (Levelt, 1989) processing of speech is seen as relatively independent of the communicative situation. Contextual theories, such as ethnography (Hymes and Gumperz, 1972) assume that communication is dependent upon context variables and thus varies strongly.

Systemic view of communication

To capture the richness of communication fully it is necessary to take a systemic view, which concurrently incorporates the concepts of information, interaction and situation. Communication is seen as a process. For the purpose of human-to-human communication there are at least two participants, who can both be senders and receivers. The communicators are seen as interacting cognitive systems distinct from, but also influenced, by their emotions and their environment; see Figure 9.4. Interactions do not necessarily have to be visible from the outside, although they are often accompanied by gestures.

> Communication is at the same time a social, a cognitive and an emotional act.

Communication is at the same time a social, a cognitive and an emotional act. It is social because different people are interacting with each other in a social and political context. Communication is cognitive because what people say and how people make sense of what they perceive depends on their mental models and prior realm of semantic, syntactic, and pragmatic knowledge. Communication also depends on how we feel about people we communicate with and about the content of the message we are trying to convey. As far as the emotional aspect is concerned, it is worth mentioning that trust and amicability, the willingness of someone to hear the thoughts of another person with good will, is often central to any communicative act (Cockburn, 2001).

As depicted in Figure 9.4, several factors influence each communication process. Because group cohesion needs to remain high, particularly if work is to be brought in on time and on budget, the choices the communicator makes will be influenced strongly by the norms and values of the project team (Maletzke, 1963). Furthermore, the individual members of the teams have different educational backgrounds and thus are entrenched in their own object world (Bucciarelli, 1994). The next layer is the organisation. Features such as the size, organisational set-up, policies and leadership style influence the communication process. A third layer is the environment, which includes society as a whole, the communicator's immediate community, the groups he or she belongs to and the individuals he or she interacts with. All these layers influence communicative behaviour.

After having outlined theoretical conceptualisations of communication and the way the concept of communication is used throughout the chapter, the following sections concentrate on those aspects of it that are specific to design, leaving aside universal factors, such as emotion.

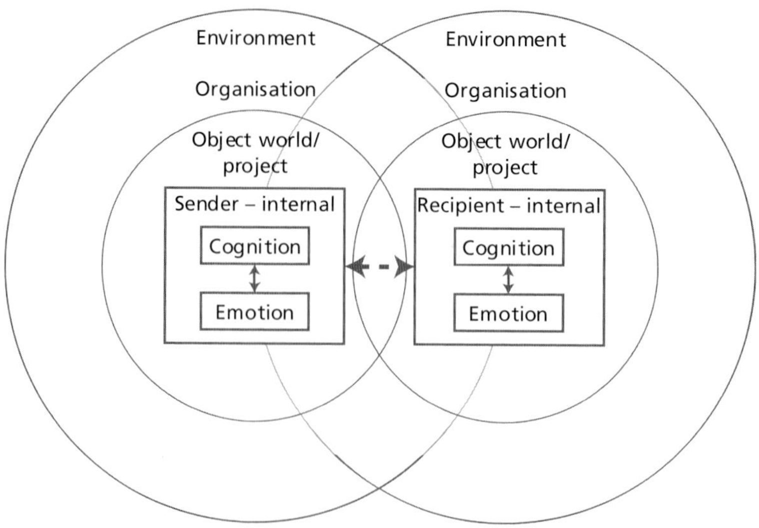

9.4 A systemic view of communication

Characterising communication situations in design

Design in its broader cultural sense includes all activities involved in the generation of a complex product, which can entail many different communication situations.

At present there is no complete taxonomy of different design tasks. Frost (1994) provides a useful classification of products, which can be used to assess characteristics of their design processes. While there have been many attempts made to describe engineering design in general in taxonomic form (e.g. Ullman, 1992), detailed taxonomies address only specific issues. For instance, Ullman (1995) classifies decision problems in design; and Kaplan *et al.* (1992) are concerned with the information requirements of tasks requiring interaction between designers.

In contrast to Kaplan *et al.* (1992), we are looking at communication activities within large design processes, where the mode of collaboration is not necessarily predetermined by the task but rather by the organisational set-up.

Product information is communicated directly according to the stage of the design development and the intended recipient of the communication.

How designers communicate, and how designers could communicate, has been studied from a variety of intellectual perspectives. But discussions of collaborative designing usually consider only a handful of activities; and support systems for cooperative design are developed for specific scenarios, whereas consideration of a wider range of uses could reveal a broader range of requirements and potential pitfalls. Product information is communicated differently according to the stage of the design development and the intended recipient of the communication.

Pahl and Beitz (1996) distinguish between original, adaptive and variant design. Each of these modes of design involves different types of communication, and very often each mode can be found in the same company, even where the company is working in an established product domain. For example, routine design work in a company may involve the rapid development of a current standard design, perhaps using parametric design or knowledge-based engineering techniques. This would be an example of variant design, and the communication issues would involve the rapid population of design automation tools and rapid communication with customer and manufacturing organisation.

With a longer term focus, the same company may incrementally adapt its design approach to improve product performance and reduce product costs. The communication issue in this case concerns the application of design evaluation tools and design for X methodologies (where X is such issues as manufacturability, maintainability, *etc.*).

Concurrently, the company may explore radical approaches to its design challenges, and will seek to be aware of disruptive technologies that will impact on its markets. Communication, in this case, concerns awareness of leading-edge technological developments, and communication within a design team to allow novel concepts to be explored.

Studies of communication in collaborative designing

Research on design collaboration has largely focused on team meetings. Many studies have given a group a design brief and analysed the resulting design activities (see Cross *et al.* (1996) for 20 detailed analyses of the same episode of collaborative design by various different researchers). Design conversations almost always employ sketches, drawings, prototypes or other visual referents, either actual or imagined (Eckert and Stacey, 2000).

Communication in joint designing is multi-modal: speech, drawing and gestures are used in combination, with each channel used to explicate and clarify what is expressed in the others (Bly, 1988; Tang, 1989; Minneman, 1991). This multi-modal communication involves the use of argumentation strategies and rhetoric and subtle modulation of the degree of commitment with which a proposal is put forward (Brereton *et al.*, 1996). Minneman (1991) points out that describing the design itself is just one aspect of design discourse. He classifies the content of design communication according to a 3-by-3 matrix (see Figure 9.5).

Communication can be about an artefact, a process, or a relation (between individuals or groups, or between people and tools, rules, representations, etc.). It can describe the state that something is now in, or how and why something got to be the way it is (making sense), or how something might or should develop (framing the future).

There has been extensive research on how using computer technology influences people's interactions in meetings. One important finding is that people will exploit ways to communicate that do not exist in conventional

Facets / Trajectories	An artefact	A process	A relation
State of			
Making sense of			
Framing futures of			

9.5 Framework for considering design communication (Minneman, 1991)

face-to-face interactions – for instance, by drawing or gesturing in the same place at the same time in a virtual workspace (Bly and Minneman, 1990). Another is that using group support systems influences what happens in meetings, but how they change what happens depends on both the technology and the purpose of the meeting; for instance, decision-making is different from idea generation (Huang and Wei, 1997).

Minneman (1991), Bucciarelli (1994) and Henderson (1999), among others, have studied large-scale engineering processes as participant observers. They report that complex designs are developed largely through social processes of argumentation and negotiation. They view designs as arising through a process of negotiation between participants, where information is actively communicated and made sense of, rather than seeing it as passively transmitted through an organisation. However, this view also downplays the role of a designer working alone communicating with himself/herself – sketching, modelling, *etc.* and then needing to communicate externally to pass on the results of the work.

A significant aspect of many design processes is the handover of information, where one designer has generated a specification that another member of the team is supposed to implement. In these cases designers do not wish to enter a negotiation process. Henderson (1999) shows that graphic representations play a critical role in structuring the design process and conveying information between people with different knowledge and responsibilities.

Eckert (2001) has analysed the communication breakdown during design handover and showed that remarkably little conversation takes place to resolve ambiguity in specifications when designers are not aware of multiple interpretations. In a handover situation, ambiguity in representations can seriously decrease the efficiency of a process (Stacey and Eckert, 2003), while ambiguity can be a driver for creativity in joined design situations, as argued in Minneman (1991). For example, creativity is enhanced by allowing designers to reinterpret sketches. Schön (1983) views this as interacting with the sketches as in a conversation: the designers see more in their sketches than they put in when they drew them, and these insights drive further designing.

Designers communicate through reference to objects.

A similar phenomenon occurs when designers communicate through reference to objects (Eckert *et al.*, 2003), when the listeners might pick up on a different aspect of a reference design than that intended by the speaker, or might break out of a mental fixation (Jansson and Smith, 1991) by being provided with an alternative frame of reference. The effectiveness of communication through references to objects varies with the objects and intentions of the speaker and the knowledge of the recipient (Eckert and Stacey, 2000). This discussion of ambiguity shows how difficult it is to understand the characteristics of communication in general. It is necessary to differentiate between different communication scenarios, modes of interaction and intentions.

The following section provides a classification of different communication situations according to the dimensions of variability, the purpose and the content of the communication act (Figure 9.6).

Interaction scenarios

The situations in which designers interact vary in a large number of ways. The dimensions of variation listed in Figure 9.7 are not orthogonal: common situations have related values along a number of the dimensions. This classification from Eckert *et al.* (2001) has been derived from industrial observations, and thus has more of a cognitive and social bias than those of others who are also considering the management of information. For example, Ostergaard and Summers (2003) started from communication between intelligent agents in a computer program.

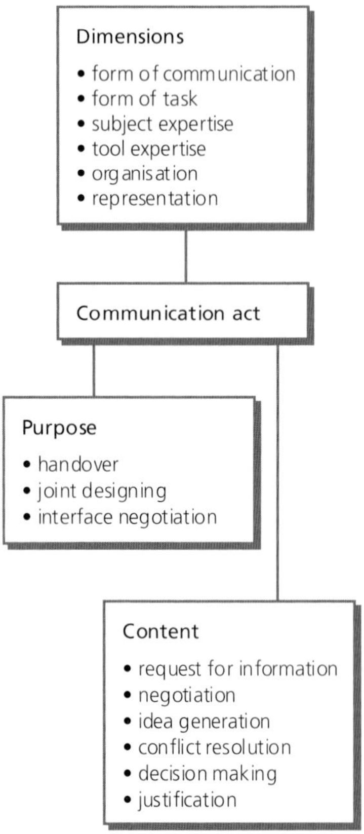

9.6 Classification of communication acts

These different situations can create different types of communication behaviour and, therefore, breakdown. For example, it is intuitively obvious that communication between people with the same expertise, who work jointly on a problem in the same room, is quite different from communication between people from different countries who have never met and come from very different lines of work.

Just a few of the dimensions of the communication scenarios listed in Figure 9.7 determine most of the characteristics of an interaction situation. They define common interaction scenarios, which turn up in many different industries. These scenarios reflect typical work situations, requiring their own support tools and methodologies. One way to classify scenarios is by the way that inputs to the tasks of the participants are related.

Handover

Handover situations are scenarios in which a person undertakes a design task and finishes it as far as possible, then passes on the design to another specialist, through a written or oral specification. The expectation is that the next person will do what is required within the specification rather

Form of communication

- *Place*. Participants are face-to-face <-> participants are geographically remote
- *Time*. Communication is interactive in real time <-> communication is asynchronous
- *Size*. Interaction between pair <-> interaction between many
- *Identity*. Recipients are known (conversation, private notes) <-> recipients are unknown (record keeping, subcontractors to be found, open audience)

Form of task

- *Objective of task*. Generation of ideas or alternative solutions <-> convergent problem solving *vs* decision making from alternatives *vs* acquisition or imparting of pre-existing information
- *Division of decision-making*. Joint problem solving <-> negotiated handover <-> sequential problem solving
- *Hierarchy of decisions*. Different participants' tasks are of equal importance <-> some tasks are subordinate to others
- *Duration*. Interactive or communicative activity is brief <-> activity is extended.
- *Information type*. Facts, proposals, specifications <-> opinions or judgements or prognoses <-> problem-solving strategy advice
- *Time pressure*. Task is time critical <-> task is not urgent

Subject expertise

- *Equality of expertise*. Participants have equal levels of expertise <-> Some participants are more knowledgeable than others (one important interaction type is apprentice consults more experienced colleague)
- *Balance of Expertise*. Participants have shared expertise (and use the same concepts and can interpret each other's terms and representations) <-> participants have complementary expertise
- *Mental representations*. Participants conceptualise topic in similar terms <-> participants conceptualise topic in different terms
- *Familiarity*. Participants know each other <-> participants cannot make assumptions about others' knowledge
- *Context*. Participants share contextual information <-> participants have different (or no) knowledge of the context

Tool expertise

- *Competence with groupware*. Experienced frequent user (skilled at and comfortable with using the medium) <-> novice or infrequent user of medium.

Organisation

- *Hierarchy*. Participants at same level of hierarchy <-> participants have different status
- *Interest*. Participants from same company <-> participants working for different companies
- *Security*. All information can be shared <-> some information must not be shared (for instance in dealings with suppliers, or with people without security clearance)

Representation of information

- *Medium*. Speech, gestures, hand-drawn sketches, hardcopy printouts of text files or CAD models, Web pages, shared files, physical objects such as prototypes...
- *Form of information*. Text, data plots, tables, diagrams, code, photographs...
- *Notation*. Some fields have alternative notational conventions for the same information

9.7 Dimensions of communication scenarios

than advancing the design by changing the specification. The participants are often collocated but communication is asynchronous.

Later tasks are often seen as subordinate, so that two-way negotiations are excluded. For example, knitwear designers give their technicians specifications, without much discussion unless problems occur (Eckert, 2001). Such over-the-wall sequential design processes are still quite common in engineering, especially when designs are handed over to suppliers or contractors.

Joint designing

Joint designing refers to scenarios in which a group of people work on one problem together. Typically they work at the same time in the same room. Individuals might work on parts of the problems, but they have easy access to each other and discuss issues as they occur. Joint designing is typically done by groups of people with similar expertise, who are solving a problem that concerns all.

Over-the-wall sequential design processes are still quite common in engineering, especially when designs are handed over to suppliers or contractors.

The team members usually share a lot of background knowledge and awareness of context, and often get to know each other well. They can talk to each other spontaneously and get rapid feedback. For example, knitwear designers work out colour schemes as a group, because they all use the same scheme. In engineering, designers often work jointly during conceptual design, when even a complex problem is addressed by a small group.

Interface negotiation

In concurrent design, there are different scenarios in which people from different fields of expertise work on a design at the same time. Their tasks have mutually dependent inputs. To achieve full concurrency, they need to work with estimates of parameter values to achieve mutually consistent solutions to their individual problems. In reality, most processes give priority to some tasks and decisions, and stagger the beginning of the tasks. It is well recognised that concurrent design processes work best with collocated project teams. Communication occurs informally, through one-to-one conversations as well as in meetings.

Episodes of interaction can have a variety of purposes, even within a meeting with a different primary purpose. The types of discussion listed below can be about most of Minneman's nine classes of subject matter (Figure 9.5).

Request for information

Designers frequently find they need more information, and usually their main source is their colleagues. A pure information request is more likely to occur in design handover or concurrent situations than in joint design sessions.

Negotiation for clarity and negotiation of constraints

Participants in a discussion must make sure that they understand each others' positions – that is, achieve compatible interpretations of the situation. This often requires understanding the constraints that the others must meet in order to understand what the constraints on their own activities should be.

Thus, negotiation for clarity often leads to a negotiation over constraints. This is particularly important when designs are handed over (not necessarily in a linear process) from one specialist to another who is doing an equally important task independently.

Designers frequently find they need more information, and usually their main source is their colleagues.

Idea generation

In many design processes that are essentially sequential, idea generation is undertaken as a joint activity in a meeting, because designers need each other's input before committing time and resources to any particular solution. Designers often reuse ideas from past designs or other sources; how much they refer to visual props depends on how much they need to explain ideas with reference to their sources.

Conflict resolution

Meetings are often set up to resolve conflicts between elements of a design, typically through real-time discussion. Conflict resolution situations vary according to whether there is an authority capable of arbitrating or imposing a decision on conflicting parties.

Decision making

Much design comprises an exploration of possibilities followed by a decision on which avenue to follow. Decisions need to be made about what trade-offs are necessary, and often about conflict resolution, as well as about concepts. If individuals make decisions on their own, then they have to justify them (see below). In meetings, decisions can be made jointly or by individuals higher up in the hierarchy.

Justification

Designers must often justify their solutions or decisions, either orally, in meetings, or in reports. The recipient cannot be assumed to have the same knowledge as the person who has to justify the solution. Justifications may be made to colleagues, bosses or outsiders; and the explanations must be pitched to the recipients' understanding. Specific justifications are often necessary in handover activities.

Each individual engages in most of these communication situations as part of their normal work. Designers use different channels on different occasions to convey different kinds of design information. For example, a designer might engage in joint problem solving with his boss in a face-to-face meeting involving conversation and sketches, when they are negotiating over the constraints on a particular problem. The designer then works on his own using a CAD system. When he has a question he sends an e-mail message or picks up the telephone. Later he has to return to the boss to justify in another face-to-face meeting the design that he has come up with.

Communication breakdown

Communication breakdown can have multiple causes.

All the communication situations discussed in the previous section carry their own problems and difficulties. In many practical design situations it is difficult to identify communication problems as such or find their root causes, because they are so strongly interwoven with other process issues. Even if a communication problem is detected, companies often struggle to see where it comes from; sometimes it is the effect of factors such as management structures, at other times the problem is purely personal.

This section discusses causes for communication breakdown, which might be worth considering when a problem is encountered (Eckert and Stacey, 2001).

Constructing meaning

At the start of the chapter, a systemic view of communication was discussed in a theoretical manner. This section discusses in more practical terms the stages a designer goes through to make sense of design information. Successfully constructing an understanding of what to do in a new or changed situation, such as a modification to a design, comprises obtaining the information needed and making sense of it.

Making sense of what you see or are told has three aspects which are inseparable in practice, shown in Figure 9.8: interpreting this information

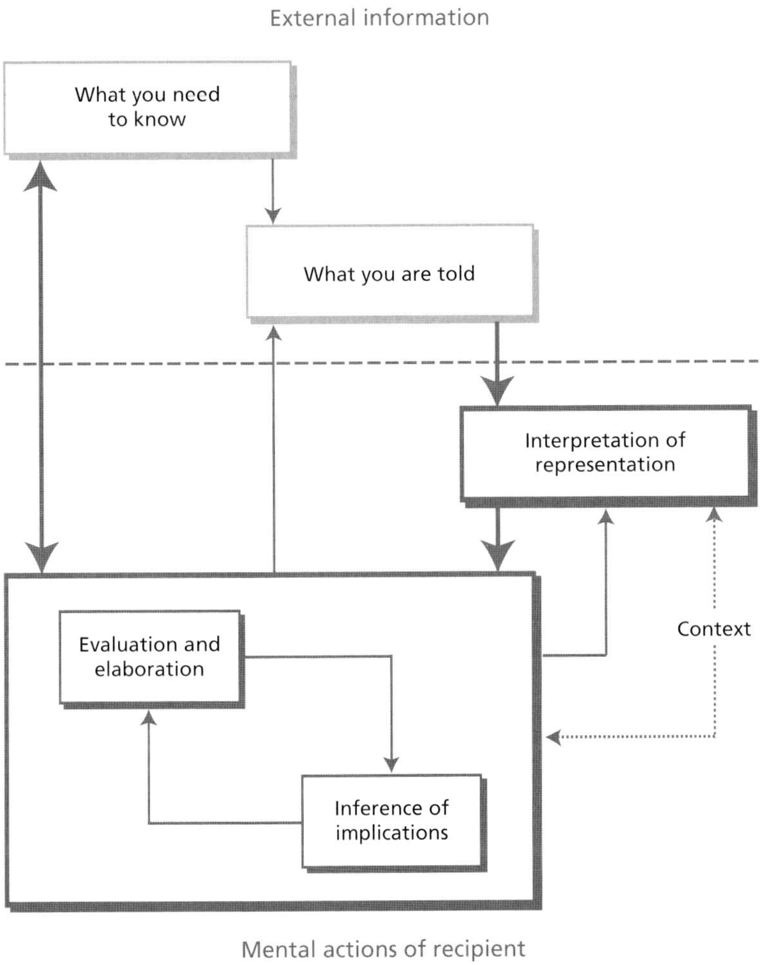

9.8 Recipient's perspective of information transmission (Eckert et al., 2001)
Reproduced by permission of the Council of the Institution of Mechanical Engineers

from the form in which it is represented; integrating it into one's understanding of the situation by elaborating it and evaluating its quality with contextual knowledge; and inferring its implications for one's own tasks and responsibilities, and how to apply it.

This necessarily involves learned interpretation skills, background knowledge and awareness of context, which are different for each participant. A representation of design information might be incomplete, ambiguous or inconsistent, or might obscure aspects of the design. Missing information must be filled in from context, typically with conventional assumptions or default values, which might or might not be right for the problem. If the recipient realises that the information is incomplete or inadequate, he or she will try to find the missing or correct information, either by going back to the person who has provided the information or by looking for other ways to find it.

Problems often arise simply because designers are not told what they need to know.

Causes of communication breakdown

Communication breakdown can have multiple causes. They are not independent, but they are listed separately here because their causal connection is not unique. For example, a lack of overview of the product can lead to designers not recognising that they are missing information, so that they do not follow it up. Alternatively, missing information leads to a lack of overview.

Not understanding the big picture

It is extremely difficult for an individual designer to fully understand a complex product or the process by which it is generated. Of course, complex products are decomposed as far as possible into modules with relatively simple interactions, to minimise the complexity of the design process. However, connectivity inevitably remains an important issue.

Designers and managers often have only localised knowledge of the processes they are involved with (*i.e.* processes of the teams they currently work with and processes they have worked with in the past). This lack of overview of the design process means that designers may not understand the context of the information that they are using. In particular, there is often a lack of awareness of:

- tasks that need to be done;
- information history;
- how information is applied;
- changes to processes.

Communication is profoundly influenced by what is communicated and how it is expressed.

Missing information provision

Problems often arise simply because designers are not told what they need to know. Others often do not know what information somebody else requires or do not have the time to talk to their colleagues. Designers often have:

- no feedback on information provided;
- no status information – people therefore often assume that values are exact and put great effort into meeting a seemingly exact target, even though the values could be just estimates or placeholders for other information;

- restricted viewpoint owing to power structure – contractors and suppliers are often excluded from decision-making processes, because they have no official standing in the company hierarchy;
- insufficient information due to confidentiality concerns – contractors or suppliers are often deliberately not given information that might be useful for their tasks, because it is considered confidential (Henderson, 1999).

Communication is a two-way process whereby at least two parties interact with one another according to their own rules.

Information distortion

In complex organisations information is often passed on via several other people before it reaches the recipient. The generator of the information may not know the ultimate recipients, or does not know the recipients' needs, tasks and background, and can thus do little to ensure accurate transmission. The following problems can occur:

- information is oversimplified or corrupted ('Chinese whispers');
- hierarchical communication paths leads to distortion and interpretation of information;
- expertise of intermediary puts a spin on the information.

Interpretation of representation

Communication is profoundly influenced by the subject of the communication act and the form chosen to express it. Any design descriptions only capture part of the object that they are describing, because of the complexity of the product and the richness of the context in which it is used and created. Any description is inevitably also selective, thus remaining ambiguous, leaving scope for interpretation. The representations that designers use to express design ideas and other information, and the representation-understanding skills they possess, have a powerful influence on design communication, because:

- interpretation of ambiguous information is based on context;
- recipients may be unable to extract the required information from the representation.

Communication strategies should be clear, engaging and sustainable.

Managing communication

With a complex phenomenon like communication it is often impossible to fully understand what bears on it; however, in practice it is often useful to think about a situation in the following terms. It is easier to recognise that something is part of a pattern than to see the pattern itself in the first place.

Communication often fails because designers lack awareness of the design process, the tasks and competencies of other designers and the interfaces between them.

Looking for different causes of communication breakdown can help to see it in a less personal way, which might ease the tension involved in difficult communication situations.

While it is hard to come up with universal ways to overcome a particular problem, it often sufficient to draw attention to a specific issue to work out a possible solution. The following sections give an overview of topics which need to be addressed in order to arrive at a necessary and satisfactory level of understanding in managing communication.

To recapitulate, communication is a two-way process whereby at least two cognitive systems interact with one another according to their own rules. The ultimate goal of research on design communication in academia and industry is to improve the design process. It is very difficult to direct and control communication in order to achieve intended results.

There is no definitive solution, partly because communication problems are often closely intertwined with process issues and partly because communication is a multifaceted concept. Yet, a setting can be provided to facilitate and encourage communication.

Improving internal communication

There are several techniques and devices available for improving internal communication which can be part of a communication strategy. The term communication strategy is used to denote a set of methods applied to realise short- and long-term objectives, and can be applied at several levels within the company and at several phases during the design of a product.

Management cannot dictate friendship, but it can create the necessary conditions.

A communication strategy for design does not solely refer to documentation and reporting at the end of phases in the design process, it should also encompass the points mentioned below. Communication strategies should be clear, simple, engaging and sustainable. A carefully thought-through communication strategy does not guarantee, but rather increases, the likelihood of successful communication. In this section, emphasis is put on necessary conditions for improving person-to-person contact.

General awareness

Major sources of information breakdowns have been discussed in the previous section. They need to be counteracted through positive measures. As was argued, communication often fails because designers lack awareness of the design process, the tasks and competencies of other designers and the interfaces between them. This understanding can be enhanced with computer tools showing product or process connectivity (Eckert and Clarkson, 2003; Flanagan *et al.*, 2003). Furthermore, it takes a certain mind set to be a good communicator. Designers must be educated to take responsibility for their general awareness of process, and also for the information needs of others.

Trusting atmosphere

Communication problems often arise from tensions between individuals or groups. While it is difficult to erase deep-rooted personal antagonisms, organisations can actively work on introducing a culture of open exchange of knowledge and ideas and can reward people actively for their willingness to communicate, both formally and informally. As Allen (1977) has shown, engineers keep abreast of their field and get a significant amount of design-related information by contacting their co-workers. He produced evidence that critical information leading to genuine innovation came from outside the immediate work group but from within the organisation.

Since proprietary information must be protected from competitors, bouncing of ideas with peers outside the company is rarely possible. Hence, management should make sure that each individual does not feel inhibited by status or other factors within the company. The design manager needs to ensure that there is an open 'no-blame' culture where team members can express their experience and knowledge freely. Management cannot dictate friendship, but it can create the necessary conditions. Social interactions serve the function of developing interpersonal understanding. The encouragement of social interactions outside the work environment is a mechanism to promote communication within the team.

Buildings and office layout can play an active role in facilitating interaction patterns, and thus communication within the work environment.

Team composition

This chapter has concentrated on the cognitive and social characteristics of communication and the root causes of communication behaviour. Some of these factors can be overcome or improved through suitable team size and team composition; indeed, overall team performance is itself quite dependent on these factors. For a detailed analysis, see e.g. Belbin (1991) and Hurley (1995). Even though the composition of a team is strongly influenced by the organisational structure and the nature of the product and the design process, product design managers should still pay attention to the way the individual team members interact and should use team communication as a factor in selecting team size and composition.

Interface management

The product architecture influences technical communication and interaction among design teams. To illustrate this, Sosa *et al.* (2003) conducted a study in an aerospace company in which they identified the impact of modular and integrative systems on design team interactions. Modular systems are those whose design interfaces with other systems are clustered among physically adjacent systems. Integrative systems are systems whose interfaces are physically distributed or functionally integrated across all or most other systems.

The conclusion of the study was that team interactions between design teams that develop integrated systems are more likely to be predicted by design interfaces than are team interactions between design teams that develop modular systems. As was expected, system boundaries impose architectural knowledge barriers, which inhibit design experts' understanding of certain design interfaces. This results in some team interactions that are not predicted by design interfaces. This work highlights the importance of identifying design interfaces during the project planning stage so that corresponding design team communication is managed efficiently during project execution.

Many companies would benefit from a careful assessment of their communication processes.

Design of office space

Buildings and office layout can play an active role in facilitating interaction patterns, and thus communication within the work environment (Allen, 1977). Penn *et al.* (1999) have found that patterns of space use and movement generated by spatial configuration have a direct impact on the frequency of contact between employees within office-based organisations. The underlying assumption of this study was that spatial patterns affect movement patterns and that movement patterns bring people past other people's workstations. Within existing buildings it is unlikely that one can change the overall structure and distribution of floors. One can, however, directly influence the layout of the office and strategically place interaction-promoting facilities, such as printers and water-coolers, so that they can be shared by several groups whose physical separation might otherwise hinder face-to-face communication. Team managers need to be alert and make the best use of the space available.

Organisational settings

Organisational settings are rarely changed solely to improve design communication and are usually beyond the control of individual design managers. However, an awareness of conflicts that might arise from the organisational settings is important in a project. For example, managers need to acknowledge that designers in a matrix organisation are often required to communicate along both lines of report and in doing so may fail to satisfy both parties.

Another example is the trade-off between project and functional teams: a project-based organisation may have deficiencies in communication amongst functional groups and *vice versa*.

Communication audit

Many companies would benefit from a careful assessment of their communication processes. A communication audit will produce a clearer understanding of how communication really works and the degree to which it satisfies the needs of the organisation. From this, ideally, flow a number of possibilities, such as improved productivity, potential discovery of hidden sources of information, better context awareness, more efficient use of time, transparency of processes, connectivity of tasks and improved morale.

In addition to the analysis of communications media, patterns, flow, channels, and technologies, a communications audit examines content clarity and effectiveness; information needs of individuals, work groups, departments and divisions; non-verbal communications and corporate culture issues; and communication impacts on motivation and performance. A communication audit could range from an informal internal study to a formal process undertaken by internal or external experts. In a more or less structured form it would go through four stages (Figure 9.9):

1. At the planning and design stage, the audit's scope and goals, unit of analysis, types of communication to be audited, methods to be used and timeframe and budget are determined.
2. The fact-finding stage begins with informal exploratory research and often moves to formal, scientific methods of gathering information. The two informal, exploratory research methods used most often are in-depth interviews and focus groups. The formal, scientific measurement method used most often for primary source research is a survey. Another method would be to conduct observations.
3. The analysis and reporting stage establishes how well the communications satisfy the needs of the organisation and the stakeholder groups today and how well these communications will serve changing needs in the operational future (1–2 years).
4. Based on the findings, the recommendation stage suggests guidelines and recommendations on how to improve communication.

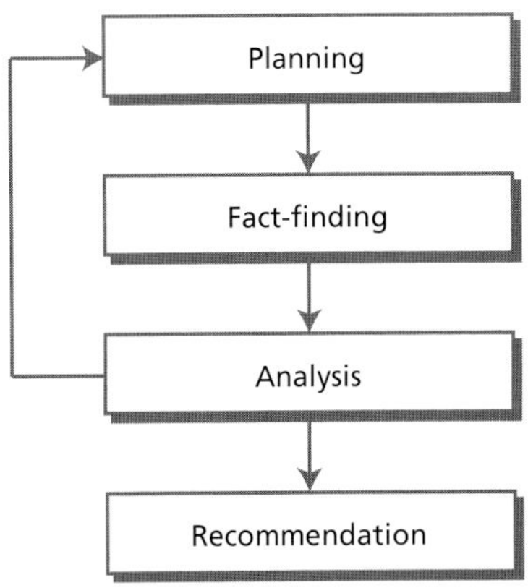

9.9 Stages of a communication audit

A communication audit is not always a linear and straightforward process. Iterative loops can occur between the stages, especially while analysing the acquired facts.

Conduct of a communications audit is usually performed by outside consultants because of their professional experience, expertise and objectivity. In addition, an independent third-party's guarantee of confidentiality often produces a higher level of trust from employees and other stakeholders in in-depth interviews, focus groups and surveys. This often produces more open, candid, real-world information than that which can be acquired by in-house research efforts.

Understanding specific communication situations

The same techniques are used in the academic community to understand how communication works as a social and cognitive process, so that tools and techniques can be developed to improve communication or aid the process it is part of. Many studies of communication involve a combination of methods. These fall essentially into three different categories (see Patton (1990) for a discussion on qualitative research and evaluation methods).

Technology plays a major role as an enabler of communication.

Observations

Observations allow for the study of the social basis of communication. Observers can see how groups and individuals act in their own context of work. Observations come in different guises. Ethnographic studies (Bucciarelli, 1994; Agar, 1996) try to look at cultures from outside, but at the same time try to understand the insider's view point. Action research, on the other hand, involves active participation in a process with reflection afterwards.

Experiments

Experiments allow a previously specified hypothesis to be tested. In the psychological tradition, context is made explicit and controlled as far as possible. Experiments can give insights into design cognition and those universal aspects of communication which are fairly independent of a specific context, such as the role of gestures. Design researchers often set up experiments, in which individuals or groups of designers are given a brief and recorded while they are designing, where the record is later analysed. These situations are, however, somewhat artificial, because design communication is very strongly influenced by the objects and terminology that designers have encountered during their previous working life.

Interviews

Interviews can be a short and efficient way to gain access to people's perception of communication behaviour. People are often happy to explain what goes on in an organisation, especially how and when communication breaks down. It is often difficult to get the real story from an individual's perception; however, a series of interviews can be one of the fastest and most efficient ways to find out what is going on in a company.

In addition to the suggestions made above, which focus on the human aspects of communication, technology plays a major role as an enabler of communication.

Supporting communication with technology

The generic term for many of the information and communications technologies that are used to support communications in design is computer supported cooperative work (CSCW – note that the term 'collaborative work' is also used). The term is frequently used synonymously with groupware, defined by Ellis *et al.* (1991) as "…computer-based systems that support groups of people engaged in a common task (or goal) and that provide an interface to a shared environment".

Figure 9.10 presents a variant of the space and time categorisation of CSCW originally presented by DeSanctis and Gallupe (1987) and refined by Johansen (1989). In design communication terms we can identify technologies that simply support the development of distributed communities

		Same time	Different but unpredictable time	Different and predictable time
		Synchronous	*Asynchronous*	
Same place		Face-to-face meetings and discussion aids	Physical bulletin and notice boards	Team meeting rooms and discussion areas
Different but predictable place	*Distributed*	Voice/video conferencing, virtual meeting rooms; shared applications	Messaging systems e.g. e-mail	Multi-user editors and collaborative writing tools
Different and unpredictable place	*Distributed*	Interactive multicast seminars	Virtual bulletin and notice boards	Workflow systems

9.10 Approaches to CSCW

and the sharing of encoded knowledge by the community, including electronic communications systems (mail systems, facsimile transfer, voice and video conferencing) and shared workspace systems (virtual meeting rooms, remote screen sharing and electronically aided intelligent whiteboards (shared applications). These are technologies that have already led to significant practical applications (e.g. Lotus Notes, 2003; Microsoft Exchange, 2003).

Mail systems, mail directories and workflow systems are now used routinely. Video conferencing is now employed in many companies, and low-cost hardware capable of transmitting highly compressed video images along telephone connections or packet-switching networks between PC computers is available. High-speed digital communication allows designers on different sites to work simultaneously on the same CAD model, and at the same time to have video and audio communication as well as the use of a shared whiteboard for drawing sketches and posting images. Research programmes demonstrated this capability in the mid 1990s (SMAC, 1995), and more recently there have been a number of programmes of shared distributed design work in academia and industry (Gomes *et al.*, 2001; Thomson *et al.*, 2001). The topic is likely to be of increasing importance as design is distributed between collaborating companies that are located throughout the world.

The design of any complex product is an inherently social process in which communication plays a vital role.

The key issue in the successful application of these CSCW technologies is the extent to which they provide a satisfactory alternative to direct, face-to-face communication, as studied for example by McGregor *et al.* (2001) and by Kunz *et al.* (1998). The emerging view appears to be that present-generation systems for handling text (e-mail, message boards) are becoming *de facto* mechanisms of working even for quite closely collocated teams. Voice and video communication is satisfactory for routine working, but for critical situations involving groups of people, and in particular where there are cultural differences, face-to-face communication is preferred.

E-mail and conferencing systems are entirely passive transmitters of information. There are, however, a number of CSCW techniques that themselves begin to incorporate encoded knowledge. Important amongst these are group activity support systems – including workflow systems that enable electronic documents to be sent on predefined routes through organisations (*i.e.* pushed), co-authoring tools for the joint writing of documents, decision-support tools to help group decision-making, and idea-generating and prioritising tools to help group creativity.

In the design context, workflow techniques are beginning to be applied in highly structured design tasks such as document sign-off (in particular associated with commercial product data management systems), and research programmes are addressing their application in less-structured parts of the design process, in particular where dynamic reconfiguration of processes is necessary (Clarkson and Hamilton, 2000).

Techniques of the World Wide Web are becoming increasingly important for the sharing of information within design organisations. Documents are routinely organised into company Intranet pages and engineering information portals. Such approaches tend, however, to require rather centralised creation and management of the content. Tools that support a more collaborative approach to content creation and management that may be more suitable for design team use include Web logs and Wikis – server software that allows users to freely create and edit Web page content and organisation using any Web browser and on the fly (Wiki, 2003).

Conclusions

The design of any complex product is an inherently social process in which communication plays a vital role. Communication is a multi-faceted phenomenon that can be characterised in many different ways. This chapter provides a characterisation and classification of communication, together with an overview of methods to improve it and to provide computational support for collaborative working. Its overall aim is to provide practitioners with a conceptual understanding of what happens in a communication process, so that they can draw their own conclusions and find a solution for problems in their context.

References

Agar M (1996) The professional stranger: an informal introduction to ethnography. University of Maryland Press

Ahmed S, Blessing L, Wallace K (1999) The relationships between data, information and knowledge based on a preliminary study of engineering designers. ASME DTM'99, Las Vegas, NE, USA

Alavi M, Leidner DE (2001) Review: knowledge management and knowledge management systems: conceptual foundations and research issues. MIS Quarterly, 25(1): 107-136

Allen TJ (1977) Managing the flow of technology. MIT Press

Austin JL (1962) How to do things with words. Clarendon

Baldwin AN, Austin SA, Hassan TM, Thorpe A (1999) Modelling information flow during the conceptual and schematic stages of building design. Construction Management and Economics, 17: 155–167

Belbin RM (1991) Management teams. Butterworth-Heinemann

Bly SA (1988) A use of drawing surfaces in different collaborative settings. CSCW'88, Portland, OR, USA

Bly SA, Minneman SL (1990) Commune: a shared drawing surface. COIS'90, Boston, MA, USA

Brereton, MF, Cannon DM, Mabogunje A, Leifer LJ (1996) Collaboration in design teams: mediating design progress through social interaction. In: Analysing design activity. John Wiley

Bucciarelli LL (1994) Designing engineers. MIT Press

Clark HH (1992) Arenas of language use. University of Chicago Press

Clark HH (1996) Using language. Cambridge University Press

Clark HH, Brennan SE (1991) Grounding in communication. In: Perspectives on socially shared cognition. American Psychological Association

Clark HH, Carlson TB (1982) Hearers and speech acts. Language, 58: 322–373

Clark HH, Murphy GL (1982) Audience design in meaning and reference. In: Language and comprehension. North-Holland

Clark HH, Schreuder R, Buttrick S (1983) Common ground and the understanding of demonstrative reference. Journal of Verbal Learning and Verbal Behaviour, 22: 245–258

Clarkson PJ, Hamilton JR (2000) 'Signposting': a parameter-driven task-based model of the design process. Research in Engineering Design, 12(1): 18–38

Cockburn A (2001) Agile software development. Addison Wesley

Cross NG, Christiaans HHCM, Dorst K eds (1996) Analysing design activity. John Wiley

Dahlbom B, Mathiassen L (1995) Computers in context: the philosophy and practice of systems design. Blackwell

Delia JG (1977) Constructivism and the study of human communication. Quarterly Journal of Speech, 63: 66–83

DeSanctis G, Gallupe B (1987) A foundation for the study of group decision support systems. Management Science, 33(5): 589–609

Eckert CM (2001) The communication bottleneck in knitwear design: analysis and computing solutions. Computer Supported Cooperative Work, 10(1): 29–74

Eckert CM, Clarkson PJ (2003) If only I knew what you were going to do: communication and planning in large organizations. In: Methods and tools for co-operative and integrated design. Kluwer Academic
Eckert CM, Stacey MK (2000) Sources of inspiration: a language of design. Design Studies, 21: 523–538
Eckert CM, Stacey MK (2001) Dimensions of communication in design. ICED'01, Glasgow, UK
Eckert CM, Clarkson PJ, Stacey MK (2001) Information flow in engineering companies: problems and their causes. ICED'01, Glasgow, UK. Figure 9.8 has been reproduced from Design management - process and information issues (ICED01), 2001, pp 43-60, Figure 1
Eckert CM, Stacey MK, Earl CF (2003) Ambiguity is a double-edged sword: similarity references in communication. ICED'03, Stockholm, Sweden
Ellis CA, Gibbs SJ, Rein GL (1991) Groupware: some issues and experiences. Communications of the ACM, 34(1): 39–58
Flanagan T, Eckert CM, Clarkson PJ (2003) Parameter trails. ICED'03, Stockholm, Sweden
Frost R (1994) A suggested taxonomy for engineering design problems. Journal of Engineering Design, 5(4): 399–410
Gomes S, Eynard B, Magnon L (2001) A multi-site design experience using a computer supported collaborative work tool. ICED'01, Glasgow, UK
Goodwin C (2000) Action and embodiment within situated human interaction. Journal of Pragmatics, 32: 1489–1522
Habermas J (1981) Theorie des kommunikativen Handelns. Suhrkamp
Henderson K (1999) On line and on paper. MIT Press
Huang W, Wei KK (1997) Task as a moderator for the effects of group support systems on group influence processes. European Journal of Information Systems, 6: 208–217
Hurley R (1995) Group culture and its effect on innovative productivity. Journal of Engineering and Technology Management, 12(1/2): 57–75
Hymes D, Gumperz JJ eds (1972) Directions in sociolinguistics: the ethnography of communication. Holt, Rinehart and Winston
Jansson DG, Smith SM (1991) Design fixation. Design Studies, 12: 3–11
Johansen R (1988) Groupware: computer support for business teams. Free Press

Kaplan SM, Tolone WJ, Bogia DP, Bignoli C (1992) Flexible, active support for collaborative work with conversation builder. CSCW'92, Toronto, Canada

Kunz JC, Christiansen TR, Cohen GP, Jin Y, Levitt RE (1998) The virtual design team: a computational simulation model of project organizations. Communications of the Association for Computing Machinery, 41(11): 84–92

Levelt WJM (1989) Speaking: from intention to articulation. MIT Press

Lotus Notes (2001) http://www.lotus.com

Maletzke G (1963) Psychologie der Massenkommunikation. Verlag Hans Bredow Institut

McGregor S, Thomson AI, Juster NP (2001) Information sharing within a distributed, collaborative design process: a case study. ASME CIE'01, Pittsburgh, PA, USA

Merten K (1976) Reflexivität als Grundbegriff der Kommunikationsforschung. Publizistik, 21: 171–179

Merten K (1999) Einführung in die Kommunikationswissenschaft. 1/1: Grundlagen der Kommunikationswissenschaft. LIT

Microsoft Exchange (2003) http://www.microsoft.com/exchange/

Minneman SL (1991) The social construction of a technical reality: empirical studies of group engineering design practice. PhD thesis, Stanford University, CA, USA

Ostergaard KJ, Summers JD (2003) A taxonomy for collaborative design. ASME DAC'03, Chicago, IL, USA

Pahl G, Beitz W (1996) Engineering design: a systematic approach. Springer

Patton MQ (1990) Qualitative evaluation and research methods. Sage

Penn A, Desyllas J, Vaughan L (1999) The space of innovation: interaction and communication in the work environment. Environment and Planning B: Planning and Design, 26: 193–218

Reddy M (1979) The conduit metaphor: a case of frame conflict in our language about language. In: Metaphor and thought. Cambridge University Press

Schön DA (1983) The reflective practitioner: how professionals think in action. Basic Books

Searle JR (1969) Speech acts. Cambridge University Press

Shannon CE, Weaver W (1949) The mathematical theory of communication. University of Illinois Press

SMAC (1995) Suppliers and manufacturers in automotive collaboration. European Community RACE Research Project R2112
Sosa ME, Eppinger SD, Rowles CM (2003) Identifying modular and integrative systems and their impact on design team interactions. Journal of Mechanical Design, 125: 240–252
Stacey MK, Eckert CM (2003) Against ambiguity. Computer Supported Cooperative Work, 12(2): 153–183
Tang JC (1989) Listing, drawing and gesturing in design: a study of the use of shared workspaces by design teams. PhD thesis, Stanford University, CA, USA
Thomson AI, Ion WJ, Temple BK, Allan M, Kernohan N, Davidson S (2001) VIDEEO - virtual development enterprise for Europe. ICED'01, Glasgow, UK
Tuomi I (1999) Data is more than knowledge: implications of the reversed knowledge hierarchy for knowledge management and organizational theory. Journal of Management Information Systems, 16(3): 103–117
Ullman D (1992) A taxonomy for mechanical design. Research in Engineering Design, 3: 179–189
Ullman D (1995) A taxonomy for classifying engineering design decision problems and support systems. AI EDAM, 9: 427–438
Wiki (2003) http://wiki.org/

Chapter 10
Engineering change

Timothy Jarratt, John Clarkson and Claudia Eckert
University of Cambridge

Change or adaptation has always been a fundamental part of engineering design; the vast majority of product design activity consists of taking a current product, concept or solution and adapting it to meet a new set of requirements. This view, whilst seldom emphasised in text books on design, is supported by a number of authors, for example:

> *...most designing is actually a variation from or modification to an already-existing product or machine.*
>
> (Cross, 1989)

> *History matters – no design begins with an absolutely clean sheet of paper.*
>
> (Bucciarelli, 1994)

" History matters—no design begins with an absolutely clean sheet of paper."
(Bucciarelli, 1994)

From a business perspective, changes to a design are "a fact of life" in taking a product from concept, through design and manufacture and out into the field (Nichols, 1990); they are the rule and not the exception in product development processes in all companies and in all countries (Clark and Fujimoto, 1991). From a high-level viewpoint, changes are made for two reasons: to remove errors from a product (rework) or to improve/enhance/adapt it in some way.

As an example of the importance of engineering change, a survey of German engineering businesses found that approximately 30% of all work effort was due to engineering changes (Fricke *et al.*, 2000); this included rework as well as the adding of functionality to a product. Terwiesch and Loch (1999) reported that engineering changes consumed between a third and a half of the engineering capacity at the firm they examined, along with 20–50% of tool costs (Figure 10.1).

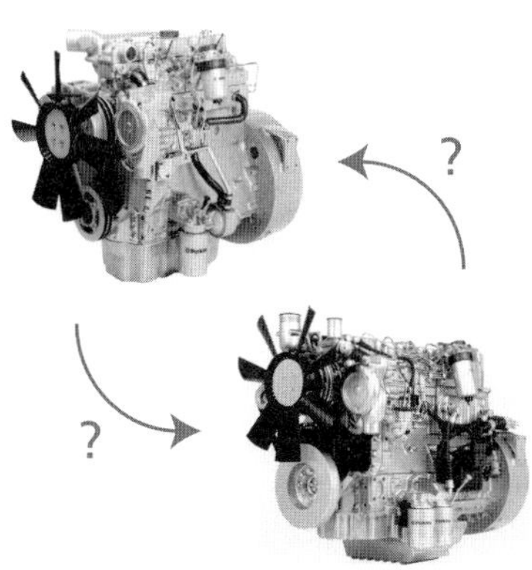

10.1 Change is most often a planned activity
Based on images © Perkins Engines Company Limited

The attitudes of engineers and managers towards engineering changes are important, as the ability of a company to implement changes effectively and efficiently is hugely dependent upon the people carrying out the task, and the way they communicate. Engineering changes are often perceived negatively because they can cause schedules to slip and budgets to overrun, but they can also be regarded as an opportunity for well-organised companies to meet the requirements of demanding customers rapidly and compete successfully with their rivals (DiPrima, 1982).

The issue of engineering changes has been gaining prominence in industry over the past two decades due to dramatic changes in markets. Maull *et al.* (1992) state that the move from the seller-dominated markets of the 1970s and early 1980s to the buyers' markets of today has led to a situation of greater diversity in products, smaller production runs and shorter product life-cycles. An increasing volume of engineering change is the inevitable consequence of such an environment (Coughlan, 1992).

Markets are now fragmented and populated by sophisticated customers who demand individualised offerings (Clark and Fujimoto, 1991). Today, there is also much more competition because of the increased globalisation of industries, such as automotive, aerospace and electronics. "The time when an innovatory product could be launched with confidence and remain unchallenged has passed" (Inness, 1994).

In order to maintain or increase market share, companies must be constantly prepared to improve and update existing products, and rapidly introduce new ones. Engineering change has always been an important part of the product design and development process, but today it is an essential aspect. For businesses to survive and compete, gaining a thorough understanding of all the issues involved is a vital design research activity for industry in conjunction with academia. This situation may be summed up by the following statement:

" ... it's absolutely necessary to understand changes and to have a good grip on them as the entire product development process can be described as a continuous change management process."
(Fricke *et al.*, 2000)

> *... it's absolutely necessary to understand changes and to have a good grip on them as the entire product development process can be described as a continuous change management process.*
>
> (Fricke *et al.*, 2000)

This chapter first takes a general look at engineering change and configuration management, as currently practised in industry. This is followed by definitions of change. The change life-cycle and a change process are then introduced. Finally, the impact of change and its relationship to a product's architecture are discussed. The purpose of this chapter is to define what is meant by an engineering change, to show when in the product life-cycle engineering change processes occur and discuss what their typical elements are.

Engineering change and configuration management

The attention that is now being paid to the management of change processes has in part been driven by the needs of companies to comply with configuration management and quality management standards such as ISO10007 (ISO, 1995) and ISO9001 (ISO, 2000), which demand clearly documented processes for all key business activities. Defining configuration management is difficult. Probably the clearest official definition comes from ANSI/EIA 649 (ANSI/EIA, 1997), which states that configuration management is

> *a management process for establishing and maintaining consistency of a product's performance, functional and physical attributes with respect to its requirements, design and operational information throughout its life.*
> (ANSI/EIA, 1997)

Change management is a formal discipline that allows complex products to be designed and produced concurrently by several business units or separate businesses separated by thousands of miles (Lyon, 2001). It is used throughout the product life-cycle from the selection of a concept to the wind-down of production. One of the key aspects of configuration management is the control of engineering changes, because uncontrolled changes will have a dramatic impact upon a product's performance and its functional and physical attributes. The engineering change process is the core process of the larger configuration management process. Each change of the product or its documentation causes a change in product configuration (Pikosz and Malmqvist, 1998).

Change management is a formal discipline that allows complex products to be designed and produced concurrently by several business units or separate businesses separated by thousands of miles.

Although originally developed for electro-mechanical goods, most recent literature on configuration management has focused on software products (Huang and Mak, 1998). The main focus is on document control and the administration of product options; the more-technical issues involved in making changes are either ignored or covered in little depth.

Configuration management is practised with differing intensities in different industries. It is a key process for the design and manufacture of complex mechatronic products such as cars and aeroplanes. As such, configuration management is a vital issue in such industries and for the companies that supply them. For example, it is doubtful whether a company such as Airbus, which has a widely distributed design and manufacturing capability, would be able to design new aeroplanes effectively and efficiently without the discipline of configuration management. Configuration management can also assist communication; it provides a framework to support contacts between groups, especially if they are geographically spread (Leech and Turner, 1985).

10.2 Configuration management is a key process for the design and manufacture of complex mechatronic products

Approximately 95% of UK firms that design and manufacture products have adopted a formal approach to engineering change management (Huang and Mak, 1999). However, it must be noted that although all companies that adopt robust configuration management procedures must have a formal engineering change process, this does not mean that all companies that have a formal approach to engineering changes must be following configuration management practice. Although the two issues are highly interrelated, they are not the same.

Defining engineering change

"An Engineering Change is a modification to a component of a product, after that product has entered production."
(Wright, 1997)

"[engineering changes are] the changes and modifications in forms, fits, materials, dimensions, functions, *etc.* of a product or a component."
(Huang and Mak, 1999)

"Engineering change Orders—changes to parts, drawings or software that have already been released."
(Terwiesch and Loch, 1999)

It is important to distinguish engineering change from the general concept of change in a business/organisational context. Change management is a term that is common in management and business literature, especially that concerning business process re-engineering (e.g. Kettinger *et al.*, 1997). It refers to the administration and supervision of corporate or organisational transformation, be it the results of merging two firms or implementing a new business process.

Engineering change management refers to the organisation and control of the process of making alterations to products. In this chapter, any mention of change refers to engineering change. It is important to establish what is meant by an engineering change or an engineering change order (ECO). Many authors use the terms interchangeably as they are approaching the issue from a management perspective, but most do not attempt to define terms, making the tacit assumption that the reader has a clear understanding of the situation.

Authors often use slightly different terms such as 'product change' (Inness, 1994), 'design change' (Ollinger and Stahovich, 2001), 'product design change' (Huang and Johnstone, 1995) and 'engineering design change' (Leech and Turner, 1985). Close inspection of these authors' work indicates that they are all referring to the same phenomenon. Throughout this chapter the term 'engineering change' is used.

On the occasions when a definition is supplied there are subtle differences which are helpful to highlight and discuss. Three definitions from often cited papers are as follows:

> *an Engineering Change (EC) is a modification to a component of a product, after that product has entered production*
> (Wright, 1997)

> *[engineering changes are] the changes and modifications in forms, fits, materials, dimensions, functions, etc. of a product or a component*
> (Huang and Mak, 1999)

Engineering Change Orders (ECOs) – changes to parts, drawings or software that have already been released
(Terwiesch and Loch, 1999)

Wright's (1997) definition restricts engineering change to the production stage and in doing so ignores the whole range of alterations that can occur during the design and development of a product. This has been the common approach in much of industry, with engineering change being regarded solely as a manufacturing issue that must be addressed to ensure product quality and to meet delivery deadlines; change before manufacture is regarded as a natural iteration of the design process. This approach creates an artificial division between engineering change and 'normal' product design and development.

The other two descriptions are more general and support the view that engineering change is an integral part of all design activities. These could range from changes made to a prototype during the development phase to an old product being updated to extend its life. Both definitions are much more suited to an environment of concurrent engineering. The definition of Huang and Mak (1999) is too general, in that it makes no mention or reference to the administration or management of design.

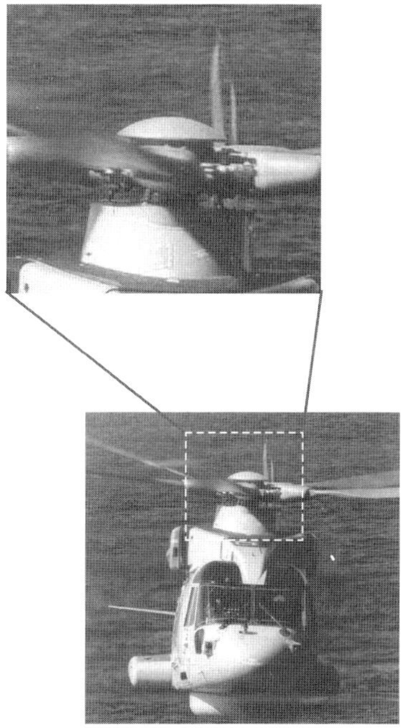

10.3 Change may be localised or apply to the whole product
© AgustaWestland

Terwiesch and Loch (1999) specifically mention the issue of software design, a vital aspect of modern mechatronic product design, which the other two ignore or at least fail to mention explicitly. By using the term ECO they are clearly approaching engineering change from a management point of view, where it is the management of change that is the big issue, especially when many changes are 'live' at the same time. They also imply that changes only occur once design details have been formally released. This links in with the formal processes for engineering change which are prescribed by configuration management standards.

It is important to appreciate that none of the definitions discussed above mention the size, scope or origin of the change. An engineering change can be anything from a small revision of a diagram taking one engineer a few minutes to a major redesign operation involving a large team of engineers working over a period of many months or even years. Designs are modified for a variety of reasons: to remove errors that have become apparent (through testing, manufacture, *etc.*); to adapt the device to open a new market sector; or to respond to customer demands.

In response to these issues, the definition of engineering change used in this chapter is based upon that given by Terwiesch and Loch (1999), but has been modified to include reference to the magnitude of the change:

An engineering change is an alteration made to parts, drawings or softare that have already been released during the design process. The change can be of any size or type, can involve any number of people and can take any length of time.

(Jarratt *et al.*, 2003)

"An engineering change is an alteration made to parts, drawings or software that have already been released during the design process. The change can be of any size or type, can involve any number of people and can take any length of time."

(Jarratt *et al.*, 2003)

Engineering change in the product life cycle

Virtually all texts on product development discuss the concept of product life cycles (e.g. Otto and Wood, (2001)). Inness (1994) describes moving from the 'birth' of a product idea, through design and development to production and shipping. Eventually, after a period of growth, the product matures; finally, its position can no longer be maintained and so it is phased out: product 'death'. Obviously, engineering change activity varies significantly depending upon which phase of its life-cycle a product is in.

An engineering change can be triggered at any point in the product life cycle once the concept for the design has been selected and defined, since at this point the design data and information start to be formally released to design teams, suppliers, potential customers, *etc.* Any changes to this data, as the product evolves, must be regarded as an engineering change. Figure 10.4 illustrates this point by using the generic product development process proposed by Ulrich and Eppinger (2003).

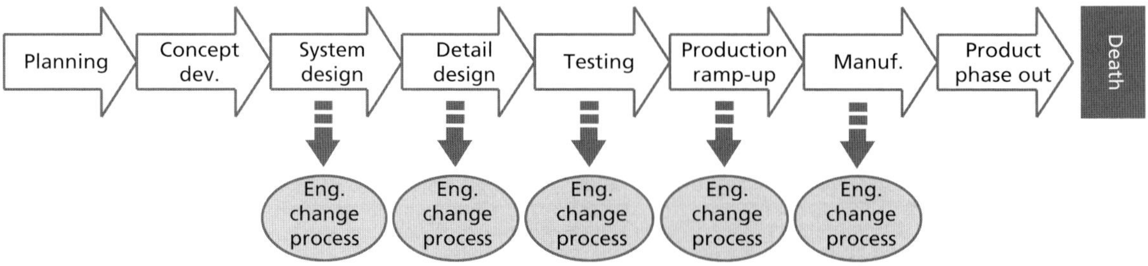

10.4 Engineering change processes can occur during the design and production life of a product – based upon the generic product design process proposed by Ulrich and Eppinger (2003) in *Product design and development*
© McGraw-Hill – reproduced with permission of The McGraw-Hill Companies

Design research, especially that which attempts to model the design process, often gives the impression that the design and development phase of a product's life has a definite end point at which the finished product is handed over to production and marketing. Although many of the original designers and engineers will move on to new projects, the product can still be developed and enhanced, engineering changes will still occur and engineering change processes will need to be controlled and managed. Thus, for the sake of completeness, two extra phases have been added to Ulrich and Eppinger's model of the product design process: manufacturing and product phase-out.

Companies will often use different terminology to describe the change processes that occur at various points in the product lifecycle (although in Figure 10.4 only the term engineering change process is used for simplicity). For example, the authors have witnessed the following terms being used in different companies: 'product change process' used to describe changes during production ramp-up and manufacture; 'prototype change process' for changes during the testing phase; and 'design changes' for changes made during the system and detail design phases.

Although different terminology can be used, the basic engineering change process is the same whenever it is triggered in the design process. It is important to realise that there are two lifecycles connected with any product: the in-production lifecycle and the in-service lifecycle. For a number of products, especially those with medium to long in-service lives, a situation can arise where production will have ceased long before the last product is retired from service and decommissioned. Examples of such products are automotive vehicles, aeroplanes, helicopters, ships, military equipment and industrial plant.

Change processes in different companies may refer to:

- product changes;
- prototype changes; or
- design changes.

The engineering change process

Most authors refer to the engineering change process, but only a few actually outline the elements or phases within it. This section will discuss some of the different engineering change processes proposed in literature and outline a generic process.

Engineering change processes

All of the engineering change processes suggested in literature and used in industry contain most of the same ideas/themes irrespective of the industry or product involved. This is because the proposed processes are similar at a macro level.

Pikosz and Malmqvist (1998) investigated the engineering change processes in three Swedish engineering companies: an automotive manufacturer, a supplier to the defence industry and a supplier of test equipment for military aircraft. They discovered that, whilst companies may perform similar tasks when examined at a high level, organisational, market and product issues lead to significant differences when the processes are investigated in greater detail. For example, if the company produces a safety-critical product, the engineering change process is focused much more on quality than on timescale or costs.

The change process is a mini, highly constrained design process or project.
(Leech and Turner, 1985)

Perhaps the clearest description of the engineering change process is provided by Leech and Turner (1985), who state that the process is a mini, highly constrained design process or project and "like any project, is only worth undertaking if its value is greater than its cost".

Different authors split the engineering change process into different numbers of elements, for example:

- Dale (1982) – (i) procedure to approval; (ii) procedure on approval.
- Huang and Johnstone (1995) – (i) before approval; (ii) during approval; (iii) after approval.
- Rivière *et al.* (2002) – (i) engineering change proposal; (ii) engineering change investigation; (iii) engineering change embodiment.
- Maull *et al.* (1992) – (i) filtration of engineering change proposals; (ii) development of solution to proposal; (iii) assessment of impact of solution; (iv) authorisation of change; (v) release and implementation of change.

Another element that is highlighted is that of review. DiPrima (1982) places an emphasis on following up any change to learn lessons. A month gap is suggested from implementation of the change to a review session. The review should examine whether everything is functioning as expected.

Learning from previously implemented changes is one of the key strategies proposed by Fricke *et al.* (2000) to cope with engineering changes, "Changes should be accepted as a chance, first, to improve the product and second, to do it better the next time".

A generic engineering change process

Figure 10.6 shows a generic high-level engineering change process based upon the elements outlined above. The process is initiated by a change trigger: this is a reason for change. Eckert *et al.* (2004) describe changes as emerging from the product (*i.e.* errors) or being initiated from outside (*i.e.* customer requests, legislation, *etc.*). Once the need for change is identified the six-phase process begins:

"Changes should be accepted as a chance, first, to improve the product and second, to do it better the next time."
(Fricke *et al.*, 2000)

1. A request for an engineering change must be made. Most companies have standard forms (either electronic or on paper) that must be completed. The person raising the request must outline the reason for the change, the priority of the change, type of change, which components or systems are likely to be affected, *etc.* This form is then sent to a change-controller who will enter it onto an engineering database.
2. Potential solutions to the request for change must then be identified, but often only a single one is examined. This can be for a variety of reasons:

time pressures, the fact that the solution is "obvious" or because engineers stop investigating once one workable solution is found.

3. The impact or risk of implementing each solution must then be assessed. Various factors must be considered: for example, the impact upon design and production schedules; how relationships with suppliers will be affected; and will a budget overrun occur? The further through the design process a change is implemented, the more disruption is caused.
4. Once a particular solution has been selected, it must be approved. Most companies have some form of Engineering Change Board or Committee, which reviews each change, making a cost–benefit analysis for the company as a whole and then granting approval for implementation. The Engineering Change Board must contain a range of middle to senior ranking staff from all the key functions connected to the product: for example, product design, manufacture, marketing, supply, quality assurance, finance, product support, etc. A thorough list of suitable functions to consider is provided by DiPrima (1982).
5. Implementation of the engineering can either occur straight away or be phased in. The option followed will depend upon various factors, such as the nature of the change (for example, if it is a safety issue, then immediate implementation must occur) and when in the product lifecycle it occurs. Paperwork must be updated. "One of the major problems frequently associated with engineering change, is that of ensuring that only current documentation is available to manufacturing areas" (Wright, 1997).
6. Finally, after a period of time, the change should be reviewed to see if it achieved what was initially intended and what lessons can be learnt for future change processes. Few companies carry out such a review process.

There are possible iterations within the process, two of which are marked by arrows in Figure 10.6. For example, a particular solution may be too risky for the company to implement and so the process will return to phase two, in order that other possible solutions can be identified. At the approval stage, the Engineering Change Board may feel that further risk analysis is required (maybe in the form of more testing) and so the process will return to phase three.

There are other possible iterative loops, but they are not marked for the sake of clarity. The most extreme loop would be when it was realised during the review phase that the solution implemented had been ineffectual or made matters worse. In that instance the process would return to the start with a new change request being raised.

10.5 Each aircraft will have a unique, and changing, build description

"One of the major problems frequently associated with engineering change, is that of ensuring that only current documentation is available to manufacturing areas."

(Wright, 1997)

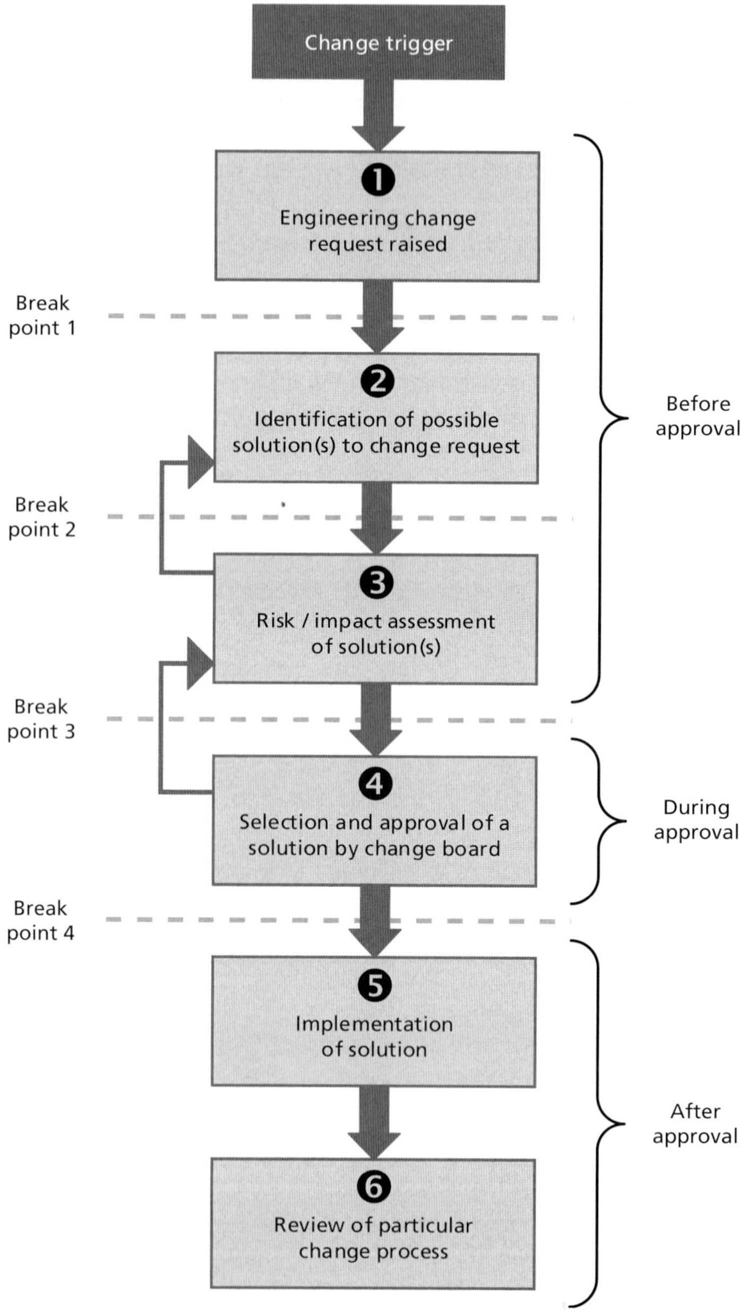

10.6 A generic engineering change process

So far, it has been tacitly assumed that the process will eventually progress to the end point of an implemented change being reviewed for lessons learnt. Only those changes that actually provide an overall benefit to the business must be allowed to proceed to the end of the process. Sometimes there is no choice if the change is as a result of a safety issue or legislation, but the majority of changes faced by a company are not so clear cut.

Fricke *et al.* (2000) state that, in their study of German manufacturing firms, only 40–60% of engineering changes were technically necessary. They report that, in the cases where a change was not technically necessary, the final decision came down to the experience and knowledge of the company members involved. As Clark and Fujimoto (1991) stress, it is important to differentiate between meaningful and meaningless changes.

Break points in the change process

There are four break points in the engineering change process shown in Figure 10.6. At each of these points the change process can be brought to a halt. They can be likened to the 'stage-gate' points used by many businesses in evaluating progress during new product development projects.

The first break point comes after the request for change has been raised. As Maull *et al.* (1992) point out, there must be a filtration of the change requests so that those which are truly impractical can be removed from the process early.

Employees must be encouraged to raise change requests, but, as many employees may not appreciate the full ramifications of their suggestions, there must be a mechanism to filter the proposals.

Employees must be encouraged to raise engineering change requests as part of continuous improvement, but, as many employees may not appreciate the full ramifications of their suggestions, there must be a mechanism to filter out the totally impractical proposals. Boznak (1993) states that effective screening can enable a company to identify improvement opportunities effectively while avoiding unnecessary change costs.

The second break point comes after the search for possible solutions. Although the request may have been suitable on initial inspection, further investigation may reveal that there are no sensible solutions.

The third break point comes after the impact/risk assessment phase. Analysis and testing may show that the proposed solution(s) are far too risky for the company to consider. The final break point comes when the Engineering Change Board meets to consider the proposed solution. Board members may feel that, given the risk analysis, the interaction of the product with other products and processes, end where the proposal is being raised in the product life cycle, the proposal is not worth proceeding with.

Engineering change process paperwork terminology

Several terms are used by different authors and companies to describe the paperwork that accompanies the engineering change process. These include engineering change request (ECR), engineering change notice (ECN), ECO and engineering change proposal (ECP). As with the definitions and processes discussed above, there is some contradiction depending upon which author's work is read or which company's process is examined. In the majority of cases ECRs and ECPs are synonymous, as are ECNs and ECOs. Definitions of these two groups are taken from Monahan (1995):

> *[the Engineering Change Request is] a form available to any employee used to describe a proposed change or problem which may exist in a given product;*

> *[the Engineering Change Order is] a document which describes an approved engineering change to a product and is the authority or directive to implement the change into the product and its documentation.*

The impact of engineering change

The assessment of the impact of a change is at the core of the engineering change process.

The assessment of the impact of a change is at the core of the engineering change process. As a result, the effects of making a change are a subject that has received much coverage in academic literature. In general, changes affect planning, scheduling and project costs.

Several authors refer to a 'Rule of 10' (e.g. Clark and Fujimoto, 1991; Anderson, 1997): the cost of implementing a change increases on average by a factor of 10 between each phase of the design process. Thus, a change made during manufacture would be 1000 times more expensive than making the same change during the detail design phase.

Terwiesch and Loch (1999) break down the costs of engineering changes into three categories:

- design;
- changes in prototype tools;
- changes in production tools.

One change that they tracked in an automotive company affected production tooling and cost by approximately $190,000. Another change to the same component cost less than $10,000 because the change was implemented before any tooling was manufactured. Changes that occur late on in the design process also affect far more people than those triggered early on. Once manufacturing, suppliers, marketing, *etc.* are involved, the number of people who must be notified of a change increases dramatically.

Engineering changes during the design process result in 'information deficiencies' for other development teams, whereby decisions about the product may be made without up-to-date data (Fricke *et al.*, 2000). This situation is increasingly common with the compressed development schedules that are now required in most markets.

Changes can propagate, *i.e.* a change can spread from the initially affected component or system to impact upon other parts of the product. The change can also spread to other products (for example, other members of the product family), processes (for example, manufacturing) and businesses (for example, suppliers, partners, etc.). Terwiesch and Loch (1999) have identified three key couplings that can lead to propagation:

- between components and manufacturing;
- between components within the same subsystem;
- between components in different subsystems.

Two other authors (Fricke *et al.*, 2000; Eckert *et al.*, 2004) have identified propagation as a key potential impact of implementing an engineering change. In particular, Eckert *et al.* (2004) have identified two different types of propagation event (see Figure 10.8):

- *Ending change propagation* – consists of *ripples* of change, a small and quickly decreasing volume of changes, and *blossoms*, a high number of changes that are nonetheless brought to a conclusion within the expected timeframe.
- *Unending change propagation* – characteristic of this type are *avalanches* of change, which occur when a major change initiates several other major changes and all of these cannot be brought to a satisfactory conclusion by a given point. Fricke *et al.* (2000) also talk of an avalanche of engineering change, whilst Terwiesch and Loch (1999) refer to 'a snowball effect'.

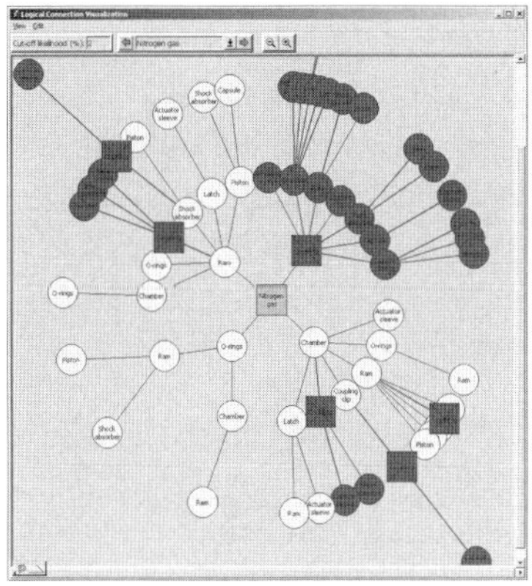

10.7 Changes may propagate via a number of different routes from an initiating change to an affected sub-system

Product architectures and change

How change affects a product is fundamentally linked to the product architecture, which is defined as:

(1) *the arrangement of functional elements;*
(2) *the mapping from functional elements to physical components; and*
(3) *the specification of interfaces among the interacting physical components.*

(Ulrich, 1995)

Changes can propagate, *i.e.* a change can spread from the initially affected component or system to impact upon other parts of the product.

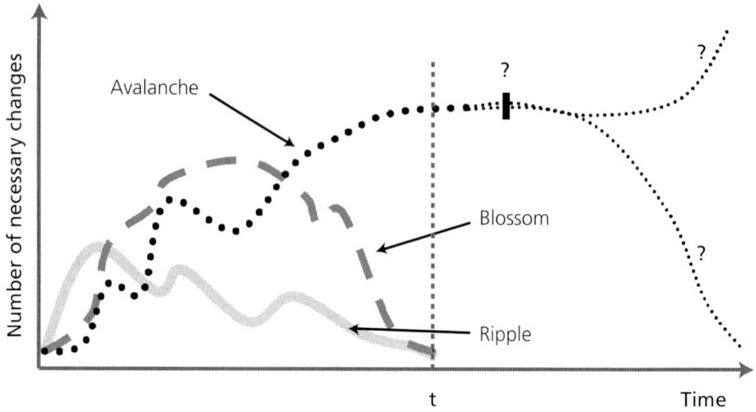

10.8 Types of change propagation (Eckert *et al.*, 2004)
© Springer-Verlag

There are two main types of product architecture:

- *Modular* – where each physical component of the product carries out only one element in the function structure and the interfaces between the components are decoupled – two components are said to have a coupled interface if a change to one causes a change to the other;
- *Integrated* – where each physical component carries out more than one functional element – this is termed function sharing (Ulrich and Seering, 1990).

In practice, most products are situated somewhere in the spectrum between full modularity and full integration. Indeed, whether a product is deemed modular or integrated depends upon the level at which it is examined. Products can be composed of subsystems that are modular in the way that they link together, but each one is highly integrated. For example, when considering a car, the radio can be considered as modular in relation to the rest of the vehicle, but when examined in isolation it is extremely integrated with high connectivity between components.

There are cost implications associated with product architecture. Without function sharing, many items, e.g. cars, would become prohibitively expensive (Ulrich and Seering, 1990). Modular designs generally cost more to manufacture and assemble than integrated ones, and this is why most mass-produced products, e.g. white goods, possess an integrated architecture. However, savings are possible through modularisation when a particular subassembly can be used on a variety of products.

It must be noted that few products are truly modular, especially as the complexity of the device increases. A good example comes from the automotive industry, where the same engine is used in a variety of car types.

Although it may appear a simple process of inserting a new module, actually the process requires a great deal of adaptive work; often this is to make the engine fit into the slightly different space offered by the new automobile. Successful modularisation allows the possibility of mass customisation, where individual products are tailored to individual customers.

Mass customisation

The assumption has always been that increased variety equates to increased costs for the manufacturer, but this is being challenged by concepts such as mass customisation (Pine, 1992). Three factors are making mass customisation possible:

- The designing of products with variety in mind (e.g. Martin and Ishii (2002)).
- Having flexible manufacturing facilities based on intelligent automated plant – for example, advances in rapid manufacturing mean that batch sizes as low as one are now economically feasible (Burton, 2003);
- Having the capability for effective and efficient product change.

Here an understanding of change propagation is critical if product architectures are to be developed that enable economic mass customisation. Early identification of those parts of a product that can vary, and those that must be kept unchanged, reduces the possibility of change avalanches (or blossoms) during customisation. Conversely, the limits of mass customisation may be set through consideration of the likely changes required.

The assumption has always been that increased variety equates to increased costs for the manufacturer, but this is being challenged by concepts such as mass customisation.

Modularity

The trend in many industries has been to promote modularity, and this, as well as creating adaptable and competitive products, has had the effect of promoting innovation, as specialist companies are able to concentrate all their expertise and resources on one particular module (Baldwin and Clark, 1997). Nowhere has this been more apparent than in the personal computer industry.

A linked trend is the concept of platform development, which is now seen widely within the automobile business. A platform is defined as:

> *a relatively large set of product components that are physically connected as a stable sub-assembly and are common to different final models*
> (Muffatto, 1999)

The main advantages of following a platform strategy are that it can lead to reduced production costs and, perhaps more importantly, it allows for delayed product differentiation, which enables producers to meet the requirements of increasingly demanding customers more efficiently (Lee and Tang, 1997).

In terms of change, modular designs can be adapted much more easily to changing requirements if the interfaces between the modules are able to remain the same. However, once the interfaces between modules need to be altered, the magnitude of the change issue will increase dramatically. Lindemann *et al.* (1998) talk of 'local change', which just involves one component or system, and 'interface-overlapping change', which involves many components and is especially common in complex products with high connectivity between parts.

Components and change

Successful mass customisation and modularity rely on the designer's ability to minimise the changes required to modify a product. In turn, the level of change required is defined by the architecture of the product and the ability of the parts of the product to 'absorb' change. Hence, a product's components or subsystems may be categorised into three approximate types with regard to their change properties (Eckert *et al.*, 2001):

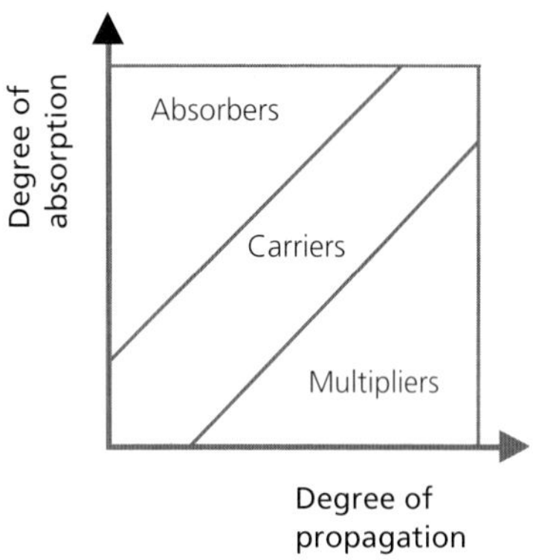

10.9 Component change properties (Eckert *et al.*, 2004)
© Springer-Verlag

- *Absorbers.* These can be either 'partial' or 'total', where a total absorber causes no further change whilst accommodating a number of changes (a rare situation), and a partial absorber contains many changes and passes on only a few.
- *Carriers.* These neither reduce nor add to the change problem – they merely transfer the change from one component to another.
- *Multipliers.* These expand the change problem making the situation more complex – such components may lead to an 'avalanche' of change.

These categories are illustrated in Figure 10.9. It is critical to appreciate that components can change between the three roles depending upon the size of the change. A component may be an absorber of small changes, but when a large alteration is necessary, it may develop into being a carrier or, worse, a multiplier. Two factors affect whether a change can be absorbed (Eckert *et al.*, 2001): the initial specification of the component and the tolerances designed into it. When reporting on the specific case of helicopter design, they comment:

> *the designers observed typically added a 25% safety margin to the specification of many components, which was gradually used as the design was put together.*
>
> (Eckert *et al.*, 2001)

Once the safety tolerances are all used up, the component will switch to being a carrier or multiplier. Successful design under these conditions requires the use of a robust design change process.

Strategies and methods to cope with engineering change

Engineering change will always be associated with engineering products. Equally, such change is likely to cause major upheavals during design and manufacture. As a result, many authors (e.g. Nichols, 1990; Terwiesch and Loch, 1999) have suggested strategies to cope with it. These help to reduce some of the negative aspects whilst maximising the positive. The most comprehensive list is from Fricke *et al.* (2000), who suggest five:

- prevention;
- front loading;
- effectiveness;
- efficiency;
- learning.

Before examining each of these strategies in detail, it is worth quoting a passage from Clark and Fujimoto's (1991) examination of the automotive industry. One aspect they identified that differentiated Japanese firms from their Western counterparts was how engineering changes were handled. Although the past decade has seen huge changes and consolidation in this industry (especially in North America and Europe), it is still worth quoting, as it covers all the main issues involved in the successful handling of engineering changes.

> *...the typical Japanese project has almost as many changes as its Western counterpart. The differences in approach lie not in numbers, but in patterns and content. Procedures are less bureaucratic and orientated more towards fast implementation than towards checks and balances. In effect this approach emphasises early versus late, meaningful versus unnecessary and fast versus slow. Engineers make changes earlier, when the cost of change and time pressure are still relatively low. They reduce the number of changes due to careless mistakes and poor communication so that changes that are made add value to the product.*
>
> (Clark and Fujimoto, 1991)

Strategies to cope with change include:
- prevention;
- front-loading;
- effectiveness;
- efficiency;
- learning.

(Fricke at al., 2000)

"The typical Japanese project has almost as many changes as its Western counterpart."
(Clark and Fujimoto, 1991)

Prevention

This strategy aims to reduce (or eliminate) the number of emergent changes that occur. Saeed *et al.* (1993) found that changes to correct errors accounted for 58% of engineering changes in the company they studied. However, examination of the sources of error in the design process in three aerospace companies found that it was difficult, if not impossible, to clearly identify the point of introduction of an error (Cooke *et al.*, 2002). In all the cases examined:

> *the one common theme was the failure to correctly identify when the uncertainty in the design was becoming unacceptably large so that high levels of risk were introduced into the project.*
>
> *Ignorance of the limits of one's own knowledge is perhaps the most dangerous [factor] of all*
>
> (Cooke *et al.*, 2002)

"Ignorance of the limits of one's own knowledge is perhaps the most dangerous [factor] of all."
(Cooke *et al.*, 2002)

Initiated changes, which enhance the product or its production, are important and "efforts to eliminate them entirely are both undesirable and unrealistic" (Clark and Fujimoto, 1991). Smith and Reinerstein (1998) state that early freezing of the design specification is a 'foolhardy' method of reducing errors, as this does not fit with reality; the initial specification is rarely accurate and the market may alter during development.

A more sensible approach would be to reduce unnecessary specifications and focus on the core customer requirements. Techniques such as quality function deployment (Otto and Wood, 2001) and the separation of technology development from product development (as proposed by Clausing (1994)) are recommended to achieve this (Fricke *et al.*, 2000).

Front loading

This strategy is proposed by a number of other authors (e.g. Nichols, 1990; Lindemann and Reichwald, 1998; Terwiesch and Loch, 1999). Early detection of required changes will result in a lower overall impact and cost, as discussed above (*i.e.* with the 'Rule of 10'). Good concurrent engineering practice, such as early involvement of suppliers and customers, coupled with techniques such as "failure mode and effects analysis" and "design for manufacture and assembly" will help bring changes forward in the design process.

Fricke *et al.* (2000) discuss in detail the front loading strategy. Although much literature promotes it, certain markets are changing so fast that following this strategy dogmatically could lead to companies losing out to their competitors by not reacting to customer wishes. Fricke *et al.* (2000)

conclude that the 'Rule of 10' must be broken and they propose Design For Changeability as a means to do this by moving away from 'single-point design'. At the heart of their proposal are the concepts of flexibility, agility, robustness and adaptability.

10.10 Effective change can contribute to commercial success
© AgustaWestland

Effectiveness

This strategy emphasises the making of effective 'effort versus benefit' analysis for each proposed change. Not all engineering changes are immediate or mandatory, as described above; in the study of Fricke *et al.* (2000) only 40–60% of changes were technically necessary. It is essential for engineers and managers to differentiate between the meaningful and meaningless, but the study showed that assessments of "possible effects of changes and the evaluation of change requests are mostly based on the experience and knowledge of the employees" (Fricke *et al.*, 2000).

Avoiding unnecessary changes, by getting the initial release right, is one of Terwiesh and Loch's (1999) four principles of change management. Analysing the effects of historic changes could be used as a method to support current change evaluation, but none of the companies surveyed by Fricke *et al.* (2000) did this.

Efficiency

Essential changes should be implemented as efficiently as possible by making best use of resources such as time and money. Essential changes should be communicated as soon as possible to all affected people and sections. Although change processes may be standardised (due to ISO 9000, *etc.*), this is not optimal for all kinds of changes; flexibility is needed. The reality of the situation is that people will often go out of process in order to improve the speed of implementation (Fricke *et al.*, 2000). They also highlight the impact of architecture on efficiency:

> *the design of the product, requirements and process, or the design of the entire project, should be of a kind that changes can be realised easily. Unfortunately, most companies focus only, if at all, on improving the administrative change process*
>
> (Fricke *et al.*, 2000)

Ways of speeding up the change process have been proposed by several authors. For example, Loch and Terwiesch (1999) examined and proposed methods of removing bottlenecks in the process.

"The design of the product, requirements and process, or the design of the entire project, should be of a kind that changes can be realised easily. Unfortunately, most companies focus only, if at all, on improving the administrative change process."
(Fricke *et al.*, 2000)

Reviewing and critiquing engineering changes offers a chance to improve the design of a product, the product design process and the engineering change process.

Learning

Reviewing and critiquing engineering changes offers a chance to improve the design of a product, the product design process and the engineering change process. However, few companies actually carry out consistent, continuous analysis (Fricke *et al.*, 2000). Another aspect of such a review process is increased awareness of the importance of engineering change and the issues amongst employees that affect it. A review and critiquing process, in a company studied by Fricke *et al.* (2000), led to a significant reduction in the average number of changes per item. Linked to this, the visibility of the engineering change process and employees' understanding of it are vital for success. However, Saeed *et al.* (1993) found that the process was very complex and few people understood it well.

Conclusions

Engineering changes allow companies to enhance and adapt their products, and to remove errors from them. Changes are a fact of life for all companies that design and manufacture products and they are a topic that is growing in importance as product lifecycles shorten and markets fragment. The engineering change process is a vital part of any product's life and it links into all the major business functions, such as manufacturing, purchasing, marketing and aftersales support.

The impacts of making changes to products can be surprising; occasionally, dramatic propagation from the initially affected component or system can occur. A key factor in whether propagation takes place is the product architecture and the interactions between components and systems. Careful design of the architecture can help minimise the negative effects of change and also allow for more product flexibility, which can be used to follow a business strategy such as mass customisation.

This chapter has highlighted five strategies to improve the handling of engineering changes. By both appreciating the importance of change and efficiently and effectively managing the process of making alterations to products, companies can gain a significant advantage over their rivals.

References

Anderson DM (1997) Agile product development for mass customisation. Irwin Professional Publishing

ANSI/EIA (1997) ANSI/EIA 649: national consensus standard for configuration management. American National Standards Institute

Baldwin CY, Clark KB (1997) Managing in an age of modularity. Harvard Business Review, 75(5): 84–93

Boznak RG (1993) Competitive product development: a quality approach to succeeding in the 90s and beyond. ASQC Quality Press

Bucciarelli LL (1994) Designing engineers. MIT Press

Burton M (2003) Design for rapid manufacture. Presentation given at Process/Design Improvement Workshop. Loughborough University

Clark KB, Fujimoto T (1991) Product development performance: strategy, organization and management in the world auto industry. Harvard Business School Press

Clausing D (1994) Total quality development. ASME Press

Cooke JA, McMahon C, North MR (2002) Sources of error in the design process. IDMME'02, Clermont-Ferrand, France

Coughlan PD (1992) Engineering change and manufacturing engineering deployment. In: Integrating design and manufacturing for competitive advantage. Oxford University Press

Cross N (1989) Engineering design methods. John Wiley

Dale BG (1982) The management of engineering change procedure. Engineering Management International, 1(3): 201–208

DiPrima MR (1982) Engineering change control and implementation considerations. Production and Inventory Management, 23(1): 81–87

Eckert CM, Zanker W, Clarkson PJ (2001) Aspects of a better understanding of changes. ICED'01, Glasgow, UK

Eckert CM, Clarkson PJ, Zanker W (2004) Change and customisation in complex engineering domains. Research in Engineering Design, 15(1): 1–21.(Pages 13, 17 and Figures 13,17)

Fricke E, Gebhard B, Negele H, Igenbergs E (2000) Coping with changes: causes, findings and strategies. Systems Engineering, 3(4): 169–179

Huang GQ, Johnstone G (1995) CMCEA: change mode, cause and effects analysis – a concurrent engineering approach to cost-effective management of product design changes. ICED'95, Praha, Czech Republic

Huang GQ, Mak KL (1998) Computer aids for engineering change control. Journal of Materials Processing Technology, 76: 187–191

Huang GQ, Mak KL (1999) Current practices of engineering change management in UK manufacturing industries. International Journal of Operations and Production Management, 19(1): 21–37

Inness J (1994) Achieving successful product change. Financial Times/ Pitman Publishing
ISO (1995) ISO 10007 quality management – guidelines for configuration management. International Organisation for Standardisation
ISO (2000) ISO 9001 quality management systems – requirements. International Organisation for Standardisation
Jarratt T, Eckert CM, Clarkson PJ (2003) A taxonomy of change. Technical Report CUED/C-EDC/TR123, University of Cambridge
Kettinger WJ, Teng JTC, Guha S (1997) Business process change: a study of methodologies, techniques and tools. MIS Quarterly, 21(1): 55–80
Lee HL, Tang CS (1997) Modelling the costs and benefits of delayed product differentiation. Management Science, 43(1): 40–53
Leech DJ, Turner BT (1985) Engineering design for profit. Chichester
Lindemann U, Reichwald R (1998) Integriertes Änderungsmanagement. Springer
Lindemann U, Kleedorfer R, Gerst M (1998) The development department and engineering change management. In: Designers: the key to successful product development. Springer
Loch CH, Terwiesch C (1999) Accelerating the process of engineering change orders: capacity and congestion effects. Journal of Product Innovation Management, 16(2): 145–159
Lyon DD (2001) Practical CM – best configuration management practices. Butterworth-Heinemann
Martin MV, Ishii K (2002) Design for variety: developing standardised and modularised product platform architectures. Research in Engineering Design, 13(4): 213–235
Maull R, Hughes D, Bennett J (1992) The role of the bill-of-materials as a CAD/CAPM interface and the key importance of engineering change control. Computing and Control Engineering Journal, 3(2): 63–70
Monahan RE (1995) Engineering documentation control practices and procedures. Marcel Dekker
Muffatto M (1999) Platform strategies in international new product development. International Journal of Operations and Production Management, 19(5/6): 449–459
Nichols K (1990) Getting engineering changes under control. Journal of Engineering Design, 1(1): 5–15
Ollinger GA, Stahovich TF (2001) Redesign IT – a constraint-based tool for managing design changes. ASME DETC'01, Pittsburgh, PA, USA

Otto K, Wood K (2001) Product design – techniques in reverse engineering and new product development. Prentice Hall

Pikosz P, Malmqvist J (1998) A comparative study of engineering change management in three Swedish engineering companies. ASME DETC'98, Atlanta, GA, USA

Pine BJ (1992) Mass customization: the new frontier in business competition. Harvard Business School Press

Rivière A, Dacunha C, Tollenaere M (2002) Performances in engineering change management. IDMME'02, Clermont-Ferrand, France

Saeed BI, Bowen DM, Sohoni VS (1993) Avoiding engineering changes through focused manufacturing knowledge. IEEE Transactions on Engineering Management, 40(1): 54–59

Smith PG, Reinersten DG (1998) Developing products in half the time. John Wiley

Terwiesch C, Loch CH (1999) Managing the process of engineering change orders: the case of the climate control system in automobile development. Journal of Product Innovation Management, 16(2): 160–172

Ulrich KT (1995) The role of product architecture in the manufacturing firm. Research Policy, 24: 419–440

Ulrich KT, Eppinger SD (2003) Product design and development. McGraw-Hill

Ulrich KT, Seering WP (1990) Functional sharing in mechanical design. Design Studies, 11(4): 223–234

Wright IC (1997) A review of research into engineering change management: implications for product design. Design Studies, 18: 33–42

Chapter 11

Risk in the design process

Chris McMahon
University of Bath

Jerry Busby
Lancaster University

Uncertainty pervades engineering design. There is variation in all materials and processes, in all engineering parts and assemblies. The use (and abuse) of engineering artefacts differs from user to user and there are large unknowns in the impact on the natural environment. Our understanding of the factors that influence artefact performance is incomplete, and our analytical and predictive methods are imperfect. We cannot predict all of the ways in which a process or an artefact might fail. We cannot completely replicate on the test bed or in prototype development the loads to which our designs will be subject in use. For these and for many other reasons engineering design is an uncertain activity, and thus a source of risk – of the possibility of an undesirable event or outcome.

11.1 The Paddington rail disaster
© PA photos

Undesirable outcomes in engineering can include poor technical or commercial performance of an artefact, danger to life and limb for a user of an artefact, or impact on the environment or some third party. Such outcomes have existed throughout the history of engineering, but today have acquired a particular importance because of the high cost and timescales and distributed nature of many engineering projects, the complexity and inherent danger of some engineering artefacts and systems, and the aversion of many people to personal and commercial risk.

The present importance of risk has led to a great deal of recent interest in its active management. This involves a number of techniques, ranging from general approaches to risk identification, assessment and monitoring through to analytical methods that represent and manipulate uncertainty in design parameters. Risk management has become a standard engineering technique, contractually required in many engineering projects. But while qualitative approaches to risk management have had some success, quantitative risk assessment has had a much lower impact except in very risk-sensitive domains such as nuclear engineering and aerospace.

It is also apparent that public and private attitudes to risk are not strictly informed by rational judgements of likelihood and impact, but also by perception, and in particular that risk perception is strongly influenced by dread and by dangers imposed by others. For these reasons perception has become an important factor in the engineer's consideration of risk.

This chapter will review all the aspects of risk and uncertainty in engineering that have been noted above. It will first provide an overview of the nature of risk and uncertainty in engineering, and will distinguish between different aspects of risk from the point of view of the engineer. It will then review current approaches to risk in engineering – first through an overview of approaches to risk assessment and management and then through a brief exploration of quantitative approaches to the evaluation of risk and uncertainty. It will finish with an overview of the impact of perception on risk in design, and a note of some aspects of risk management in practice.

The nature of risk and uncertainty in engineering

Risk in engineering design encompasses a variety of issues for a wide range of stakeholders. It encompasses risk to organisations in the product supply chain – manufacturers of parts, assemblies and integrated systems, maintainers and recyclers – to the customer or user of artefacts, and to the wider community both in the present day and in the future. It also involves a variety of concerns, which include:

- *Technical risk* – i.e. risk that the artefact will not perform as intended. Technical risks include, for example, the possibility that an aircraft will not reach its payload/range targets or that components of an automobile engine will fail prematurely.
- *Project risk* – i.e. risk that a project will fail or will overrun in cost or time. Examples of adverse outcomes in project risk include a military procurement contract that exceeds budget and a civil engineering occupation of a railway track that exceeds an allocated time period.
- *Risk to life and limb* – i.e. risk that someone will be killed or injured as a consequence of use or even abuse of the artefact. Examples include the risk of injury from failure of transportation devices or production equipment and also long-term hazards to health from asbestos insulation.
- *Risk to the environment, or to future generations*. Examples include risk of pollution from a manufacturing process or of depletion of scarce materials.

Risks exist in all aspects of life, but those associated with the manufacture or construction and use of engineering artefacts are often particularly acute.

Risks exist in all aspects of life, but those associated with the manufacture or construction and use of engineering artefacts are often particularly acute. The artefacts are in continual use in very large numbers: we all spend many hours of each day interacting with them (to the extent that they may be so familiar to us that we fail to show them the respect that they deserve), and the artefacts themselves often have a high propensity to cause injury or death as a result of the energies involved in their construction and use.

Complexity

Engineering artefacts are also often characterised by complexity in a number of respects. Many artefacts themselves are both complex and complicated, involving very many component parts and requiring significant skill and knowledge in their construction and use. In modern aviation systems, for example, individual aircraft may have in the order of a million component parts, and they interact with other aircraft, with airport and air traffic control systems (Figure 11.2) and so on. The number of potential failure modes is enormous, as is the number of modes of interaction between components and subsystems.

11.2 Air traffic control, part of a complex modern aviation system

Complexity in engineering also extends to the number and geographic distribution of the people and organisations involved in the design and construction of engineering artefacts. A design team can today be spread between three continents, as can the companies in the supply chain. This geographic distribution is necessary because the cost of large design and development programmes, such as those for aircraft or automobiles, is now so great as to require firms to collaborate in order to spread the development costs and achieve the necessary economies of scale. These costs also mean that the number of new product programmes in some areas is small, and therefore the implications of failure for the organisations concerned (including governments where these are the customers) can be severe.

A further aspect of complexity and coupling in engineering concerns the interactions between the engineered artefact and the natural environment. In this regard, hazards such as those imposed by extreme events including earthquakes, large waves or high winds are well known, but an emerging understanding is developing of the implications for the natural world of long-term use of engineering artefacts, owing to the interaction of man-made materials with the environment, the impact of pollutants and so on.

Human factors

Finally, and of considerable importance, people have a huge impact on risk in design. Many failures and uncertainties in the engineering process are due to human error, and there are many uncertainties in the way in which people may interact with an artefact, ranging from areas such as market acceptance of a new product and, in particular, unforeseen abuse of the artefact itself.

The subject has perhaps been investigated most widely by those concerned with the consequences of design error resulting in structural failure, and these have tended to concentrate on the nature and effect of human error. For example, Stewart (1992) suggests that reviews of statistical data indicate that up to 75% of structural failures are human errors, and suggests that human error also accounts for much of the discrepancy between estimated and actual probabilities of artefact failure. Petroski (1991) argues that human error is the most likely cause of fundamental errors made at a conceptual stage, which can be the most serious and elusive of design errors. Cambell (2002) suggests that some 30% of construction failures are due to design error, and emphasises the importance of education and quality systems that ensure all aspects of the design are thoroughly and independently checked.

Human error is also very significant in accidents and other undesirable outcomes resulting from the use of engineering artefacts. For example, it is estimated that 70% of aircraft accidents involve pilot error (and error by maintenance and other ground staff will contribute further), while 80% of shipping accidents involve human error (Hawkins, 1993; Lucas, 1997). Such bald statistics may, however, obscure the contribution that can be made by other factors even in cases that are ostensibly due to human error. Bennett (2001) argues that bad design, poor training, unrealistic rosters, substandard maintenance and other factors outside the control of the flight crew may often be significant in aircraft failures.

A similar picture may be found in UK National Health Service hospitals, where it is estimated that adverse events, in which harm is caused to patients, occur in around 10% of admissions – or at a rate in excess of 850,000 a year; and that these cost the service an estimated £2 billion a year in additional hospital stays alone. It is thought that human error may sometimes be the factor that immediately precipitates a serious failure, but there are usually deeper, systemic factors at work which, if addressed, would have prevented the error or acted as a safety net to mitigate its consequences (DOH, 2000).

Approaches to risk management

Although risk pervades engineering, designers have traditionally used very limited tools to assess the likelihood and impacts of risks.

Although risk pervades engineering, designers have traditionally used very limited tools to assess the likelihood and impacts of risks. Engineering calculations have generally been deterministic, with uncertainty taken account of through so-called "factors of safety". Project risk has often been dealt with simply by trying to identify likely risk factors and to take steps to mitigate them.

The past 20 years have, however, seen a significant change in attitude to risk for the reasons noted above: the complexity of modern engineering projects is such that the investment in time and money in new product development is large; and a single product failure may have a major impact on a company. Projects are often distributed between companies and often between countries, and risk has to be formally managed within the frameworks for collaboration. There is a much more widespread use of fixed-price contracts, especially by government. Consumer awareness has put an increased emphasis on safety and reliability, and customers and others impacted by products have become increasingly litigious: we are living in a "risk society" (Lupton, 1999). There has also been an increasing awareness of the impact of artefacts on the environment, and of other external impacts such as that on national economies (Kammen and Hassenzahl, 2001). There exists also risk relating to everyday interactions, particularly within the work place (Bloor, 1995).

The changes in attitudes have been reflected in developments both in design practice and in research in design and in the social sciences. Formal risk management has become a requirement for a significant number of projects, in particular those financed from public funds (MOD, 1996a). Many more companies incorporate risk management in their procedures, both for project and technical risk, although not contractually required to do so (Crossland *et al.*, 1998, 2003). New techniques have been developed for project and technical risk assessment and management. These include a number of risk management methodologies (Carter *et al.*, 1994; Simon *et al.*, 1997; ICE, 1998; Patterson *et al.*, 1999), and software tools for risk management and assessment (@Risk; Monte Carlo; CIRIA; BSI, 1991; Kletz, 1992).

Risk management is defined as "co-ordinated activities to direct and control an organisation with regard to risk."
(ISO/IEC, 2002)

In the ISO guide to risk management vocabulary (ISO/IEC, 2002), risk management is defined as "co-ordinated activities to direct and control an organisation with regard to risk". There are many published methodologies prescribing an idealised generic process for risk management, including that published by the Risk Special Interest Group of the Association for Project Management (Simon *et al.*, 1997), Chapman and Ward's (1997) nine-phase generic risk management process structure, and the Riskman methodology (Carter *et al.*, 1994). All of these are intended to provide a framework for risk analysis and control, rather than a detailed prescription of techniques. Nevertheless, the four key phases (Figure 11.3) in all such risk management processes are (MOD, 1996b, c; DOD, 2000; ISO/IEC, 2002):

- *Risk identification* – the process of finding, listing and characterising elements of risk.
- *Risk assessment* – the overall process of risk analysis and risk evaluation.
- *Risk treatment* – the process of selection and implementation of measures to modify risk.
- *Risk monitoring, review and communication* – a continual process of re-examining assumptions, reviewing developing risk and communicating likely impacts to stakeholders.

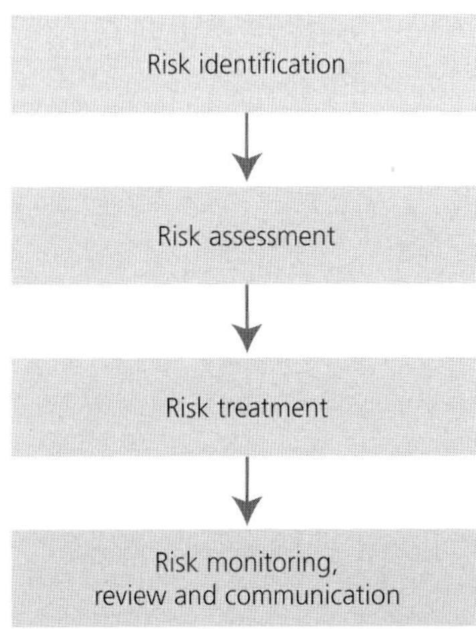

11.3 The four key phases of the risk management process

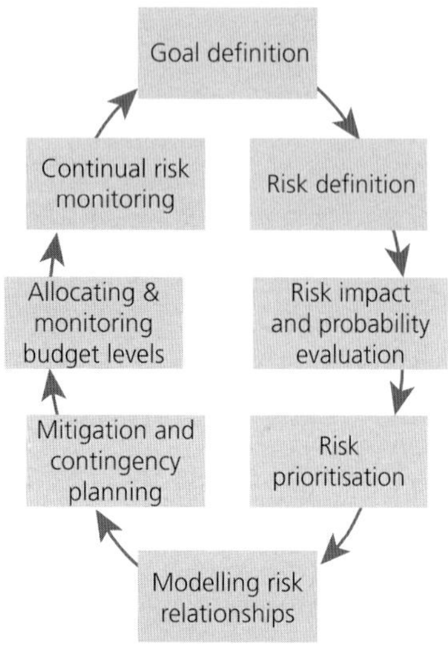

11.4 A general risk management process (Crossland *et al.*, 1998)

The risk management cycle

The four key phases identified above may be expanded into a cyclic sequence of risk identification, prioritisation, monitoring and review, representing a plan for risk management action, as shown in Figure 11.4. The stages of this cycle are broadly as follows.

Goal definition: identification of measurable control parameters and determination of a base plan (the planned structure of project elements if no risk events occur) and risk management plan. The identified and recorded risks represent deviations from this plan.

Identification of both risks and opportunities, and of the members of the project team who are most closely concerned with those risks (the "owners"). The tools and techniques used for risk identification include questionnaires, checklists, prompt lists, expert interviews, formal risk review procedures, workshops, brainstorming, risk response analysis (Cooper and Chapman, 1987) and knowledge-based systems (KBS) (Niwa, 1989; Cailleaud *et al.*, 1999). Identified risks are recorded in a risk register (Carter *et al.*, 1994) or risk list (CCTA, 1995).

Risk impact and probability evaluation: the impact and probability of risks is identified and recorded. Numerical evaluations are given wherever possible, and recorded in the register. Techniques for analysing and evaluating the probability and impact of identified project risks include schedule-specific techniques such as the critical-path method, Gantt charts and the program evaluation and review technique (PERT) (Moder and Phillips, 1970; Starkey, 1992), qualitative techniques such as probability/impact matrices and use of high/medium/low categories for probability and for impact (Carter *et al.*, 1994; Coppendale, 1995). Equivalent techniques for technical risk include failure mode and effects analysis (FMEA), hazard and operability (HAZOP) and preliminary hazard analysis (PrHA).

Risk prioritisation: the evaluated impact and probability for each identified risk are used to determine which risks should be included in the risk model.

Modelling relationships: relationships are modelled in terms of time, cost, performance or other measures. Some methodologies reduce everything to cost.

Mitigation and contingency: the base plan is changed to reduce probability or impact. Contingency plans are triggered and trade-offs identified.

Budgets are allocated and monitored for measurable/controllable parameters.

Risk monitoring of the identified risks takes place. Probabilities and impacts are updated. New risks arise. Existing risks are eliminated. Trigger events are monitored. The risk monitoring activity in turn contributes to the next cycle of risk identification, prioritisation and monitoring, so closing the loop.

Risk management in practice

A number of industries have been at the forefront of developments in risk management. A good deal of the early focus was on risk to life and limb, especially in high-impact industries such as nuclear, aerospace and construction, and these industries have remained a strong focus of risk research. So far as project and technical risk are concerned, a good deal of work on design project risk management has concentrated on the design of software systems (Boehm, 1991; Ould, 1999), which seems to be inherently more technically risky than many other kinds of design. The defence and construction industries (Edwards, 1995; Godfrey, 1995) have also been at the focus of formal project risk management methods, owing to the sheer size of their projects. Issues of technical risk have also been particularly important in defence programmes, owing to the rapid pace of technical change combined with long programme timescales (MOD, 1996a – c). Technical risk is also at the forefront of concerns in aerospace, nuclear and medical engineering, where the impact of failure is particularly high (Health and Safety Executive, 1992; FDA, 2000; Ward and Clarkson, 2004), and in construction programmes such as the design of flood and coastal defences owing to the unpredictable nature of natural forces and the long timescales involved (Godfrey, 1995). Both uncertainty and risk issues are paramount in the oil and gas sector, where a single decision determines massive financial investment. There are huge uncertainties regarding what lies beneath the ground and there are huge health and safety issues, for example Piper Alpha and Exxon Valdez (Heising and Enzenbach, 1991; Aven and Pitblado, 1998; Bea, 1998).

With the increasing use of analysis and simulation techniques in engineering it is very important for engineers to understand the uncertainties and risks inherent in the use of such techniques. Computer models in engineering design are representations of products or processes that may be prone to uncertainty, variability or error. There is a need for approaches that help engineers understand the nature of such variability and identify whether models are appropriate for specific uses. In this regard, a number of approaches for the evaluation of the suitability of techniques have been devised – for example, Rajabally *et al.* (2003) propose a methodology that uses Bayesian belief nets to capture the reasoning associated with justifying model trustworthiness and Balci (2001) proposes a systematic approach for the evaluation of hierarchies of direct and indirect indicators and the aggregation of indicator scores. These approaches depend on expert assessment of techniques – often there is a lack of well organised verification data – the organisation and accumulation of such data is an important future research issue.

With the increasing use of analysis and simulation techniques in engineering it is very important for engineers to understand the uncertainties and risks inherent in the use of such techniques.

A workshop on issues in engineering risk assessment and perception held at the University of Bristol in 2002 (McMahon *et al.*, 2002) suggested that identifying potential risks is seen as particularly important in an industrial context. The consensus was that where a risk is identified, then the assessment and mitigation carried out are generally effective. The identification of a potential risk in the first place is the weakest part of the process. One problem is that those who experience failures of the product (for example, disgruntled users, seriously injured or relatives of deceased) are often not on good terms with those who make or specify the product. There is reluctance to contact such people to gather information, and yet they often have unique stories to tell.

Crossland **et al**. (1998, 2003) carried out a survey of risk management practice in UK engineering companies. In one part of the survey, respondents were asked to identify the difficulty of dealing with different sources of technical risk. As shown in Figure 11.5a, the aspects considered "difficult" or "extremely difficult" to deal with by the most respondents were (in descending order) aggregate budget overruns (where budgets include cost,

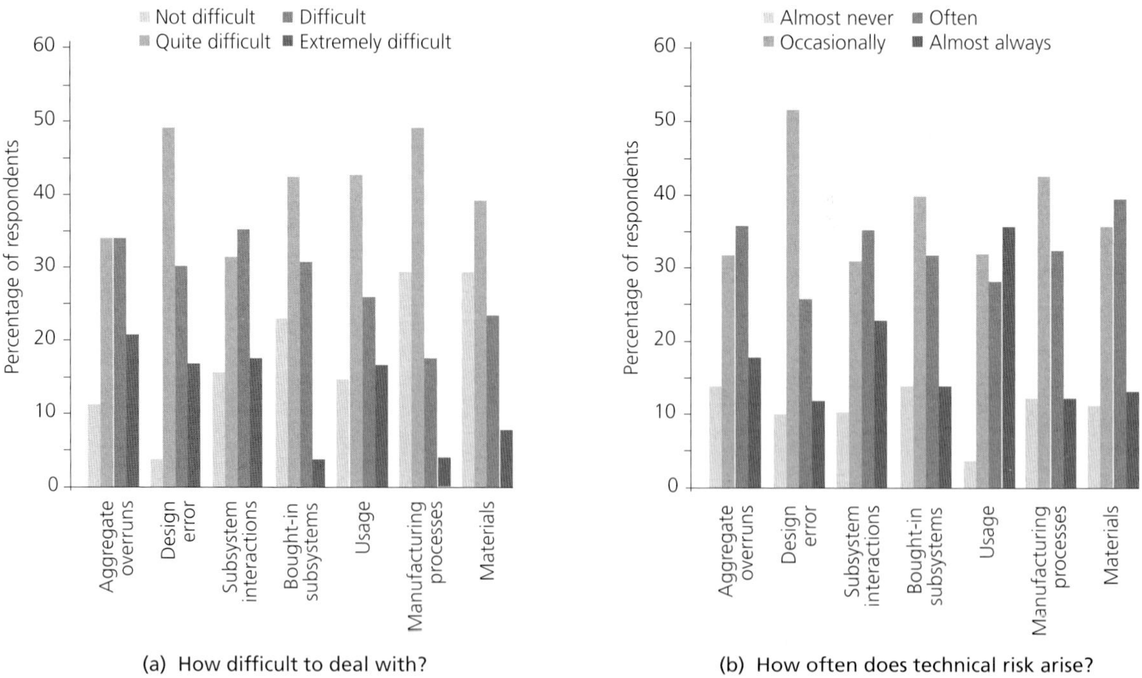

11.5 Risk management survey results

weight, *etc.*), design errors, subsystem interactions and product usage (understanding the loads and usage that a product will be subjected to during its life). Respondents were also asked how often technical risk arises in each area, and here usage, subsystem interactions, materials and aggregate overruns were the most important areas, as shown in Figure 11.5b.

The survey covered a number of other topics, and is reported in full in Crossland *et al.* (1998). Perhaps the most important conclusion from the work from the point of view of future design methodologies is that, while many companies collect data about risk, the incorporation of quantitative models into risk management is rare. Improved techniques are needed to link together data collection with predictive and modelling methods.

Risk in teams

We have noted that engineering is more than ever carried out by large teams, usually distributed between several organisations and often separated by substantial distances. Many of the difficult aspects of engineering risk come from the complexity associated with these large teams. We have also noted that risk is difficult to assess and control where it arises from subsystem interactions, from interactions between participant groups in a project, and from aggregate budgets – where, for example, the responsibility for the weight budget or cost budget for an artefact is spread amongst many participants in a project. Understanding of the risks and uncertainties in a project or in the performance of an artefact will also be distributed amongst the members of a team – in this case the issue is one of communicating this understanding to those responsible for decision making.

Many of the difficult aspects of engineering risk come from the complexity associated with large teams, usually distributed between several organisations and often separated by substantial distances.

In all of these cases, a major issue in risk assessment and management concerns the provision of methodologies that allow members of a team to collaborate in building a shared understanding of risks and uncertainties. Examples of research issues include:

- How can the team accumulate an understanding of the risks and uncertainties associated with the processes and activities that they undertake, particularly to accumulate evidence about the uncertainties inherent in analytical and simulation methods?
- How can the team record its view of the risks and uncertainties arising from subsystem and group interactions and emerging aggregate budgets?
- Can an environment be provided that allows team members to flag up and record their concerns in a confidential manner?

The quantitative evaluation of risk and uncertainty

Although a good deal of risk management is still qualitative, the quantitative assessment of risk is a significant engineering objective, and a number of techniques have been developed to support this. These techniques are also closely allied to the development of more general approaches to design analysis under uncertainty. The most widely used approaches to quantitative analysis of risk, and of uncertainty more generally, are firmly grounded in probability, although fuzzy systems have had an impact, as have some other techniques. For all approaches, introduction has been facilitated by vastly improved computing capabilities.

The most widely used approaches to quantitative analysis of risk are firmly grounded in probability.

The main quantitative risk assessment techniques applied in risk assessment include (Andrews and Moss, 2002):

Fault tree analysis (Schneeweiss, 1999a, b). This is a graphical technique in which occurrences in a system which can result in an undesirable outcome are described in the form of an inverted tree. The most serious outcome, such as explosion, toxic release, *etc.*, is selected as the top event of the tree, and then the remainder of the tree, constructed by considering the sequence of events which individually or in combination could lead to the top event. The construction of the tree allows the probability of contributory events and the logic of event combination to be considered.

Event tree analysis. This is again a graphical technique, used to analyse the consequences arising from a failure or undesired event. An event tree, by contrast, begins with an initiating event, such as a component failure, and then considers consequences of the event through a series of possible paths, where each path is assigned a probability of occurrence. In this way the probability of the various possible outcomes can be calculated.

Decision tree analysis. As the name implies, decision trees are again a graphical technique, but in this case the branching of the tree reflects both choices of action that may be taken and chance events, and the numerical values assigned to the branches reflect probabilities and values of outcomes.

Influence diagrams use more general graphs, in which the nodes represent variables or decisions, and the edges indicate the path or direction in which one node can influence another. Influence diagrams can be used as a basis for decision trees, but can also model more subtle and sophisticated relationships and are perhaps the most general of the diagrammatic techniques.

FMEA. This is a technique that aims to identify potential ways in which a product or process might not meet expectations and any possible causes of such failure, and to rank failures and their causes to indicate where

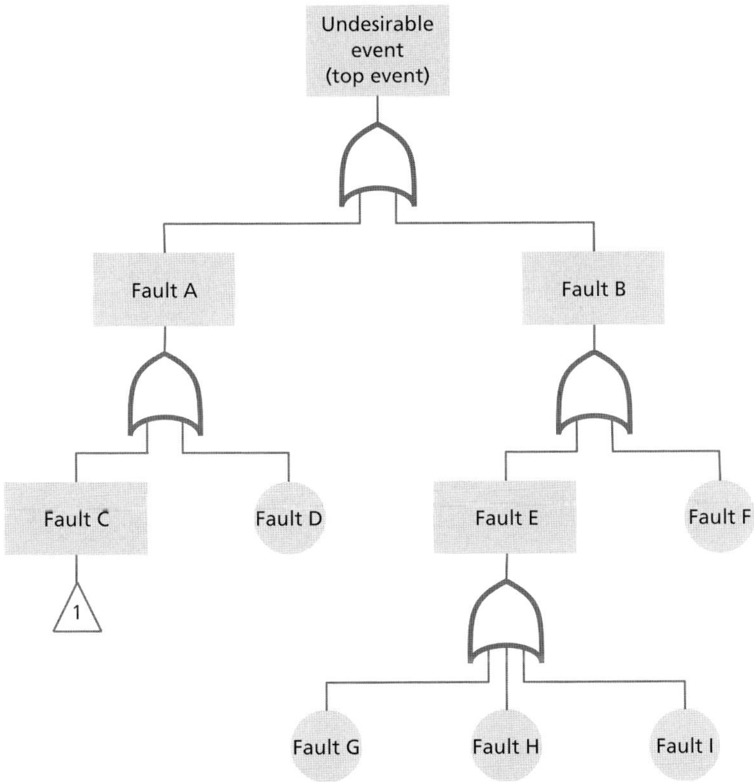

11.6 Fault tree analysis, a graphical tool for risk assessment

engineering effort should be expended to reduce failure likelihood and severity. The basis of FMEA is to try to identify and list all possible ways in which an assembly, a part or a process might fail. For each possible failure mode an assessment is made of the severity should failure occur and possible causes of the failure. For each cause, assessment is made of the likelihood of its occurrence and the likelihood of detection. The three assessments – severity, occurrence and detection – are then multiplied together for each failure mode/cause to give a risk priority number (RPN) which is used as an aid to indicate the priority of action for each mode.

Technical risk assessment tools include all of the techniques mentioned so far, as well as safety factors and a number of reliability techniques, in particular based on limit state analysis and the first- and second-order reliability methods (FORM and SORM) (Hasofer and Lind, 1974; Fiessler *et al.*, 1979). Limit state theory provides the framework within which the performance of engineering components can be assessed against various limiting conditions, e.g. a condition of load exceeding resistance in a structure such that the component is no longer able to fulfil its intended function. In the FORM, the limit state is linearised around the design point, the point on the limit state with the highest probability. FORM has the advantage of simplicity, but in highly non-linear situations and as the degrees of freedom of the problem increase it may be subject to increasing error. The SORM is constructed by fitting a parabolic surface (as opposed to a plane surface in FORM) to the limit state function at the design point. The information about the curvature of the limit state function is utilised in SORM, therefore improving results from FORM.

Monte Carlo analysis is extensively used in technical risk assessment for simulations involving extensive computation.

Monte Carlo analysis (Hammersley and Handscomb, 1964) is extensively used in technical risk assessment and probabilistic analysis for simulations involving extensive computation. Where the performance function is computationally expensive (for example, with finite element analysis), techniques such as the response surface method (Bucher and Bourgund, 1990), in which an approximate mathematical function of the performance function is used to avoid computations of the actual performance function, minimise the computation required. Advanced mean value (Wu *et al.*, 1990) and fast probability integration (Wu and Wirsching, 1987) are further approximate techniques designed to achieve good results for computationally intensive situations.

Other methods

There are many other quantitative methods for risk and uncertainty analysis in design. Traditionally, designers have often used deterministic analysis combined with safety factors; in the absence of information about statistical probabilities for design variables, techniques such as interval analysis (for example, applied in tolerance stack analysis) and the absolute worst-case variation (in which the variables are either set to the lowest or largest expected value) are used. Fuzzy theory has had some application in risk assessment, but the use of fuzzy methods is most appropriate in manipulating design imprecision in earlier design phases, whereas probabilistic design is most suited to problems with stochastic uncertainty (Wood and Antonsson, 1989).

Traditionally, designers have often used deterministic analysis combined with safety factors to manage risk.

Industrial application of quantitative methods

The Society of Automotive Engineers (SAE, 2003) reports the following barriers to implementation of probabilistic methods:

1. the methods are a radical departure from existing practices;
2. they are not compatible with existing tools;

3. they are too difficult to use and take too long;
4. they take too much data;
5. the results from probabilistic methods cannot be verified and output data is difficult to interpret;
6. the complexity of multiple failure modes is an issue.

They also note the following limitations of probabilistic methods:

1. lack of guidelines for dealing with remote probabilities;
2. lack of guidelines for data adequacy;
3. lack of guidelines for model adequacy;
4. difficulty in validation;
5. required deterministic calculations can be too expensive;
6. failure modes are often poorly identified;
7. difficulty in negotiating risk limits.

Our experience in exploring the use of probabilistic methods in component life assessment is that many of these issues are important, but by far the biggest difficulties, at present, concern the lack of sufficiently complete data (and associated data and model guidelines) for the application of the method – for example, in automobile engineering the necessary data would include that on road conditions, driver behaviour, material properties and the effects of treatment (for example, on residual stresses), the behaviour of tyres and bushes and so on. And even if a full set of data were available on all aspects of the design problem, there would still be limitations in our understanding of the uncertainty inherent in the analytical techniques. This suggests the need for a database framework that would allow information to be collected and collated for use in risk and uncertainty evaluation.

Even if a full set of data were available on all aspects of a design problem, there would still be limitations in our understanding of the uncertainty inherent in the analytical techniques.

Risk perception

Risk to life and limb has always been of particular concern to engineers, and many of the quantitative approaches to risk and much of the legal and regulatory emphasis on risk have concerned such hazards. However, it is now increasingly recognised that the separation between the objective and subjective in risk is difficult to maintain – it is also accepted that all knowledge of risk has an element of subjective judgement. The subjective is particularly important in judging the societal impacts of hazards.

A central problem, however, lies in the discrepancies between the analytical frameworks used by designers to determine risk, and the qualities of a risk that actually influence risk bearers. Design risk analyses assess probability and impact, whereas lay people appear to perceive risk on the basis of a variety of factors that give them a richer picture of what a risk means to them. These factors include dread (lack of control, catastrophic potential, inequitable distribution, *etc.*) and the extent to which a risk is unknown (being new to society, being delayed in its effects, *etc.*) (Slovic, 1987). They seem to be influenced by various cultural biases (Adams, 1995) and the information they receive about risk is mediated by a range of social mechanisms (Kasperson *et al.*, 1988).

An important impact of risk perception is that people often overestimate the risk associated with very low probability events, and underestimate that associated with high probability events. As an illustration of this issue, consider Lomborg's (2001) observation that "if we drink water which contains pesticides at the EU limit value for a whole lifetime, we face the same death risk as if we smoke 1.4 cigarettes, cycle 15km, live two months in a brick building or drink a half litre of wine – just once". If we asked people what they perceived to be the risk from these various sources, we would surely get a very different view of the relative risks inherent in the different activities.

People often overestimate the risk associated with very low probability events and underestimate those associated with high probability events.

There is a basic question about whether design, in the service of society, should replace society's inexpert risk assessment with its own conception of what is rational – or whether it should incorporate in its own risk assessment models some of the dimensions that influence risk bearers. If the former, then designers need to communicate and influence users more effectively, and there are basic questions as to how to do this. If the latter, then there are some difficult questions about how qualities like dread should be incorporated in risk analyses in sensible ways.

Risk perception is intimately associated with attitudes to risk and acceptance of risk, and has been the subject of study from a number of perspectives, including the psychology and sociology of risk and the economics of risk (Pidgeon, 1999; Slovic, 2000). Perception is part of the management of risk – people think of risk management as risk reduction, but this is not always possible (Sandman **et al.**, 1997). It is associated with risk communication: through the supply chain, right through to honesty with the public. The issue is how to communicate the residual risk. Psychology and issues of the man–machine interface also have a strong place in studies of error and hazard – human and organisational factors cause up to 80% of risks – and in their impact on health and safety issues.

Conclusion

A number of factors have contributed to the present emphasis on risk in engineering design. We live in a world of complex, interacting engineered systems. The design process is itself often complex, with many, distributed participants working over long time periods to bring products to market. The cost of the process may be high, and the financial implications of failure significant. And both the users of engineering products and the wider community are much more averse to risk arising from engineering design than before – in particular to life and limb, but also commercial and technical risk.

This article has reviewed some of the responses that have been made to the need to manage risk actively. It has introduced the nature of the risk management cycle, has outlined some of the qualitative and quantitative techniques that can be applied in risk assessment and monitoring, and has given an overview of their impact in practice. From this review it has been noted that, while many approaches have been developed, the application of quantitative risk management in practice is limited, and human error, both in designers and in users of their products, remains a significant issue. Furthermore, there is a limit to the extent to which quantitative approaches can be applied owing to the importance of societal attitudes to risk and to acceptance of risk. The engineering designer must take an approach that considers both the formal assessment of risk and the implications of societal risk perception.

References

@Risk Palisade Europe.

Adams J (1995) Risk. UCL Press

Andrews JD, Moss TR (2002) Reliability and risk assessment, 2nd Edition. Professional Engineering Publishing

http://www.palisade-europe.com/html/decisiontools_suite.html

Aven T, Pitblado R (1998) On risk analysis in the petroleum activities on the Norwegian and UK continental shelves. Reliability Engineering and System Safety, 61(1/2): 21–29

Balci O (2001) A methodology for certification of modelling and simulation approaches. TOMACS, 11(4): 352–377

Bea R (1998) Human and organisational factors in the safety of offshore structures. Reliability Engineering and System Safety, 61(1/2): 109–126

Bennett S (2001) Human error – by design. Perpetuity Press

Bloor M (1995) The sociology of HIV transmission. Sage

Boehm BW (1991) Software risk management: principles and practices. IEEE Software, 8(1): 32–41

BSI (1991) Reliability of systems, equipment and components. BS 5760: PART 5. http://www.bsi-global.com

Bucher CG, Bourgund U (1990) A fast and efficient response surface approach for structural reliability problems. Structural Safety, 7: 57–66

Caillaud E, Gourc D, Garcia LA, Crossland R, McMahonC (1999) A framework for a knowledge-based system for risk management in concurrent engineering. Concurrent Engineering Research and Applications, 7(3): 257–268

Cambell P (2002) Learning from construction failures: applied forensic engineering. John Wiley

Carter B, Hancock T, Morin J-M, Robins M (1994) Introducing RISK-MAN methodology: the European project risk management methodology. NCC Blackwell

CCTA (1995) An introduction to managing project risk. HMSO

Chapman CB, Ward S (1997) Project risk management: processes, techniques and insights. John Wiley

CIRIA RiskCom. http://www.ciria.org.uk/

Cooper DF, Chapman C (1987) Risk analysis for large projects: models, methods and cases. John Wiley

Coppendale J (1995) Manage risk in product and process development and avoid unpleasant surprises. Engineering Management Journal, 5(1): 35–38

Crossland R, McMahon CA, Sims Williams JH (1998) Survey of current practice in managing design risk. University of Bristol

Crossland R, McMahon CA, Sims Williams JH (2003) The practical application of design risk assessment models. IMechE, Part B, 214: 227–234

DOD (2000) Military standard 882D – standard practice for system safety. US Department of Defence. http://www.reliasoft.org/mil_std/mil_std_882d.pdf

DOH (2000) An organisation with a memory – report of an expert group on learning from adverse events in the NHS, chaired by the Chief Medical Officer. UK Department of Health. http://www.doh.gov.uk/orgmemreport/

Edwards L (1995) Practical risk management in the construction industry. Thomas Telford

FDA (2000) Medical device use-safety: incorporating human factors engineering into risk management. http://www.fda.gov/cdrh/humfac/1497.html

Fiessler B, Neumann H, Rackwitz R (1979) Quadratic limit states in structural reliability. Journal of Engineering Mechanics Division, 105(EM4): 661–676

Godfrey PS (1995) Control of risk: a guide to the systematic management of risk from construction. CIRIA

Hammersley JM, Handscomb DC (1964) Monte Carlo methods. Methuen

Hasofer M, Lind NC (1974) Exact and invariant second-moment code format. Journal of Engineering Mechanics Divisions, 100(EM1): 111–121

Hawkins FH (1993) Human factors in flight. Ashgate

Health & Safety Executive (1992) The tolerability of risk from nuclear power stations. HSE Books

Heising C, Enzenbach W (1991) The ocean ranger oil rig disaster: a risk analysis. Risk Analysis, 9(1)

ICE (Institute of Civil Engineers and Institute of Actuaries) (1998) Risk appraisal and management for projects (RAMP). Thomas Telford

ISO/IEC Guide 73 (2002) Risk management – vocabulary – guidelines for use in standards. International Organisation for Standardisation

Kammen DM, Hassenzahl DM (2001) Should we risk it? Princeton University Press

Kasperson R, Renn O, Slovic P *et al.* (1988) The social amplification of risk: a conceptual framework. Risk Analysis, 8(2): 177–187

Kletz TA (1992) Hazop and Hazan identifying and assessing process industry hazards. Rugby Institution of Chemical Engineers

Lomborg B (2001) The skeptical environmentalist. Cambridge University Press

Lucas D (1997) The causes of human error. In: Human factors in safety critical systems. Butterworth-Heinmann

Lupton D (1999) Risk (key ideas). Routledge

McMahon CA, Bennett S, Busby JS, Barr G (2002) Risk perception and assessment in design, research review and priority setting exercise. University of Bristol

MOD (1996a) Ministry of Defence: initiatives to manage technical risk on defence equipment programmes. Report HC 361, UK National Audit Office

MOD (1996b) Defence standard 00-56(PART 1)/2 – safety management requirements for defence systems - Part 1: requirements. UK Ministry of Defence. http://www.dstan.mod.uk

MOD (1996c) Defence standard 00-56(PART 2)/2 – safety management requirements for defence systems – Part 2: guidance. UK Ministry of Defence. http://www.dstan.mod.uk

Moder JJ, Phillips CR (1970) Project management with CPM and PERT. Van Nostrand

Monte Carlo for Primavera systems Inc. http://www.primavera.com/products/p3_montecarlo.html

Niwa K (1989) Knowledge-based risk management in engineering: a case study in human–computer cooperative systems. John Wiley

Ould M (1999) Managing software quality and business risk. John Wiley

Patterson FD *et al.* (1999) Managing the risks within automotive manufacturing. Risk Management, 1(3): 7–23

Petroski H (1991) Paconius and the pedestal for Apollo: a case study of error in conceptual design. Research in Engineering Design, 3: 123–128

Pidgeon NF (1999) Social amplification of risk: models, mechanisms and tools for policy. Risk, Decision and Policy, 4(2): 145–159

Rajabally E, Sen P, Whittle S, Dalton J (2003) Combining evidence to justify the appropriate use of models in engineering design. ICED'03, Stockholm, Sweden

SAE (2003) Society of Automotive Engineers. G-11 Probabilistic Methods Committee. http://www.sae.org

Sandman PM, Weinstein ND, Hallman WH (1997) Communications to reduce risk overestimation and underestimation. Risk Decision and Policy, 2: 1–16

Schneeweiss WG (1999a) Advanced fault tree modelling. Journal of Universal Computer Science, 5(10): 633–643

Schneeweiss WG (1999b) The fault tree method (in the fields of reliability and safety technology). Hagen

Simon P, Hillson D, Newland K eds (1997) Project risk analysis and management guide (PRAM). Association for Project Management

Slovic P (1987) Perception of risk. Science, 236: 280–285

Slovic P (2000) The perception of risk (Risk, Society and Policy Series). Earthscan Publications

Starkey CV (1992) Engineering design decisions. Edward Arnold

Stewart MG (1992) Simulation of human error in reinforced concrete design. Research in Engineering Design, 4(1): 51–60

Ward JR, Clarkson PJ (2004) An analysis of medical device-related errors: prevalence and possible solutions. Journal of Medical Engineering and Technology, 28(1): 2–21

Wood KL, Antonsson EK (1989) Computations with imprecise parameters in engineering design: background and theory. Transactions of the ASME, 111: 616–624

Wu YT, Millwater HR, Cruse TA (1990) Advanced probabilistic structural analysis method for implicit performance functions. American Institute of Aeronautics and Astronautics Journal, 28(9): 1663–1669
Wu YT, Wirsching PH (1987) New algorithm for structural reliability estimation. Journal of Engineering Mechanics, 113: 1319–1336

Chapter 12

Design for X

Harald Meerkamm and Michael Koch
Universität Erlangen-Nürnberg

To be successful, companies must develop, produce and deliver products which fulfil a multitude of different requirements. The challenge for the companies' product development engineers is to deliver products that not only meet the customers' requirements, but also respond to constraints imposed on the design process.

Trade-offs are inevitably required across a range of engineering disciplines. As a result, decision making, by evaluating and weighing the various requirements in order to find out those most relevant and important, is a difficult task for the designers.

Company resources also need to be managed in response to other expectations, such as growing business volume and rising profit. Traditionally, such resources have included people, time, material and energy. In addition, environmental issues are increasingly important. Figure 12.1 illustrates this situation, highlighting the importance of optimising the use of such resources and impact on the environment.

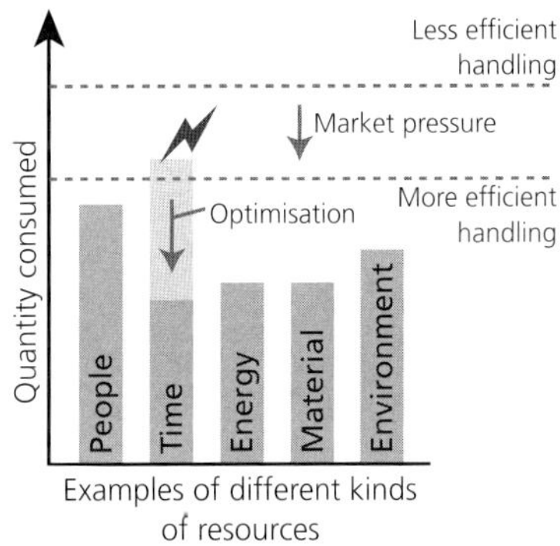

12.1 Need for optimisation due to the competition of the market

Against this background, 'design for X' (DfX) has emerged as a promising approach to identify an efficient way to manage resource optimisation throughout all the different stages of the product development process.

DfX offers strategies for supporting fundamental decisions at the planning stage of the developmental process and can also provide guidance during the later stages. In addition, DfX can assist the broader goal of distributing limited people, time, energy, material and environmental resources in an optimal way in response to market pressures.

'Concurrent engineering' (CE) has also been identified as an important approach for successful product development. CE intends to improve quality, reduce cost, compress cycle times and increase flexibility:

> *through cooperative teamwork between multiple disciplinary functions to consider all interacting issues in designing products, processes and systems from conception through production to retirement.*
>
> (Huang, 1996)

DfX in combination with CE is seen as a functional strategy for the designer to manage the development of a product within the above boundary conditions. To give an impression of 'DfX' itself and the possibilities for the user applying this approach, a few fundamental definitions follow, along with an insight into various DfX implementations.

Definitions

DfX is a systematic approach for making decisions in product development related to products, processes and plants, where the Xs may be in conflict.

The field of DfX consists of fundamentally different ideas of what DfX is or does. For some people DfX is a tool, helping the designer to find the best detailed design. Others think of it as procedures with different check-lists to tick off demands. In this chapter the main focus on DfX is as a 'way of thinking' for the designer, offering:

- a kind of a management tool to define the fundamental boundary conditions for a product during planning and conceptual design;
- detailed descriptions for a designer to support the shaping of the product during embodiment design and detailed design.

To clarify the different terms used in this chapter, some essential definitions are presented:

Design for X approach

The concept of DfX draws together all the tasks that are necessary in order to form a product with respect to the diverse goals and restrictions which apply to that product. Huang (1996) defines the DfX 'approach' in general as "making decisions in product development related to products, processes and plants". This definition presents DfX as an holistic approach, through which the total product development process is supposed to be influenced.

Design for X criteria

All production-oriented characteristics that are conceivable in the frame of a product development are designated as DfX 'criteria', where the 'X' is changed as appropriate for each criterion. Design for assembly (DfA) and design for manufacture (DfM) are but two examples out of an immense number of possible criteria. Herein lies the core problem: the number of imaginable DfX criteria is not limited, leading to difficulty in the management of the approach.

Design for X method

A DfX 'method' is the procedure by which the product developer selects and weighs different DfX criteria.

A DfX 'method' is the procedure by which the product developer selects and weighs the different DfX criteria for the respective product. It can be seen as a way of navigating through a large number of DfX criteria.

Design for X strategy

A DfX 'strategy' is a collection of DfX criteria which results from using the DfX method.

Design for X tools

DfX 'tools' support the product developer in the realisation of a DfX strategy. They enable the synthesis and analysis of one or more DfX criteria. Different implementations of DfX tools will be described later in this chapter.

An approach for structuring design for X criteria

As has been shown, DfX means that the designer has to follow many guidelines during the whole product development process. Starting from the conceptual stage up to the embodiment and detail design, these guidelines and rules are constraints on his or her path to the optimal design solution.

Hubka (1984) defined several "categories of characteristics" according to geometry, kinematics, mechanics, acoustics, *etc*. Using this classification, the first criteria for DfX were developed. The terms DfM and DfA were originally used by Boothroyd and Dewhurst (1983) to encompass their approach to ensuring that a product is both manufacturable and simple to assemble. Since then the expression DfX has emerged to cover a wide range of approaches to product design and a diverse collection of tools, techniques and philosophies.

DfX has emerged to cover a wide range of approaches to product design and a diverse collection of tools, techniques and philosophies.

The problem in adopting different DfX criteria is the rapidly growing complexity of the design process. This results from the fact that every single criterion can be more or less comprehensive in detail, and that there is usually a mix of a large number of criteria from several (sometimes totally different) areas of design. Furthermore, all the different criteria can interact, such that in some instances they support each other (complementary criteria) and in others they contradict each other (competing criteria). Criteria which do not interact at all (indifferent criteria) are possible as well. This diversity requires the designer to compromise between competing criteria.

In the literature there is no useful structure for classifying DfX criteria. They are generally presented in no particular order and there is no difference made according to their location in the design process or to the hierarchy between them. An example of this unstructured usage is to compare design for production with design for cost. At first sight the subtle difference between these criteria is confusing and difficult for the novice to understand.

To reduce the confusion, a new structure for different DfX criteria has been developed (Bauer, 2003). The criteria are classified and hierarchical dependencies are defined, based on the fundamental aims of a company, namely: design for profit, design for resources and design for staff (Figure 12.2).

Guidelines for these aims should represent the restrictions from outside the company (for example, laws) as well as the limitations from inside (for example, limited options in production). These guidelines affect the entire process of design and are boundary conditions which cannot be optimised or changed. Therefore, they are not part of the DfX approach, but have to be considered when using every single DfX criterion, since all known DfX criteria can be seen to be hierarchically subordinate to them.

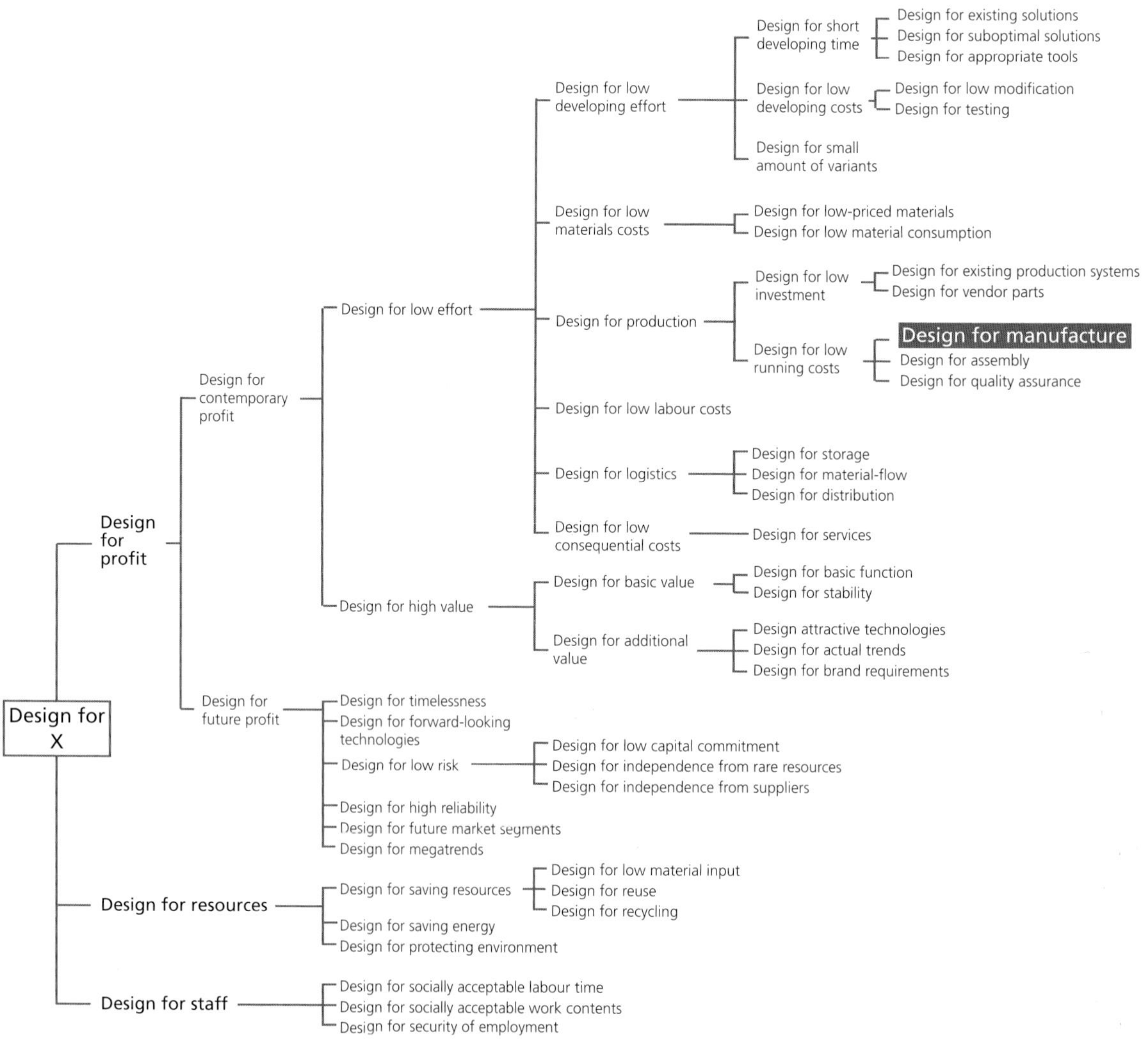

12.2 An approach for structuring DfX criteria (Bauer, 2003)

Each criterion can be split again. For example, the criterion DfM can be broken down into its sub-criteria (Figure 12.3). This particular example shows both the variety of DfX and the fact that different criteria are spread over a wide range of hierarchy levels. Hence, many interconnections and dependencies arise inside this hierarchical structure, leading to the complexity mentioned earlier.

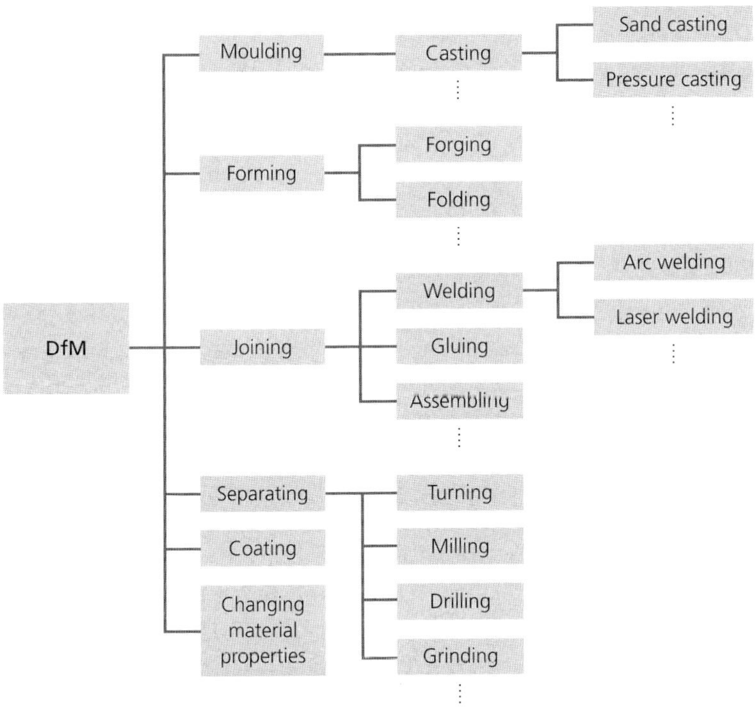

12.3 Complexity of the singled out criterion "design for manufacture" (DIN 8580)

Every company chooses, either intentionally or implicitly, a DfX strategy which includes some of the above-mentioned main aims. This strategy can vary from product to product. To realise different kinds of strategies (for example, one for extended business volume, one for quality products) the weighting of the DfX criteria has to be modified. How each criterion is weighted depends on the specific strategy in the context of the company's general business aims and objectives. According to Miller (1956), seven (plus or minus two) criteria are best. As soon as these elements are chosen, the strategy can be fixed and the criteria than used for realising the product. Different approaches can be considered to support the design process, but CE has proved to be the most suitable and effective.

Integrated product development as the basis of design for X

In order to meet the challenge of DfX, four key areas must be considered:

- methods;
- organisational issues;
- attitude of the designers (considering the complexity of DfX);
- tools (for example, computer systems).

These four areas are included within the framework of 'integrated product development' (IPD), which is understood to promote both an integrated way of thinking and collective, interdisciplinary behaviour. IPD influences both the product and the processes leading to it, as a result encouraging:

- A uniform methodical approach for all product characteristics in the context of the whole product life cycle: function, security, DfM and DfA, cost, design, ergonomics, *etc*. In doing so the customer and user (if not the customer) are involved in determining the desired and actual characteristics of the product.
- An holistic (*i.e.* interdisciplinary) view of the development of products whose components originate from different areas of engineering: for example, mechanical and electrical engineering, mechatronics, thermodynamics and hydrodynamics.
- A methodical procedure, adapted not only to the thinking and working behaviour of the designer, but also to the needs of business, in terms of timescale and communication.
- The functioning of an interdisciplinary team in accordance with the principles of simultaneous or concurrent engineering, fulfilling the above-mentioned demands (in terms of quality, time and cost) in an optimal way.

CE is considered to be an ideal environment for the DfX approach to product development.

It is a well-known fact that the correcting of a design fault is more expensive the later it is detected. Therefore, using DfX, it is necessary to show the designer the consequences of his or her decisions as soon as possible, ideally during the design process, but certainly not after assembly. This can be assisted by the adoption of CE.

Concurrent engineering

CE is considered to be an ideal environment for the DfX approach to product development. The objective of CE is to reduce the system or product development cycle time through better integration of activities and processes. A more formal definition is provided by the Institute for Defense Analyses (IDA, 1988):

Concurrent Engineering is the systematic approach to the integrated, concurrent design of products and related processes, including manufacturing and support. This approach is intended to cause the developers to consider all elements of the product life cycle from conception through disposal, including quality, cost, schedule, and user requirements.
(IDA, 1988)

CE provides an integrated, parallel approach to design. The connection between DfX and CE is shown in Figure 12.4. By paying attention to all aspects of the design during each phase, errors are more likely to be detected prior to being implemented in the product. This integrated design process must include a strong information sharing system, an iterative process of redesign and modification, trade-off analysis for design optimisation and documentation of all parts of the design.

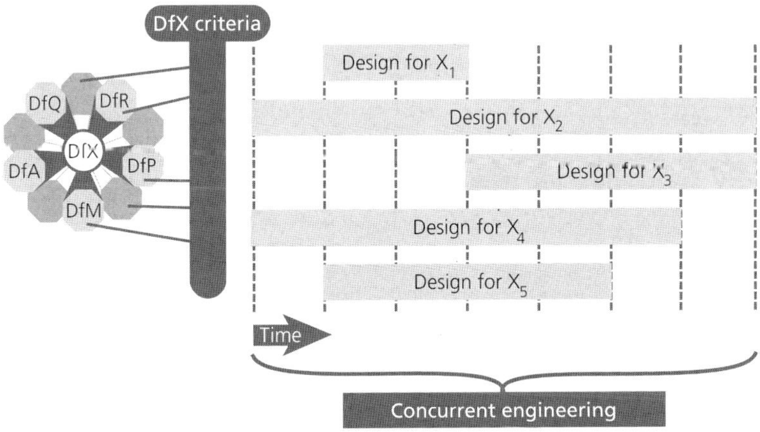

12.4 Connection between DfX and CE

Conventional support for the designer

The simplest kind of support for the designer using DfX is to use the basic 'rule' – to design "unambiguously, simply and securely" (Pahl and Beitz, 1995). This rule is derived from the general objectives of a technical system, i.e. "fulfilment of the technical function" ... "economic realisation" and ... "security for people and environment". Compliance leads to a better chance of realising the product; whereas non-compliance can lead to problems, mistakes and damage. However, this only represents a very high-level view of design, and the use of other rules remains the responsibility of the designer. In addition, there exist high-level principles and strategies for design which are already tailored to different design contexts.

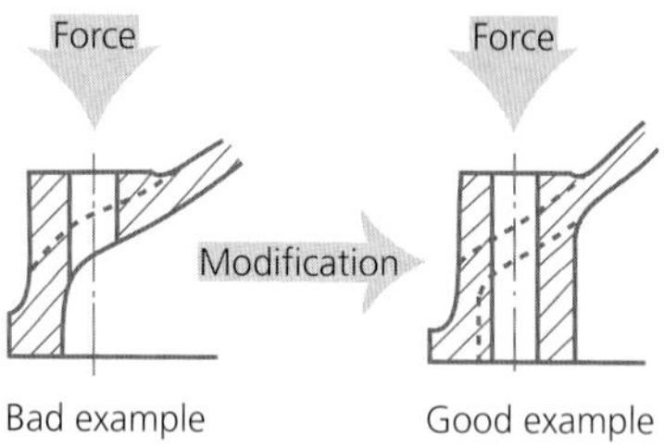

12.5 Good/bad figures as a kind of conventional guideline-based support

Design 'guidelines' are a lot more focused on technology and more problem-oriented, representing instructions for the appropriate design of technical products. They can also be used to review the design retroactively with respect to various DfX criteria. Guidelines are most often represented by illustrations of good and bad practice, supplemented by text describing the benefits of using them (Figure 12.5). The advantage of this approach is that they are easily understood, practical and simple to apply.

'Checklists' are a special form of guidelines. They can act as a stimulus when starting a design task and can be used on completion of the task for systematic checking. Good checklists need to be generated with reference to the product as well as to the company structures. Furthermore, they must be continuously updated to minimise the presence of out-of-date or irrelevant information that could impede their use.

The disadvantage of design guidelines and checklists is that they may appear to be too generic, making it difficult to identify aspects relevant to the current design. Translation of the general rule into useful recommendations is left to the designer. In addition, guidelines very seldom quantify the effects that result from their use. Currently, most design guidelines are documented in the technical literature as collections of solutions. They may also be found in construction catalogues. Computer-based representations are only used in a few individual cases and for limited areas of design.

The disadvantage of design guidelines and checklists is that they may appear to be too generic, making it difficult to identify aspects relevant to the current design.

Stand-alone information technology-based supporting tools

In this section 'stand-alone' software tools which address various independent aspects of DfX are discussed. Many of these tools are based on the finite elements method (FEM), focusing on the analysis of the mechanical characteristics of a design. Other tools focus on examining the manufacturability of a product, e.g. the simulation of the injection moulding of pressure castings, on the basis of digital models (Digital Mock-Up), is state of the art. In this case, three-dimensional geometry models generated in computer aided design systems provide the most important source information and are imported over special interfaces into the simulation program. By their use, casting processes can be simulated (for example, form filling and/or behaviour of solidification) to help determine real process conditions (for example, the profile of injection speed).

A further group of 'stand-alone' tools support the calculation of costs. With these tools each component is subdivided into its production features for which set-up time, primary processing time and auxiliary processing time and thereby the detailed costs of production are acquired. The links between the manufacturing features and the production costs must be stored *a priori* in the knowledge base of the cost-calculation -program. In order to calculate the material costs, the ratio between product geometry/material and the cost for the necessary semi-finished part also has to be defined in advance.

Methods borrowed from investigating DfA in existing products represent a further interesting stand-alone tool. The evaluation of DfA results from the breakdown of the product into its hierarchical building blocks, estimating the time for assembly and developing an 'assembly efficiency' factor on the basis of experience metrics (ratio of the theoretical to the true time for assembly). The impact of using this method is a reduction in the product part count and the optimisation of the assembly and handling processes.

In summary, stand-alone tools enable a range of investigations which deliver good results, with respect to limited (individual) criteria, for well understood tasks. The results are also very dependent on elaborate data modelling. Furthermore, the different tools are concentrated only on the one aspect of DfX.

Stand-alone tools enable a range of investigations which deliver good results, with respect to limited criteria, for well understood tasks.

A problem with all these tools is the transfer of the required geometrical information from existing CAD models. This requires the use of standard interfaces which frequently cause the loss of information. Therefore, the extraction of relevant features often comes back to the designer. Finally, stand-alone tools only test out weaknesses in a design, leaving the interpretation and improvement of the work to the designer.

An integrated approach (Engineering Workbench mfk)

The different facets of and problems with DfX were discussed above. Owing to the complexity of the problems, an integrated approach is required. In response, the Engineering Workbench mfk (KSmfk) has been created; it is a system which offers a range of tools to the designer to support the use of different DfX criteria. The fundamental requirements for such a tool are detailed below, followed by examples of specific DfX implementations.

Architecture and mode of operation of the Engineering Workbench

The system consists of an information-generating synthesis part and an information-processing analysis part. The link between these and the main part of the system is the product model, which contains all the data required to design the product.

The Engineering Workbench (Figure 12.6) is connected to a CAD system through an interactive interface. The CAD system is provided as a user interface and for visualisation of the design.

The synthesis part of the system offers an object-oriented description of the components to the designer by allowing access to design elements through a design module. These descriptive elements are structured hierarchically. With basic elements, the user can construct hierarchically higher design elements, together with design building blocks (solution related). These three hierarchical levels are sufficient to describe components.

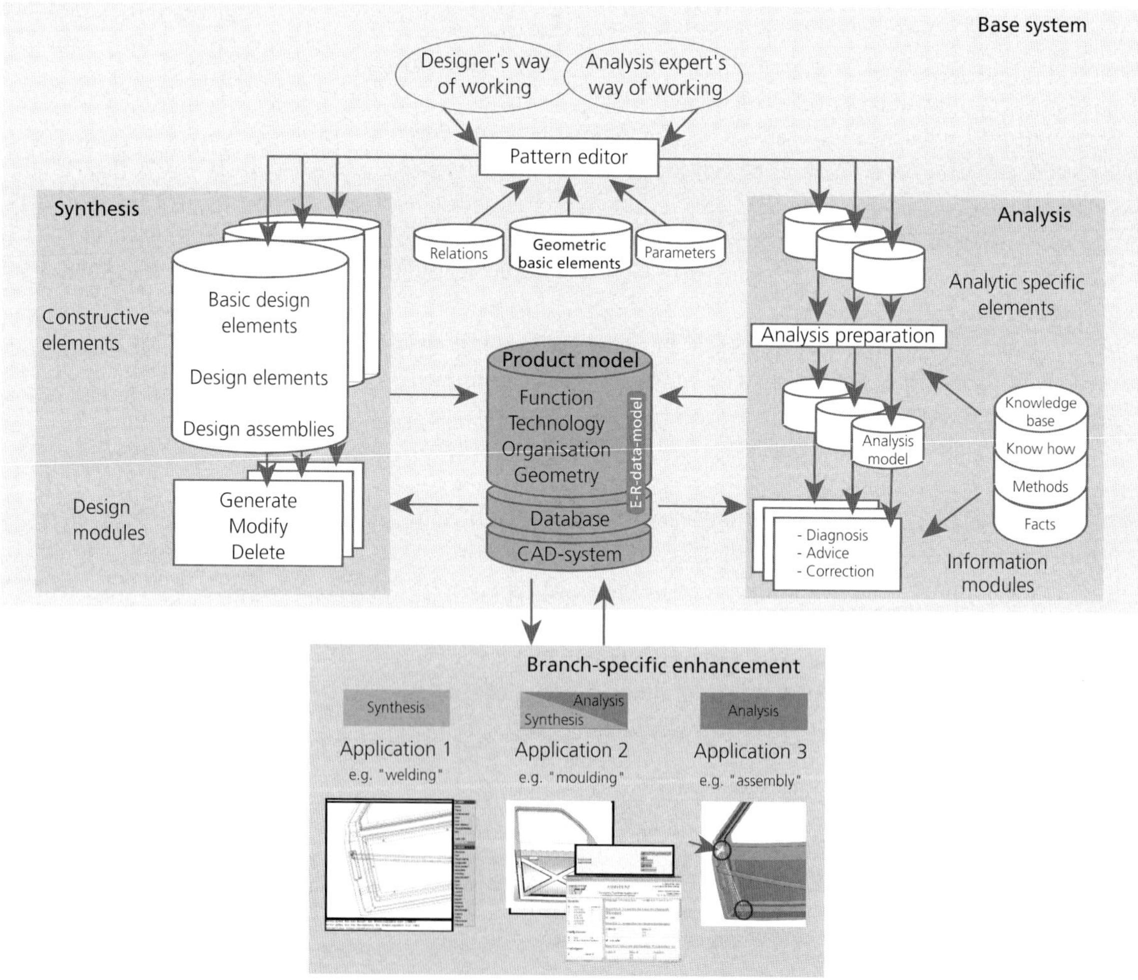

12.6 Engineering Workbench KSmfk

The complete description of components calls for different types of basic design element. It is important to differentiate between geometric elements, technology elements, function elements and organisational elements, all objects from the designer's language, with regard to their information content. The functions for handling the design elements are available in the design module. In addition to the shape functions (generating, changing) there are also a range of management functions.

In the information-processing part of the design system, the information modules allow access to an extensive knowledge base. It contains – in the form of facts, methods and experience – the knowledge that is necessary for the completion of a number of individual problem-oriented analyses.

The functionality of this analysis part enables the assessment of the previously generated design (from the synthesis part) and facilitates different levels of analysis. The simplest is diagnosis, with which the system only asserts that the rules have been violated, e.g. rules of design for production. In addition to this, the system can advise the designer, based on the diagnosed errors, of several alternative suggestions for eliminating the diagnosed problem. At the most complex analysis level, the system can automatically perform the suggested correction, an action that only has to be confirmed by the designer.

Product model

The interface between the synthesis and analysis parts of the system is the product model. It contains all the product-defining data specified by the designer as regards geometry, technology, function and organisation. Product model data are generated automatically during the synthesis procedure. Recent developments have changed the conventional product model into a 'hybrid' version based on relational structures. This offers much potential for the support of the early design stages, because accurate geometric data and any desired semantic information can both be stored in the product model. By using a commercial database (Oracle™) it is possible to save and retrieve all the accruing or required information at every point of the design process. Moreover, accessing the database over standard interfaces (SQL) makes it possible to supply the product model with data from different software tools. Most often this link is used by CAD systems, but other stand-alone modules can also access the product model. Efficient support for the widespread design process increasingly relies on accessing the product model in this way.

The most interesting relationships, with respect to the CAD systems, are the topological ones, particularly in connection with groups of components, e.g.

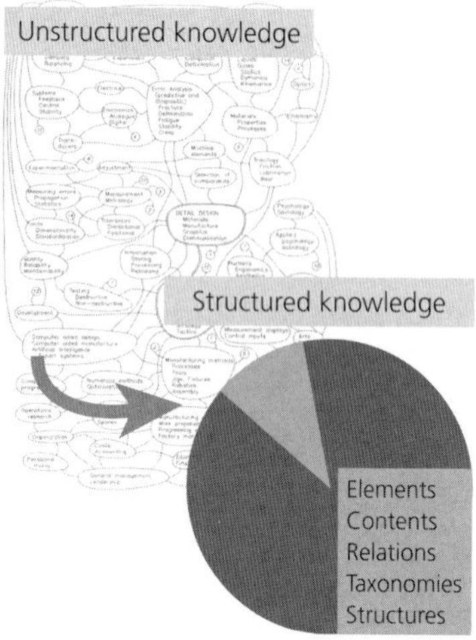

12.7 Categorising unstructured knowledge

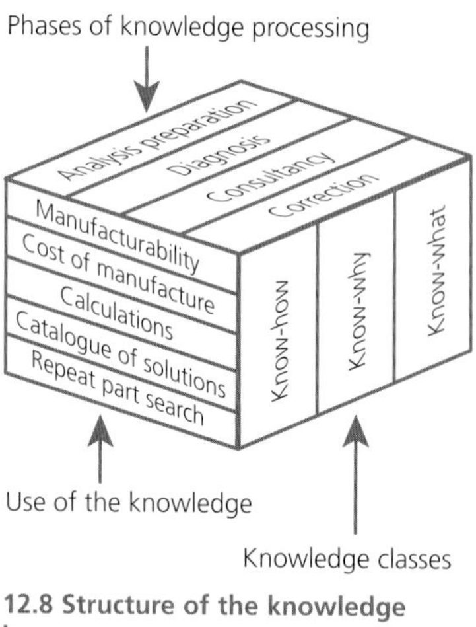

12.8 Structure of the knowledge base

problems relating to tolerancing or assembly. For a completely new product, the designer moves from the abstract to the concrete. The Engineering Workbench, therefore, should support the designer from the early stages. Starting from the functions within a function structure, the user can design the operating structure of the whole concept in detail and non-solution-specifically. For this purpose, the above-mentioned function elements serve to describe information concerning part functions, loads, qualitative and quantitative constraints as well as relationships.

Knowledge processing in the Eengineering Workbench

The information modules of the Engineering Workbench present DfX methods in the form of analysis functions that enable the assessment of design results. Thus, they allow simultaneous access to an extensive knowledge base. Quick access to the relevant knowledge requires the structuring of the knowledge base, according to the different criteria. The complexity of the knowledge relevant for design is shown diagrammatically in Figure 12.7.

As a consequence, one finds comparable complexity when using DfX to explore relationships between a number of DfX criteria, *i.e.* properties. So it is vital to find the right structure for the knowledge designers need to work with. Otherwise, they will not succeed in using the tools currently available, e.g. databases, expert system shells, hypermedia systems, *etc.* Figure 12.8 shows the rough structure of the knowledge base used for the engineering workbench. First of all, a structure is required for knowledge classes:

- The know-why knowledge class, for example, includes design rules which can be described in the form of if – then relationships.
- The know-how can be set up in the form of feasible programs. With this know-how, facts are determined that can be used as operators for the 'condition' part of the rules.
- The know-what contains information that can be processed from the above-mentioned programs of the know-how.

The structuring in different knowledge classes must be universal as well as factory specific.

The designer needs different types of knowledge for different jobs. That is why structuring the knowledge base according to individual needs is necessary (for example, for calculations, analysis of the production rules, manufacturing cost calculations, *etc.*). It must be possible to get specific views from the knowledge base.

DfX with the Engineering Workbench

Several design modules, in combination and cooperation with a complex but well-structured knowledge base, provide good support for the designer in the difficult field of DfX.

Design for stress

The prerequisite for the integration of dimensioning and the selection of calculations is the provision of all relevant information. The above-described component model fulfils this condition in most cases. As well as the structured shape information, forces, moments and different possible views can be shown. Furthermore, it contains material information necessary for the required detailed calculations.

The applied calculation packet, BETSYAX, is a tool developed by a research group at the Department for Technical Mechanics (University of Erlangen, Germany). It is useful for the calculation of notch stresses in axi-symmetrical components. The component model generated in the synthesis part of the system is the starting point for the calculation. The designer describes the component's shape, load and position. The models for the different calculation methods (FEM and boundary element method) are automatically read from the component model. This ensures that inconsistencies between the different models are excluded. For the preparation of the data for the calculation model, the knowledge base has to be reaccessed. The visualisation of the calculation results (diagnosis) follows within the CAD system using traditional post-processors.

The designer needs different types of knowledge for different jobs. That is why structuring the knowledge base according to individual needs is necessary.

This integration allows the designer to use numerical calculation methods for frequently recurring calculations during shape optimisation (for example, the calculation of notch stresses). For such standard analysis, this way of working can replace the calculation engineer. However, the designer is still responsible for keeping all rules and data up to date.

Design for production

The Engineering Workbench enables many types of analysis to be carried out and is especially suitable for supporting the designer during design for production. One way is by checking manufacturability in general. During this process, the system informs the designer whether the component generated in the synthesis part of the tool can be manufactured using the tools available in the factory. Semi-finished products, tools, machines and fixtures, whose data are stored as facts (know-what) in the knowledge base of the Engineering Workbench, can all be examined in this way.

12.9 The designer should be able to test, by checking against the design rules, whether the design is feasible for production
© Airbus

Over and above that, the designer can test whether the design is feasible for production by using the analysis part of the tool and checking against the design rules (Figure 12.9). For example, a component can be checked for tool run-outs, minimum wall thickness, material accumulation, *etc*.

To cover the area of design for production fully, a large number of design and information modules are necessary, e.g. for the design of turned components, sheet metal components and cast components. In this way, the individual design modules make application-specific, basic design elements available. Information about the production process is stored in the knowledge base. It is important that all such modules are compatible with one another to enable the designer to change quickly from one application to another.

The system starts with a design object which provides the outline of the workpiece. Material information is specified in the technology element (Figure 12.6). Using this information and comparisons with a database of semi-finished products, the system orders the right semi-finished product for the workpiece.

By comparing the semi-finished dimensions with the maximum allowable within the available workspace, the designer can check whether the workpiece can actually be produced on a machine in the factory. This is possible since details of the production equipment are also stored in the knowledge base. Finally, elements are assigned for detail design, where recognised errors are shown graphically and alphanumerically on the workpiece. The designer also receives suggestions for the correction of the error.

Design for casting

It is important that any desired component geometry should be able to be analysed in terms of design for casting, independent of its complexity. Here, the analysis is based on the component's shape. Critically, design for casting must identify areas of material accumulation, wall thickness difference and undercuts. For this purpose, functions were developed which analyse all areas of the cast part with respect to these issues.

The starting point for these analysis functions is the basic concept that undercuts in the CAD model may be determined by 'sending out' an analysis ray. In contrast to previous research work, in this approach a strategy is pursued to determine an analysis point for each surface of the cast part, thereby dramatically reducing the volume of data generated and the corresponding time required for the analysis. The procedure described above is best suited to the analysis of undercuts. If wall thicknesses, material accumulations or hot spots need to be detected, the designer can take advantage of simulation tools which, in combination with appropriate computer hardware, enable nearly every desired parameter to be analysed at an early stage in the design process.

Design for tolerancing

Tolerance analysis is very important in the design process, and choosing the right kind and size of tolerances has a large influence on product cost (affecting manufacturing, assembly, *etc.*). As a result, the Engineering Workbench includes a tool for tolerancing, which presents the necessary tolerance information to the designer as required (see Figure 12.10).

Tolerance analysis is very important in the design process, and choosing the right kind and size of tolerances has a large influence on product cost.

The tool enables tolerances to be examined at different stages of the design process. This is made possible by the component or product module, which stores tolerances as technological elements in the form of independent data sets. The tool allows for different types of tolerance processing, ranging from the calculation of dimensional variations for a component to the automatic generation of a numerical control program for manufacture.

Cost estimation

The designer has a basic responsibility to be aware of costs, and this requirement means that cost estimates should be available as early as possible in the design process. However, until now the coupling of design and cost calculation has been prevented by the use of different databases for these processes and the shortage of bi-/unidirectional conversion programs for their contents.

The necessary technical information required for cost estimation, e.g. cost models for step drilling and fittings, *etc.*, are not found in a normal CAD structure. They are, however, available in the above-mentioned component model. Hence, for the first time, the Engineering Workbench can now be coupled with a program, the HKB package (see Figure 12.11), for the evaluation of cost estimates (Wartzack, 2001).

The designer has a basic responsibility to be aware of costs, and this requirement means that cost estimates should be available as early as possible in the design process.

The product model is first of all searched for manufacturing characteristics, *i.e.* objects associated with the cost calculations that can be defined by the HKB user. These objects contain a combination of geometrical, technological and organisational information, against which unambiguous costs can be allocated. Each new manufacturing characteristic must be defined by the Engineering Workbench as a combination of design elements in the form of a mask. Analysis of the data structure then identifies such characteristics, enabling cost estimations to be carried out using standard manufacturing cost estimation programs.

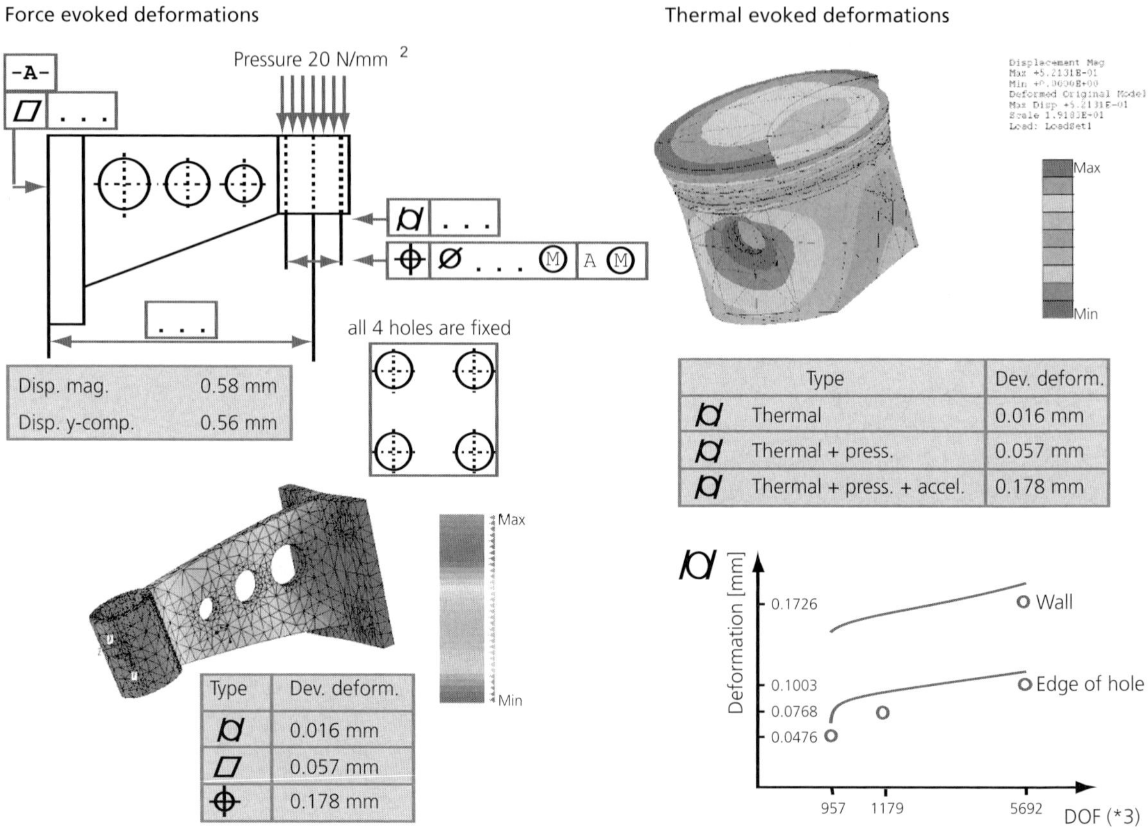

12.10 Results of two examples of the tolerancing module of the KSmfk

Design for environment/recycling

Design for recycling, which represents a small part of the approach to design for environment, encourages the designer to look at the end of the normal product lifecycle. It questions what can be done after the product has reached the end of its life and looks at the opportunities for reusing parts or recycling materials, *etc*.

The challenge in this area is that the designer has to make actual decisions about products which will probably only be recycled after 10 or 20 years. Ideally, the designer should be supported in these decisions by a comprehensive knowledge base containing the relevant material data, and disassembly and recycling strategies, *etc*. It is possible to integrate such a knowledge base in the design system as a module. In this way the designer can be provided with a tool for analysing a design with a view to recycling.

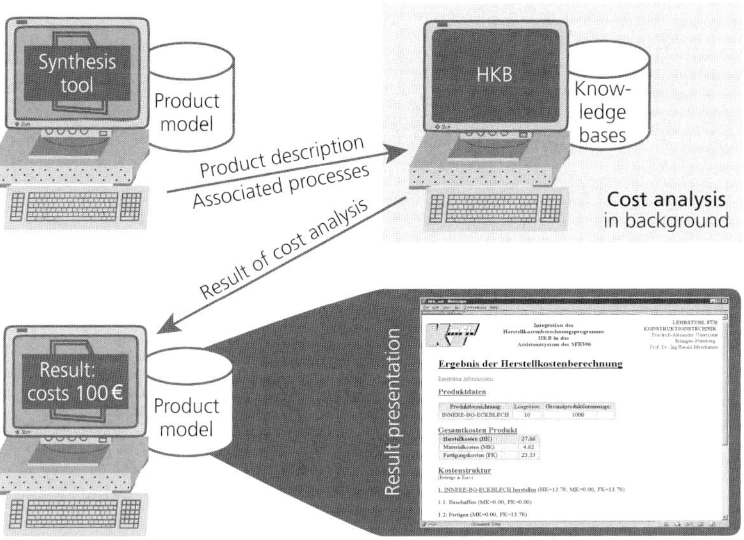

12.11 Integrated cost analysis

Conclusion

In this chapter the field of DfX has been introduced and discussed. Even though approaches for structuring the different DfX criteria exist, DfX remains a complex area: it is not easy for the designer to consider all the issues throughout the design process. Accordingly, a number of different implementations have been described which are able to support the designer by giving him an appropriate methodological base and suitable computer tools based on the latest hardware and software.

References

Bauer S (2003) Design for X – Ansätze zur Definition und Strukturierung. 14th Symposium DfX, Neukirchen, Germany

Boothroyd G, Dewhurst P (1983) Design for assembly handbook. Boothroyd Dewhurst Inc.

Huang GQ (1996) Design for X. Chapman and Hall

Hubka V (1984) Theorie Technischer Systeme. Springer

IDA (1988) Institute for Defense Analyses. Alexandria, VA, USA

Miller GA (1956) The magical number seven, plus or minus two. Psychological Review, 63: 81–97

Pahl G, Beitz W (1995) Engineering design: a systematic approach. Springer

Wartzack S (2001) Predictive engineering. VDI Fortschrittsberichte 336

Design management

Chapter 13

Engineering knowledge management

Ken Wallace, Saeema Ahmed and Rob Bracewell
University of Cambridge

The need to improve engineering knowledge management (EKM) is driven by the current challenges facing manufacturing organisations in the emerging global economy and, in particular, by the important role knowledge plays in the engineering design process. Industrial organisations are facing increasing international competition, and in response their engineering products are improving in terms of performance, reliability and cost of ownership. Current pressures on industry include:

- need for improved product quality, shorter lead times and reduced costs
- difficulty of managing large, multidisciplinary design teams that are not necessarily collocated
- increasing complexity of both products and processes
- rapid rate of change of technology
- problem of retaining knowledge and experience
- increasing complexity of computer-based tools
- requirement for sustainable development
- increasing risk of product liability litigation.

The aerospace industry is subject to all the pressures listed above with its extremely complex products, which are often designed and developed by large multidisciplinary teams as part of a multinational collaboration. Our recent research in the aerospace industry has focused on the problem of retaining design knowledge for future use as experienced engineers and technology experts move to other organisations or retire.

The product development process involves a number of stages, including feasibility study, design, development, production, distribution, operation and recycling. Starting with a product idea or market need identified during the feasibility study, the decisions taken by the design team define the product that will eventually be produced and sold. The correctness, or otherwise, of these decisions, therefore, has a fundamental bearing on the future commercial success of an organisation.

The importance of an effective and successful design process to the commercial viability of an industrial organisation cannot be overstated.

The earlier in the product development process that an error is identified, the lower the cost to rectify it. Expenditure rises rapidly as a product goes into production, and errors identified at this stage are costly and time consuming to rectify, and can delay a product's launch. Worse still, if a fault occurs once the product has gone into operation then not only are the costs of rectification extremely high, but also the reputation of the organisation can be seriously damaged. The importance of an effective and successful design process to the commercial viability of an industrial organisation cannot be overstated (Moore and Rayson, 1996).

Traditionally, large amounts of knowledge and experience are never written down and are only stored in the heads of individuals.

Traditionally, large amounts of knowledge and experience are never written down and are only stored in the heads of individuals. Empirical studies have shown that engineering designers spend around a quarter of their working day gathering information and that the most common source of this information is to ask colleagues (Court, 1995; Marsh, 1997). Changes taking place in manufacturing organisations, along with increasing globalisation, mean that those with the required expertise are either not going to be so readily available to consult in the future or, if they are, not necessarily face to face. The reason for this is the increasingly transient nature of modern industry. When individuals move to another part of the organisation, leave or retire, they take their knowledge with them, and in many cases this knowledge is lost forever.

EKM is a broad field, so only a number of selected issues will be addressed. We draw on our experience of EKM research in the area of aerospace engineering design. However, we believe that the issues discussed are equally relevant to other areas.

A key issue is how to capture knowledge and experience from the memories of individuals, store and then retrieve it, so that it can be reused in future design processes.

Design process

Designing involves *people*, a design team with the appropriate expertise, undertaking a *process*, a sequence of activities arranged into phases and steps, to define a *product*, its configuration, components, materials and construction. This activity takes place within a particular *organisation*, which provides the necessary infrastructure and resources. For the members of the design team, it is *knowledge* that links everything together and enables them to take the actions and to make the decisions that direct the process and determine its outcome (Pahl and Beitz, 1996).

The engineering design process is a knowledge-intensive activity. Each step in the process involves members of the design team identifying the knowledge that defines a particular sub-task and then using their expertise to process that knowledge into a state that defines the selected sub-solution. The final product definition is an appropriate combination of all the selected sub-solutions. The quality of human expertise and the ability to retrieve and use knowledge throughout the design process are crucial to the outcome.

Human expertise comprises personal ability and knowledge. Ability includes intelligence, talent, creativity, judgement and skills. Some of these qualities, such as intelligence and talent, are innate, while others, such as judgement and skills, can be developed through exposure and training. Knowledge includes knowing about, knowing how and knowing why and is acquired through exposure, reasoning and education.

Experts have a high level of expertise in a particular field and are recognised by their peers. Although exposure to a variety of design projects always increases experience, it does not necessarily make an individual an expert. Novices, by definition, cannot be experienced, but they may have sufficient knowledge and ability to be experts, particularly in fast-moving new fields.

A design team can start a design task with different levels of prior knowledge and information (Bohn, 1994). At the lowest level, the team might have no prior knowledge of the product area and simply be given a task, e.g. 'devise a means of shortening grass'. Here, the members of the team must rely on their existing knowledge and experience along with what they can retrieve. This starting point has the advantage that it can avoid fixation on past design solutions and open the door to novel ideas. At the next level, the design team might be given the task and several physical examples of previous products, e.g. lawnmowers, to take to bits and analyse. At the third level the team might have available some documentation, traditionally drawings and reports, that provide insights into the rationale behind the design decisions. Finally, at the fourth level, the design team might also have access to the designers who worked on the previous products. In many engineering design situations, including aerospace, the fourth level is the most common.

Observational studies by Marsh (1997) in the aerospace industry in 1996 produced the result that in no less than 90% of information requests another person was approached rather than documentation. Observational studies by Aurisicchio and Wallace (2004) in the same aerospace company in 2003 showed that this figure had dropped to around 75%. The reduction may be due to the improvements in information technology (IT) that have taken place in the intervening period. However, the message is clear: members of a design team rely very heavily on being able to consult colleagues and experts.

Members of a design team rely very heavily on being able to consult colleagues and experts.

The reason for using human sources for acquiring information is that they provide:

- rapid, accurate, up-to-date and trustworthy information;
- answers based on the context of the situation rather than on the questions asked;
- the rationale behind the process and the product not available elsewhere;

- strategic guidance on how to tackle a particular stage in the design process;
- guidance about who to ask if they cannot provide the answers themselves;
- support and confidence that the right approach is being adopted.

An important issue is how to help designers acquire expertise, in particular the relevant product and process knowledge, as quickly as possible.

Knowledge

When considering future EKM systems, and in particular design support systems, it is important to start by considering: (1) what we understand by knowledge; and (2) the knowledge requirements of engineering designers.

The exponential growth in computer processing speed and storage capacity means that almost unlimited amounts of information and data can now be stored, processed and searched. It might, therefore, be expected that more progress would have been made in capturing design knowledge, and in particular design rationale, in an effective design support system. However, knowledge based and expert systems have, so far, failed to live up to their initial promise. Some of the reasons for this are discussed by Studer *et al.* (1998) in their comprehensive review of knowledge engineering. Knowledge is generated in large quantities throughout the design process and has to be captured either during the process, *i.e.* in real time, or after the process has been completed, *i.e.* retrospectively. It is clear that capturing the required knowledge as the process proceeds is best, but designers are always under considerable pressure to meet project deadlines and will, quite naturally, resist any additional burden. Some argue that, with almost unlimited computer storage capacity, all the knowledge generated should be captured. How this knowledge should be captured is seldom discussed, but the assumption is that it will be possible to search this knowledge using sophisticated search engines. However, a more pragmatic approach would be to aim to capture and store only the knowledge that is likely to prove useful in the future.

When individuals capture knowledge in their minds, they structure and index it (Akin, 1990). One of the problems with indexing knowledge as it is stored is that one does not know how, where or when that knowledge might be useful in the future. If a subsequent search is based on an indexing system that is too rigid, the required knowledge might not be retrieved. In contrast, the human mind is very flexible, and by assimilating context can jump across mental indexes, which is one of the reasons why so much knowledge is currently sought and retrieved from human sources.

One of the problems with indexing knowledge as it is stored is that one does not know how, where or when that knowledge might be useful in the future.

There are many ways of defining and classifying knowledge, information and data (Ahmed *et al.*, 1999). Whereas data are clearly distinguished, the distinction between knowledge and information is blurred and the two terms are frequently used synonymously. Our particular view is that what is stored and transferred outside the human mind is information. Knowledge only exists when information is interpreted. One classification of the types of process and product knowledge, along with some examples, is shown in Table 13.1.

	Stored externally	Stored internally in human memory		
	Information	Explicit knowledge	Implicit knowledge	Tacit knowledge
Process	Descriptions of the design process (information)	Explanations about the process (rationale)	Understanding about the process (strategies)	Intuition about the process (insights)
Product	Descriptions of the product (information)	Explanations about the product (rationale)	Understanding about the product (relationships)	Intuition about the product (insights)

13.1 Classes of knowledge and information

Explicit knowledge can be articulated, *i.e.* 'written down' and stored externally in the form of information, e.g. in external repositories such as physical media, paper-based media and electronic media.

Implicit knowledge cannot be articulated by the person possessing it. However, it is possible to articulate it and store it externally after it has been extracted through knowledge elicitation methods.

Tacit knowledge is knowledge that, by common definition, cannot be articulated. However, its influence on the design process can be researched.

When designing, designers use explicit knowledge, including their own, as well as that gained from colleagues and from external repositories, along with their own implicit and tacit knowledge, to generate new knowledge. The information retrieved will come from that stored about the current project, past projects, the domain and general knowledge.

Attempts to identify the knowledge requirements of designers have focused largely upon characterising the types of knowledge designers require and include the research of Baya (1996). His study focused upon questions asked explicitly by the participants, *i.e.* it was assumed that the participants were aware of their knowledge needs. However, research carried out by Ahmed and Wallace (2004) found that designers are not always aware of their knowledge needs. In this study, a total of 633 queries made by novice designers were analysed to identify their knowledge needs and the results suggested that they were aware of the specific knowledge that they needed to know in only about one-third of all their queries. The main conclusion from this particular study was that novice designers need guidance in forming questions to identify what they need to know. This highlights an important shift in knowledge management systems, which should prompt designers about what they need to know, *i.e.* supply them with questions as well as answers.

Novice designers need guidance in forming questions to identify what they need to know.

The study showed that novices were confident asking others for 'know about' knowledge, but were less inclined to seek 'know how' and 'know why' knowledge, e.g. knowledge about design strategies and design rationale. These two areas will be addressed in later sections.

Important issues are how to capture, store and retrieve more explicit and implicit process and product knowledge in external information repositories; how to structure it so it can easily be interpreted by those who need to re-use it; and how to understand more fully the role of tacit knowledge in the engineering design process.

Process and product knowledge

Designers need a wide range of knowledge, including knowledge about the structure of the organisation and the people in the organisation, but we will focus on the process and product knowledge, referring again to Table 13.1.

Process knowledge

Explicit knowledge about the process can be articulated and stored *externally* as information. It includes descriptions about how to undertake the stages and steps of the design process; when to apply the methods and tools; and where to obtain missing knowledge. This information is stored in reports, standards and manuals and is easily retrieved.

Explicit knowledge about the process, which is stored *internally*, includes knowledge that experts have about undertaking a process. If asked, they can provide explanations of why the process is the way it is, *i.e.* the rationale underpinning the process. This rationale can be articulated and stored, although it seldom is.

Implicit knowledge about the process includes the understanding that experts have about how to undertake a process or procedure in practice.

This know-how demonstrates itself in the strategies adopted by experts when tackling design tasks and the questions they ask themselves when searching for additional knowledge.

Tacit knowledge about the process expresses itself as the intuitive feel experts have for a process, observable through the actions they take.

Product knowledge

Explicit knowledge about the product stored *externally* includes descriptions about a product's functions, its configuration, its components, its materials, and its construction. This type of information is well captured and stored for both current and past projects, traditionally in the form of drawings and documents, but now increasingly in the form of CAD files, product data management systems and databases. It can easily be retrieved using modern search engines that employ indexes, links and keywords.

Explicit knowledge about the product, which is stored *internally*, includes explanations of why the product is the way it is, *i.e.* designers understand the rationale underpinning the product. They are encouraged to record the reasons for their decisions in reports at the end of a design project, but these reports are usually done retrospectively and often in a hurry as the next project starts. Acceptance decisions are recorded but the rejection decisions seldom are.

Implicit knowledge about the product includes the understanding of why a product is the way it is and the relationships that make it work. This know-how expresses itself in the ability to create sound designs without apparent discursive reasoning.

A key issue is how to capture knowledge and experience from the memories of individuals, store and then retrieve it, so that it can be reused.

Tacit knowledge about the product expresses itself as an intuitive feel for the working of a product, which experts cannot articulate but is observable through the judgements they make.

Two specific issues for supporting designers are how to improve the capture, storage and retrieval of: (1) design strategies; and (2) design rationale.

Supporting designers – design strategies

Expertise, which is built up through exposure to problem-solving situations, takes time to acquire. It is argued that it takes at least 10 years of exposure before one can be considered an expert in a particular field (Sonnentag, 1998). In the future it is likely that designers with less experience will have to tackle new design tasks, so any means of helping designers to acquire the required expertise faster would be beneficial.

The expertise possessed by designers plays a significant role when solving problems, as does the nature of the situation,e.g. degree of urgency and working conditions, as well as the social aspects, e.g. established team and supportive colleagues (Weth, 1999). Experts rapidly assess the complexity of problems based upon the knowledge available to them and adjust their problem-solving strategies accordingly. Experts clearly have a larger body of internal knowledge gained from past projects to draw on and intuitively know what questions to ask when retrieving that knowledge. In order to assist novices, it is important to understand more about the strategies experts adopt when tackling design tasks and how they retrieve knowledge.

Until recently, research into the differences between novices and experts focused on problems where constraints and contexts are well defined, and a limited number of rules apply, e.g. chess. The problems tackled in engineering design are ill-defined as they do not have a set of rules to achieve a solution. There is no clear goal state for design problems and there is no one definitive solution. When tackling design problems experts tend to reason forwards, whereas novices tend to reason backwards (Zeitz, 1997). Forward reasoning progresses from the given information to the unknown, whereas backward reasoning progresses from a hypothesis about the unknown back to the given information. When solving complex problems, experts alternate between forward and backward reasoning. Kavakli and Gero (2001) suggest that the forward reasoning employed by experienced designers suggests that they possess and apply strategic knowledge, but they do not identify what this strategic knowledge is. If this is indeed the case, identifying this strategic knowledge and employing this as part of a design support system would have the additional benefit of encouraging forward reasoning, *i.e.* expert behaviour.

When tackling design problems experts tend to reason forwards whereas novices tend to reason backwards.

Observations of novice and experienced engineering designers attempted to characterise some of the strategic knowledge that experienced designers possess (Ahmed *et al.*, 2003). Twelve observations of novice and experienced designers carrying out design tasks were undertaken. The observations identified eight strategies that experienced designers adopted. Novice designers had not developed such strategies and were found to be unaware of them. The eight design strategies have been combined in a method named C-QuARK and are summarised below (see Figure 13.2). The arrows represent the most common moves between strategies made by experienced designers.

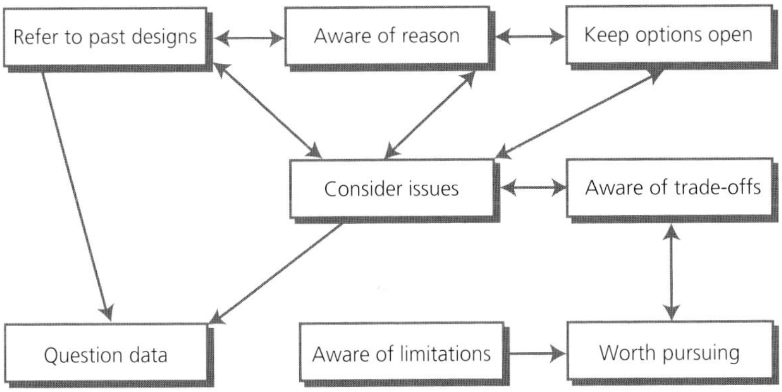

13.2 C-QuARK: design strategies employed by experienced designers

Consider issues: experienced designers tended to consider several relevant issues, and decided which were the most important. They were also aware when issues were not relevant.

Aware of reason: experienced designers were often aware of the reasons behind the use of a particular design solution or manufacturing process.

Refer to past designs: experienced designers referred to past projects to find similar designs, similar environmental and functional conditions, and similar problems that had been encountered and resolved.

Worth pursuing: experienced designers asked themselves how much they could expect to achieve if they continued a particular approach and if it was worthwhile.

Question data: experienced designers questioned data they obtained from any source. They questioned the accuracy of the data, how components were modelled or tested, how much accuracy was required, customer specifications, and the applicability of standards.

Keep options open: experienced designers rejected an option or delayed a decision on an option if it limited later options in the design task. They were aware of what needed to be considered further on in the design process.

Aware of trade-offs: experienced designers were aware of the relationships between issues. They were aware that many decisions were based on compromises and when they were aware of the trade-offs, they would question whether it was better to pursue the task or to implement a decision.

Aware of limitations: experienced designers were aware of the limitations of the current design task and hence of the amount of time to spend on it.

Using the C-QuARK method has been shown to encourage novices to begin to adopt these strategies and to encourage them to ask more appropriate questions when seeking knowledge from others.

Supporting designers – design rationale

The potential for design-rationale-capture tools to improve the design process is very great. Lee (1997) lists the advantages that such tools could offer, if they could be used naturally by designers without impeding their work. For example, they could provide better support for redesign, reuse, maintenance, learning, documentation, collaboration, and project and dependency management.

When individuals move to another part of the organisation, leave or retire, they take their knowledge with them and, in many cases, this knowledge is lost forever.

The rationale-capture research field is well covered in the book edited by Moran and Carroll (1996), which contains contributions from most of the major groups. It is now acknowledged that rarely have such capture techniques been successfully applied in industry, except perhaps in the context of facilitated meetings (Conklin *et al.*, 2001). The various classes of design-rationale tool will now be described, along with an assessment of the aspects that inhibit their use by engineering designers.

Research into capturing and mapping the rationale for complex decisions can be traced back over 30 years to the issue-based information system (IBIS) of Kunz and Rittel (1970). The basic concept of IBIS is simple. It is a tree or directed graph, with nodes representing issues to be solved that are linked to child nodes representing alternative solutions. These, in turn, are each linked to further children representing arguments for or against. As can be judged by the number of its derivatives (Moran and Carroll, 1996), including gIBIS, itIBIS, PHI, QOC, DRL, Questmap (Conklin *et al.*, 2001), and Compendium (www.compendiuminstitute.org), the simplicity and expressive power of IBIS holds strong intellectual appeal. Additionally, a study of a number of existing design reports written by designers in an aerospace company showed that much of their contents would map well into an IBIS structure (Bracewell and Wallace, 2003). However, the same study suggested that the practical use of these existing IBIS tools, by the designers who authored these reports, would be severely hampered by a number of problems. The first is their use of single-line labels to represent rationale elements, whereby the full text of the element can only be viewed or edited by double-clicking to pop up a window. Thus, for every issue, solution or argument captured, the user needs to summarise it meaningfully into no more than five or six words, which is likely to prove an intolerable burden.

A second problem is that these tools have no clear and consistent way of representing element status, so the user is forced to adopt text conventions in labels to represent status information. For example, in Questmap it is conventional to prepend the label of accepted elements with a star, while enclosing the labels of rejected elements in brackets. These are difficult to pick out at a glance and make the rationale difficult to follow.

There are two main classes of IBIS tools, mainly differentiated by their strategies for managing large rationale spaces. These are planar graph-based tools, such as Questmap or Compendium, and collapsible outline or tree-style tools, such as itIBIS, Égide/DRAMA (Bañares-Alcántara *et al.*, 1995), or R-Objects Pepper (Ernst, 2002). There are further practical problems, specific to each class, for their use in engineering design.

A major problem with planar graph-based tools is the unavailability of clear, comprehensive hard-copy output that can easily be related to what the designer sees on the screen. This is because, while the graphs can easily be printed, the arguments are unclear because the text is hidden within the rationale elements. Perhaps an even more serious problem is the use of what is known as transclusive linking to distribute a single large rationale across multiple views. This simply means that the same node appears in more than one view. The fundamental problem with this is that answers must sometimes be separated from the issue that they are addressing, and arguments divorced from whatever they are arguing about. The result again is a loss of clarity of the design arguments. Additionally, the only way to appreciate the complete context of any element is to visit in turn all views in which that element appears – a laborious procedure requiring much information to be retained in short-term memory.

Collapsible outline style tools, on the other, hand allow a large rationale to be captured in a single view, thus avoiding the problems of transclusive links. The user chooses what to display at any time, by expanding or collapsing levels at will. Additionally, unlike graph-based tools, comprehensive hard copy is not a problem. The outline is simply printed fully expanded, and the troublesome text hidden in elements displayed by one further level of expansion. However, these advantages come with two even more serious problems. First, while many design and decision spaces are largely tree structured, adequately representing those parts that are not, in a tree- or outline-based tool, invariably leads to serious difficulties.

The quality of human expertise and the ability to retrieve and use knowledge throughout the design process are crucial to the outcome.

The second, more subtle problem, is the lack of clarity with the outline style presentation, where it is difficult to determine which levels need to be collapsed and which need to be expanded to appreciate a particular point. It is easy to miss a crucial argument simply because it is hidden in a collapsed branch, or is scrolled off the top or bottom of the screen because unimportant branches have been expanded.

Design Rationale editor (DRed) – a new IBIS-based tool

DRed is a new graph-based IBIS design-rationale-capture tool that addresses the problems listed in the previous section, and as a result allows a much clearer view of the rationale structure and content than has previously been possible. The rationale is contained in nodes of various types arranged on multiple linked workplanes, with dependencies between them represented by directed links (Bracewell and Wallace, 2003). Figure 13.3 shows a screen shot of the visual representations of all of the available node types and their statuses.

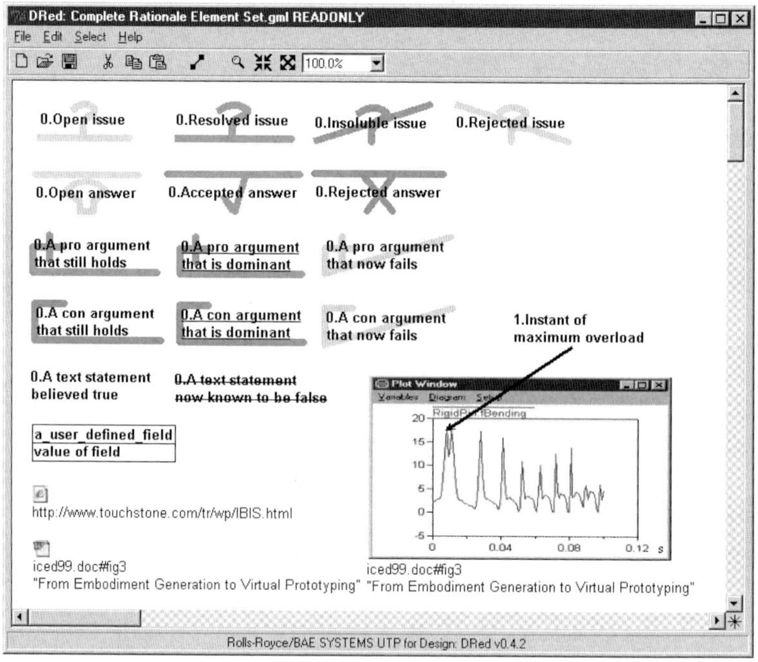

13.3 DRed element types and their statuses

Distinguishing features of DRed are now briefly described. Tunnelling links appear to tunnel into the workplane, to reappear elsewhere, either on the same or a different workplane. Such links permit large and complex rationales to be distributed across multiple workplanes, and laid out legibly, while still allowing easy navigation. Double clicking the tunnel mouth, represented by a small circle, carries the mouse pointer through the tunnel to the opposite mouth (see Figure 13.4). Double clicking that mouth carries the pointer back again. Unlike transclusive linking, each element appears on just a single workplane, where its meaning is clear. DRed uses text elements overlaying background graphics that automatically resize with the text displayed, of arbitrary width and number of lines. The type of node is represented by the background shape. This allows the whole contents of the node to be viewed at a glance together with a surrounding knot of related elements, without the need to open additional windows. Support for embedded graphics with anchored call-outs means that sketches, screen dumps from other tools such as CAD, and other graphical elements can easily and seamlessly be embedded in the rationale structure.

Rationale dependency links can point to and be anchored to particular feature locations within the graphical elements. References to external files and Internet URLs are easily included. All elements have a small set of allowed statuses, clearly displayed by suitable combinations of changes in colour and geometry of the background shape and font style of the text. These statuses are changed by the designer as work progresses from, for example, unresolved to resolved. Thus, the tool provides clear prompts of what needs to be done, combining the functions of a notebook and a 'to do' list. Additionally, dominant arguments are emphasised clearly. Clear, unambiguous, comprehensive hard copy is always available, with no information hidden in the tool that is not visible in the printout.

An evaluation of DRed in industry has shown it to be easy and intuitive to use. This makes the design process faster overall and more rigorous. The view of the rationale structure is comprehensive, clear to see and understand, both by users and others. Tunnelling links simplify the preparation and presentation of large rationale structures in a way that no other system provides.

An evaluation of DRed in industry has shown it to be easy and intuitive to use.

Tunnelling links provide dependency linking between views by: (1) showing clearly that a dependency on an element of a particular type on a particular workplane exists; (2) allowing easy navigation through the tunnel to the view in which that element was defined and where its context, meaning and status is clear; and (3) allowing easy navigation back again through the tunnel. A particular advantage is the natural and intuitive interface, which means that the system is easy to learn, easy to use and clear as an archival method. The method of displaying the rationale element type and status, using resizable coloured background shapes overlaid by text, saves screen area and allows more of the linked nodes, which provide the context of a particular node of interest, to be viewed at a glance along with that node. It also makes it easier to grasp immediately the type and status of a particular statement in a complex rationale.

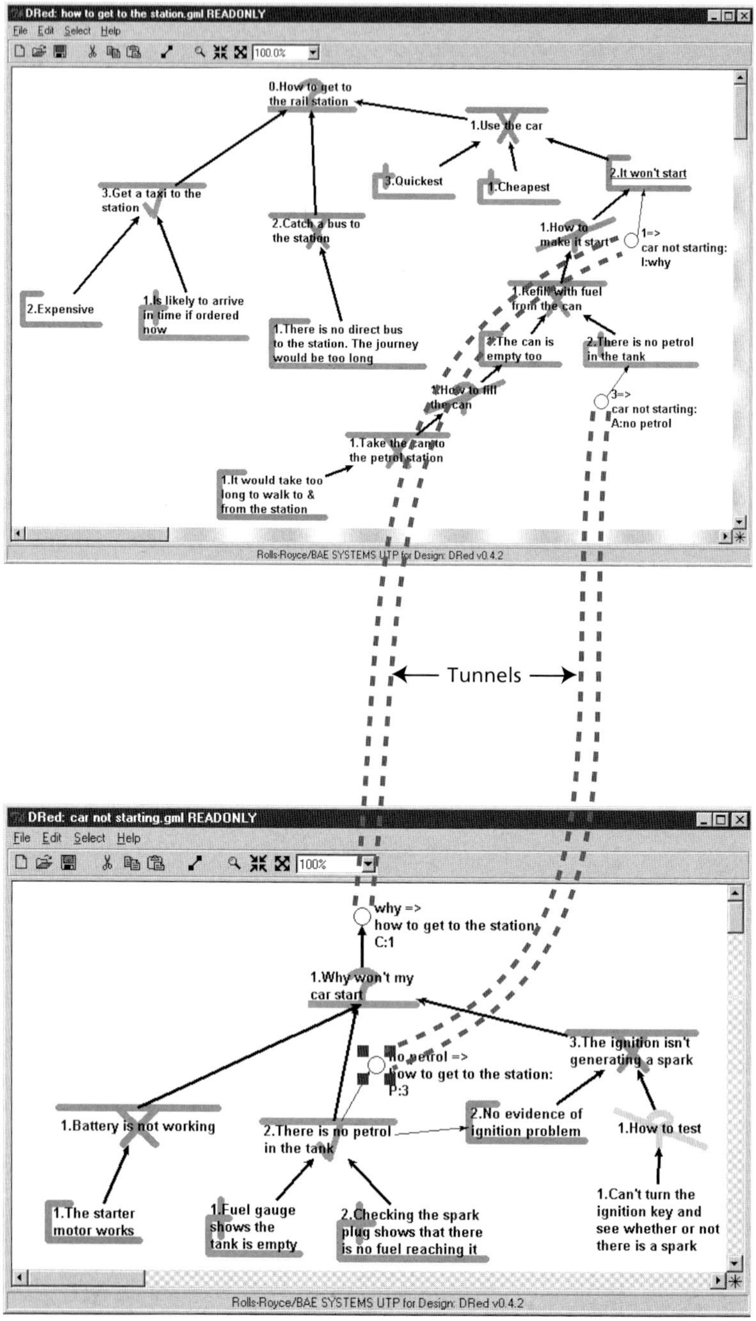

13.4 Example DRed graph distributed between workplanes using tunnelling links

Figure 13.4 shows an example DRed graph, simple enough to be self-explanatory, that explores the best way of getting to the railway station when the car will not start. It is distributed across two workplanes by means of tunnelling links.

Conclusions

Industrial organisations are facing many pressures and must continually improve their products to stay competitive. Effective engineering design is key to their survival and is a knowledge intensive activity. Engineering designers acquire most of their knowledge from colleagues and experts. However, due to the current transient nature of modern industrial organisations, experienced designers and technology experts are not going to be so readily available to consult in the future. A key issue is how to capture knowledge and experience from the memories of individuals, store and then retrieve it, so that it can be reused in future design processes.

Engineering designers with less experience and novices will have to undertake more design projects in the future. An important issue is how to help designers acquire expertise, in particular the relevant product and process knowledge, as quickly as possible.

For future EKM systems, and in particular design support systems, an important issue is how to capture, store and retrieve more explicit and implicit process and product knowledge in external repositories, and to understand more fully the role of tacit knowledge in the engineering design process. Two specific issues for supporting designers are how to improve the capture, storage and retrieval of: (1) design strategies; and (2) design rationale.

It has been shown that experts adopt eight basic strategies and by applying these strategies, supported by a set of generic questions, novice designers can move more rapidly towards adopting expert behaviour.

Most current design rationale capture systems have proved unsatisfactory and too cumbersome, but a new IBIS-based tool called DRed has been shown to overcome many of the problems. Trials have shown that DRed is easy and intuitive to use and actually helps designers structure their design process, *i.e.* it assists rather than hinders. The fact that it captures a considerable amount of design rationale effortlessly as the design process proceeds, and reduces the need for formal reports, is seen as a tremendous advantage.

There is a need for industrial organisations to be more process-based rather than product-based, and, in particular, to address the issues of how to capture, store and retrieve design knowledge independently of human sources.

There is a need for industrial organisations to be more process-based rather than product-based, and, in particular, to address the issues of how to capture, store and retrieve design knowledge independently of human sources.

References

Ahmed S, Wallace KM (2004) Understanding the knowledge needs of novice designers in the aerospace industry. Design Studies, 25(2): 155–173

Ahmed S, Blessing LTM, Wallace KM (1999) The relationship between data, information and knowledge based on an observation of engineering designers. ASME DETC'99, Las Vegas, NV, USA

Ahmed S, Wallace KM, Blessing LTM (2003) Understanding the differences between how novice and experienced designers approach design tasks. Research in Engineering Design, 14(1): 1–11

Akin Ö (1990) Necessary conditions for design expertise and creativity. Design Studies, 11(2): 107–113

Aurisicchio M, Wallace KM (2004) Information requests and consequent searches in aerospace design. Design 2004, Dubrovnik, Croatia

Bañares-Alcántara R, King JMP, Ballinger GH (1995) Égide: a design support system for conceptual chemical process design. AI System Support for Conceptual Design Workshop, Lancaster University, UK

Baya V (1996) Information handling behaviour of designers during conceptual design: three experiments. PhD thesis, Stanford University, CA, USA

Bohn RE (1994) Measuring a technological knowledge. Sloan Management Review, 36(1): 61–73

Bracewell RH, Wallace KM (2003) A tool for capturing design rationale. ICED'03, Stockholm, Sweden

Conklin J, Selvin A, Buckingham Shum S, Sierhuis M (2001) Facilitated hypertext for collective sensemaking: 15 years on from GIBIS. Hypertext'01, Århus, Denmark

Court A (1995) The modelling and classification of information for engineering designers. PhD thesis, University of Bath, UK

Ernst J (2002) Collaborative decision making and personal knowledge management with R-Objects Pepper. WWW2002, Honolulu, HI, USA

Kavakli M, Gero JS (2001) Strategic knowledge differences between an expert and a novice. In: Strategic knowledge and concept formation III. University of Sydney, Australia

Kunz W, Rittel HWJ (1970) Issues as elements of information systems. Working Paper 131. Berkeley, CA, USA

Lee J (1997) Design rationale systems: understanding the issues. IEEE Expert Intelligent Systems and their Applications, 12(3): 78–85

Marsh JR (1997) The capture and utilisation of experience in engineering design. PhD thesis, University of Cambridge, UK

Moore CJ, Rayson P (1996) The electronically defined aero engine. CALS'96, London, UK

Moran TP, Carroll JM eds (1996) Design rationale: concepts, techniques, and use. Lawrence Erlbaum

Pahl G, Beitz W (1996) Engineering design: a systematic approach. Springer

Sonnentag S (1998) Expertise in professional software design: a process study. Journal of Applied Psychology, 83(5): 702–715

Studer R, Benjamins VR, Fensel D (1998) Knowledge engineering: principles and methods. Data and Knowledge Engineering, 25: 161–197

Weth RVD (1999) Design instinct? The development of individual strategies. Design Studies, 20: 453–463

Zeitz CM (1997) Some concrete advantages of abstraction: how experts' representations facilitate reasoning. In: Expertise in context. AAAI Press

Chapter 14
Quality management

Graham Thompson
UMIST

The subject of quality in design process improvement can be considered from a number of perspectives. Indeed, it is quite possible that a survey of engineers could elicit views of the subject that differ markedly. Therefore, the treatment given here does not represent any specific school of thought or narrow perception of the subject. Rather, we cover a diverse range of topics that are influential on our theme of design process improvement and of interest to design practitioners.

Design activities and their timing, the use of appropriate design tools, human factors and, most importantly, satisfying the clients are discussed. No narrow definition of the client is assumed. The word *client* is understood here to include consumers who purchase products, internal clients within a company, services to another company, society at large or the environment. Thus, design activities may pertain to products and systems both large and small.

To achieve high-quality design, the design team must understand what is required by the client and what will best fulfil the client needs. Such a simple statement belies the difficulties that may be encountered. For example, does the client really know what he wants, has he thought through his needs in depth, does the design team believe mistakenly that they know best, and in a long project will the client perspective be lost?

To achieve high-quality design, the design team must understand what is required by the client and what will best fulfil the client needs.

Given a clear understanding of the client needs, the design team must undertake their work effectively and efficiently to create the desired outcome. This can involve a diverse range of activities ranging from conceptual to detailed design. Different thinking skills must be employed; for example, there will be times when divergent thinking and solution finding will be at a premium, whereas at other times convergent analytical ability will be essential to produce a quality product. The literature contains many design methods (tools) which are highly valuable when used competently and at the correct time. But used wrongly or inappropriately, the same design tools will achieve a poor quality outcome. Finally, it is people who will carry out the design work and, in addition to their ability to perform certain tasks, they have certain attributes that are highly influential on the outcome of any design activity. High-quality people who are unduly influenced by their personal traits will not achieve quality design outcomes.

The following sections explore the above factors, but it is not possible to provide a fully comprehensive treatment. The aim is to cover the topics noted in sufficient depth to be of use to practising engineers and to give references to enable further in-depth study if the reader is so minded.

Design activities

14.1 A novel and useful product

There are many publications that describe in-depth design activities and their relationship to each other (Pugh, 1991; Roozenburg and Eekels, 1991; Pahl and Beitz, 1996; Cross, 2000). Although authors differ in their descriptions, there is no fundamental variation between them, and the basic design activities may be described as follows:

- understanding the needs of the client (including market analysis);
- writing a design specification that defines requirements;
- generating and evaluating concepts;
- developing schemes (embodiment of concepts) and evaluation;
- detail design to enable manufacture to take place.

For a detailed discussion of what is involved in each of the above activities, the above references may be consulted.

Typically, these activities are presented in a sequential manner with iteration between stages, the iteration being stated either explicitly or implicitly. A design engineer could undertake each activity in turn and progress from the first contact with a client through to detail design. In fact, engineering design teachers will often refer to a 'design process' from analysis of need through to detail design. However, many design engineers in industry will not practice the process in the course of their work because their company departmental organisations are based on tasks, e.g. a body department in a motor company. Within each department, parts of the idealised design process will be carried out and the relationships between people within the same, or different, departments may well be recognised within the context of the overall process. The classical descriptions serve well to identify the principal design activities. But, for the objective of achieving high-quality design, a different approach can be taken, albeit within the context of the generalised design process.

Creativity can be defined as a combination of novelty and usefulness.

It is useful to turn to the field of creative problem solving (CPS) to help understand how each individual design activity might be undertaken to achieve high-quality outcomes. Creativity can be defined as a combination of novelty and usefulness. It is a practical subject for engineers (Figure 14.1). Briefly, the CPS process starts, continues and ends with the client needs (Isaksen *et al.*, 1994; Fox and Fox, 2000). It can be summarised as:

- need finding, making sure the requirements of the client are understood;
- problem finding, ensuring that the correct problem is solved;
- idea finding, finding potential solutions and evaluation using client focused criteria;
- acceptance finding, exploring how to implement solutions with the client.

The quality of the CPS process does not lie in the identification of activities and their sequencing. Of course, the correct sequencing of activities is necessary, but of more importance is *how* each activity is undertaken. The success of the process lies in the disciplined divergent and convergent thinking that takes place at each stage and the type of tools that are employed. Divergent thinking involves the suspension of judgement as one investigates a problem and searches for information, formulates problem statements, generates ideas, searches for solutions, *etc*. Convergent thinking involves the imposition of value judgements, analysis and decision making.

Divergent thinking involves the suspension of judgement as one investigates a problem and searches for information, formulates problem statements, generates ideas, searches for solutions, *etc*.

Superficially, we could map CPS process onto the design process and talk about equivalence, but it is more important to recognise that each design activity requires divergent–convergent thinking. The use of rigorous divergent–convergent thinking during each design activity is a significant step to achieve high-quality outputs for each activity. Also, it is important to use appropriate design methods (tools) at each stage, which will be considered later in this chapter.

Therefore, each design activity should involve both divergent and convergent thinking:

- The identification of client needs and derivation of a design specification involves exploration of diverse factors (the market, environment, the aims of the client) and the analysis of the whole to clarify specific requirements.
- Concept design involves generation of ideas, their evaluation, development of ideas to create feasible concepts and finally a choice of preferred concept – this is a set of divergent–convergent activities.
- Schematic and detail design involves identification and consideration of alternative, particular detail solutions, and their evaluation and choice.

Figure 14.2 illustrates the divergent–convergent nature of design in the context of CPS.

If any stage of design omits, or does not treat effectively, the divergent thinking activity, then unsuitable, unimaginative, uncompetitive solutions will be produced. This is especially true if divergent thinking is lacking in the early stages of design. Engineers tend to be good at analysis and criticism, but exploration and suspending judgement is harder. At the heart of any consideration of quality design lies the creation of competitive solutions that are fit for purpose, solutions that satisfy the real needs of a client. Particular design methods may be used to create reliable products with high performance, but there is no point in creating an excellent solution to the wrong problem.

Convergent thinking involves the imposition of value judgements, analysis and decision making.

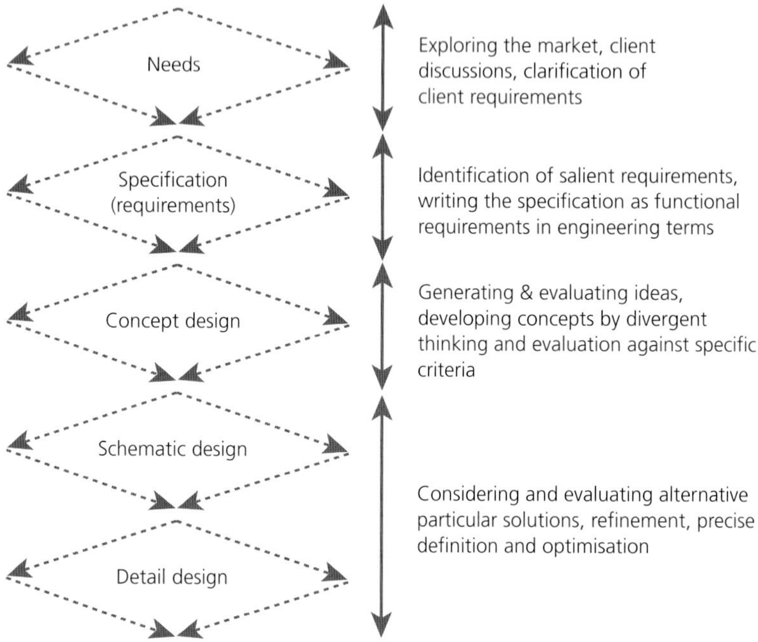

14.2 The divergent–convergent nature of design

People – cognitive style

The outcomes of design activities, processes and methods depend on people. In addition to abilities that may be described variously as academic, common sense, technical know-how, *etc.*, there are certain personality traits that are highly influential in determining the outcomes of design work. There are a number of personality factors that can be described with confidence, e.g. the Myers–Briggs type indicator (Isaksen *et al.*, 1994; Fox and Fox, 2000). A highly significant factor is that of personal preference for a particular problem-solving style, described by Kirton (2003) on a measurable continuum from *adaptor* to *innovator*.

The adaptor prefers to work within the paradigm, to improve existing solutions, and to achieve practical outcomes. When ideas are sought, the adaptor will generate a limited number of ideas but they will be practical. Adaptors can handle detail and prefer to work on a limited number of projects simultaneously.

Innovators will readily break the paradigm and look for solutions with a high degree of novelty. They prefer to change rather than seek to improve. When required to generate alternatives, innovators will produce many ideas and a number will be highly impracticable. The term 'innovator' is used here in a particular context with respect to problem-solving style, it does not equate to the word 'innovation'.

Of course, the above descriptions of innovators and adaptors are the extremes. Most people lie somewhere on the continuum between the extremes, but the differences between people are noticeable. Some engineers clearly prefer to improve on existing solutions whereas others have a marked tendency to look for a high degree of novelty.

The significance of problem-solving style is the influence it has on the outcomes of all types of design activity. If the workplace is dominated by adaptors, then that culture will prevail long term because innovators will be uncomfortable, perhaps not be appreciated, and may well leave. The converse is also true. Innovators and adaptors will tend to prefer design methods that best suit their style. For example, innovators will not be attracted to detailed methods that tend to improve solutions. Conversely, adaptors will not be comfortable with extreme divergent methods that generate abstract concepts. The choice of design method affects the type of solution generated; for example, brainstorming will tend to produce innovative solutions.

Therefore, it is important that the *solution requirements* shape the choice of solution. It is the client needs that have to be satisfied, not those of the designer. Whist this may appear to be an obvious point, too many times one can see problem solving style preferences reflected in the solutions opted for by designers. In all design activity, the personal traits of individuals will be influential. With suitable training, one may flex between styles to suit solution requirements. However, prolonged working outside personal preferences induces stress. Thompson and Lordan (1999) give a discussion of CPS principles and their applicability to engineering design.

It is important that the *solution requirements* shape the choice of solution. It is the client needs that have to be satisfied, not those of the designer.

Design methods (tools)

The literature contains many design methods (tools), many of which can be used effectively to improve the design process and achieve high levels of quality. Those methods that have specific uses, e.g. finite element analysis, are used widely and need no further mention here. However, there are a number of general design methods, particularly concerned with simulating divergent thinking, convergent analysis, and especially multi-objective decision making, that do not find wide application even though they have much to offer. A study of the use of design methods in industry (Lopez-Meza and Thompson, 2003) reveals typical results:

- the number of methods used in industry is relatively small and those that are used are used in a non-systematic way;
- the wrong implementation of methods and their *ad hoc* selection prevents companies from using a wide range of methods;
- the way the methods are delivered to engineers today does not suit their needs.

Incorrect application of a method invalidates its results. One typical failing is to use evaluation methods that rely on detailed information too early in the design process. Engineers can be found 'guessing scores' for performance criteria in concept evaluation when there is absolutely no justification. Also, whilst brainstorming may potentially be the single most beneficial method to generate ideas, it is often practised badly with scant regard for suspended judgement, the group dynamic and the use of extended effort.

There are many actions that could be taken in order to enhance the use of methods in industry, and there are also many levels at which those actions could be taken, e.g. management level, product development level, university level, *etc*. Some authors have pointed out that an important factor for the successful implementation of design methods in industry is the availability of easy-to-use software tools (e.g. Killander, 2001). Others have pointed to the need for support teams or help desks in industry and that management should encourage the use of methods (Ernzer and Birkhofer, 2002). An Internet-based integrated learning, information, and training environment is being developed to train learners in validated methods (Birkhofer *et al.*, 2001).

For divergent thinking, the most effective methods are brainstorming, brainwriting and the use of a morphological chart.

One of the earliest texts that described design methods was by Jones (1970), in which numerous methods were discussed. Over the last three decades new methods have been introduced and old ones developed and refined. Interestingly, there are certain methods that have stood the test of time and which have been found to be very effective. Cross (2000) gives a concise account of engineering design methods for product design, as do the other main text books (Pugh, 1991; Roozenburg and Eekels, 1991; Pahl and Beitz, 1996).

Methods for divergent and convergent thinking

For divergent thinking, the most effective methods are brainstorming, brainwriting and the use of a morphological chart. Brainstorming is often misused and should be practised with a high degree of discipline with respect to the group dynamic and suspension of judgement. Brainwriting is practised in groups where each person has a sheet on which they describe, say, three ideas at the top of columns. The sheets are exchanged and the next person develops the idea in each column, or possibly produces a new idea stimulated from the first idea. Brainstorming and brainwriting are, therefore, a valuable pair of methods. Brainstorming tends to produce a diverse range of ideas that can well be paradigm breaking, whilst brainwriting tends to encourage adaptive change. In early concept design, paradigm breaking may be advantageous, but later in concept development a more adaptive style of change in which improvements are sought is the better way to progress.

The evaluation of any aspect of design depends upon two factors: the criteria used to evaluate the proposals and the appropriate choice of method.

A morphological chart is very useful if a problem can be broken down into particular functions, e.g. in the case of a motor vehicle there is the power generation, power transmission, body, suspension, *etc*. The functions are set out in the first column of a matrix. Alternative solutions are sought for each function and are described briefly in the row defined by each function. The objective is then to determine the optimum permutation of solutions to produce the best overall design. Note, the best solution for each row may not be the best overall solution. For example, in a manufacturing system it may be preferable to choose either pneumatic or hydraulic equipment rather than a hybrid to reduce spares and ease maintenance.

The evaluation of any aspect of design depends upon two factors: the criteria used to evaluate the proposals and the appropriate choice of method. For convergent thinking, two particular methods deserve mention: one for conceptual design and one for detail design. It is important that methods are used at the appropriate stage.

The method advocated by Pugh (1991) has proved highly effective for evaluating concepts. In this method, the concepts under consideration are described briefly along the top row of a matrix. The evaluation criteria are listed in a column to the left of the matrix; see Figure 14.3. One of the concepts is selected as the reference or datum (a preferred choice or an existing solution) and the other concepts are compared with it.

The method proceeds as follows. For each criterion individually, a concept is compared to the reference concept and a decision of 'better than', 'worse than', 'same as' or 'don't know' is made. Note that it always possible to compare two things with respect to one criterion in this way. For 'better than' a '+' is inserted into the matrix, for 'worse than' a '–' is inserted, and for 'same as' or 'don't know' an 'S' is used. Note that no numerical scores are used at any stage. In Figure 14.3, concept 1 is judged to be better than the

		Concepts							
		1	2	3	4	5	6	7	8
Assessment criteria	A	+	+	–	S	D	S	S	+
	B	+	+	+	+	A	+	S	S
	C	–	+	S	+	T	–	S	S
	D	–	–	S	–	U	+	–	–
	E	S	S	S	–	M	–	+	–
Sums	+	2	3	1	2		2	1	1
	S	1	1	3	1		1	3	2
	–	2	1	1	2		2	1	2

14.3 Concept evaluation (Pugh, 1991)

datum with respect to criterion A; therefore, a '+' is inserted under the concept 1 column against criterion A. After the matrix is completed, the total '+', '–' and 'S' ratings are added. Each concept (perhaps the top few if there are many concepts) is reviewed to determine if any '–' or 'S' ratings can be improved to '+'. Eventually, the preferred concept emerges. The strength of the process is that it forces the designers to think about the evaluation criteria and the appropriate assessment of concepts with respect to the criteria.

For detailed design evaluation, a numerical scoring method can be used as follows. First the evaluation criteria are defined. For each criterion, two performance levels are prescribed and a performance defined: the minimum acceptable level below which a product or system is unacceptable and a level of performance which, considering engineering feasibility and market aspirations, would be perfectly acceptable. The unacceptable performance is scored 0 and the perfectly acceptable score is 10. For each criterion, a calculation of the expected performance of the proposed design is undertaken. Using a linear function, a score is determined for that criterion between 0 and 10 (see Figure 14.4). A total score can then be calculated by combining the components as follows:

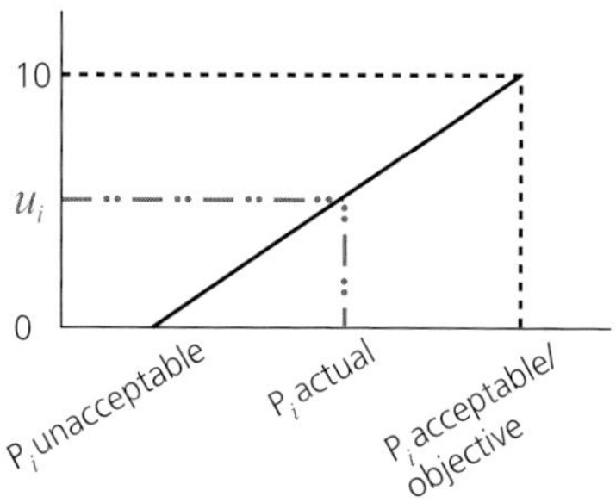

14.4 A numerical scoring method

$$\text{Total\_score} = N\left[\left(1/_{\text{score1}}\right)+\left(1/_{\text{score2}}\right)+\left(1/_{\text{score3}}\right)+\ldots\left(1/_{\text{scoreN}}\right)\right]^{-1}$$

where N is the number of criteria.

Such a method avoids estimating (guessing) scores and the use of an inverse calculation avoids the problem of addition in which a very low score can be compensated by high scores. Multiplying by N simply brings the total score to a range 0–10. Scale factors can also be used (Thompson, 1999).

For a comprehensive treatment of methods to stimulate divergent thinking see Pahl and Beitz (1996), Pugh (1991) and Thompson and Lordan (1999). Pugh's concept evaluation method is described fully in Pugh (1991). Pahl and Beitz (1996), Pugh (1991) and Roozenburg and Eekels (1991) describe a wide range of methods.

Specific design methods to improve quality

Whilst the objective of all design methods is to improve the quality of engineering solutions, there are specific methods that require particular elaboration. They are quality function deployment (QFD) and Taguchi methods. As will be seen below, QFD is very useful in the early stages of design and Taguchi methods are applicable to improvements through testing and development.

Quality function deployment

Total quality management (TQM) is concerned with the continual improvement of company processes and activities with the aim of achieving a high level of customer satisfaction. The focus, throughout all company activity, is the customer and all activities are evaluated with respect to their contribution to the achievement of customer satisfaction. In the context of this general picture, an important method in engineering design is QFD.

Total quality management (TQM) is concerned with the continual improvement of company processes and activities with the aim of achieving a high level of customer satisfaction.

QFD is a specific design tool that links the customer requirements to an engineering specification. The method is well described in Pugh (1991), Cross (2000), Roozenburg and Eekels (1991) and Bergman and Klefsjo (2003), and in numerous other texts; therefore, only a brief outline is given here.

The basic objective of QFD is to relate customer needs to engineering characteristics. For a particular product, the attributes of the product that a customer will perceive are first identified. Using a matrix, particular engineering characteristics are then related to customer attributes so that it is clear which engineering characteristics influence which customer attributes.

For example, in the case of an electric hover-type lawnmower, a customer may use the following criteria to judge competing models: appearance, switch feel, responsiveness, noise, cut width, cut quality, ease of cut adjustment and number of times the bag needs emptying. The engineering characteristics could be defined as: motor torque and speed; blade diameter, toughness, adjustability of blade assembly; bag volume and pressure in the bag; force to operate the switch and switch travel; structural stiffness weight, and surface finish; soundproofing. Figure 14.5 gives a matrix that shows the particular engineering characteristics that determine the product attributes, *i.e.* cut quality is determined by the torque and speed of the motor and the toughness and hardness of the blade.

The matrix can then be developed further to determine design objectives or target values for each engineering characteristic. The target values, including those of competitors if required, are shown placed at the base of the matrix (Figure 14.6). A further development is to show which engineering characteristics influence other characteristics. The interactions between characteristics are indicated on the top of the matrix (Figure 14.6), which is the 'roof' of the so-called 'house of quality'.

QFD does provide a clear link between the customers' perception of a product and the engineering characteristics that will be the objects of designers' attention throughout the project. The customers' needs are translated into specific engineering requirements. It is clear that QFD is an attractive method in product design. For engineering system design, such as process plant or other manufacturing systems, customer requirements may be specified more directly in engineering terms (for example, plant throughput, product purity, availability), therefore, there may be less need to apply the QFD method.

Engineering characteristics

Product attributes	Motor torque	Motor speed	Blade diameter	Blade hardness	Blade assembly	Bag volume	Bag pressure	Force to operate switch	Switch travel	Structural stiffness	Weight	Surface finish	Soundproofing
Appearance												*	
Switch feel								*	*				
Responsiveness										*	*		
Noise										*	*		*
Cut width			*										
Cut quality	*	*		*	*								
Cut height adjustment					*								
Time to fill bag						*	*						

14.5 Engineering characteristics and product attributes

The above description of QFD deals with the salient principles as applied to the early stages of design concerning customer needs. A more comprehensive approach can be taken to apply the method at four stages: customer needs vs. engineering attributes, engineering characteristics vs. parts characteristics, parts vs. process operations and key process operations vs. production requirements. Bergman and Klefsjo (2003) gives a comprehensive description of QFD applied through design and manufacture.

Taguchi principles

The Taguchi principle is based on the achievement of precise product performance values that relate to customer values. A quality improvement programme is used to ensure that deviations from target product performance values are minimised. Such deviation is seen as bringing about serious deleterious effects on product quality. Emphasis is placed on specific product performance

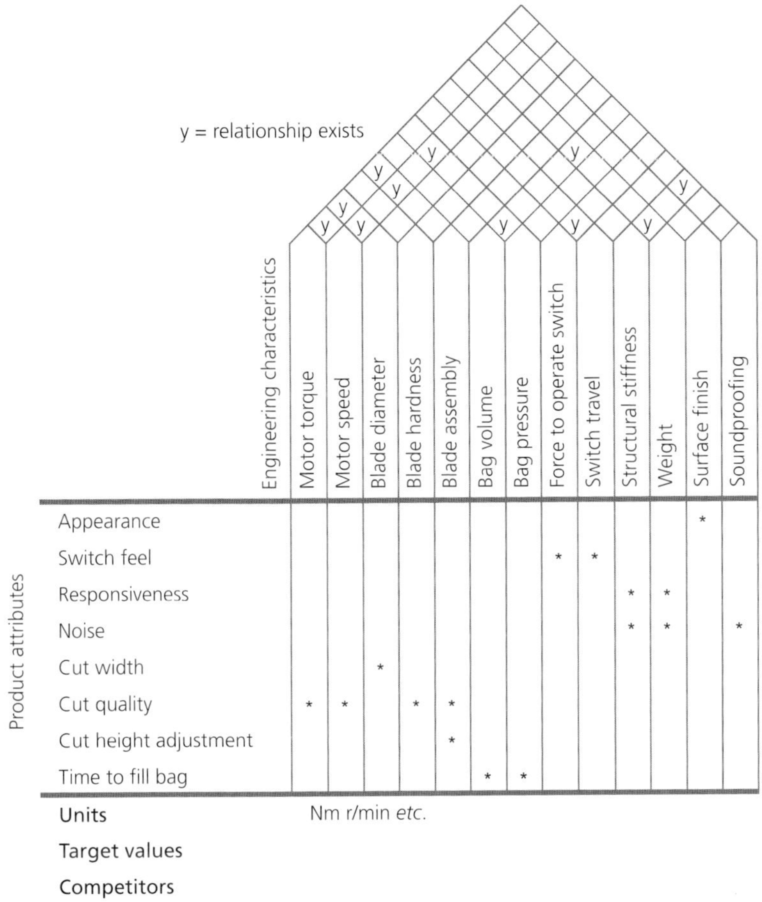

14.6 The 'house of quality'

targets rather than tolerance bounds of acceptable performance. Product quality is understood to deteriorate rapidly if precise targets are not achieved.

The process is one of continual improvement. The objectives are to achieve precise product performance targets that determine quality and to drive down costs. To this end, the method encompasses design and manufacturing. Manufacturing has a highly significant role to play in Taguchi principles. However good the design process has been, the potential product quality may not be attained if the manufacturing process is poor. The focus of attention is that set of parameters, especially parameter combinations, that relate to customer quality, not just general manufacturing tolerances. Statistical design of experiments is used to link design parameters with product characteristics;

for further information see Roy (2001). There has been some controversy regarding Taguchi methods, but undeniably the Taguchi approach has led to many improvements in quality. For a good discussion of Taguchi principles see Bergman and Klefsjo (2003).

The perceived quality of many products depends upon their reliability.

In design, there is a clear link between the objectives of the QFD method and Taguchi. The QFD method creates the relationship between the customers' perception of a product and the engineering characteristics that determine product performance. The outcome of the QFD method is a set of precise product performance targets. The Taguchi principles are based on such targets and are concerned with minimising deviation from them.

Reliability and maintainability

The perceived quality of many products depends upon their reliability. Failures are perceived by many clients as an indication of poor quality, and quite rightly so. Many designers are reluctant to consider failure; indeed, some will even claim that they produce designs that work rather than designs that fail. Therefore, it is important to adopt proactive methods at all stages of design, and in all activities, that will improve reliability.

Similar arguments apply with respect to maintainability. The client may, or may not, undertake maintenance and repair operations, but he certainly pays for the maintenance of products throughout the product lifecycle.

Failure rate prediction and the 'reliability case'

Reliability is the probability that a device, system or component will continue to perform a specified duty under prescribed environment and loading conditions for a given time.

Reliability is the probability that a device, system or component will continue to perform a specified duty under prescribed environment and loading conditions for a given time. Failure rate prediction has been of interest to engineers since the 1970s and is best carried out using failure rate data from equipment operated under similar environment and loading conditions to the case under examination. When this is not possible, a nominal failure rate (obtained from such data sources as are available) is modified by two factors: one for environment and one for loading. The failure rates for all components are added (assuming no redundancy in the system) to obtain a failure rate for the system. This 'component count' method has found more success for electronic equipment than for mechanical components.

However, in recent times there has been a shift away from an attempt to predict a precise failure rate that may not be accurate. The emphasis today is on identifying the key factors that affect system or product reliability and setting down a clear strategy, with detailed actions, to ensure that a high standard of product reliability is achieved. Thus, reliability is achieved through control of the product design and manufacturing process.

In *engineering design*, reliability and maintainability (R & M) considerations should be, and can be, included in every stage and in all design activities:

- when the specification is derived;
- during concept generation and evaluation;
- in detail design.

An overview of these is given next; for a full discussion see Thompson (1999).

The design specification

If the design specification does not include appropriate clauses pertaining to R & M then the client has no contractual comeback if poor R & M is provided. But the client may then not purchase again, or a design team's reputation may be sullied in the eyes of others. Either way it is the designers who lose in the long term. Therefore, it is important that the designer takes a proactive stance with respect to R & M (Figure 14.7).

R & M may be included both quantitatively and qualitatively into the design specification. Quantitatively, this may be done by the inclusion of mean corrective repair time and mean time to failure objectives. Mean corrective repair time (or mean active repair time) is the mean time that is required to return a machine or system to operation given that spare parts and manpower are available. It may be estimated by a simple calculation and demonstrated before contract completion. The mean-time-to-failure calculation is subject to greater error since it relies on absolute values of failure rate data, which can be erroneous and cannot be demonstrated in the short term.

Qualitative inclusion of maintainability criteria can refer to the skill levels required for maintenance and repair; for example, is multi-tasking an option, is maintenance to be carried out by a highly skilled, or otherwise, workforce? Specific skills should be cited. In the case of reliability, the specification should describe clearly and precisely the operating environment of the machine or system. Also, the skill level of operators should be described realistically. The environment in which a machine will work and the way it is operated will also occasionally have a highly significant affect on reliability.

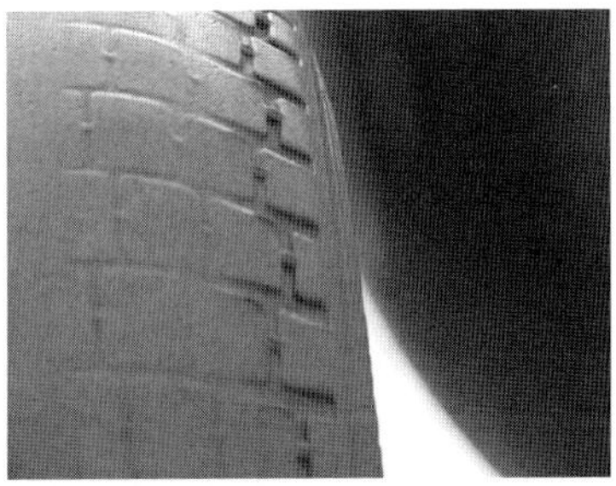

14.7 Reliability and maintainability should be considered at every stage of the design process

The specification may require that certain design methods be used. For example, the specification could state that a top set of critical items that would be most influential on system reliability must be identified by, say, a system reliability model or a HAZOP study plus fault tree analysis (FTA). Then, a detailed reliability assessment should be carried out for each critical machine. A failure mode effect analysis could be required for a critical system to identify the areas of high risk.

Qualitative statements that refer to R & M in the design specification are extremely valuable. They capture the operating climate of the system and involve the client thinking deeply about the design. However, qualitative statements should be made very specific. It is largely worthless to include terms in a specification such as 'good maintainability', 'maximum reliability must be achieved', 'full attention should be given to maintainability and reliability', *etc.* They mean nothing and add little value to the document; for example, how good is 'good', how can maximum reliability be achieved – at what cost or expense to other parameters? Often R & M are linked; for example, joints may be introduced to improve maintainability but their presence reduces reliability.

Concept design

14.8 Products should be designed for ease of maintenance

It is quite possible to include R & M considerations in concept design. One attractive way is to undertake a specific R & M evaluation using Pugh's method. Specific R & M criteria can be defined for use in Pugh's evaluation method; for example:

- simplicity and elegance;
- minimum number of parts;
- suitability for modular construction;
- accessibility;
- sensibly sized components;
- ease of adjustments;
- precise definition of maintenance skill levels;
- minimum number of moving parts.

The above criteria appear rather general; more specific criteria can be defined for particular cases (Figure 14.8).

Detail design

There are certain analysis methods that are suitable for detail design that integrate R & M considerations with other performance parameters. However, the simple checklist remains one of the most cost-effective ways of ensuring that the client will be satisfied. Taken from Thompson (1999), typical examples include:

- *Spares*. Is the variety of spares required reasonable and not excessive? Is the future availability of spares assured?
- *Ergonomics*. Can the forces/torques required for maintenance be provided by persons of average physique?
- *Faults*. Are people safe in the event of mal-operation? Is other equipment protected in the event of mal-operation? Can the operator readily detect if the machine operates out of specification?
- *Condition monitoring*. Is provision made for hand-held condition monitoring devices to be used? Is provision made for installed condition monitoring instrumentation if required?
- *Corrosion*. Are the components, and especially fasteners, resistant to external corrosion? Are the materials selected to resist the internal corrosion of any parts in contact with fluids?

Specific design methods for failure analysis

Failure mode and maintenance analysis (FMMA), FTA and failure mode effect and criticality analysis (FMECA) are precise methods that specifically consider the consequences of failure.

FMMA is a simple approach in which, on completion of a piece of design, the principal failure modes are listed and, with reference to the design work carried out, the ways in which failure is corrected are written down for each failure mode. It is best done by someone other than the designer and is suitable for a design review of critical machines. Such a study may lead to redesign and/or the introduction of condition monitoring equipment.

Failure mode effect and criticality analysis is a 'bottom up' approach to failure analysis.

FMECA is a 'bottom up' approach based on a risk assessment. Firstly, the components of a system are listed, and for large systems the study will sometimes begin at an intermediate level. For each component, the failure mode is defined and the consequences of failure are considered. The likelihood of failure can be predicted using failure rate data if available, but more commonly an estimate is made on a 0–10 or 0–5 scale (high number equates to more likely).

The consequences of failure may be estimated in real terms, e.g. serious injuries/year, or more often as a severity rating on a scale similar to the likelihood of failure. The most critical items are identified by the product of the likelihood of failure and consequence ratings, *i.e.* maximum risk, and a decision taken whether or not the risk is acceptable or whether remedial action is called for. FMECA studies can be carried out for products and processes. In some cases, the analysis is extended to include the likelihood of detecting a failure mode and a risk priority number calculated as the product of the three factors: likelihood of occurrence, consequence and likelihood of detection.

The FMECA approach is a simple method that is widely used. Its drawbacks are that it can be time consuming and consequences of combinations of individual failure events can be undetected.

Fault tree analysis is a top-down approach to failure analysis.

FTA is a top-down approach to failure analysis and begins with a clear statement of a system failure. The events that must occur to create that top failure mode are then identified and arranged under the top event in a tree with logic gates (the most common ones used are 'and' gates and 'or' gates). Then, for each failure on the second level, the events are identified that need to occur in order for that particular event to occur and the events drawn on a tree using logic gates. Thus, a comprehensive fault tree is created showing all the failure events and their dependencies that need to occur if a certain top-level failure is to transpire.

The tree can be analysed to give a figure for the top-level failure rate by inserting failure rates for each event. However, for the designer, especially when failure rate data are not available, the fault tree can be used to identify the key components or subsystems that are likely to lead to a system-level failure. A fault tree can become quite complicated, but, by using modern software, fault trees for the diverse parts of complex systems can be derived separately and linked with common elements if required.

BS 5760-0 reliability of systems, equipment and components

British Standard 5760-0 gives a description of reliability concepts, processes and methods.

British Standard 5760-0 (1986) gives a description of reliability concepts, processes and methods. In addition to precise definitions in the field of reliability, the standard covers business organisation to achieve reliability and descriptions of particular methods and their use in practice. It is a very useful standard and is used by many companies. It guides the reader in the practical application of reliability principles and methods to industrial practice. In this way, it is more useful to the practitioner than certain text books.

There are links between the quality standard (BS EN ISO 9000-1 (1994), etc., see below) and BS 5760-0. For example, BS CECC00804 (1996) is concerned with a harmonised system of quality aspects for electronic components: Interpretation of 'ISO 9000:1994' – reliability aspects for electronic components. There are many such helpful examples, too numerous to list here, and the BSI Web site should be consulted to search for appropriate standards and guides.

Design review

A design review is a quantitative and qualitative examination of a proposed design to ensure that it is safe and has optimum performance with respect to maintainability, reliability and those performance variables needed to specify equipment.

Maintainability and reliability are included in the definition to ensure they are considered, but they should be dealt with in conjunction with other parameters. An effective design review will ensure that design proposals are fit for purpose. A design review is much more than a perusal of drawings and calculations. It should be a systematic procedure that is integrated with normal design activity, which is outlined below. For a full description see Thompson (1999).

A design review is a quantitative and qualitative examination of a proposed design.

A systematic procedure

The review should begin from the derivation of the design specification and continue through to detail design.

Specification. The objective is to ensure that all salient points in the design specification are understood by the design teams, including maintainability and reliability requirements and influential factors. This is especially important when design work is put out to sub-contract.

System level review. The aim is to identify critical areas that are most sensitive to the achievement of client needs. The outcomes of a QFD analysis would be useful in this respect. The design review team might also comment on the need to follow high-risk options, say the introduction of new technology, to satisfy certain system requirements.

Functional unit level. Particular designs of equipment are reviewed in detail, say by a applying multi-criteria, quantitative assessment method. Such detailed evaluations would only be carried out for critical items. Checklists could be used more generally as a cost-effective solution.

Detail level. Generally a major project cannot be reviewed at a detailed level due to time constraints, nor would one ever expect so to do. Rather, one would expect the design review team to identify particular areas for detailed scrutiny, say the seals in part of a chemical plant.

The review should begin from the derivation of the design specification and continue through to detail design.

Design review team

In some cases a design review team is formed that remains together for the duration of a project and which has an important management role to play. From the above, it can be seen that an experienced multi-disciplinary team is required comprising design personnel, particular technical experts, maintenance engineers, production staff and safety specialists.

Design review teams may be formed from personnel within a company, from an external consultancy or from internal and external sources. The advantage of in-house personnel is that they must 'live' with the outcome long term. If internal expertise is limited, then external consultants have benefits and the consultants can point to best practice elsewhere.

EN ISO 9001 quality systems

BS EN ISO 9001: 1994 (formerly BS 5750) is the international standard on quality systems that includes product design. It is the model for quality assurance in design, development, production, installation and servicing. The quality system requirements include:

- management policy;
- quality system;
- contract review;
- design control;
- document and data control.

Other sections of the standard refer to aspects of purchasing, customer-supplied product, inspection, measurement and testing.

The standard unequivocally states that the responsibility of the management of the product supply company is to define and document its quality policy, including specific goals and the expectations of the customer. Therefore, the client needs are again at the heart of design activity. The particular responsibilities and authority of management personnel should be stated clearly. The quality system includes planning that defines how quality requirements will be met and the controls that will be put in place. At the tender stage, the contract requirements must be reviewed thoroughly.

A quality system includes planning that defines how quality requirements will be met and the controls that will be put in place.

Design control is a significant part of the standard and the following activities are covered:

- design and development planning for each design activity;
- organisation of the technical interfaces between different groups;
- design inputs that refer unambiguously to requirements;
- design outputs that specify terms that can be verified and validated with respect to the requirements;
- design review at each stage of design;
- verification and validation;
- design changes.

To achieve good quality in design, reliability and maintainability, the requirements in the design specification are important and have been discussed above. Importantly, in order to satisfy the quality standard a design review must be undertaken. The design review is a very important means by which R&M may be included in the design process in order to achieve high levels of quality.

Another important aspect of the quality standard which, perhaps, does not receive the prominence it deserves is organisation of the technical interfaces between groups. Figure 14.9 shows the range of inputs to a major project. *All projects start and finish with the client.* On the way there are many inputs to the project, including mainstream designers, sub-contract designers, technical specialists, regulatory authorities (municipal authorities and specific government departments), consultants, *etc.* If the interfaces between all these groups are not managed then the output of the design activity will suffer badly.

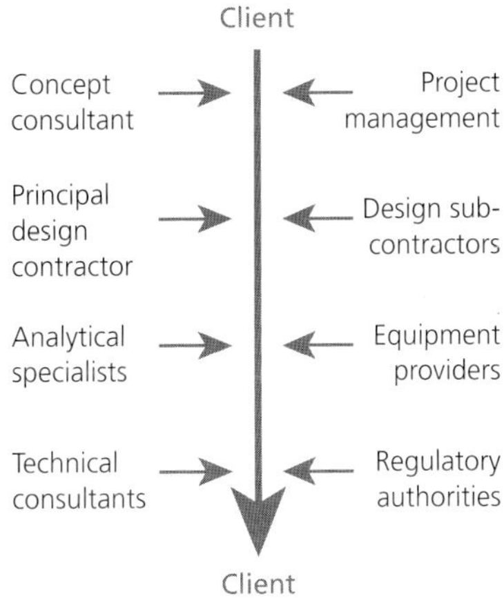

14.9 The range of inputs to a major project

Conclusion

Quality is perceived in many different ways, but the only view that counts is that of the client or customer. Unless the client perceives good quality then sales will suffer and a company will not survive. The client will perceive quality in terms of performance, and in reliability terms, *i.e.*: Does the product continue to perform satisfactorily without failure or without deterioration in performance?

In order to achieve good quality in design, design activities need to be carried out using appropriate methods that will yield solutions meeting the problem requirements, especially with respect to innovative or adaptive change. The people undertaking the design must select solutions that are client focused and not be unduly influenced by their personal cognitive style. There are many methods that can be used in different design activities. Reliability, QFD and Taguchi methods are very significant in the achievement of high reliability.

Reliability in design can be achieved by considering apposite reliability parameters at each design stage, from specification through to detail design. Qualitative and quantitative methods can be used in all design activities and a comprehensive design review is an important contribution to quality improvement. The building up of a reliability case in this way by including reliability in all design activities is preferred to relying on a precise, but possibly inaccurate, failure rate prediction calculation.

Quality is a major subject that encompasses much more than design. The above sections have outlined briefly certain aspects that are highly pertinent to design. No claim to exhaustiveness can be made, and different authors will place different emphasis on topics. The wide subject of quality is covered in ISO 9000 and its associated parts.

References

Bergman B, Klefsjo B (2003) Quality: from customer needs to customer satisfaction. Pub Studentlitterature, Lund, Sweden

Birkhofer H, Lindemann U, Albers A, Meier M (2001) Product development as a structured and interactive network of knowledge – a revolutionary approach. ICED'01, Glasgow, UK

BS 5760-0 (1986) Reliability of systems, equipment and components – introductory guide to reliability. http://www.bsi-global.com

BS CECC00804 (1996) Harmonized system of quality assessment for electronic components: interpretation of ISO 9000:1994 – reliability aspects for electronic components. http://www.bsi-global.com

BS EN ISO 9000-1 (1994) Quality management and quality assurance standards – guidelines for selection and use. http://www.bsi-global.com

Cross N (2000) Engineering design method: strategies for product design. John Wiley

Ernzer M, Birkhofer H (2002) Selecting methods for life cycle design based on the needs of a company. Design 2002, Dubrovnik, Croatia

Fox JM, Fox RL (2000) Exploring the nature of creativity. Kendall/Hunt

Isaksen SG, Dorval KB, Treffinger DJ (1994) Creative approaches to problem solving. Kendall/Hunt

Jones CJ (1970) Design methods. John Wiley

Killander AJ (2001) Why design methodologies are difficult to implement. International Journal of Technology Management, 21(3/4): 271–276

Kirton MJ (2003) Adaptors and innovators: styles of creativity and problem solving. Routledge

Lopez-Meza B, Thompson G (2003) Exploring the need for an interactive software tool for the appropriate selection of design methods. ICED'03, Stockholm, Sweden

Pahl G, Beitz W (1996) Engineering design: a systematic approach. Springer

Pugh S (1991) Total design: integrated methods for successful product engineering. Addison-Wesley

Roozenburg NFM, Eekels J (1991) Product design: fundamentals and methods. John Wiley
Roy RK (2001) Design of experiments using the Taguchi approach: 16 steps to product and process improvement. John Wiley
Thompson G (1999) Improving maintainability and reliability through design. Professional Engineering Publications
Thompson G, Lordan MA (1999) Review of creativity principles applied to engineering design. IMechE, Part E1, 213: 17–31

Chapter 15

Workflow for design

Sándor Vajna
University of Magdeburg

Within a company, engineering appears the most important area. In this area all relevant characteristics of a product and its manufacturing, distribution, usage, service and recycling processes are fixed. It is well known that engineering determines up to 75% of the expected product costs (Wiendahl, 1970). Engineering consists of marketing, product development, production process planning, prototyping and testing processes. This is a complex set of processes of which some run serially and some in parallel, and the process participants are usually distributed over different locations.

Within engineering, different technologies are used, for example, CAx (CAD, CAM, *etc.*) systems, electronic document management (EDM) systems, product data management (PDM) systems, virtual reality, digital mock-up, Internet applications, and office software applications. Usually it is hard (if not impossible) to find the right documents, data or information at the right time. Therefore, it is difficult to finish work quickly and to an appropriate level of quality.

Furthermore, in the light of decreased development time, customers usually are not aware of all their requirements and thus change them often during a product development process (creating so-called 'running targets'). In summary, it is difficult to manage schedules, resources and costs of a project with traditional approaches. New approaches like process modelling are required.

One of the most suitable ways to model highly complex issues (like engineering processes), with limited expenditure, is to use the approach of the morphological box.

(Zwicky, 1966)

Process modelling

A comprehensive process modelling approach has to take into account the high complexity and the dynamics of today's processes. In order to handle this complexity without losing both process rationale and context, the decomposition approach, *i.e.* the reduction of a process into phases (or even into its generic component 'atoms') is most commonly used (Dym, 1994). In engineering design, the decomposition into phases is applied by the majority of 'standard' literature (e.g. Pahl and Beitz, 1996), but also for quality assurance and manufacturing (e.g. Ploetz and Biehl, 1999; Aurich and Wagenknecht, 2003). Most of the different decomposition sources can be traced back to the 'Therblig' approach of Lillian M Gilbreth and Frank B Gilbreth from 1904 (Gilbreth and Gilbreth, 1924; and quoted after Ferguson, 2000). Therbligs were created to define standard and enclosed activities within manufacturing in order to improve the worker's situation and to minimise fatigue by reconfiguring and recombining Therbligs. They exist at a low level of any manufacturing process.

In order to handle the combination diversity of these generic process 'atoms' and to increase the transparency of their possible relations, one of the most suitable ways to model highly complex issues (like engineering processes), with limited expenditure, is to use the approach of the morphological box (Zwicky, 1966). Although this approach was originally developed for product configuration purposes, it may be applied to any kind of configuration activity. Within this approach, it is assumed that the high variety of possible solutions (which is one of the key factors of today's strategy of mass customisation) can mostly be realised by a high variety of combinations of a limited number of (mostly generic) building blocks.

The ability to address a variety of combinations of a limited number of building blocks is one of the key factors of today's strategy of *mass customisation*. The term 'mass customisation' was derived as a link between *mass production* and *customisation*. Its aim is the development, manufacture, marketing, sales and service of products, which are to be supported in such a way that almost all customers can get exactly the product they are really looking for, at an acceptable cost, and with enough variability, so that they can use the product in the specific way they intend.

All necessary activities, from the first idea to the start of production, are performed within the engineering process.

For process modelling purposes, these building blocks are converted into enclosed activities that are called *process elements*. Their description is based on a predefined structure that is suitable for application in a computer system.

With the definition of appropriate process elements, the approach of the Gilbreths can be unambiguously transferred into engineering and applied to process modelling purposes using the morphological box approach (Vajna and Freisleben, 2002). The combination of these two approaches allows the setting-up, running, and evaluation of any type of process, where the resulting process model may be hierarchical, or in the form of a net, line or matrix, *etc*.

The term 'workflow' (*i.e.* the smooth flow of work) is quite commonly used to describe activities within manufacturing, controlling, and administration processes. It is also used to describe activities associated with preparing an organisation for the adoption of computer support (Conger, 1999), *i.e.* for rigidly coupled activities that are completely predictable and reproducible. Within product development, however, there are usually very few rigidly coupled activities (mainly for release management and change management); rather, there are mainly dynamic and non-predictable processes. In order to distinguish clearly between processes within product development and processes outside of product development, the term 'workflow' in this context is used in its original sense for rigidly coupled activities, whereas the terms process and process element are used to describe activities in product development.

Definitions

In order to ensure a common understanding, frequently used terms in this area besides 'process element' are defined as follows (Vajna *et al.*, 2001):

- *An activity* is the 'molecule' of a working environment, e.g. product development. An activity is a logically enclosed operation. It contains at least one or more working steps in a specific configuration. It is started by one or several events and ends in one or several events. It needs certain inputs and creates certain outputs whose format and structure can usually be fixed.
- *A working step* is the smallest subset of an activity, its 'atom'.
- A *process* is a meaningful set of activities or sub-processes to solve a class of possible tasks. The combination of activities and/or sub-processes is always flexible and can be adapted dynamically to a specific task. A process is a virtual object that describes how tasks may be solved;
- *A workflow* is a dedicated, rigid sequence of working steps, process elements or sub-processes,e.g. a release workflow, which is not changed.
- *A sub-process* is a subset of a process and is also a set of activities or other sub-processes. Sub-processes are usually applied for repetitive or standardised processes that may be used in different contexts. By contrast with a workflow, a sub-process may be adapted to a given situation.
- *A project* is a process with an actual and real task to address. A project defines the initial conditions of a process, e.g. requirements, starting time, due dates, budget limitations and resources. It initiates one or more processes and/or workflows.

The interrelationships between these terms are illustrated in Figure 15.1.

The most important characteristics of the product are specified early in the engineering process.

Modelling requirements

The reasons for creating a new product (either a physical object or non-material service) are orders, customer requests, general market needs or spontaneous innovative ideas. It is well known that the most important characteristics of the product are specified early in the engineering process. All necessary activities, from the first idea to the start of production, are performed within this process. Systematic and methodical support of these activities can dramatically increase the productivity of the company.

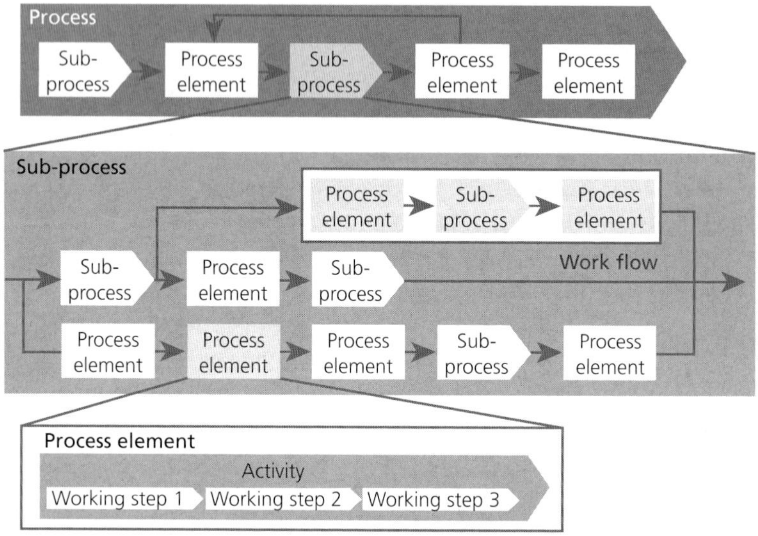

15.1 Hierarchy of definitions of process components

However, which requirements should a support system fulfil?

- The transparency of the engineering process within every project should be improved, especially if several projects run in parallel. If everybody knows which activities follow or come before their own, they can come to better decisions or can ask the right person for relevant information.
- The response time for customer requests should be reduced (thus increasing flexibility), because it is well known that customers often change their requirements during the project. The support system should respond to these 'running targets' by changing dates, milestones, and resources and their allocation appropriately and in real time.
- Problems and possible bottlenecks must be identified before they occur. A simulation of process variants before the start of a real project would use resources more efficiently.
- The optimisation of processes to meet different priorities, e.g. time, cost or maximum application of a specific tool, should be possible.
- With better project coordination the system will reduce throughput time, 'time to market', and 'time to money' of a product.
- Following a successful process it should be possible to store details for efficient reuse of the process knowledge on the next similar project.
- Every user should be freely guided through a project or process. The system has to support the daily work of the user.

Regarding these requirements, it does not seem to be possible to manage engineering processes with traditional workflow systems such as are provided, e.g. by enterprise resource planning (ERP), EDM/PDM, or project management systems. These systems are based on a rigid process model of manufacturing, controlling or administration departments and their predictable processes. Table 15.2 shows the main differences between process in engineering and processes in the other areas of a company.

Engineering processes need an approach that:

- continuously monitors the process and predicts possible bottlenecks;
- creates and evaluates potential process flow alternatives to overcome possible changes (for example, new requirements of a customer, a failed resource, a missed deadline, unforeseen disturbances) in real time;
- offers these alternatives to the user and allows him to select the alternative he would prefer, then re-evaluates the process.

The behaviour of such a system is best described as 'navigation', since it always leaves the control and the decision competence with the user, whereas in process control, the users become mostly 'production means', strictly guided by the system. Another metaphor for navigation is its description as a game of chess between the user in a company and the customer. In response to possible customer changes and resource shortfalls during project execution, and in order to keep within time and budgetary constraints, the user has to update the project strategy continuously.

(Business) processes in manufacturing, controlling and administration	Engineering processes
• Processes are fixed, rigid, have to be reproducible and checkable to 100%	• Processes are dynamic, creative, chaotic; many loops and go-tos
• Results have to be predictable	• Results are not always predictable
• Material, technologies, and tools are physical (e.g. in manufacturing) and/or completely described (e.g. in controlling)	• Objects, concepts, ideas, designs, approaches, trials (and errors) are virtual and not always precise
• Possibility of disruptions is low, because objects and their respective environments are described precisely	• Possibility of disruptions is high because of imperfect definitions and change requests
• No need for a dynamic reaction capability	• There is a definitive need for dynamic reaction capabilities
→ Process control	→ Process navigation

15.2 Differences between processes in a company

Modelling with process elements

Approximately 50 generic process elements that are independent of any specific type of industry, branch, or product (Freisleben, 2001) have been defined and stored in a process element library (Figure 15.3). They were derived from engineering process descriptions and methodologies (VDI Richtlinie 2221, 1986; Hubka and Eder, 1992; Eder, 1996; Pahl and Beitz, 1996; Girard and Merlo, 2001; Marle and Bocquet, 2001) as well as from consulting experiences in different industries. The elements cover 'classical' engineering activities (for example, conceptual design) as well as organisational (for example, gating) and structural aspects (for example, serial/parallel activities, loops, and alternatives).

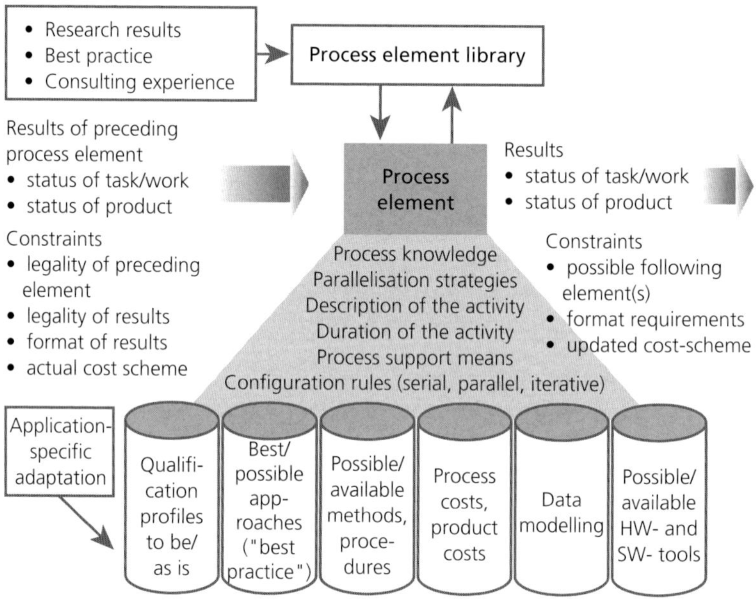

15.3 Basics of a process element

Each process element includes predefined knowledge, which in this context is defined as consisting of data, information, rules and meta-rules, where meta-rules describe application and combination possibilities of the other components. Hence, each process element comprises:

- A unique process element name.
- A description of the activity to be performed, based on the results of best practice within the respective company.
- The required skills to handle the activity using a qualification profile template (Figure 15.4).

- Possible and meaningful methods, working techniques and tools. If more than one support possibility exists, the process element will propose the best possible means appropriate to the actual activity.
- Allowed and excluded preliminary and following process elements with input and output information regarding content, structure and format of data.
- Structural information,e.g. combination possibilities with other processes (for example, sequential, parallel, or iterative).
- Cost development (product and process costs), driven by rules.
- Estimated and allowable duration of the activity, subdivided into value-adding time, waiting time, transportation time, and interruption time.

The general procedure of process modelling is shown in Figure 15.5 as a knowledge-based engineering process model.

A given engineering process (that comes mostly in a serial configuration, top left of Figure 15.5) is modelled with the above-mentioned process elements (top right), thus creating the *as-is* process topology. Another possibility is that a neural network creates a first proposition for a topology of process elements, based on given process requirements.

Qualification	Level of knowledge: Beginner (B)	Advanced (A)	Expert (E)
Draughtsman			x
CAD designer		x	
Technician			x
EDM/PDM user	x		
Controller	x		
Calculator (FEM, ...)		x	
:			

15.4 Qualification profile template

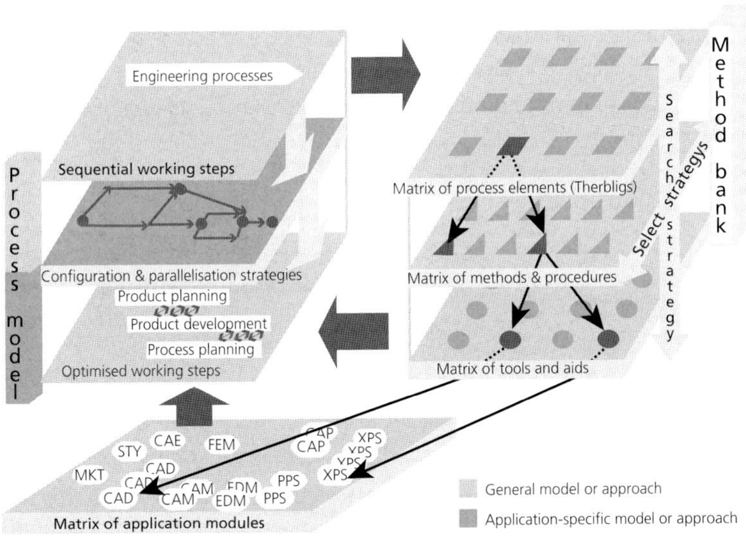

15.5 Working with the knowledge-based engineering process model

According to the situation, the appropriate methods and tools are assigned to each process element (the black arrows in the three-level -structure middle right), using both a process-to-methods matrix and a methods-to-tool matrix.

Search strategies within the matrices provide the right links, whereas selection strategies supply the best possible support for a given activity in a given environment. Both strategies are based on application knowledge gained from literature reviews as well as from research and consulting experiences.

Optimisation approaches change the as-is topology to an improved model with optimised working steps, using as many configuration and parallelisation possibilities as possible (low middle left). Finally, the actual application modules (low left) are activated and linked to the improved process element topology. The process model is now ready to support a corresponding project.

Process optimisation

An important goal for improving engineering is to optimise its processes and their appropriate activities.

An important goal for improving engineering is to optimise its processes and their appropriate activities. It should be kept in mind that the early phases of product development are of great importance, since a large part of the later costs of a product, about 75%, are fixed by conceptual and strategic considerations (Wiendahl, 1970). Because the costs incurred during these early phases of the product development are rather low, optimisation of these activities at this stage will improve the subsequent activities (for example, in manufacturing) and will lead to higher efficiency. Targets for optimisation are requirements fulfilment, process quality and time and budgetary requirements. The first step towards improvement is the simulation and testing of modelled processes and process structures. In this way it is possible to identify:

- resource bottlenecks;
- problems with dates and milestones;
- sequences of activities that might not work well in practice.

In general, four subsequent steps are then applied to maximise the potential of the optimisation process.

Qualification balancing

This step covers the rearrangement of people and resources (methods, procedures and tools) in order to find the most suitable relation between the available people and resources and the process elements of a given process. First, the required qualification profiles for each process element are extracted, thus generating the *to-be* qualification profile for the given process. This to-be profile is compared with the profile of the collaborators available for this process, where there will usually be mismatches between exiting profiles and to-be profiles. Using the template in Figure 15.4, people are given the most suitable process element to work on, thus assuring the realisation of the 'actual one best way' approach for each process activity.

This first step within qualification balancing has some interesting side effects. It provides an overview of the qualification profile of the staff and provides hints as to the areas where people's qualifications should be increased. It may even lead to a rearrangement of departments according to necessary and existing qualification profiles.

An analogous approach is carried out for resources, using a large knowledge base. As every process element contains generic knowledge of which resource would be best applicable to a given activity, the best possible method, procedure, and/or tool is proposed. Examples are the application of analogies from evolution for product optimisation (Vajna *et al.*, 2003) or, for product modelling, the use of a 3D CAD/CAM system instead of a 2D system. In case of unavailability, this may lead to further investment in new resources. Their possible benefit may be comfortably predicted by using holistic evaluation approaches, for example, the behaviour and process modelling (BAPM) approach (Schabacker, 2002).

As a result, the utilisation of resources is improved whilst the process elements and process element topology are not changed.

Simultaneous engineering

In principle, simultaneous engineering is defined as the parallelisation of unequal activities, which traditionally are processed in a sequential manner (for example, development, design and process planning), thus putting significance on shortening the throughput time. The main challenge in parallelising activities is to find the right moment when interim results, created by earlier activities, are mature enough to be handed over to and be processed by a formerly following, now parallelised, activity. In this context, 'mature' means that the probability of the results changing is low, and that, if a change occurs, the costs of this change are lower than possible penalties, which arise when a deadline is exceeded. Note that, during the overlap of the parallelised activities, a continuous permanent exchange of results between the two activities is necessary.

Simultaneous engineering is defined as the parallelisation of unequal activities which traditionally are processed in a sequential manner.

In this context, the output data of each process element ('results') are continuously compared with the input data and *vice versa*. One approach is to check the degree of fulfilment of each activity, *i.e.* a measure of the respective maturity of each result. The other is to take a closer look at the different components of the throughput time (see Figure 15.6):

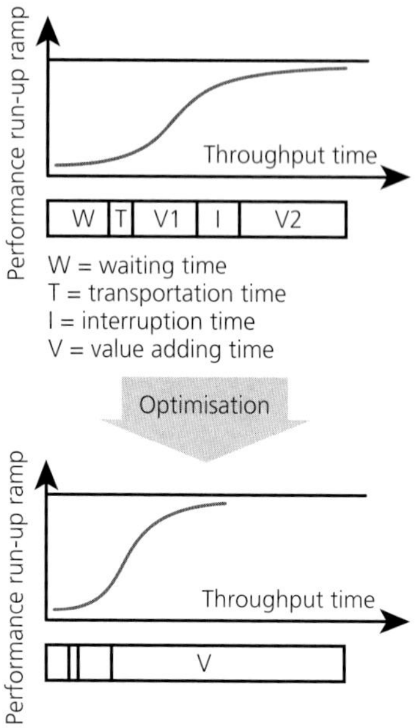

15.6 Components of throughput time

- *value adding time* where the 'real work' happens;
- *waiting time* before work can start;
- *transportation time* to provide the necessary input (for example, product models or necessary methods) for the work;
- *interruption time* lost when the work has to be interrupted due to unforeseen circumstances.

The performance of the process element shows a run-up ramp, which can be shortened by minimising the non-value-adding components of the throughput time. Possible means to shorten transportation time are company-wide data distribution systems (for example, EDM/PDM systems). Means to shorten waiting time and interruption time or, respectively, shifting the start of the value-adding time of the parallelised element in order to reach the appropriate point in time are dynamic process navigation systems, as described in this contribution.

In order to simplify over-complicated process nets which result from simultaneous engineering, an organisational process element, the 'gating' process element, can be included additionally in the process model. This element requires a specified input at a specified time from several process elements. Its main activity is to harmonise the different inputs and to come to a common agreement on the results, thus defining interim results as a firm '*status quo*' in the process.

As a result, the way of working on a process element is changed. The process topology is changed to include more parallel elements.

Concurrent engineering

A process element usually consists of either several working steps and/or of several loops or iterations of the same set of working steps. Therefore, it can be decomposed into smaller portions of equal activities to be processed in parallel by several employees, possibly at different locations. In advance, competencies have to be fixed and both the working areas and their respective interfaces (both physical and temporal) have to be clearly defined in order to keep the work consistent. Along with linear optimisation methods, evolutionary methods can be applied to this step (Vajna and Freisleben, 2002), changing the topology within a process element.

Time concentration

The whole process topology can be reconfigured using evolutionary methods to achieve the shortest possible throughput time. This usually results in different working sequences and maximum parallelisation. Time concentration includes all preceding optimisation steps. The main difference compared with the other steps is the focus on the whole process, whereas the other steps take a more 'localised' view. The resulting process topology is completely changed. All the optimisation steps are shown in Figure 15.7.

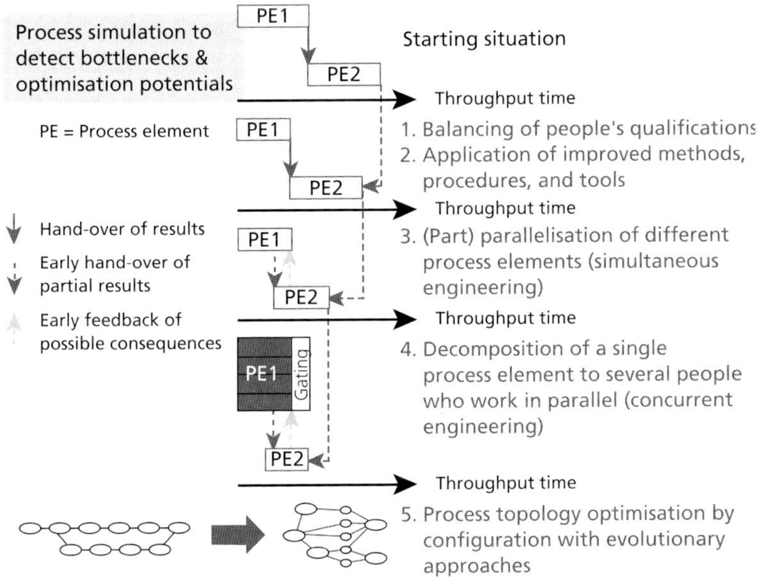

15.7 Summary of process optimisation steps

Process performance review

In business theory there are no suitable benefit evaluation procedures for the performance of engineering processes or of their respective tools. Evaluation systems from business theory lack focus for use with unpredictable processes (see Figure 15.2). It is typical of engineering tools that the necessary costs for their implementation and application appear in cost categories that are not the same as those for the benefits of the application. Another problem is the missing process orientation, as well as an inadmissible mix of quantifiable and qualitative benefits. Hence, the results are difficult to comprehend.

The process review approach used here applies the so-called Benefit Asset Pricing Model (BAPM®) (Schabacker, 2002). This portfolio of engineering benefit classes is based on the portfolio theory of Markowitz (1952). It also takes into account the viewpoints of the balanced scorecard approach. Methods and procedures for yield and risk evaluation of capital market investments can be applied to benefit evaluation. It can be shown that the BAPM® portfolio, which consists of different benefit classes with their appropriate risks, behaves in a similar way to a portfolio in the capital market containing shares, bonds and zero bonds.

With this evaluation approach, BAPM® provides detailed information on the expected return on process performance and investment in tools, estimates the investment risk, and provides an easy way to understand and overview potential benefits. For the evaluation, it uses the process structures and other information from the process model. Results include the yield and risk portfolios for each benefit class (Schabacker, 2002).

Process navigation with proNavigator

The approaches described here have been implemented as the proNavigator (2003) software suite, which has been successfully applied in several major industrial companies, creating amazing results in improved process work.

This system is able to model existing engineering processes, to navigate users through a project or a product development process, to respond dynamically to disturbances during the product's development (for example, when a customer changes product requirements during order processing or a resource fails), to monitor processes and their improvement, to evaluate benefits, and to document the actual execution of the project.

The proNavigator software maps the processes within product development using predefined process elements from its library. As shown in Figure 15.3 (top left), the initial package of process elements has to be customised to the actual situation in the given company, e.g. by adding the available qualification profiles, methods and tools. This is usually done by examining the departments in question. From experience, the cost of adaptation (in person days) is approximately equal to half the number of employees in the product development departments.

Within proNavigator, process elements are treated as objects, thus allowing the use of object-oriented modelling. The system offers executable processes and creates a holistic view of the whole project. During execution proNavigator offers documents, working techniques, design methods and tools to the user, ensuring that he does not forget any of the necessary activities when developing the product. It documents all executed processes, supporting distributed and network-based development.

Processes are evaluated by proNavigator with regard to costs and benefits, and can be executed forwards in time (prospective) and backwards in time (retrospective). It is, for example, possible to evaluate the benefit of using a new technology or a new software tool in general or within a specific process. The general system structure and the information flows are shown in Figure 15.8.

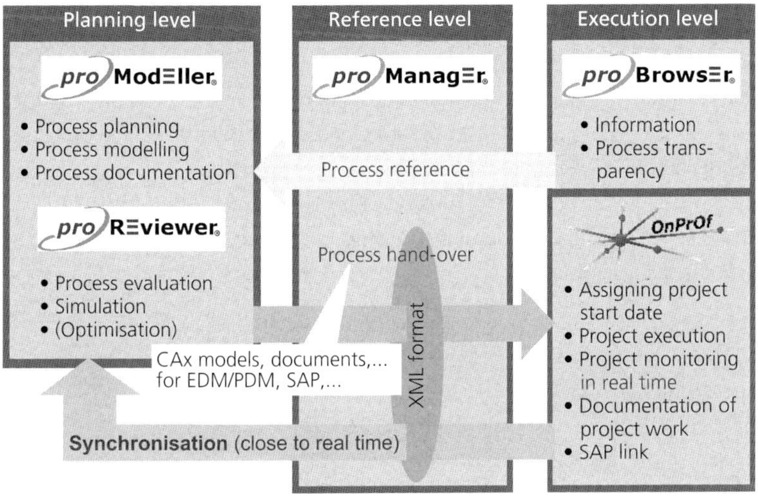

15.8 The system structure of the proNavigator software suite

The system structure was designed with regard to two distinct approaches within process navigation. The so-called planning level covers all necessary steps to plan, model, document, evaluate and simulate and to optimise a process. These are performed with the proModeller and the proRe viewer tools respectively. The result is a complete process model, optimised to fulfil actual goals within a given environment.

This process and all adjacent objects (for example, documents and procedures) are handed over to the project management system using standard interfaces (*i.e.* in XML format). Since proNavigator provides several interfaces, most standard project management systems can be linked to it. However, these systems have to fulfil certain requirements, especially in project monitoring.

In the current system, the OnPrOf (2003) system from Freudenberg FAW was selected. Assigning the project start date initiates the change from process to project status. During its run, the project is continuously monitored, providing the project team members and the respective managers with a complete overview. OnPrOf is multi-process oriented, *i.e.* it supports the execution and monitoring of parallel projects.

If any problem occurs during execution (for example, resources shortfall or change of requirements), the project is stopped and returned to the planning level, where any necessary adaptations and evaluations are performed. The adapted process is handed back to OnPrOf, where the project continues from the same process element where it was stopped.

Process modelling

The proModeller module user interface is shown in Figure 15.9. It supports efficient modelling of a process by allowing users to drag-and-drop process elements from a customised library. There is no preferred modelling direction, i.e. processes can be modelled top-down or bottom-up due to the object character of the process elements. During modelling, process knowledge is automatically captured.

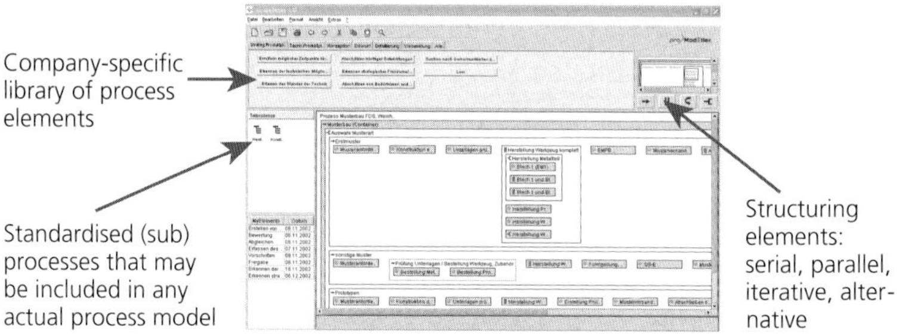

15.9 proModeller user interface

Any process modelling state can be evaluated regarding times and costs using the proReviewer module. As soon as modelling is required, process models can be automatically handed over to the project management software.

Processes can be stored and can be included as standardised (sub) processes in an actual process model. This capability assures the reuse of proven and consistently documented processes and the adjacent knowledge, e.g. as reference elements.

Process execution

As soon as the process is handed over from proNavigator (see Figure 15. 8) and thus becomes a project, OnPrOf starts the execution and monitoring of the process. The project team member is presented with the following interface shown in Figure 15.10.

The main issue is to ensure that the team member cannot forget any project step, whilst they have the freedom to choose that project element (or

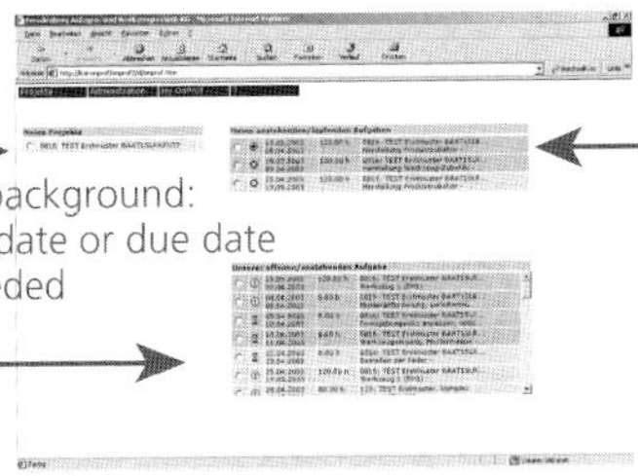

15.10 Interface for the team member showing all current related activities

'task') they want to work upon. OnPrOf provides a complete and dynamic task list for every team member, which is updated every minute. This list includes all tasks that are assigned to a certain member, in either processing or waiting status. Through the use of different colours, the member sees the actual state of every task.

The team member selects the task he wants to process next from the task list. This selection is based purely on the member's decision. He is not forced by the task list to proceed in any particular order. As soon as the selection is confirmed, OnPrOf provides the necessary data and tools to work with, based on proNavigator's process model. If, by this selection, other tasks become critical, these are highlighted immediately, both to the team member and to the project manager.

The project manager is provided with an actual overview of all parallel running projects, which is updated every minute (Figure 15.11). Using this overview, the project manager (or any other authorised person) is able to discover quickly not only already critical projects, but also possible critical projects. In both cases, he can link into a detailed view of the project in question, identify the critical activity, then stop the project and return to proModeller (Figure 15.9). He can then resolve any problems (usually together with team members) before continuing with the revised project.

Application examples

In an automotive supplier company, 30 employees in a specific department process 1,400 projects per year. The average throughput time of each project is between 60 and 70 days, i.e. each employee is involved in about 140 projects that run in parallel. As a result, employees forgot to process critical tasks, applied wrong methods and tools, or lost process threads in this multi-project environment.

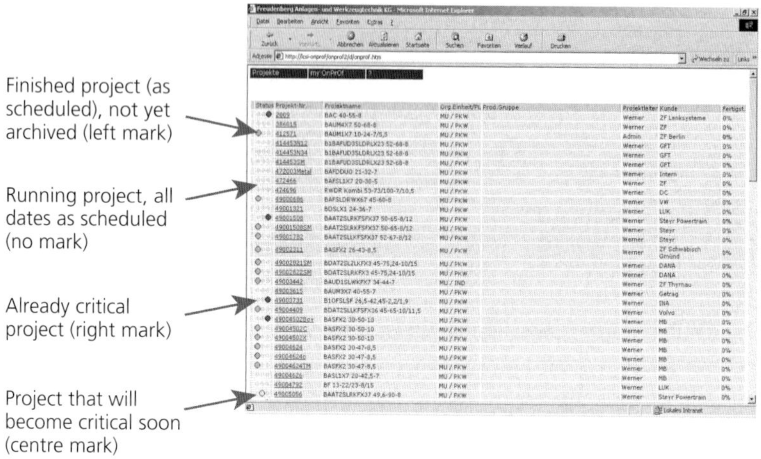

15.11 Overview of all active projects

After the proNavigator was introduced, employees regained process transparency. Overtime was significantly reduced. The throughput time dropped on average by 5%. From the investment point of view, a minimum decrease of only 0.5% was required, *i.e.* the application was economically successful. These results were achieved within 4 months after implementation of the proNavigator.

Figure 15.12 shows the results of another industrial application at a communication company. After the processes were modelled with the pro-Modeller module, the proReviewer module, which applies the BAPM approach, was used to evaluate costs, benefits and risks of a possible ERP support system for certain process elements and their combinations.

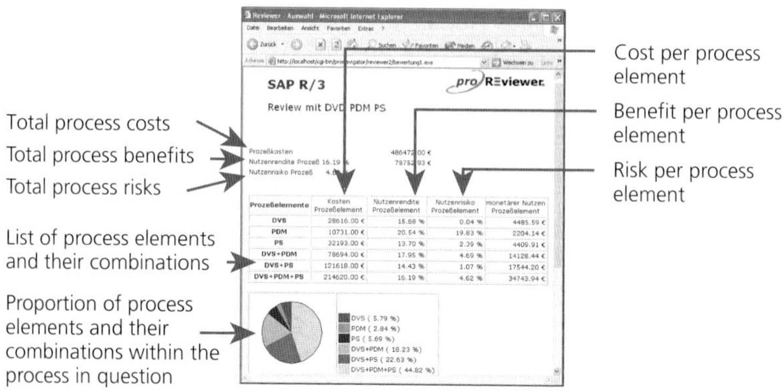

15.12 Review of process elements supported by an ERP system

Not only were the total costs, benefits, and risks involved in achieving the benefits provided, but also a view per element. This was especially helpful when a decision was required to determine at which process element the implementation of a tool would be started in order to achieve a quick return on investments.

Conclusion

Processes within engineering can largely be modelled by configuration and combination of quasi-standardised process elements. These elements may be regarded as building blocks, which can be combined by using the rules of a morphological box to build up a process topology. The standardised process elements are furnished with corresponding knowledge (which has to be adapted to a given situation). The content and structure of this approach enable real-time modelling, evaluation and dynamic navigation through any type of process or project. Based on this approach, a software system has been developed that has been successfully implemented in several companies of different types and sizes.

The emphasis of the actual research and implementation work is on the realisation of optimisation capabilities, using approaches and tools from artificial intelligence (for example, neural networks and evolutionary algorithms) for both the improvement of one process and its extension to a multi-project environment. The latter will be an analogous application of an already existing approach of product optimisation (Vajna *et al.*, 2003).

In the future, pattern recognition approaches will be included to enable and simplify the retrieval and the reuse of existing processes, thus partly shifting the process modelling work from original design to adaptive design.

References

Aurich J, Wagenknecht C (2003) Bausteinbasierte Modellierung unternehmensübergreifender Produktionsprozesse. ZwF, 98(12): 661–664

Conger M (1999) The emergence of workflow management in enterprise-class business solutions. Web Publishing, Inc. http://sunsite.uakom.sk/ sunworldonline/swol-09-1999/swol-09-itarchitect.html

Dym C (1994) Engineering design: a synthesis of views. Cambridge University Press

Eder WE ed. (1996) Engineering design and creativity. WDK 24. HEURISTA
Ferguson D (2000) Therbligs: the key to simplifying work. The Gilbreth Network. http://gilbrethnetwork.tripod.com/therbligs.html
Freisleben D (2001) Gestaltung und Optimierung von Produktentwicklungsprozessen mit einem wissensbasierten Vorgehensmodell. Integrierte Produktentwicklung, 2
Freisleben D, Vajna S (2002) Dynamic process navigation: modelling, improving, and review of engineering processes. ASME'02, Montreal, Canada
Gilbreth FB, Gilbreth LM (1924) Classifying the elements of work. Management and Administration in Manufacturing Industries, 8(2): 151–154
Girard P, Merlo C (2001) An agent environment to support co-ordination between design actors. ICED'01, Glasgow, UK
Hubka V, Eder WE (1992) Engineering design: general procedural model of engineering design. HEURISTA
Markowitz HM (1952) Portfolio selection. Journal of Finance, 3: 77–91
Marle F, Bocquet J-C (2001) A multi-project management approach for increased planning processes. ICED'01, Glasgow, UK
OnPrOf (2003) Online project office. Freudenberg FAW. http://www.onprof.de
Pahl G, Beitz W (1996) Engineering design: a systematic approach. Springer
Plötz M, Biehl M (1999) Intelligent kombinieren. Qualität und Zuverlässigkeit, 44(5): 580–584
proNavigator (2003) Process Navigation Technologies GmbH. http://www.pronavigate.com/home.htm
proNavigator (2004) http://www.pronavigate.com
Schabacker M (2002) Benefit evaluation of new technologies. ASME'02, Montreal, Canada
Vajna S, Clement St, Jordan A (2003) The autogenetic design theory: an evolutionary view of the design process. ICED'03, Stockholm, Sweden
Vajna S, Freisleben D (2002) Dynamic project navigation: modelling, improving, and reviewing engineering processes. The Quest, 6(2002): 2
Vajna S, Freisleben D, Schabacker M (2001) Improvement of engineering processes. ICED'01, Glasgow, UK

VDI Richtlinie 2221 (1986) Methodik zum Entwickeln und Konstruieren technischer Systeme und Produkte

Wiendahl H-P (1970) Funktionsbetrachtungen technischer Gebilde: ein Hilfsmittel zur Auftrags-abwicklung in der Maschinenbauindustrie. Thesis, RWTH Aachen, Germany

Zwicky F (1966) Entdecken, Erfinden, Forschen im morphologischen Weltbild. Droemer Knaur

Chapter 16

Integrated new product development

Jonathan Cagan and Craig M Vogel
Carnegie Mellon University

The creation of high-valued products and services requires a process that is driven by anticipating the needs, wants and desires of key stakeholders, and Cagan and Vogel (2002)provide a strategy and series of tools to help companies navigate the earliest stages of product development. This is the portion of the process that is uncertain and undefined, often referred to as the 'fuzzy front end' of the process.

We argue that it is no longer sufficient to approach product development through the "form follows function" cost-driven process of much of the latter half of the 20th century. Instead, the mantra for the 21st century is that "form and function must fulfil fantasy". Fantasy, in this case, is the anticipation of an optimum consumer experience based on the value system of a particular market segment. When a product meets the anticipated desires of a customer a company can often generate greater profits. In some product categories this level of design has become the cost of doing business.

"Form follows function" is no longer sufficient. Products must capture the users' imagination.

In our research, we have discovered that while many companies say they are customer focused they often fail to maximise the fuzzy front end. Product development teams often focus on the wrong issues too early. Decisions are made with a concern for manufacturing quality and efficiency, overanticipating the needs of downstream processes. This concern for the back end of product development takes resources and attention away from the creation of user-centred product attributes.

It is important to use the front end to lay out a strategy that will connect to the desired market and help to establish or extend brand equity. The idea is not to ignore downstream quality issues, but instead to focus on development innovation in the early stages. If used properly the fuzzy front end establishes the innovation for the product and allows teams to focus on implementing quality processes downstream without costly overruns and changes.

The secret of great product development is to gain significant insight into the needs, wants and desires of the key stakeholders. This requires the use of a variety of qualitative methods that complement existing quantitative processes used by most marketing groups. This chapter highlights our integrated New Product Development (iNPD) process (Cagan and Vogel, 2002) and some of the tools that help a product development team understand the value needs of the product and work in an integrated way to achieve them.

The next section reviews some frameworks for product development found in the literature and introduces the iNPD process. An approach for dividing value into discrete attributes, called Value Opportunities (VOs), is presented, followed by a description of their application to product development through the use of VO Analysis. A case study helps to illustrate the process. Finally, the difference between hard and soft quality and the need for both in any successful product is discussed.

Product development processes

There are a variety of presentations of the product development process that have been discussed in the literature. All tend to take a particular discipline viewpoint, rather than a complete integrated team approach. However, each offers insights or organisation to help the discipline understand or improve the process. Many also discuss the challenges and the importance of addressing the early stages of product development, though few offer techniques to understand the true value that the end stakeholders seek.

One approach often used in industry is the Stage-Gate® (Stage-Gate® is a registered trademark of the Product Development Institute, Inc.; www.prod-dev. com) process of Cooper (2001). In that process, detailed requirements are specified and met at various stages (in terms of timing and development) throughout the process. There are principally five gates (see Figure 16.1) and five stages:

- scoping;
- building the business case;
- development;
- testing and validation;
- launch and post-launch review.

The process is effective and has helped many companies become more organised in their product development process; most companies that have criteria to move through design reviews are using the principles of the Stage-Gate® process. What the Stage-Gate® process does not do is tell you *how* to get through the stages themselves, especially early in the process, or how to enter the process itself, *i.e.* what is an opportunity for a new product.

There have been several presentations of an engineering-based process as found in Pugh (1990), Otto and Wood (2001), and Ullman (1996).

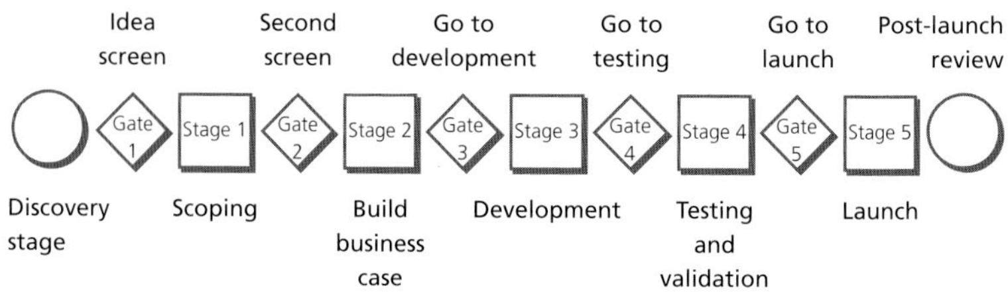

16.1 The Stage-Gate® process
The Stage-Gate® is a registered trademark of the Product Development Institute Inc.; www.prod-dev.com – image reproduced with permission of Dr. Robert G Cooper

Pugh is most noted for the development of what have become known as "Pugh charts", weighted matrices that help a product development team qualitatively compare, differentiate, and filter out competing design concepts.

Otto and Wood (2001) give a thorough discussion of the road map for the engineering design process. One of the most interesting chapters is the first, which presents the design processes, many with a Stage-Gate® flavour, of several design firms and corporations, including Ford, Raychem, Design EDGE, Raytheon, and Motorola. The chapter also presents a list of significant design theory developments from ancient Egypt through to today. The book is a good resource for the process from an engineering perspective once a product opportunity is understood and the conceptualisation process is to begin.

Ullman (1996) is a more succinct presentation of the same process, with particular emphasis on the downstream activities once the engineering specifications are ready to be articulated. In each of these approaches the process begins with product specification, followed by conceptualisation, then detailing, manufacturing specification and quality, and then production.

Ulrich and Eppinger (2003) present a broader view from both an engineering and business perspective (see Figure 16.2). Management and the economics of the process complement many of the engineering techniques included in the previous set of references. Ulrich and Eppinger discuss some of the benefits of industrial design as a player in the process, though they maintain a technical and business approach for the core methodology. Like the above engineering methods, and the Stage-Gate® process, the process assumes an understanding of the product opportunity and direction for product specification, but nicely leads the user through planning, specification, concept generation and selection, refinement, and design for manufacturing and cost assessment.

The variety of books available represent a wealth of information and guidance to help companies work through the product development process.

Other books focus on the business case. For example, Wheelwright and Clark (1992) detail specification and feature requirements, project and team management, and development timing, efficiency and acceleration. Smith and Reinertsen (1998) argue that time, rather than cost, is the critical factor in product development. They focus on the management of the process and teams to help move through the early stages quicker and more effectively.

Most of these books begin once a product focus is understood. They also take a discipline-specific focus and tend to represent the process and product in terms of marketing, manufacturing or functional goals alone. Together they represent a wealth of information and guidance to help companies work through the product development process.

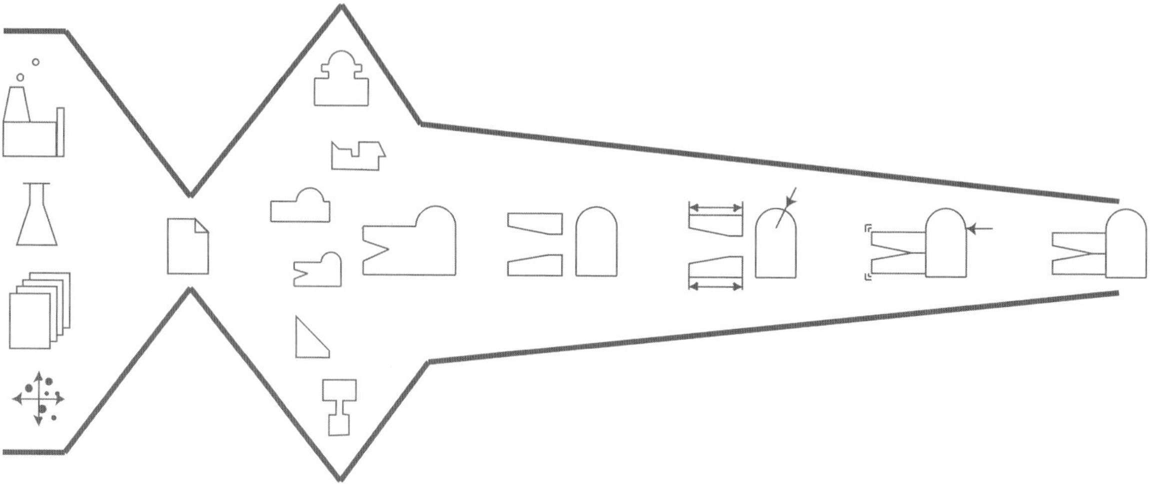

16.2 An engineering and business perspective (reproduced from *Product design and development* by Ulrich and Eppinger, 2003)
© McGraw-Hill – reproduced with permission of The McGraw-Hill Companies

Integrated new product development

In each of the processes presented above, there is little in the way of explanation or tools to help navigate the earliest stages of product development, what has been called the 'fuzzy front end' of the process. The use of the term fuzzy is not arbitrary – to many product developers the uncertainty and fast-evolving/chaotic nature of the early product definition stage is uncomfortable and to be avoided.

Most product development processes begin once it is known what technology a company wants to design and why it wants to design it. Engineers, in particular, are very comfortable taking a product definition and quickly moving it from function to mechanism, which often prematurely dictates product form and interaction.

Unfortunately, all too often, the early definition and purpose are not well understood, which leads to an ineffective or at least sub-optimal solution downstream. Engineering analysis and parameter optimisation tools then take a bad idea and work to make it acceptable.

In contrast, our iNPD process emphasises the earliest stages of product development with a focus on identifying and understanding product opportunities and the value required by stakeholder needs, wants and desires. In addition, equal participation is expected from engineering/manufacturing, marketing/finance, and industrial design/interaction, *i.e.* all major participants in the process. Moreover, the concept of VOA gives everyone involved a set of value targets that engineering, design and marketing can share from the beginning of a program, which has deeper ramifications in how resources are allocated to the process (Cagan and Vogel, 2002).

If the timing and cost allocation of all parts are considered as equal, as is often done in traditional engineering or marketing approaches, then when parts are not detailed for cost and manufacture early, and system costs cannot be determined upfront, production timelines are threatened and cost targets are challenged.

An alternative is to recognise that all parts are not designed alike and take into account their lifestyle impact and aesthetic integration into the overall product. This requires new tools and methods to work through the development process.

Development of the iNPD process

We have studied the very early stages of product development, the part that begins before the Stage-Gate® process when the product opportunity is just being formed as a vague description of intent. We have also studied industrial product development processes in a variety of industries, have consulted with consumer, medical, and business-to-business product and services companies.

The concept of VOA gives everyone involved a shared set of value targets.

We have taught an innovative product development class at Carnegie Mellon University for over a decade. This course requires engineers, industrial and communication designers, and marketing (MBA) students to work together to create patentable (and often patented) products in 16 weeks. It is a course that emulates the environment of the fuzzy front end.

We have developed tools and methods to help product development teams navigate through the early stages. Our approach uses four phases that bring each team from product opportunity identification through to the point of program approval where a company commits to patenting and manufacturing costs.

The methods and tools integrate with current processes within companies, or can serve as the basis for developing a new process for those companies looking to develop their own. In the next section the basic iNPD process is reviewed and then, in the subsequent sections, the part of the process most foreign to many engineers, namely understanding and articulating true customer value, is discussed.

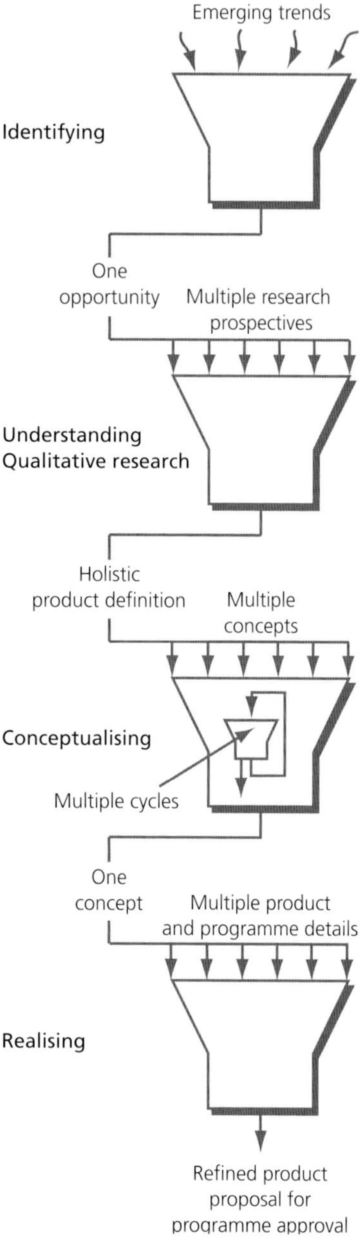

16.3 Four-phase product development process (adapted from Cagan and Vogel (2002)) © 2002 Prentice Hall PTR

The iNPD process

The iNPD process is made up of four phases, namely: identifying; understanding; conceptualising; and realising. These are described below and summarised in Figure 16.3.

Identifying

Product opportunities are identified. We describe social, economic, and technology (SET) factors, which interact in a dynamic way to create product opportunities. By constantly scanning these factors, trends in culture and lifestyle can be identified. This reveals gaps in the marketplace otherwise known as Product Opportunity Gaps (POGs). Initial customer-based and secondary literature-based research lends credibility and insight to the opportunity, transitioning to the second phase.

Understanding

Qualitative research focused on a deep understanding of the key stakeholders leads to actionable insights that provide a framework for product form and feature development. This phase is what differentiates an insightful process that can break through existing solutions from the standard approach of minimal change and innovation. The challenge is to identify, understand, and articulate the key attributes of value to be developed in the product. In the next section, customer value is discussed along with VOA. The end result of this phase is an initial product description that indicates who the target market is, and what characteristics the product will articulate.

Conceptualising

A more traditional part of product development that takes the insight from phase II as a basis for generating concepts and resembles the second stage of the Stage-Gate® process, for example. The difference between iNPD and other methods, however, lies in having already conducted the research in phase II. This makes the conceptualisation more effective and meaningful; the initial product criteria developed in that second phase, in addition to serving as the point of departure, also serve to direct and confirm each concept developed. In order to reach an optimum conclusion at the end of phase III, it is important to use an iterative conceptualisation process. This requires multiple cycles of quick, interactive prototypes tested or discussed with the key stakeholders to help direct the process. At the end of this phase the basic product is now designed, setting up the fourth phase.

Realising

The concept is detailed to the point that the company can make a go/no-go decision as to whether to move the product to production. In the Carnegie Mellon class, for example, students have a complete and accurate form model, technical proof of concept often shown through a functional prototype, a marketing plan with complete financial and roll-out strategy, and a manufacturing plan. Even in phase IV the basis for success lies in the eyes of the stakeholders as identified by the team. Very often this phase can be compromised by internal groups feeling that the product is a success and rushing to judgement without customer feedback.

After the fourth phase the product goes into the stage of refinement toward production and launch. These steps are well understood, but the challenge is to protect the innovation created in the earlier phases. Because the product was developed with a good understanding of the customer, that knowledge provides the rationale to protect the features and to prevent cost reduction from reducing feature quality as well. We present tools to help the team carry out deep qualitative research on the customer. In addition, we argue for integration of an inter-disciplinary product development team, and introduce tools to help bridge the natural perceptual gaps between disciplines. In the next section the concept of product value and VOs, a tool that makes phase II of our process so effective, is presented. The iNPD approach is compatible and complementary with each of the processes discussed earlier in the chapter, giving effective guidance to the earliest stages of product development and completing the discipline perspectives.

VO characterise the impact of a product or service.

Value opportunities

It used to be that value was equated with having the most features in a product for the lowest price. For products that are highly desirable, value is not the number of features you can get for the least money; rather, it is how effectively the features meet the expectation of usefulness, usability and desirability of the desired market segment. Value is represented through impact of the product or service on the user's lifestyle, use of the product or service through enabling features, and meaningful ergonomics.

We have broken value into seven categories, each with distinct attributes. These categories are called VOs. A product development team can use them to assess the current state of products in each category and to determine where improvement is possible. Each of the seven VO classes (emotion, ergonomics, aesthetics, identity, impact, core technology, and quality) contributes to the overall experience of the product. We map the VO attributes onto a VO chart.

The VO chart forms the basis for the VOA tool, which helps a product development team analyse the current state of products on the market, the ideal state of a product, or the realistic expectations of what attributes of value a new or next-generation product can achieve. These value categories make sense for all disciplines involved in the product development process and help teams to develop a shared understanding of their goals and to develop new products. Feedback from users indicates that they find the tool useful in structuring the qualitative goals of product programs. The seven VO classes and their attributes are now described.

The VO chart helps a product development team analyse existing and future products.

Emotion

This is closely related to a user's fantasies, and can be broken into six attributes:

- *Sense of adventure* – the product promotes excitement and exploration.
- *Feeling of independence* – the product provides a sense of freedom from constraints.
- *Sense of security* – the product provides a feeling of safety and stability.
- *Sensuality* – the product provides a luxurious experience.
- *Confidence* – the product supports the user's self-assurance and motivates him to use the product.
- *Power* – the product promotes authority, control, and a feeling of supremacy.

Ergonomics

The core of physical interaction, ergonomics is broken down as follows:

- *Ease of use* – product must be easy to use from both a physical and a cognitive perspective. It should function within the natural motion of the human body. The size and shape of components that a person interacts with should be logically organised and easy to identify, reach and grasp.
- *Safety* – product must be safe. Moving parts should be guarded.
- *Comfort* – product should be comfortable to use and not create undue stress during use.

Aesthetics

The aesthetic attributes are:

- *Visual* – the visual form must relate shape, colour, and texture to the context of the product and the target market.
- *Tactile* – physical interaction with the product, focusing primarily on the

hand but including also any other physical contact between the product and user, must enhance the product experience.

- *Auditory* – product development must determine and integrate the appropriate sounds and eliminate undesired sounds.
- *Olfactory* – product development must consider the impact of smell, providing appropriate aromas and eliminating undesirable odours.
- *Gustatory* – products that are designed to be eaten or used as a utensil, or that may otherwise be placed in the mouth (for example, a child's toy), must have an optimum flavour or no flavour at all.

Identity

Three attributes of product identity are:

- *Personality* – the two main issues in a product personality are (1) the product's ability to fit among and differentiate from its direct competition, and (2) the product's connection to the rest of the products produced by that company.
- *Point in time* – in order to be successful, a product has to capture a point in time and express it in a clear, powerful way.
- *Sense of place* – products must be designed to fit into the context of use.

VO include:
- emotion
- ergonomics
- aesthetics
- identity
- impact
- core technology
- quality

Impact

Connected to corporate brand and responsibility, and probably the least explored of all the VOs, impact has two attributes:

- *Social* – product can have a variety of effects on the lifestyle of a target group, from improving social wellbeing to creating a new social setting.
- *Environmental* – the impact of products on the environment is becoming an important issue in terms of consumer value. Design for the environment focuses on minimising negative environmental impact associated with manufacturing, resource use during operation, and disposal.

Core technology

People expect technologies to evolve rapidly and be increasingly:

- *Enabling* – core technology must be appropriately advanced to provide sufficient capabilities in a product. It may be emerging high technology or well-manufactured traditional technology, as long as it meets customer expectations in performance.
- *Reliable* – consumers expect technology in products to work consistently and at a high level of performance.

Quality

The quality VO includes two attributes:

- *Craftsmanship – fit and finish* – the product should be made with appropriate tolerances to meet performance expectations.
- *Durability – performance over time* – the craftsmanship must hold up over the expected life of the product.

We have shown that this breakdown sufficiently describes the value quotient of over 20 products from consumer goods to industrial products to services like United Parcel Service Inc. (UPS), and even the emerging retro baseball parks in the USA. Firms have also used this breakdown proactively in product development in service industries, the medical products industry, the auto industry, chemical companies and commodity manufactures of raw materials. These concepts have been introduced to electronics consumer manufacturers and the durable goods industries. In each of these cases this approach has helped their clients understand what aspects of value relate to their target customer base.

Figure 16.4 shows a complete list of VOs, where each VO can be evaluated qualitatively as zero, low, medium, and high, based on how well that attribute addresses the goal of the product. The resulting VO chart can also show the profit impact, brand impact, and extendibility of the product to other products in the company. The chart can be used to set expectations of where a new product ranks on the different attributes of value. It can be used to compare one product against a competitor. It can also be used to compare a current product to how a redesigned one should improve the value quotient.

Figure 16.5 shows a VOA, where one product is compared with another. Here, the OXO GoodGrips vegetable peeler is compared with its generic counterpart that was the standard for over 100 years prior to the OXO introduction. Visually, it is clear how much better the OXO product compares with the generic standard. The generic peeler ranks low in the emotions of independence and confidence, and meets a low level of the ergonomic attributes of comfort, safety and ease of use. The form follows function aesthetics are poor and the product makes no statement about brand identity. Although the durability is high (it will last forever), its VOA clearly indicates a missed opportunity in the marketplace.

The GoodGrips, on the other hand, excels in its ability to meet strong emotion VOs in independence, confidence, and even security, especially for the target of elderly or arthritic users. The product also excels in all aspects of

Proposed product

Emotion	adventure independence security sensuality confidence power
Ergonomics	comfort safety ease of use
Aesthetics	visual auditory tactile olfactory taste
Identity	point in time sense of place personality
Impact	social environmental
Core tech.	reliable enabling
Quality	craftsmanship durability
Profit impact Brand impact Extendible	

16.4 VOs (Cagan and Vogel, 2002)
© 2002 Prentice Hall PTR

ergonomics, core technology and quality. The form and tactile design of the product make strong aesthetic and brand identity statements of value; it is a product people want to own and are willing to spend five times the cost of the generic counterpart to possess.

The GoodGrips also has very strong social impact, stemming from the success of the handle design that enables people to hold the product with a greater sense of security. As a result, the patented GoodGrips handle has

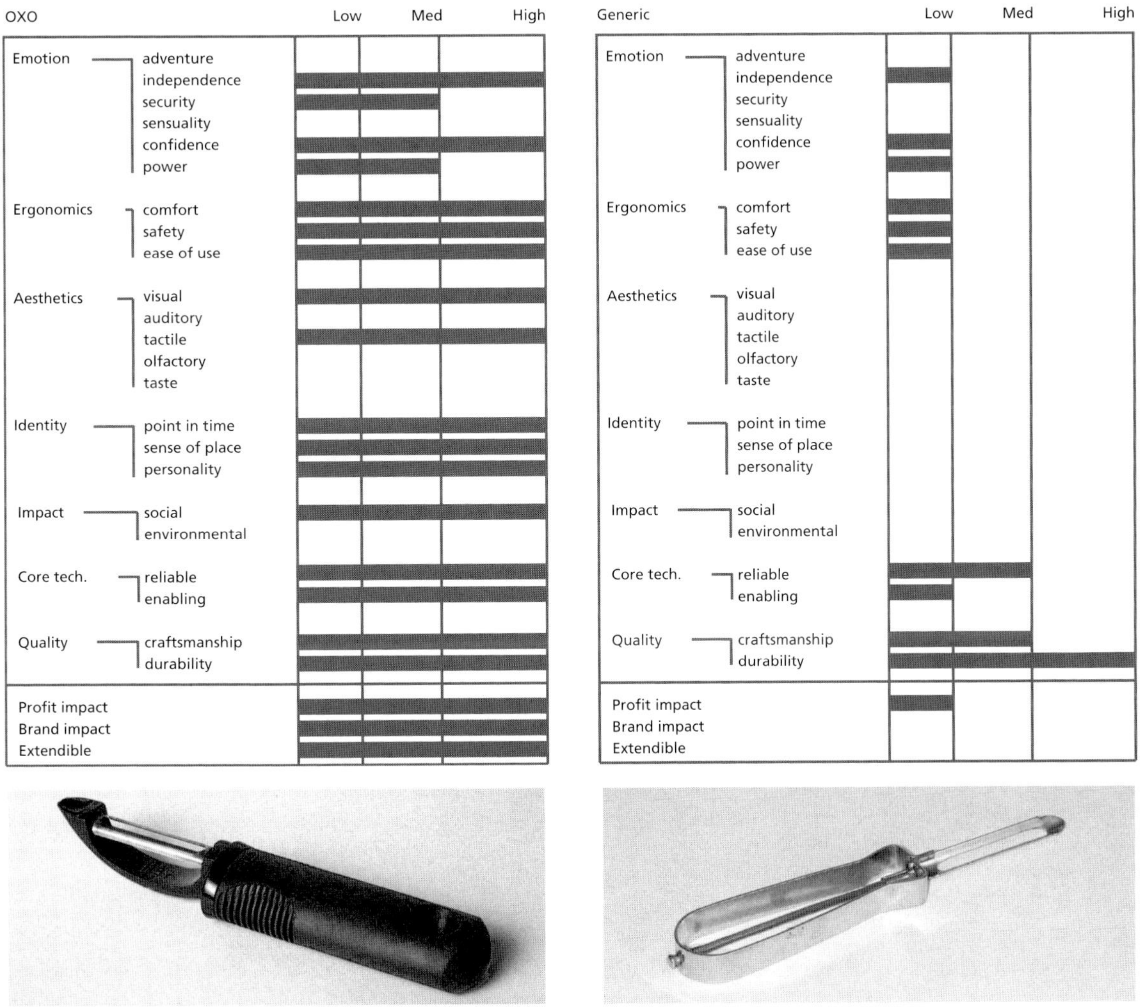

16.5 VOA of OXO GoodGrips vs generic peeler (Cagan and Vogel, 2002) © 2002 Prentice Hall PTR

helped the company launch over 350 products, including gardening tools, construction tools and other kitchen products.

VOA is a user-driven approach to product development that addresses the core value sought by the user.

The VO and VOA is a user-driven approach to product development that addresses the core value sought by the user. Understanding this value to begin with is a critical process that uses various qualitative research tools, including new product ethnography, human factors, task analysis and lifestyle reference. The VOA is just a first step. The major challenge is to convert this qualitative measure into what may be called 'actionable insights', namely goals that achieve each VO. This set of goals provides an early specification for a product well before the form or features are designed.

Case study: design of interior cleaning system for cars

In recent years the Integrated Product Development class at Carnegie Mellon has attracted corporate sponsorship where companies have had intellectual property rights to products created. Ford Motor Company sponsored the class twice, and from 12 projects they have submitted five patent applications (two of those patents have been issued so far). Most recently, two companies in the bioengineering field, Respironics and BodyMedia, have supported the course.

One project from this class, supported by Ford, focused on the design of integrated, interior lifestyle features for a small sport utility vehicle (SUV), the Ford Escape. In particular, it focused on interior update, or cleaning, of the vehicle and was designed by seven students: designers Joseph Genuardi, Jon Mayer and Lisa Villemure, engineers Andrew Birnbaum and Erika Wetzel, and marketing students Samir Kayande and Esperanza Lo.

After brainstorming over 100 possible product opportunities, they narrowed down to the opportunity of maintaining a clean environment inside the vehicle. They chose to focus on a particular market segment characterised as families with dual careers. These couples often have several young kids and lots of activities, and are always "on the go" but still with limited income. They need to maintain and clean their vehicles without spending the $100 it takes to detail an SUV at a specialty shop. The opportunity statement for this team was: keeping the interior of a car clean, as defined by the expectations of their chosen market segment.

The team then moved into the second phase and pursued multiple directions of field testing, primary and secondary research and observation to become experts in interior cleaning of vehicles. Their research with their target market gave them meaningful insights. Here are some of valuable quotes they obtained from their intended customers: "If there was some cheap, easy, quick way of cleaning my car, that would be good", "I wish I had a dedicated spot where I could put my garbage", "My husband tries to clean it, because I don't", "I eat when I'm driving…I'll make time to clean when I can't stand it anymore", "I can't vacuum, because there's no power outlet near my car".

The team developed an overview of the opportunity presented by a continuum of trash and dirt in a vehicle. They determined what portion of that continuum their solution needed to address. This led to the proposal of the need to develop a cleaning system with two main components: a handheld vacuum and a trash bin. Their research on competitive products showed no current product that addressed the problem in a way that met the lifestyle needs of the target user. The research on their target market, both primary and through access to lifestyle-based databases, led to a VOA of the way trash is currently disposed and vehicles are currently cleaned (see Figure 16.6a), and of the goals of the new product they will design (see Figure 16.6b).

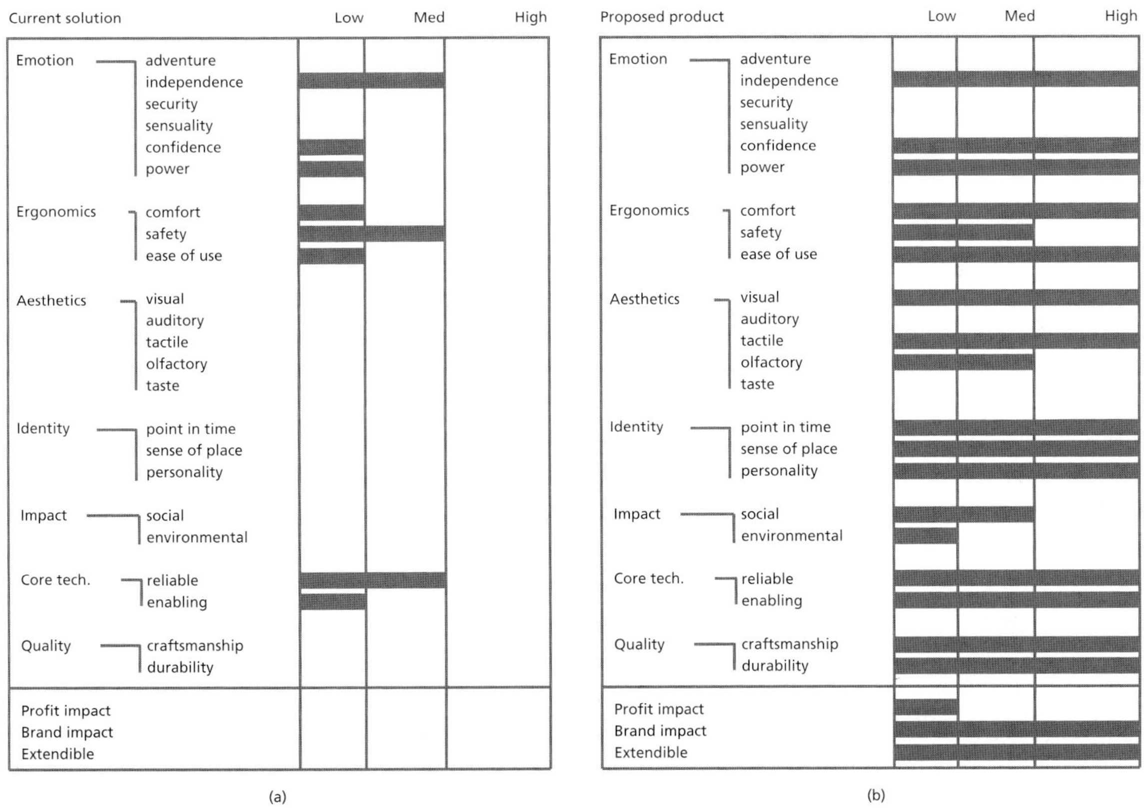

16.6 VOA of (a) current solution and (b) proposed product

The result of the second phase was a good understanding of the features and characteristics of the product opportunity. During the third phase the team focused on conceptualisation and reverse engineering. The specifications for the vacuum system were determined. The location of the system was decided (in the centre console). Because research from their users indicated that the one item currently stored in the console that they did not want to be without was their CDs, a CD holder was included in the concept. The vacuum cleaner needed to be constantly available and charged, so space became a challenge, especially with the size required to generate enough suction. Multiple vacuum forms were explored and tested with their target users.

The final phase led to the detailed design shown in Figure 16.7. The top of the centre console opened to hold a trash receptacle. Plastic shopping bags from local supermarkets were used to collect the trash, and a slit rubber cover kept the odour in. On the back of the console rested the constantly charging vacuum cleaner. The unit was specially designed to be ergonomic and meet the needs of hard to get to spaces in a vehicle. The vacuum swivelled closed to allow for compact storage. Finally, in the front was a CD case that held 10 CDs without the plastic jewel cases. The case also served to accent the Ford brand with a Ford logo and the case was portable to allow the user to take the CDs outside of the vehicle. Target user research showed that more than 75% of their market would want all or part of the system.

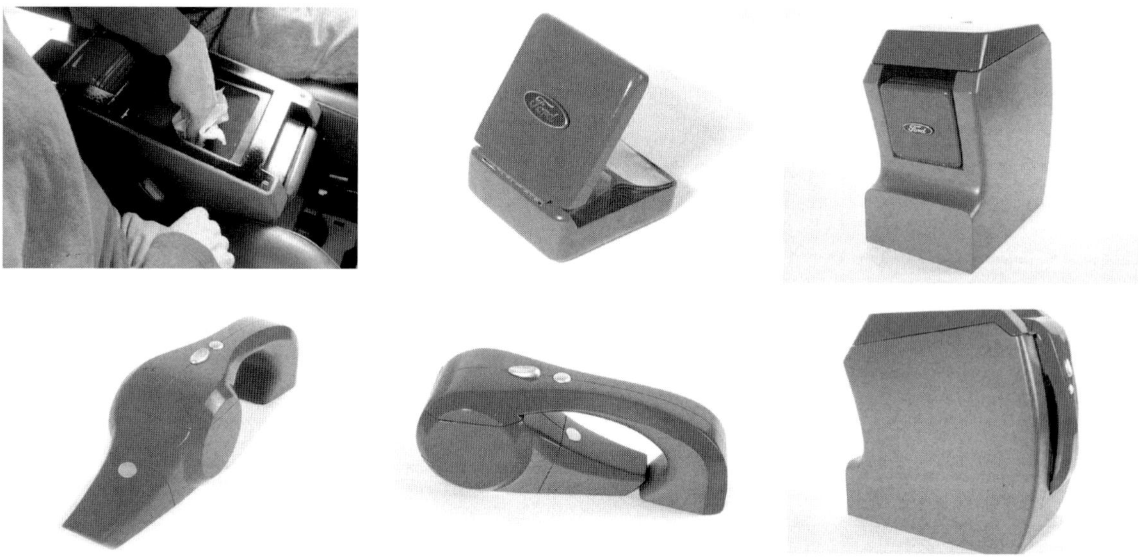

16.7 The final prototype for the SUV interior cleaning system

Resource allocation in the experience economy

The goal of new product development today is to create not just a physical product or service, but to create or enhance an overall experience for the customer. To develop highly valued products requires a new commitment from companies, especially tech-driven ones.

As shown in Figure 16.8a, traditionally tech-driven companies have focused on hard quality attributes, those attributes of manufacturing and technology development, with the form design thought of almost as an afterthought for an industrial designer to complete to finish off the product. The problem with this approach is that high-tech products that are not usable or desirable often give poorer than expected performance in the market. At best, interaction with and experience of these products is never as good as it could be. At worst, a major investment of time and money can be compromised and additional damage done to the perceived brand equity of the company.

This new product service design challenge is the result of the 'experience economy' discussed by Pine and Gilmore (1999). In product terms, we argue that products and services are interrelated and together create or enhance the overall experience that a set of stakeholders undergo when interacting with a product. In many industries, where competition, especially from companies/countries with cheap labour and costs, has driven a product toward being a commodity, design for experience is the only way to move back toward high-margin product development.

Pine and Gilmore work with many companies fighting this pressure to succumb to low-cost commodity approaches. The problem is that this approach leads to a competitive downward cycle that ends up taking the heart out of a company, destroying the potential for competitive innovation. It has been observed that only one company can be the cheapest in a given market. The challenge is to compete by identifying value and to generate greater profits through innovation that creates anticipatory solutions connected to the VOs.

As shown in Figure 16.8b, to commit to product development in the experience economy is to commit to investing in both hard and soft quality. Soft quality is the combination of lifestyle features, user interaction, aesthetic attributes, and brand identity that create the emotional attributes of the product at purchase and initial use. Soft quality creates the brand identity of the product and the initial excitement that leads to the purchase of a product. Hard quality affects the brand identity of the company and the long-term satisfaction with a product.

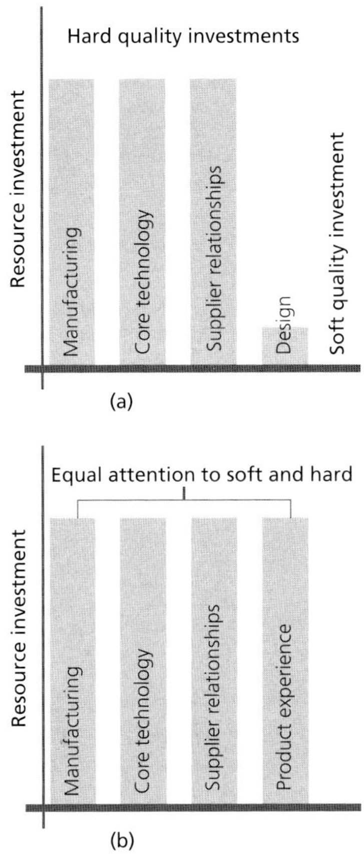

16.8 Shift of resource allocation from (a) hard to (b) both hard and soft quality in the experience economy

As shown in Figure 16.9, the VOs can be divided between hard and soft quality, emphasising the need for both in product development. What is hard for many companies to realise is that it is not just technology that constantly changes. Changes in product aesthetics and ergonomic preferences are as important as technology advances. Executives in tech-driven companies are usually sceptical about the true value of recognising that the soft quality changes are as significant as the hard quality ones.

As shown in Figure 16.8b, resources must move from treating the role of design as an afterthought to soft quality investment upfront. Soft quality investment implies not only industrial design involvement upfront, but commitment from all players.

Engineers should be active participants in the design of soft quality attributes. For example, the acoustics of a Harley-Davidson exhaust are specifically designed to support the experience of the ride. Engineers should be active participants in lead user research.

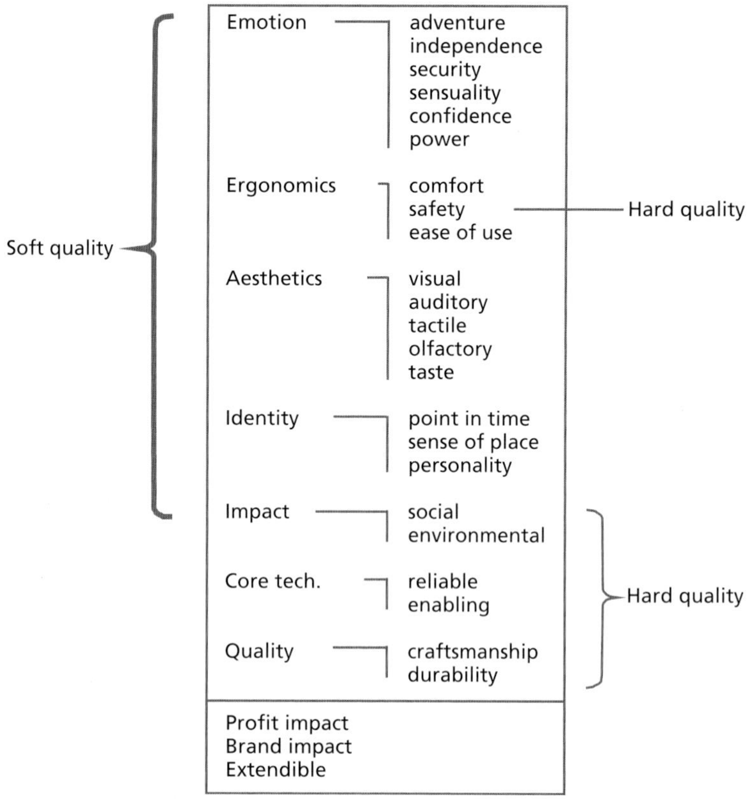

16.9 VO chart indicating both hard and soft quality attributes

Our iNPD method helps teams understand the importance of the soft quality attributes and provides tools, such as the VOA, to help them include these attributes in the product they are creating.

Conclusion

The fuzzy front end can be navigated effectively with the use of methods and tools specifically developed to maximise that part of the product development process. This process requires a commitment by management to support the use of multiple disciplines working in an integrated way, driven by valuable insights gleaned through qualitative research with intended customers.

A product development process that helps the user through the process can lead to effective and efficient downstream product development. The key is to translate the understanding of key stakeholders into VOs that can be translated into product criteria. A key attribute of the success of such a process is to recognise the advantage of both hard and soft quality features.

Devoting resources, including time, to the early stages of product development leads to a more balanced product with fewer downstream development problems. More important, it helps companies develop products and services that meet the needs and opportunities of the experience economy.

References

Cagan J, Vogel CM (2002) Creating breakthrough products: innovation from product planning to program approval. Financial Times Prentice Hall

Robert G Cooper (2001) Winning at new products, 3rd ed. New York, Perseus Books

Otto K, Wood K (2001) Product design – techniques in reverse engineering and new product development. Prentice Hall

Pine BJ II, Gilmore JH (1999) The experience economy. Harvard Business School Press

Pugh S (1990) Total design: integrated methods for successful product engineering. Addison-Wesley

Smith DG, Reinertsen DG (1998) Developing products in half the time. John Wiley

Ullman DG (1996) The mechanical design process. McGraw-Hill

Ulrich K, Eppinger S (2003) Product design and development. McGraw-Hill

Wheelwright S, Clark K (1992) Revolutionizing product development. Free Press

Chapter 17
Product portfolio management

Marco Cantamessa
Politecnico di Torino

One of the most famous citations in management literature comes from Henry Ford, who reputedly claimed that customers of his Model T could have it "in any colour, as long as it was black". Apart from the weak historical evidence that the 'father' of mass production ever did use these words, this situation did not last long. After a few years, Ford ran into problems when forced to chase the strategy of Alfred P. Sloan, who had restructured General Motors around a divisional organisation and successfully started selling differentiated motor cars. With Sloan's objective of providing a car for every taste and for every budget, product portfolio management entered the modern industrial world.

The problem of product portfolio management can be found in virtually any firm and is indeed a complex matter (Figure 17.1). If you side with marketing, their ideal would be to fit a product to each individual customer. If you listen to product development, they would talk about the nightmare of having to manage more projects simultaneously than one can even remember. If you talk to manufacturing, they would probably remind you of a technique called 'variety reduction program' that was quite successful a few years ago.

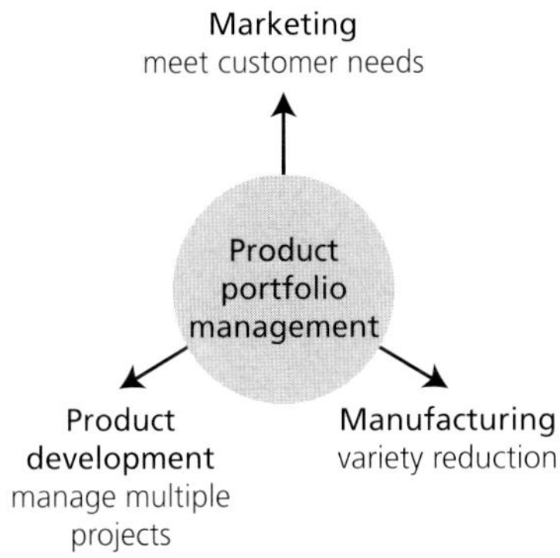

17.1 The pressures of product portfolio management

In response to the implications of different organisational functions, this survey on product portfolio management has been based on contributions from different fields, including economics, marketing and operations management. I hope that this heterogeneity will not disrupt the thread of the discussion, which is structured as follows: the next section will discuss the 'front-end' of product portfolio management or, in other words, the marketing perspective. The second section will discuss the 'back-end', which is concerned with the design and development of multiple products. The third section will present portfolio management tools that may help bring the two perspectives together. Conclusions and open issues that ought to be matter for further research will be briefly discussed in the final section.

The front end of product portfolio management

Having stated that product portfolio management is a problem for industry, one might wonder about the reasons why firms provide multiple products for their markets. Students of industrial economics are accustomed to explaining this issue under the heading of product differentiation. According to this theory, products may be differentiated either horizontally or vertically.

Horizontal differentiation

Horizontal differentiation exists when, by changing a design variable, utility grows for some customers but decreases for others. Horizontal differentiation is, therefore, related to the particular tastes of customer segments: a car may be given more elegant or more sporty design, and some customers will prefer the former while others will favour the latter. The same happens for perfumes (subtle fragrances as opposed to stronger ones), food (mild as opposed to spicy) and many other products.

An economist would model this situation by saying that the reservation price of customer x for product y (i.e. the price at which the customer would be indifferent either to buying or not buying the product) is given by the utility they gain from their 'ideal' product, minus a function of the distance between this ideal and product y. Customers will, therefore, be willing to pay more for a product that exactly matches their taste and less for a product that is more distant. A monopolistic firm providing a single product would, therefore, be forced to lower the price substantially, while catering separately to each market segment allows a firm to keep prices higher (see Figure 17.2).

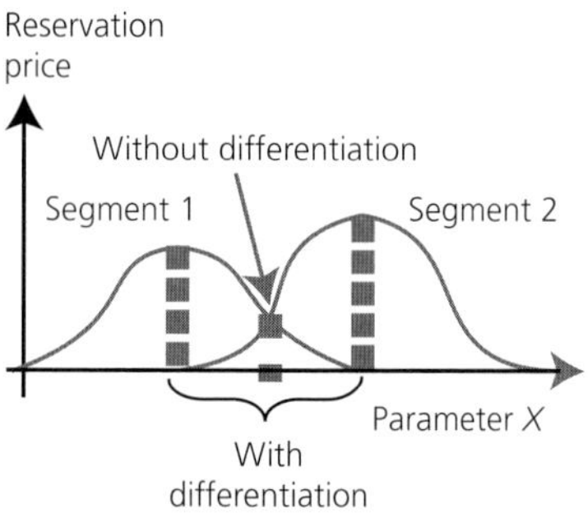

17.2 Horizontal product differentiation can increase prices

When firms are in Chamberlain monopolistic competition (i.e. when sellers are many and products are slightly differentiated), or in oligopoly (when competing firms are fewer) theory shows (Tirole, 1989) that, by aiming at separate market segments, there is less competitive interaction among firms, and this decreases downward pressure on prices. This explains why marketing, whose aim is to maximise revenue, would like to sell a distinct product to each customer.

Vertical differentiation

With vertical differentiation, changing the design variable makes utility grow or decrease for all customers in the same direction, though at a different rate. A car with a greater top speed, better fuel consumption, or more comfort will provide more utility to all customers, though some will value the increase more than others. Vertical differentiation, therefore, has to do with performance and quality and the way that this affects customers' willingness to pay.

A firm providing a single high-quality product will be forced to choose between setting a higher price and catering to the 'premium' market segment only (i.e. the one that values quality more) or setting a lower price and serving all market segments. This latter option would, however, give the premium customers a deal, since they would walk away with more 'surplus' (i.e. the difference between their reservation price, which is equivalent to the utility they gain from the product, and the price they are actually asked to pay). Alternatively, the firm might provide a single low-quality item and serve the 'basic' segment only (i.e. the customers who value quality less), but lose revenue from premium customers. In order to increase revenue, the firm could provide a high-quality item at a higher price and a low-quality item at a lower price, thus serving both segments at (or close to) their reservation prices. In doing so, the firm must be aware that it risks cannibalisation of its high-quality products, as shown in the Figure 17.3.

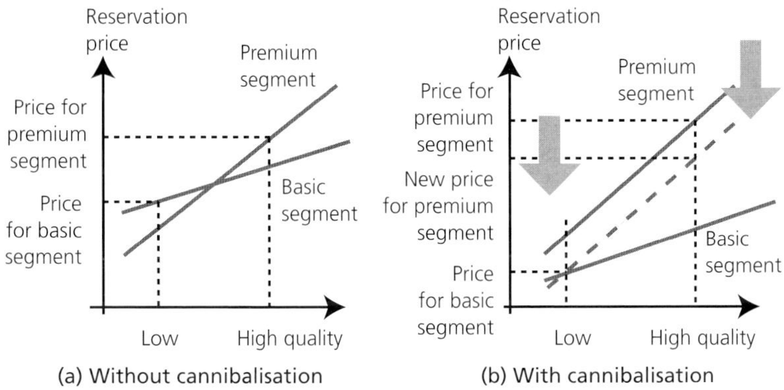

17.3 Vertical product differentiation can decrease prices

Figure 17.3a depicts a vertically differentiated firm, serving two segments without cannibalisation. The figure shows the utility of the two segments as a function of product quality, and the quality levels and prices for the two products it sells. The two utility curves cross, so that the premium segment is willing to pay the required amount for the high-quality product, whereas the basic segment is willing to buy the low-quality product. Neither segment would have any benefit in switching to the other product, since surplus would be negative for them.

Figure 17.3b shows a case with cannibalisation. While the basic segment still buys the low-quality product, the premium segment finds that by buying the low-quality product, they would gain positive surplus, with the fall in price being greater than the fall in utility. In order to avoid cannibalisation, the firm can either lower the price of the high-quality product, or keep the price fixed but increase the quality of the high-end product, or even purposely degrade the low-quality product, so as to place it to the left of the intersection between the two utility curves.

For example, airlines sell seats in economy class and business class at very different prices. However, many firms often save money by making their staff fly economy class and use tricks, such as buying two return 'back-to-back' tickets, in order to avoid Saturday night stayovers. In order to avoid cannibalisation, airlines can discount their business class fares (for example, back-to-back ticketing can be discouraged by pricing business class at less than double the cheapest economy return fare), or act on the parameters that determine service quality. For instance, they can increase the value of business class travel by providing more facilities at the reserved airport lounges. Alternatively, they can degrade the value of economy class travel to business people by doing away with on-board meals: while people travelling for leisure would find little discomfort in eating at a different time, this might be unbearable for someone travelling on a tight business schedule.

The degree to which cannibalisation is present in a specific market is often measured by using Moorthy and Png's (1992) index:

$$R = \frac{\text{size of premium segment}}{\text{size of basic segment}} \left(\frac{\text{valuation per unit of performance of premium}}{\text{valuation per unit of performance of basic}} - 1 \right)$$

which assumes that the market evaluates performance with a linear function. The index goes from zero (if the premium segment is very small or values quality in a quite similar way to the basic segment) to one (if the size of the premium segment is equal to the reciprocal of the ratio between valuations) and tends to infinity (when the basic segment is very small).

The previous discussion has provided the theoretical foundation explaining why, at least in terms of revenue, firms should offer differentiated products to their markets. Of course, reality is slightly more complicated. Apart from the obvious remark that high product variety comes at a cost, in a competitive environment it can also become a 'must-have' feature that all firms provide in order to serve the market, but without gaining significant competitive advantage from it.

With vertical differentiation firms can provide multiple products that will compete head-on and develop identical product offerings, each occupying a separate niche.
(De Fraja, 1996)

With vertical differentiation, as discussed by De Fraja (1996), firms can provide multiple products that will, in the absence of cooperative agreements, compete head-on and develop identical product offerings instead of specialising and each occupying a separate niche. Competition at the same quality levels will, therefore, force price reductions and decrease profits. This behaviour can be observed in most industries (for example, personal computers and cameras), which are generally dominated by companies providing very similar, broad product lines.

This discussion also suggests that niche players, who may reap very good profits from their positioning, cannot emerge out of competitive manoeuvring, but must base their existence on truly inimitable assets or competencies.

A further result is that, when the number of competing firms increases, product differentiation tends to be lower. With an infinite number of firms, the product becomes ever more the commodity at the highest quality level, and price decreases until it reaches marginal cost. On the empirical side, Bayus and Putsis' (1999) study of the personal computer industry shows that product proliferation has not led to reduced competition, and that benefits accruing from increased demand have been offset by higher costs. Though the authors admit that it is not possible to generalise these findings reliably, they note that it should at least be recognised that product proliferation is a double-edged strategy. Kekre and Srinivasan (1990) suggest that firms must handle product proliferation very carefully, so that the cost of variety is kept in control.

Costs of variety are examined in depth by Randall and Ulrich (2001) in their study of the US bicycle industry. They show how the provision of greater variety implies greater costs both in production, since it is harder to exploit economies of scale, and in 'market mediation', since managing the supply chain in order to match fragmented demand is more expensive. They find that the manufacturing technology and the structure of the supply chain chosen by a firm depend on which of the two costs is dominant.

The back end of product portfolio management

The previous discussion should lead to a more critical understanding of management literature, which has in recent years publicised the idea of broadening product lines to the point of serving each customer individually.

Mass customisation in perspective

Strategies such as mass customisation (Pine, 1993) are not *per se* a guarantee of success, since competitive advantage may only come from the capability of executing them more effectively or efficiently than other firms. For instance, one can think of the problems encountered by the now-merged computer manufacturers HP and Compaq when they set out to imitate Dell's make-to-order business model. Pine *et al.* (1993) stress that mass customisation has more to do with a complete overhaul of the internal organisation and culture of the firm than to a simple broadening of the product line. Gilmore and Pine (1997) argue that a mass customisation strategy must be carefully studied if it is to be successful, and it must match customer requirements to the firm's capabilities. In order to support the process, they propose a simple a 2×2 matrix that classifies product variety under the two axes of 'change in appearance' and 'change in product' (see Table 17.4).

Strategies such as mass customisation are not *per se* a guarantee of success, since competitive advantage may only come from the capability of executing them more effectively or efficiently than other firms.
(Pine, 1993)

17.4 Product variety for mass customisation

Change in product	Change in appearance	
	No change	Change
Change	Transparent mass-customisation	Collaborative mass-customisation
No change	Adaptive mass-customisation	Cosmetic mass-customisation

According to this framework, the highest amount of customisation occurs when both the product and its appearance vary. This is labelled *collaborative* customisation, with the producer providing tailor-made changes for customers who appreciate variety but do not find it easy to choose within a very broad offering. The opposite is *adaptive* mass customisation (low levels of change on both axes), in which the firm sells a standardised product that the user can adapt by himself. The other two categories are *cosmetic* (*i.e.* the firm sells a product that is for the most part standard, but contains some superficial variety) and, finally, *transparent* (with customisation being provided without the user even being aware of it). The four categories require a different design of both the products and of the processes that relate the firm to its customers.

Concentrating on the design aspects, masscustomised products generally require the development of a modular architecture, so that product variety may be provided at a low cost by combining components and options at the later stages of the manufacturing process, or even at the user's site. Product architecture is closely related to product variety, not only when dealing with customers individually, but also when the firm designs its products so that components are shared across a broad product line.

Modularisation

According to a widely accepted definition (Ulrich, 1995), a product architecture is modular when components are functionally independent, *i.e.* when there is a 1:1 mapping among components and functions. Functional independence has a deep impact on the supply chain, since components may efficiently be developed and manufactured separately by different organisations, as well as on the product offering, since variety may be created with greater ease by simply swapping components.

Modularity can affect core functional elements of the product (for example, when combining CPUs, hard disks and graphic cards in a personal computer) or it can be more superficial (for example, when applying covers and loading screen savers, ring tones and games on to a cellular phone). In some instances modularity requires redesign of the manufacturing process, since it is more efficient to move the phases that provide variety and flexibility to its end. For instance, it is well known that Benetton made operations reversal (Lee and Tang, 1998) a key feature of its strategy when it started to knit sweaters before dyeing them. This innovative process enabled Benetton to provide its stores with the right product mix in 'almost real time', and without having to build excessive inventory. A similar approach, discussed by Swaminathan and Tayur (1988), requires the manufacturing of intermediate semi-finished products, termed 'vanilla boxes', and the addition of components according to specific customer orders.

Modularisation is a complex phenomenon that has a wide-ranging impact on the firm and on the supply chain.

Modularisation is a complex phenomenon that has a wide-ranging impact on the firm and on the supply chain. The effects of modularisation can be beneficial, but failing to design the product architecture properly or to understand the required impact on the firm can lead to semi-finished and inconclusive results. To this purpose, Hansen *et al.* (2002) propose a framework for a better understanding of modularisation (Figure 17.5), which has been developed and tested within a number of industrial case studies.

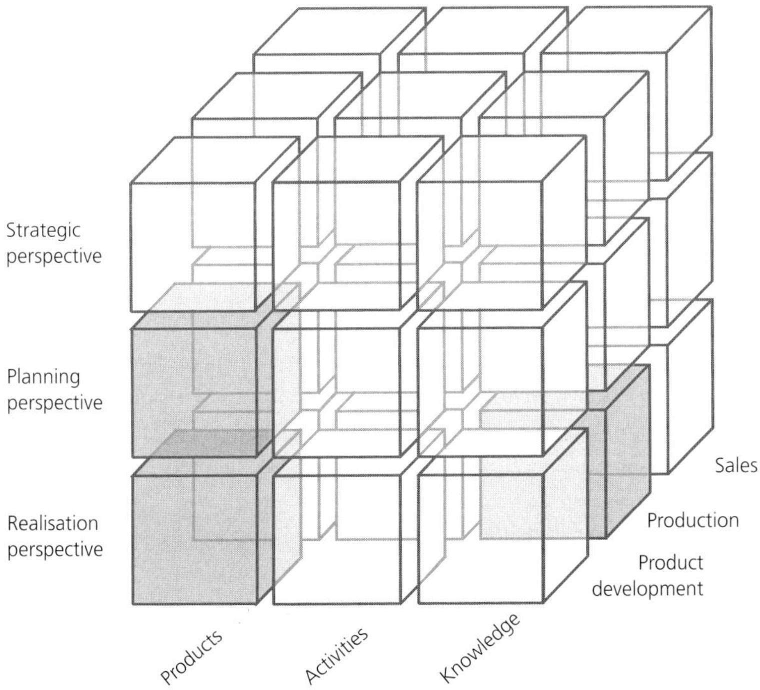

17.5 A framework for understanding modularisation (Hansen *et al.*, 2002)
Adapted with permission of the Design Society

The framework shows the three main axes on which modularisation has an impact and that should, therefore, be taken into account when dealing with this kind of strategy. The first axis deals with the temporal horizon, which ranges from a strategic level (*i.e.* defining goals and designing architectures) to a planning level (*i.e.* methods, procedures, plans, *etc.*) down to actual realisation. The second axis deals with the three corporate functions that are principally involved, *i.e.* product development, production and sales. The third axis is based on the widely accepted hypothesis that product architecture is closely related to the organisational structure of the firm, which can be described both in terms of its business processes (Henderson and Clark, 1990) and its knowledge structure (Sanchez, 2000). Accordingly, this axis represents the impact modularisation has on the product, on activities and on knowledge.

The framework in Figure 17.5 is used by Hansen *et al.* to show concisely the way the companies they studied have dealt with modularisation (for example, they insert comments on activities being observed in the appropriate cells). However, this framework could be used as a three-dimensional checklist that management might use to assess the comprehensiveness of the modularisation strategy used by the firm.

A 'module' (or 'chunk') may be viewed as a self-contained subset of components having a defined interface that connects it with other modules.

The design of modular products is an important strand of engineering design research, since it is tightly linked with the problem of embodying a functional structure in a physical assembly of components, which is in turn central to the engineering design process. For instance, Riitahuhta and Pulkkinen (2001) have developed a systematic approach enabling companies to develop highly configurable products based on modular architectures. They distinguish among four levels of modularity, which can be *assembly based* (with modules designed in different sizes, allowing a limited degree of customer-specific product configuration), *function based* (with modules designed on the basis of functionality, so that products may be customised to a greater extent), *platform based* (which introduces a separation among standard components and customer-specific ones) or, finally, there can be *dynamic* modularisation, in which modularity is also designed in view of the product family lifecycle.

In this context, each 'module' (or 'chunk') is viewed as a self-contained subset of components having a defined interface that connects it with other modules. The reasons for which a specific set of components should be selected to form a module may be disparate and are often conflicting. The analysis of these trade-offs and the consequent decision, therefore, requires attentive evaluation by the designer. For instance, modules may be formed in order to allow a wide product range to be generated through combinatorial variety, but other important criteria may be functional interdependence among components, technical issues (for example, energy efficiency, safety, and reliability), flexibility in use (for example, the ease of providing add-on accessories or component upgrades) and ease of operations (*i.e.* technological or economic aspects associated with sourcing, manufacturing, assembly, maintenance and recycling).

Modules may be formed in order to allow a wide product range to be generated through combinatorial variety.

Methods for defining modules are manifold (Breidert, 2003) and include analysis of the functional schematic of the product (Stone *et al.*, 1998; Holta *et al.*, 2003), block-diagonal rearrangement of matrices representing interactions between components (Pimmler and Eppinger, 1997; Huang and Kusiak, 1998; Lanner and Malmqvist, 1998) and algorithms operating on system-theoretic representations of component relations (Gaso and Otto, 2003). Multiple criteria evaluation of modules and the relationship between module definition and the management of technology are covered by Cantamessa and Rafele (2002).

Platforms

The provision of product variety is often based on the concept of platform-based product development, for which a fundamental reference is the textbook by Meyer and Lehnerd (1997). Platform strategy has been associated with the success of firms in many different industries, such as consumer electronics (for example, Sony's family of Walkman cassette players), watches (for example, Swatch) and automotive (for example, the strategy adopted by Volkswagen in the 1990s across its four main brands).

Product platforms have been defined as intellectual and material assets shared across a family of products (Robertson and Ulrich, 1998). This rather broad definition goes beyond the "physical" idea of a platform as a common architecture and set of components. In this way, it covers related but different interpretations that have been given to the platform concept. For instance, Clark and Wheelwright (1993) use the term platform to describe next-generation product development projects, while automotive manufacturers define as 'platforms' those organisational units that are in charge of developing component platforms.

The provision of product variety is often based on the concept of platform-based product development.

In essence, platform-based product development consists of configuring the product development pipeline in a two-tier structure. Platform projects are large-scale projects whose main goal is to create a technological basis and/or a shared set of components. For a given amount of time, the firm may then base a set of smaller (in terms of cost and development time) derivative product development projects on this platform. This arrangement has four main advantages.

The first and most obvious advantage is that platforms allow a high degree of component sharing among product versions, which can lead to significant economies of scale in manufacturing and purchasing.

Second, the development of a platform usually requires significant investment, but it allows firms to perform a stream of derivative projects at low marginal cost and with reduced time-to-market. In principle, the overall development cost of the platform project and of its derivatives should be less than what would have been spent with on an equivalent number of independent projects. By enabling quick execution of derivative projects, platforms allow firms to react more rapidly to changes in the market.

The third advantage is that alternating platform and derivative product development projects can help the firm achieve a less markedly cyclical performance. In terms of costs, a platform-based product development portfolio can be designed with a level resource utilisation profile, thus reducing the need for changes in the work force (or avoiding inefficient troughs and delay-inducing peaks in the overall work load). In terms of revenue, the competitiveness and profitability of derivative products will decline over time, since these will be based on an ageing platform. If platforms associated with different product families are staggered in time, the firm will exhibit a balanced product portfolio with respect to age and profitability.

Platform projects can be used to validate a set of new technologies, individually and with respect to their interoperability, and to create the know-how needed to deploy them in derivative product development.

The fourth and last advantage is that a platform-based product development strategy tends to keep more innovative activities separate from the less innovative ones. As shown in Figure 17.6, firms can use platform projects to transfer results from research into product development. This approach gives research projects a clearer objective ("we must finish project X by month K, so as to feed its results into platform project Y") and allows them to test new and riskier technologies within a sufficiently large-scale project that, not being directly pulled by the market, is not generally subject to an exceedingly tight schedule.

Platform projects can be used to validate a set of new technologies, individually and with respect to their interoperability, and to create the know-how needed to deploy them in derivative product development. Following terminology used at Hewlett-Packard, this is often called a 'pizza-

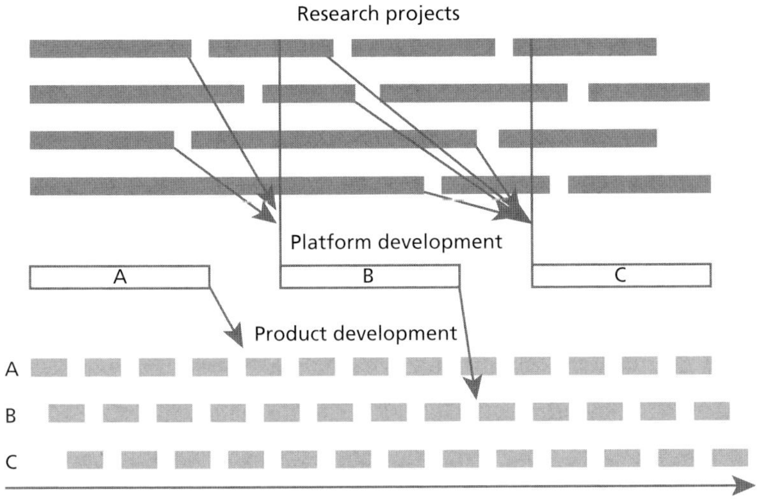

17.6 Platform projects transfer results from research into product development

bin' approach. At the same time, by taking the more innovative design tasks out of derivative product development, engineers are discouraged from over-designing individual products ("why don't we try technology X in this new product Z?"), which results in increased cost and lead time, often with dubious benefits.

This separation of the more innovative activities from product development has been studied by Krishnan and Bhattacharya (2002), who analyse and criticise the pizza-bin approach. They discuss whether basing product development only on a proven technology risks leading the firm to develop inferior products. Instead, they argue, it might be profitable to defer commitment and concurrently both develop products and validate the unproven technology. This may be done either by allowing two parallel product development processes (one per technology), or by overdesigning the product so that it may use both technological options. The choice between these alternatives depends on the added cost and on the estimate of the profitability gap shown by the two technologies.

The analytical model developed by Krishnan and Bhattacharya shows that, if the estimate of mean added profitability for the unproven technology is low, the pizza-bin approach is appropriate. If the mean estimate is high and the variance is low, they recommend the parallel approach, whereas high mean and variance make the overdesigned approach better, since this approach moves the commitment point to the latest point in time, when uncertainty regarding the new technology will be minimal.

When using platform-based product development, the evaluation of performance on *a per project* basis can be misleading, since there are dependencies among products and/or platforms and derivatives. Projects must therefore be managed with respect to the overall impact on the product pipeline and not individually. Meyer *et al.* (1997) discuss the problem of aggregate-level R&D metrics and make a distinction between the development of initial platform architectures, platform extensions (*i.e.* enhancements to subsystems that do not modify the platform architecture) and platform renewals, in which the architecture is altered. They propose measures for the efficiency of a platform (*i.e.* the degree to which a platform allows economical development of derivatives) and effectiveness (*i.e.* the degree to which derivatives produce revenue with respect to their development cost, where the use of revenue instead of profit is due to the practical concern that it is difficult to obtain reliable estimates of product-specific costs).

Platform efficiency is defined by the ratio between the average R&D cost (or development time) for derivative products over the R&D cost (or time) spent for the platform. A low value of this ratio implies that the platform is able to sustain economic development of derivatives, and *vice versa*. In the case study they present, Meyer *et al.* record values of platform efficiency around 0.1, though this figure cannot be generalised. Platform effectiveness is given by total sales of a platform and its derivatives over the total development cost. In this case, higher values of this indicator imply better performance. They recommend using these indicators both statically, in order to compare performance of different platforms, and dynamically, in order to observe the degree to which a platform is still able to sustain the low-cost development of derivative products and/or to generate meaningful revenue.

$$\text{Platform efficiency} = \frac{\text{average R\&D cost for derivative products}}{\text{R\&D cost spent for the platform}}$$

The management of multiple products through component sharing has attracted significant interest from researchers, since common sense and industrial experience make it apparent that platforms cannot be a universal answer to product strategy, for there must be trade-offs to be considered. For instance, excessive component sharing across brands in the automotive sector has often been criticised by consumers and the press, as in the case of Ford components being used in Jaguar cars, or Volkswagen's use of the same platform for widely different models.

In this context, Krishnan *et al.* (1999) present a model for the optimal design of a product family, with differentiation restricted to a single performance attribute, and in which both development cost and revenue are considered. They hypothesise that the firm develops the platform first and then starts to develop products with increasingly improved performance by progressively adding new components or by adapting/improving components from previous variants. They identify the main trade-off decision as that between the increased revenue due to a rich product line (coming from more sales and/or greater profitability) versus the development costs, which depends on both the degree of component sharing achieved by the platform and the number of product variants.

The assessment and design of a product line should be made on profits, and in aggregate and not on a *per product* basis.

Ramdas and Sawheny (2001) expand the traditional literature on product-line definition in order to discuss trade-offs when components are shared among products. They develop a mixed-integer linear programming model and discuss a few insights resulting from its application in a watch manufacturing company. The discussion of revenue effects due to product variety is particularly interesting. These effects are classified in the three categories of *demand expansion* (*i.e.* sales to customers who would not have bought a similar product at all), *competitive draw* (*i.e.* sales to customers who would have bought a product sold by a competitor) and *cannibalisation* (*i.e.* sales to customers who would have bought a different product sold by the same firm). Ramdas and Sawheny argue that assessment and design of the product line should be made on profits, and in aggregate and not on a *per product* basis. For instance, it may be unprofitable to prune out low-selling items, since these may actually be gaining sales from demand expansion and competitive draw and/or have little additional cost. Conversely, it is possible to have unprofitable high-selling items, either because of their high cost, or simply because they sell primarily through cannibalisation.

The paper by Desai *et al.* (2001) is also quite appealing because of the conceptually simple setup it is based on, which allows us to gain some interesting insights on the problem of product differentiation based on shared components. The model views a two-segment market (high and low, or H and L) and two components that determine product quality, which can be designed in two quality levels (*premium* or *basic*). For simplicity, it is assumed that the second component must be designed according to the segment being targeted, so that three possible configurations emerge (Table 17.7). They assume that customers evaluate quality through a linear combination of component quality, while the manufacturing cost of components varies quadratically with quality. The setup is modelled as a three-stage game in which the manufacturer selects the configuration and the design effort for each component, then it sets prices in order to maximise profits, and finally customers decide whether, and which product, to buy.

Segments	Configurations		
	Unique	Premium common	Basic common
High	Premium Premium	Premium Premium	Basic Premium
Low	Basic Basic	Premium Basic	Basic Basic

17.7 Product differentiation based on shared components

At first, the analysis shows some interesting facts with regard to revenues alone. If compared with the unique design, where there is no component sharing, the premium common design might not necessarily grant higher prices and revenues, even though it leads to a better basic product. This can happen because the quality gap between the two products becomes narrower and the firm must be careful to avoid cannibalisation. So, the additional revenue gained from the higher price that can be asked for the basic product might be more than offset by the lower price that must be applied to the premium product. When comparing the basic common design with the unique, the optimal price for the basic product is the same, but the lower quality premium product must be sold at a lesser price, which causes a fall in revenue.

Concerning profits, the optimal configuration depends on a number of parameters. The premium common design may be more profitable than the unique depending on the trade off between three elements: the previously discussed increase or decrease in revenue, the increase in cost due to using the more expensive component in the basic product and decrease in cost due to economies of scale. The basic common design may be more profitable than the unique depending on the interplay of two effects: the fall in revenue and the cost savings due to the lower quality of the component being used and to greater economies of scale. In addition to these insights, Desai *et al.* study the profitability of making individual components common and find that they should be ranked according to the index

$$I = \frac{\text{cost coefficient of quality}}{(\text{weight in quality evaluation function})^2}$$

In other words, components to be shared are the ones for which manufacturing cost varies more with the quality level and/or the ones that are less important to customers in their evaluation of product quality. It is interesting that the index does not depend on the way with which the two market segments evaluate product quality. This index may be applied in practice to map components on the two axes present in the ratio (Figure 17.8a). This map can be used instead of, or in conjunction with, qualitative mapping techniques that are often used in industry (Figure 17.8b). Other criteria for component sharing (Fisher *et al.*, 1999) include the assignment of components to a spectrum that ranges from the purely aesthetic (maximum variety is required) to the purely functional (maximum sharing would instead be preferable), or the distinction among components that have a strong influence on perceived quality (customer utility can be thought to vary quasi-linearly with performance) and those that do not (with customer utility more or less following a step function, with no utility below a threshold level of performance and a constant utility above).

In contrast to Desai *et al.*, Krishnan and Gupta (2001) instead take the platform and the associated list of shared components as a given and study whether this sharing is beneficial or not. They argue that platform-based product development may have benefits, but entails costs associated with the overdesign of low-end products (or the underdesign of high-end ones) and with opportunity costs that arise by delaying market launch. They develop a model in which a firm must serve the needs of two customer segments (basic and premium) with four product planning options:

- a platform-based approach (P1), where the platform matches with the low-end product, and a second project adds features leading to the high-end product;
- the independent development of the low-end and the high-end product (P2);
- the development of the low-end product only (P3), to be sold to both segments;
- the development of the high-end product only (P4), to be sold to both segments.

They argue, based on an analytical model of revenues and costs, that the optimal choice among the four options depends on two main parameters: the degree of market diversity, measured by using the previously introduced 'degree of cannibalisation' by Moorthy and Png (1992), and non-platform economies of scale (*i.e.* the degree to which components that do not belong to the platform can benefit from economies of scale).

The main findings of Krishnan and Gupta are summarised in Table 17.9. With respect to these two parameters, platform-based product development is profitable for 'intermediate' products because of revenue and cost issues. From

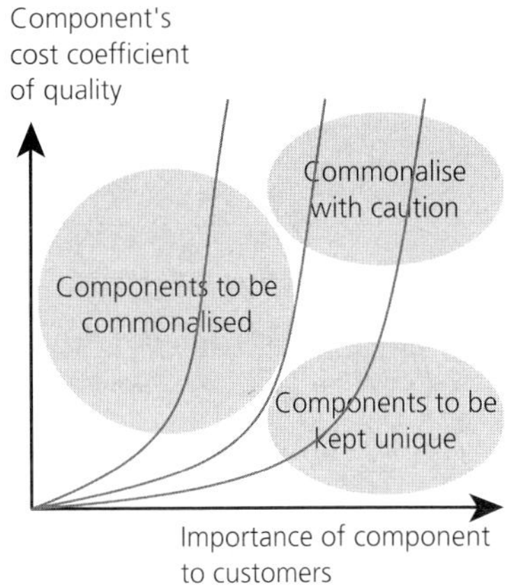

(a) Cost vs. importance

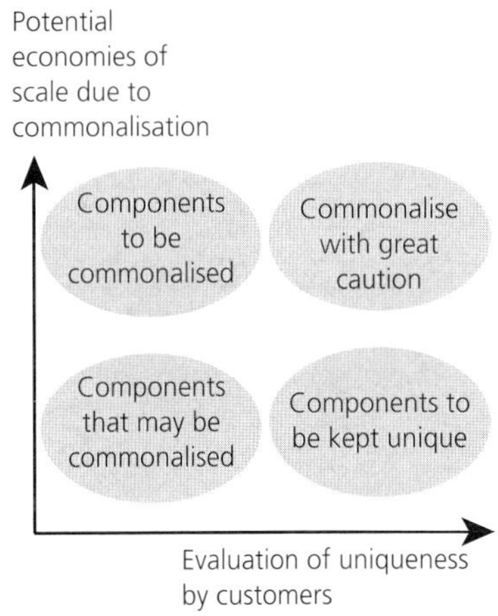

(b) Economies of scale vs. uniqueness

17.8 Mapping components

Non-platform economies of scale	Market diversity Low	Medium	High
Low	Low-end product only	Product family, with or without a platform	High-end product only
Medium	Low-end product only	Platform-based product family	High-end product only
High	Low-end product only	Low-end product only	Low-end product only

17.9 Economies of scale for platform-based development (Krishnan and Gupta, 2001)
Management Science

the side of the market, when market diversity is low the firm is better off with a low-end product only (which costs less to develop), whereas a high degree of market diversity makes it advisable to develop the high-end product only, since this caters better to the needs of the premium segment and avoids the cost of developing an additional low-end product. From the side of view of costs, high non-platform economies of scale can make it profitable to forgo the platform approach and develop the low-end product only, so as to exploit these economies of scale to a fuller extent.

Krishnan and Gupta also explore the timing of product introduction, in which the main trade off is between the delayed revenues due to the delayed launch and the reduced cannibalisation when only one product is on the market. The parameter that mostly determines the optimal choice is the ratio between the firm's discounting factor (which is related to profits) and the customers' (which instead is related to surplus). In essence, this ratio measures the relative impatience of the two agents. They show that a greater discount factor for the firm suggests the simultaneous launch of the two products, so as to speed up revenues, whereas greater impatience from customers makes sequential launch optimal (Table 17.10).

Discount factor	Market diversity Low	Medium	High
Discount factor greater for the customer	Low-end product only, or family launched sequentially without platform	Platform-based family launched sequentially	Platform-based family launched sequentially
Discount factor greater for the firm	Low-end product only, or family launched simultaneously without platform	Platform-based family introduced simultaneously	High-end product only

17.10 Discount factors and market diversity (Krishnan and Gupta, 2001)
Management Science

Robertson and Ulrich (1998) propose a practical method for planning product platforms, with the objective of finding the right trade-off between distinctiveness, which is a driver of revenue, and commonality, which instead tends to reduce cost. Their framework is based on three 'plans', namely the *product* plan (in which the firm defines a portfolio of products and variants along with launch dates and target segments), the *differentiation* plan (in which the firm identifies 'differentiating attributes', gives them a score and then defines how each product in the product plan should relate to such attributes) and the *commonality* plan (in which the firm identifies component modules, collects data on fixed and variable costs, and then assigns them to each of the products in the plan). The idea behind the approach is to revise these three plans iteratively until the decision-maker reaches a sufficient degree of consistency.

Multiple project management

Even though companies engaged in product development generally operate more than one project at the same time, multiple project management (MPM) has received scant attention from academics, probably because of its formidable difficulty. From the perspective of operations research, multiple project scheduling under resource constraints involves a very high computational complexity and for practical purposes requires heuristics to be solved, a topic that is often scarcely appealing to academics and is more likely to be found in practitioner-oriented literature.

Multiple project scheduling under resource constraints involves a very high computational complexity and for practical purposes requires heuristics to be solved.

An exception is the paper by Yang and Sum (1997), who assume a dual-level structure with a programme manager overseeing a number of projects, each led by a project manager. They study due date, resource allocation, project release and activity scheduling rules together and show that, at an individual project level, there is a very important trade-off between the due date negotiated with the customer and the resources allocated to the project. This trade-off obviously affects the project, but can have an impact on the other projects as well, since a late and underresourced project will compete desperately for extra resources, thus disrupting the scheduling.

They also show that the decision on project release dates is very important. It is often better to keep a project out of the system for some time, rather than having it compete with other projects and ending up with resources too thinly spread. This can be even more critical in the case of multiple-resource problems, because there is a greater chance that activities may be held up waiting for the right combination of resources to free themselves.

Product portfolio management is reviewed by Payne (1995) from a practice-oriented viewpoint and under five main perspectives (capacity of resources, complexity, conflict management, commitment and context). Drawing from personal experience, it can be argued that a critical issue in MPM, which is often overlooked in the literature, is the management of simultaneous projects that may be differ widely with regard to project size, skills required and urgency.

The interactions among different projects due to resource sharing have also been studied by Adler *et al.* (1995), who suggest looking at the product development function as a process, rather than as a collection of individual projects, and using queuing network theory as an analytical approach. The main issue they raise is that firms generally approach project management on a *per project* basis and do not attempt to get the 'big picture', with an objective assessment of the resources required by the active project portfolio with respect to the amount that is effectively available. Queuing network theory fits in well, since one of the main tenets of this approach is to highlight mismatches of this kind and to trace effects in terms of delays in task completion. They show that, even though it is generally considered sound management to use resources at capacity, this causes delays in task processing that may grow out of control, except for low values of task variability.

These results are fairly standard in industrial engineering, but have seldom been applied to product development. Apart from the direct application of their method, Adler *et al.* use the results to emphasise the need to monitor the existence of bottleneck resources closely, by aggregating projects and calculating resource workload profiles, and to take measures so that they are adequately staffed.

Another suggestion is to cap the number of projects in the firm to the point of mimicking a just-in-time 'pull' system in which the start of a new project is authorised only when the overall number of projects falls below a given threshold. Another measure that could be introduced in order to reduce waiting time further is cross-training and pooling of resources (conversely, one could devise an analogy of a cell-based manufacturing system and specialise resources with the criterion of assigning tasks with similar duration, in order to reduce variability).

MPM is complex because interactions between projects are not only associated with shared resources, but also with information transfer.

MPM is complex because interactions between projects are not only associated with shared resources, but also with information transfer, for instance when one project serves as the basis for a second one. Nobeoka and Cusumano (1995) discuss this problem on the basis of a survey carried out in 10 American and Japanese automotive manufacturers. They define four typologies of design interaction: carrying out a new design without any interaction, rapid design transfer from a base project to a new one (similar to concurrent engineering), sequential design transfer among projects related to different product lines (in which the new project starts when the base project is ended), and design modification (similar to the previous case, but with both projects related to the same product line).

Empirical results show that rapid design transfer is the more effective strategy for engineering man hours. Rapid design transfer, however, must meet some requirements if it is to be effective. These are part technical, related to the features of the technological platform that must serve both projects, and part organisational, associated with the definition of senior roles and responsibilities, the planning of projects in order to exploit synergies and minimise rework, and the management and sharing of design knowledge and design rationale.

It is well known that project management should be associated with rigorous methods of planning, scheduling, budgeting and controlling projects (PMI Standards Committee, 2000). Marle and Bocquet (2001), go beyond the PMI guidelines to propose a method for MPM in the context of new product development. Their approach is based on decomposition (of the program into projects, and of these into smaller activities), assignment (of resources and responsibilities) and state management by the resources endowed with responsibilities.

It is well known that project management should be associated with rigorous methods of planning, scheduling, budgeting and controlling projects.
(PMI Standards Committee, 2000)

Bringing the two together: project portfolio management

Readers may have noticed that a theme often raised in the previous sections is that the firm must be able to carefully pick the projects it engages in, so as to ensure profitability both from the side of revenue and that of cost. The management of the project portfolio is a key element of strategic decision-making in the firm.

The strategic role of project portfolio management

Project portfolio management (PPM; not to be confused with product portfolio management) determines the allocation of scarce resources across time and project scope, a concept that is represented graphically in Figure 17.11, showing how PPM sets the basis for product portfolio management and multiple project management.

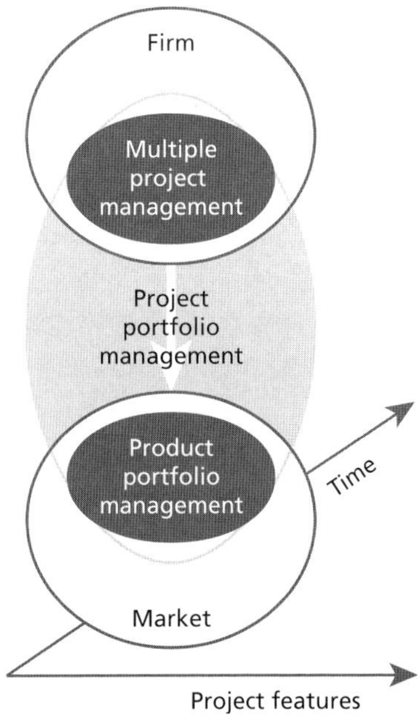

17.11 Project portfolio management

It should be noted, however, that setting the basis does not mean that the two latter are strictly *part* of the former. The strategic role of PPM (or aggregate project planning) in the product development process has been amply discussed by scholars. For example, Wheelwright and Clark (1992) make it a key element of product strategy, as in Figure 17.12.

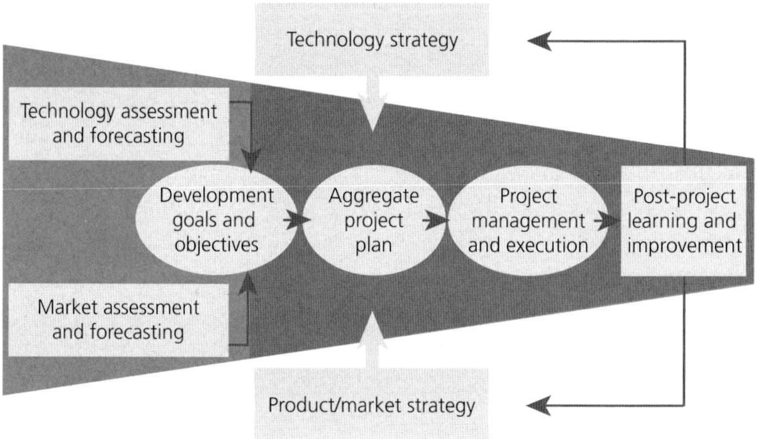

17.12 The strategic role of product portfolio management
Adapted with the permission of The Free Press, a Division of Simon & Schuster Adult Publishing Group, from REVOLUTIONIZING PRODUCT DEVELOPMENT: Quantum Leaps in Speed, Efficiency, and Quality by Steven C Wheelwright and Kim B Clark. © 1992 by Steven C Wheelwright and Kim B Clark.

Since the product development process is nearly universally operated on the basis of projects, it may safely be claimed that strategic decision making in this context is coincident with PPM. In broader terms, the implementation of an R&D strategy must go through the analysis and the redesign of the firm's project portfolio, so that projects that match with strategic needs are kept alive and assigned sufficient resource, while the ones that do not are pruned out.

As surveys have often shown, most companies do not manage their project portfolio at all, or do so informally and without a structured process. Not having a systematic approach for PPM leads the firm to have too many projects at the same time, since it is always easy to start a project ("let's try it out"), but it becomes very difficult to terminate one. Resources become overloaded and thinly spread out among projects that are widely different technical content and strategic fit. Table 17.13, after Cooper *et al.*, (1998), provides a summary of these consequences.

Techniques for project portfolio management

PPM consists in the assignment of a limited amount of resources, human, technical and financial, to a set of projects, each of which can be characterised in terms of expected economic value and risk, so as to obtain an acceptable

Not having a formal PPM approach implies	Immediate consequences	Final consequences
Reluctance in killing projects, too many approved projects, unclear objectives	Too many projects, resources thinly spread, low execution quality	Long lead times, high percentage of unsuccessful projects
Weak go/kill decisions	Too many mediocre and low-level projects. Insufficient resources assigned to important projects	Too many projects with weak innovative content. Few projects able to provide competitive advantage
No objective selection criteria. Projects selected on emotive or political basis	Wrong projects selected	New product failures
No strategic criteria for project selection	No clear direction for the project portfolio. Insufficient synergies among projects	New products do not support the firm's strategy. R and D resources used inefficiently

17.13 The consequences of no portfolio approach
Cooper *et al.* (1998) *Portfolio management for new products* – table reproduced with permission of Dr. Robert G Cooper

overall result. This definition is very similar to the problem of managing a portfolio of financial securities, even though there are important differences, such as the synergies and exclusions that may exist among projects.

PPM can be discussed both as a process and with respect to the techniques that it uses. In the first view, there is a standard classification between bottom-up and top-down PPM. In the former, projects are proposed from the lower tiers of the organisation and the program manager must decide on the acceptability and the funding of each proposal. In the latter, the program manager assigns budgets to organisational units and/or to project categories (for example, research vs. development and product vs. process) and delegates decisions.

The bottom-up approach allows top management to have a better view of the project portfolio, but requires greater effort to create a tight fit with the firm's strategy and to manage the selection process. The design of PPM as a process has been tackled by a number of authors, such as Cooper *et al.* (1998). Archer and Ghasemzadeh (1999) propose a framework process for PPM based on seven phases that bring together, in a coherent way, a number of well-known best practices. The phases are pre-screening, individual project analysis, screening, portfolio selection, portfolio adjustment, project execution and stage-gate evaluation.

Stummer and Heidenberger (2003) propose a PPM procedure that considers project interdependencies, based on the three phases of screening, multi-objective optimisation and search for Pareto-optimal portfolios and, finally, project selection. Nidamarthi *et al.* (2003) present a portfolio management methodology used at ABB for analysing cost and revenue of individual products within a product family and for optimising overall profitability.

Conversely, if one looks at PPM from the perspective of techniques, these can roughly be classified in the following categories: financial methods, optimisation methods, multi-criteria methods, and mapping methods.

Financial methods

In principle, product development projects should be evaluated according to the Net Present Value of the relevant cash flows, including development and manufacturing costs and revenue.

In principle, product development projects should be evaluated according to the net present value (NPV) of the relevant cash flows, including development and manufacturing costs and revenue. This means that it is incorrect to consider expenses that have already been allocated and cannot be reversed (sunk costs). Cash flows should be discounted at a rate appropriate to the risk inherent in the project, which is not easy to determine, although many firms incorrectly define a single internal cost of capital and apply it to all of their activities. One way out of this problem is to use indexes that separate development cost, contributions from sales and technical and commercial risk. An example is expected commercial value.

Such indexes are fairly easy to use, but neglect interactions that often exist between projects. The case where the results of a first project provide the basis for the development of a second project is particularly interesting. In this case, the decision whether to activate the second project or not can be deferred and made after having observed the results of the first. This deferral reduces risk and provides the first project with an 'option value' that has to be added to its intrinsic value (*i.e.* the value it would have if there were not further decisions to be made at its end).

The term 'option' is a reminder of the financial instruments having the same name. Specifically, there is an analogy with European call options, which give their owner the option, but not the obligation, to buy a security at a given price on a given date. The option will be exercised if it is advantageous (*i.e.* if the security is traded above the exercise price). In order to distinguish the two uses of the word 'option', the application of this concept to concrete activities is often termed 'real option'.

The evaluation of real options can be carried out by exploiting the analogy with financial options, for instance by using Black and Sholes' pricing formula.

Owing to the difficulty in correctly evaluating discount rates, more sophisticated approaches have also been developed, such as the replicating portfolio method (Copeland and Antikarov, 2001). Since projects usually have discrete outcomes, a more straightforward, though approximate, computation of real option values may be based on standard decision trees.

Optimisation methods

Mixed-integer linear programming models can be used to represent and solve PPM problems (in operations research terms these would be classified as standard 'knapsack' problems). Boolean decision variables represent the decision whether or not to activate a given project, and the objective function generally represents the sum (to be maximised) of the NPV of the selected projects. Constraints are added to ensure that activated projects do not require more resources than those available, and other constraints can model interdependence among projects.

Complex optimisation models following this approach may be found in the papers by Dickinson *et al.* (2001), who include NPV, strategic fit and project interdependence, and by Loch and Kavadias (2002), who use a dynamic programming approach and model risk aversion and interaction among products.

Multi-criteria methods

In the project selection process, projects must be compared according to heterogeneous criteria, such as economic value, risk, coherence with the firm's strategy and competencies, project complexity, *etc.* Some of these criteria are difficult to assess in economic terms, and decision makers are usually reluctant to endorse such a process, since they realise that the results would be rather unreliable. Multi-criteria evaluation techniques, such as *Electre* (Roy, 1996) or the analytical hierarchy process (Saaty, 1980), help compare projects on heterogeneous criteria in a more natural way.

In the project selection process, projects must be compared according to heterogeneous criteria, such as economic value, risk, coherence with the firm's strategy and competencies, project complexity, *etc.*

Despite their potential, these techniques are not widely diffused, partly because of their complexity and partly because managers perceive that they do not have sufficient transparency. This can be a problem when used in a process that, being subject to strong political pressure, should be as clear as possible. Firms, therefore, tend to use very crudely scored models, such as weighted sums with thresholds for screening, as shown in Table 17.14, and tolerate the fact that the attribution of weights and scores is arbitrary and that the final results can often be paradoxical.

Weight	2		3		
Threshold	**	...	***		
Criterion	Criterion 1	...	Criterion n	Passed?	Score
Project 1	***	...	*	r	r
Project 2	***	...	***	a	12
...	...	...	...	...	...
Project m	**	...	****	a	16

17.14 Using weighted sums with thresholds for screening projects

Mapping methods

PPM also requires tools for visualising, in an intuitive way, the project portfolio and related data. A popular approach is to use bubble diagrams, with two variables expressing 'project positioning' on the Cartesian axis and the bubble size proportional to some measure of project size (Figure 17.15).

The implicit message associated with mapping methods is that the firm should go for a 'balanced' project portfolio, which of course may not be the optimal one. The dimensions that can be used for these bubble diagrams include:

- strategic fit (for example, high, medium, low);
- duration of competitive advantage offered by the project (for example, short, medium and long-term);
- economic value;
- technological level (for example, from standard to breakthrough technology);
- probability of commercial and/or technical success (for example, high, medium, low);
- project complexity (for example, high, medium, low);
- market attractiveness;
- investment required for development;
- investment required for commercial exploitation;
- lead time;
- product innovation.

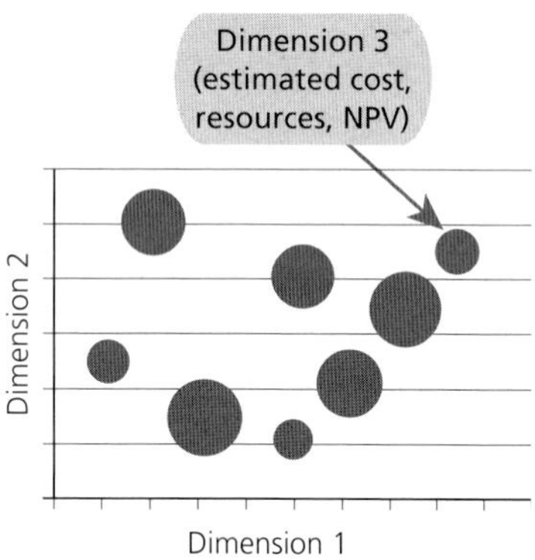

17.15 'Product positioning' using bubble diagrams

In classifying projects for mapping purposes it may be useful to follow the proposal by Shenhar (2001). On the basis of empirical research, he presents a taxonomy for development projects using the two axes of technological uncertainty and system scope. In addition, he suggests two qualitative but formally defined measurement scales (one per axis), that ensure greater fit with empirical results and, therefore, a less error-prone classification of projects.

Application of project portfolio management techniques

The previous list of techniques shows that firms have many 'building blocks' available to set up a proper PPM process. However, firms exhibit widely different behaviour in their PPM practice. Cooper *et al.* (1999) present a survey of PPM practices in more than 205 firms and find that satisfaction with PPM depends on the quality of the process and to what degree it matches management's requirements. They then find that 'benchmark' businesses (*i.e.* the ones that exhibit a PPM approach with a high degree of quality and management fit) share five main common traits:

1. the PPM methods are established, explicit, formal and with clear rules;
2. the PPM method is applied constantly;
3. the PPM method considers all projects together and pits them one against the other;
4. management follows recommendations from PPM methods;
5. PPM is based on a combination of financial methods and of tools that help evaluate the degree to which projects fit with the firm's strategy.

Conversely, they find that firms using financial methods alone derive the worst satisfaction from PPM. In a subsequent paper Cooper *et al.* (2000) warn against PPM methods that evaluate projects independently from one another and neglect resource absorption, which implies they do not consider the opportunity cost that arises when resources are committed to one project and not to another one. They suggest that PPM should be realised on the three axes of economic value, strategic orientation and balance across markets and scope (*i.e.* short vs. long term).

Defining a balanced project portfolio with respect to project scope is not easy, because of the uncertainty associated with projects in general, and especially long-term ones. In this area, literature proposes simple mapping approaches together with more complex analytical studies. Concerning the former, Mikkola (2001) proposes a mapping-based method in which projects are located depending on competitive advantage (or scope) and benefits to customers.

Defining a balanced project portfolio with respect to project scope is not easy, because of the uncertainty associated with projects in general, and especially long-term ones.

Coskun Samli (1996) proposes a process for developing breakthrough products based on three phases (generating, evaluating and prioritising ideas).

Among more sophisticated papers it is possible to mention Ding and Eliashberg (2002) and Lieb (1998), who both propose analytical models of a two-stage development process with upstream 'research' feeding into downstream 'development' after an intermediate screening. Lieb looks for the optimal 'choke' between the two phases (*i.e.* the fraction of projects that should be allowed from one to the other). The trade-off he studies arises because a tight choke leads to fewer effective projects in the development phase and/or the need to start many more research projects in order to feed development at the required rate. A wide choke means that too many possible failures are taken into the development phase. The elements that determine optimal choke are the relative cost of research vs. development projects, and the firm's ability to discriminate between good and bad projects at the review point. The discussion shows that two elements become of paramount importance: the generation of an adequate number of high-quality concepts in the research phase and a quality project screen.

Conclusions

This chapter has reviewed the topic of product portfolio management, which is of strategic importance to a firm in general and, specifically, to the processes that are tasked with developing products. Because of the breadth of the topic it has been necessary to tackle it from a number of perspectives, starting from the side of marketing. The benefits and the possible drawbacks of a broad product portfolio have been discussed by comparing the basic economics of product differentiation with recent results on product proliferation. The chapter has then covered the 'back end' of product portfolio management. First, modularisation and platforms, which are two mainstays of modern product development strategy, have been introduced and critiqued in order to highlight the trade-offs that determine their applicability. Then, a few contributions on MPM have been reviewed. Finally, the two perspectives have been brought together by introducing PPM as a process that can help determine the product portfolio by simultaneously addressing issues of supply and demand. PPM has been discussed with regard to overall methodology and related support techniques.

Product portfolio management is of strategic importance to a firm in general and, specifically, to the processes that are tasked with developing products.

Despite the hype that clouded the so-called 'new economy' at the beginning of the decade, it is undeniable that most firms nowadays operate in an environment that is more complex than the traditional linear supply chain, in which each company developed a clearly identifiable product and positioned itself between a well-defined set of suppliers and customers.

Corporate 'unbundling' (Hagel and Singer, 1999), the phenomenon in which the three processes of product innovation, customer relationship management and infrastructure management are no longer performed by the same company but demerged in different firms, is indeed happening in many industries. For instance, cellular phones are often designed by the former 'manufacturers' (for example, Samsung, Nokia, Sony-Ericsson and Motorola), produced by Far East contract manufacturers, and sold under the brand of network operators (for example, Vodafone Live!). The same may be said for the ecosystem model (Moore, 1993), in which a number of companies cooperate within a complex and dynamic network of relationships that go beyond the traditional links between suppliers and customers.

So, the natural question is, what happens to product portfolio management when the firm is unbundled, operates in an ecosystem, or provides a product-service? In principle, one might say that two complementary perspectives have been achieved. From the perspective of each unbundled firm, it is necessary to redefine the local concept of 'product' and product portfolio and use traditional techniques in order to manage it. For instance, the 'product innovator' will have to manage a portfolio of product designs, the 'customer relationship manager' will have to manage a set of customised services obtained by assembling physical 'building blocks' with service-oriented processes, while the 'infrastructure manager' will manage a portfolio of manufacturing services. This perspective must be completed with an inclusive picture of the product portfolio that end users are effectively observing and buying (in other words, one must remember that "the whole is greater than the sum of the parts"). This picture should be used to assess the profitability of all the cooperating parties, so as to ensure their commitment.

References

Adler PS, Mandelbaum A, Nguyen V, Schwerer E (1995) From project to process management: an empirically-based framework for analyzing product development time. Management Science, 41(3): 458–484

Archer NP, Ghasemzadeh F (1999) An integrated framework for project portfolio selection. Project Management, 17(4): 207–216

Bayus BL, Putsis WP (1999) Product proliferation: an empirical analysis of product line determinants and market outcomes. Marketing Science, 18(2): 137–153

Breidert J (2003) Tools supporting the development of modular systems. ICED'03, Stockholm, Sweden

Cantamessa M, Rafele C (2002) Modular products and product modularity: implications for new product development. Design 2002, Dubrovnik, Croatia

Clark KB, Wheelwright S (1993) Managing product and process development. Free Press

Cooper RG, Edgett SJ, Kleinschmidt EJ (1998) Portfolio management for new products. Perseus Books

Cooper RG, Edgett SJ, Kleinschmidt EJ (1999) New product portfolio management: practices and performance. Journal of Product Innovation Management, 16(4): 333–351

Cooper RG, Edgett SJ, Kleinschmidt EJ (2000) New products, new solutions: making portfolio management more effective. Research-Technology Management, 43(2): 18–33

Copeland T, Antikarov V (2001) Real options: a practitioners' guide. Texere

Coskun Samli A (1996) Developing futuristic product portfolios. Industrial Marketing Management, 25: 589–600

De Fraja G (1996) Product line competition in vertically differentiated markets. Industrial Organization, 14: 389–414

Desai P, Kekre S, Radhakrishnan S, Srinivasan K (2001) Product differentiation and commonality in design: balancing revenue and cost drivers. Management Science, 47(1): 37–51

Dickinson MW, Thornton AC, Graves S (2001) Technology portfolio management: optimizing interdependent projects over multiple time periods. IEEE Transactions on Engineering Management, 48(4): 518–527

Ding M, Eliashberg J (2002) Structuring the new product development pipeline. Management Science, 48(3): 343–363

Fisher M, Ramdas K, Ulrich K (1999) Component sharing in the management of product variety: a study of automotive braking systems. Management Science, 45(3): 297–315

Gaso B, Otto K (2003) Generating product architectures using partitioning algorithms. ICED'03, Stockholm, Sweden

Gilmore JH, Pine BJ (1997) The four faces of mass customization. Harvard Business Review, 75(1): 91–101

Hagel J, Singer M (1999) Unbundling the corporation. Harvard Business Review, 77(2): 133–141

Hansen PK, Andreasen MM, Harlou U, Gubi E, Mortensen NH (2002) Understanding the phenomenon of modularization. Proceedings of the 7th International Design Conference Design 2002. Ed. Marjanovic D, Dubrovnik, Croatia. Vol. 1, Figure 2 on page 155

Henderson R, Clark K (1990) Architectural innovation: the reconfiguration of existing product technologies and the failure of established firms. Administrative Science Quarterly, 35: 9–30

Holta K, Tang V, Seering WP (2003) Modularizing product architectures using dendograms. ICED'03, Stockholm, Sweden

Huang C-C, Kusiak A (1998) Modularity in design of products and systems. IEEE Transactions on Systems, Man and Cybernetics, Part A, 28(1): 66–71

Kekre SK, Srinivasan K (1990) Broader product line: a necessity to achieve success? Management Science, 36(10): 1216–1231

Krishnan V, Bhattacharya S (2002) Technology selection and commitment in new product development: the role of uncertainty and design flexibility. Management Science, 48(3): 313–327

Krishnan V, Gupta S (2001) Appropriateness and impact of platform-based product development. Management Science, 47(1): 52–68

Krishnan V, Singh R, Tirupati D (1999) A model-based approach for planning and developing a family of technology-based products. Manufacturing and Service Operations Management, 1(2): 132–156

Lanner P, Malmqvist J (1996) An approach towards considering technical and economic aspects in product architecture design. WDK-workshop on Product Structuring, Delft, The Netherlands

Lee HL, Tang CS (1998) Variability reduction through operations reversal. Management Science, 44(2): 162–172

Lieb EB (1998) How many R&D projects to develop? IEEE Transactions on Engineering Management, 45(1): 73–77

Loch CH, Kavadias S (2002) Dynamic portfolio selection of NPD programs using marginal returns. Management Science, 48(10): 1227–1241

Marle F, Bocquet J-C (2001) A multi-project management approach for increased planning process. ICED'01, Glasgow, UK

Meyer MH, Lehnerd AP (1997) The power of product platforms. Free Press

Meyer MH. Tertzakian P, Utterback JM (1997) Metrics for managing research and development in the context of the product family. Management Science, 43(1): 88–111

Mikkola JH (2001) Portfolio management of R&D projects: implications for innovation management. Technovation, 21: 423–435

Moore JF (1993) Predators and prey: a new ecology of competition. Harvard Business Review, 71(3): 75–86
Moorthy KS, Png I (1992) Market segmentation, cannibalization and the timing of product introductions. Management Science, 38: 345–359
Nidamarthi S, Mechler G, Karandikar H (2003) Product family development and manage-ment: architecting for maximum profitability. ICED'03, Stockholm, Sweden
Nobeoka K, Cusumano MA (1995) Multiproject strategy, design transfer and project performance: a survey of automobile development projects in the US and Japan. IEEE Transactions on Engineering Management, 42(4): 397–409
Payne JH Management of multiple, simultaneous projects: a state-of-the-art review. Project Management, 13(3): 163–168
Pimmler TU, Eppinger SD (1994) Integration analysis of product decompositions. ASME DETC'94, Minneapolis, MN, USA
Pine BJ (1993) Mass customization, the new frontier in business competition. Harvard Business School Press
Pine BJ, Victor B, Boynton AC (1993) Making mass customization work. Harvard Business Review, 71(9): 108–119
PMI Standards Committee (2000) A project management body of knowledge (PMBOK). Project Management Institute
Ramdas K, Sawhney MS (2001) A cross-functional approach to evaluating multiple line extensions for assembled products. Management Science, 47(1): 22–36
Randall T, Ulrich K (2001) Product variety, supply chain structure and firm performance: analysis of the US bicycle industry. Management Science, 47(12): 1588–1604
Riitahuhta A, Pulkkinen A (2001) Design for configuration. Springer
Robertson D, Ulrich KT (1998) Planning for product platforms. Sloan Management Review, 39(4): 19–31
Roy B (1996) Multi-criteria methodology for decision aiding. Kluwer Academic
Saaty TL (1980) The analytical hierarchy process. McGraw-Hill
Sanchez R (2000) Product and process architectures in the management of knowledge. In: Resources, technology and strategy. Routledge
Shenhar AJ (2001) One size does not fit all projects: exploring classical contingency domains. Management Science, 47(3): 394–414

Stone RB, Wood KL, Crawford RH (1998) A heuristic method to identify modules from a functional description of a product. ASME DETC'98, Atlanta, GA, USA
Swaminathan JM, Tayur SR (1998) Managing broader product lines through delayed differentiation using vanilla boxes. Management Science, 44(12): 161–172
Stummer C, Heidenberger K (2003) Interactive R&D portfolio analysis with project interdependencies and time profiles of multiple objectives. IEEE Transactions on Engineering Management, 50(2): 175–183
Tirole J (1989) The theory of industrial organization. MIT Press
Ulrich K (1995) The role of product architecture in the manufacturing firm. Research Policy, 24(3): 419–440
Wheelwright SC, Clark KB (1992) Revolutionizing product development. Free Press
Yang K-K, Sum C-C (1997) An evaluation of due date, resource allocation, project release and activity scheduling rules in a multiproject environment. European Journal of Operational Research, 103: 139–154

Chapter 18

The transfer of methods into industry

Ralf Stetter
AUDI AG

Udo Lindemann
Technical University of Munich

Global competition and the transition from the sellers' market to the buyers' market force industrial companies to develop products in less time, at lower cost and with higher quality. The competitive capacity of a company is ultimately determined by its product development capabilities.

A shortened development time is one of the most important success factors, but commonly results in increased process complexity. In addition, rising quality requirements, distributed development processes, and more complex products have also increased process complexity. Therefore, the need for methods, strategies and tools that support designers in their endeavour is ever increasing.

Elaborate methods, such as failure mode and effects analysis (FMEA), quality function deployment (QFD), design of experiments, design for manufacture and assembly, as well as integrated product and process development have received much attention in recent years as means to improve industrial product development processes. However, the implementation of these methodologies has not always been successful (Mai, 1999).

Circumstances in industrial companies, such as time pressure, stringent quality requirements or characteristics of existing methods, hinder the implementation of new methods, strategies and tools. As a result, the rich body of design methodology, which is the result of more than 30 years of research, is only reluctantly transferred into industrial practice and the available methods are commonly not utilised to their full extent.

The competitive capacity of a company is ultimately determined by its product development capabilities.

In recent years, a number of strategies and guidance methods for the implementation of design methods into industrial practice have been developed, which can be summarised under the heading of 'method implementation'. Method implementation should be understood as a collection of practical measures that transfer methods into practice and ensure that they are actually used.

This chapter contains:

- an overview of research into method implementation;
- the presentation of a model of method implementation;
- a list of available strategies and guidance grouped according to the model;
- a description of a current method implementation process.

The contents of this chapter are intended to be used as a 'map' that provides designers and design managers with an overview of the issues that have to be considered during method implementation, and references to the literature where they can find strategies and guidance for success.

Studies of method implementation

A number of studies are listed in this section, each of which contributes to a comprehensive description of method implementation. The results of these studies, which are grouped according to the methods looked at, are among the main sources of strategies and guidance available.

Introduction of design methods/product development methods

Academic studies are one of the main sources for strategies and guidance for method implementation.

Early research projects were aimed at analysing decision making in engineering design (Tebay *et al.*, 1984) or were carried out by means of action research with the intention of implementing information systems in design (Antill, 1986).

Several research projects concerning the introduction of systematic design methods are based on a non-specific combination of theoretical considerations, experiences and case studies. The use of systematic design methods in industry was a major research focus of leading German design researchers in the beginning of the1990s (Birkhofer, 1991, Beitz *et al.*, 1992). Similarly, Tiggesbäumker and Pingel (1992) and Lohse (1993) describe approaches to introducing systematic design methods in industrial companies.

The research of Helbig (1994) was concerned with the development and introduction of product- and company-specific guiding systems for engineering design, which are also based on systematic design methods. Moreover, the results of an investigation described by Schneider and Birkhofer (1999) were based on experiences from applying systematic design methods in 25 projects with industrial companies. Additionally, Merte *et al.* (1999) describe a project with the goal of applying an approach to systematic product development at an automotive supplier.

The introduction of systematic design methods was also investigated by means of extensive surveys. Their adoption in the Italian mechanical industry was the focus of an investigation, carried out by means of questionnaires, by Bonaccorsi and Manfredi (1999). A similar investigation in UK industry is described by Gouvinhas and Corbett (1999). Glen and Lord (1995) carried out an extensive survey investigating design practitioners' attitudes towards a formal approach to development.

Cantamessa (1997) describes an investigation with postal questionnaires, covering 98 companies in Italy. In the first stages of this research more than 20 companies were visited and interviewed (Cantamessa, 1998). Similar studies were carried out in New Zealand (Whybrew *et al.*, 2001) and Poland (Rohatynski, 2001). Further surveys are concerned with design process planning (Eckert and Clarkson, 2003).

Radcliffe and Harrison (1994) employed an action research approach when investigating a 3 year implementation process of systematic design methods in a small company producing hydraulic cylinders. The objective was a complete transformation of design practice with regard to the use of these methods.

Araujo and Duffy (1997) report on a research project with the general aim of investigating the multiple dimensions of the acquisition, assessment and selection of product development methods, tools, and strategies. Andreasen and Hein (1998) report from two comprehensive organisational innovation campaigns within product development in Danish industry. They were able to identify supportive elements in the changes of attitudes and behaviour of the parties involved. A case study was carried out by Cantamessa *et al.* (1999) in order to identify clearly whether there is truly a need for a high-level concept of 'design co-ordination', *i.e.* the planning, scheduling, representation, decision making and control of product development.

Norell (1998) describes an interdisciplinary research program in co-operation between researchers from engineering design and work psychology that, amongst others, tried to answer the question: "What characterises the successful implementation and use of development tools?" The study was performed by means of interviews with people from 12 Swedish companies, which were operationally involved in product development. In prior work, Norell (1993) showed that the introduction of the methods design for assembly, QFD and FMEA followed a two-step process model. She also used a four-level model characterising the extent of use of methods.

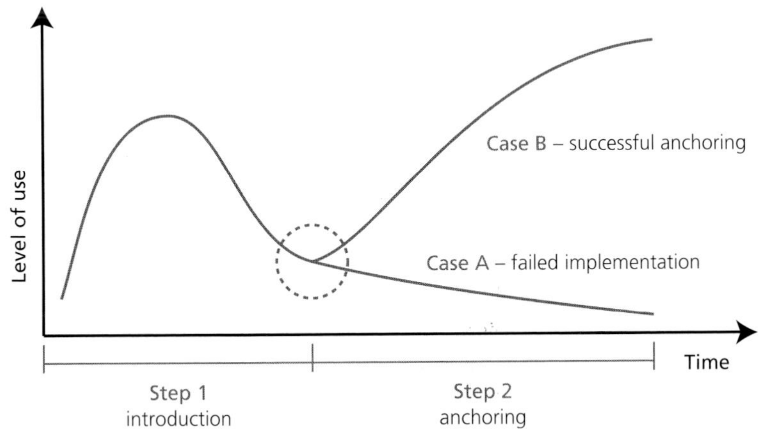

18.1 What characterises the successful implementation and use of development tools? (Norell, 1998)

Method introduction into industry needs to be carefully planned.

Connected work is focused on the introduction of QFD (Beskow *et al.*, 1998, 1999; Beskow, 2000) and the co-operation between supplier and buyer (Eneström *et al.*, 1999). The continuous improvement of the product development process is the focus of another research project of this research group (Ritzén *et al.*, 1999). The evaluation of the use of QFD was the main subject of a study carried out by means of participant observation and in-depth retrospective interviewing by Griffin (1992).

Research concerning the introduction of methods has been an important topic at the chair of product development at the Technische Universität München for a number of years. First contributions originated from Wach (1994), who developed problem-specific tools for integrated product development (IPD). The importance of method implementation and the necessity of a balanced approach was highlighted by Ambrosy (1997). Zanker (1999) developed a procedure for adapting methods based on situatedness. On the basis of this research, a structure for method implementation was developed (Stetter, 2000). Current research on method implementation is focused on the introduction of design for environment methods, mainly in small and medium-sized companies (Ernzer *et al.*, 2002; Lindemann *et al.*, 2003).

Connected research work was concerned with the development of an elaborate model for method implementation (Viertlböck, 2000). In ongoing research work, a strong emphasis towards action orientation (Lindemann and Wulf, 2001) and an understanding of methods as networks of methods is expressed (Lindemann, 2003).

Introduction of concurrent engineering

Extensive research concerning the introduction of the concurrent engineering (CE) methodology is performed in two major projects: the PACE project – a Practical Approach to Concurrent Engineering (Driva and Pawar, 1997), and the CEPRA project – Concurrent Engineering in Practice (Weber *et al.*, 1999). The main emphasis of these projects is on the development of computer tools (or 'electronic consultants'), which assist companies in improvement processes. Another large-scale project, Concurrent Simultaneous Engineering System (CONSENS), was aimed at developing and implementing integrated solutions (Bullinger and Warschat, 1996).

A series of smaller projects also concern the introduction of CE. Usher (1996) reports introductions of CE with a focus on small manufacturing enterprises. He developed an introduction strategy in a form of a cyclic approach centred around the concept of continuous improvement.

A case study with the aim of improving new product development according to the principles of CE for a medium-sized cable manufacturer is presented by Albin and Crefeld (1994). Similarly, the introduction of simultaneuous engineering (SE) in small and medium-sized enterprises was the main subject of the case study of Hindson *et al.* (1998). Additionally, Carlson-Skalak *et al.* (1997) outline a CE methodology developed specifically for small companies.

A method for the introduction of SE in product development is the result of a study by Ahrens and Beitz (1997), which was carried out in co-operation with a German company. A research method based upon focus groups was employed by Lettice *et al.* (1998) in order to understand the process of CE introduction.

Introduction of new product development

The strategic and methodical support of new product development (NPD) is the focus of several research projects. Dooley *et al.* (2002) investigated, by means of a survey with 39 respondents, whether the adoption of best practices in NPD leads to a higher NPD effectiveness; a comprehensive survey covering 383 companies in the USA was carried out by Griffin (1998).

Introduction of Total Quality Management

A series of investigations are concerned with the introduction of methods, strategies, and tools that are part of the Total Quality Management (TQM) methodology. In an early research approach, Lewis and Samuel (1991) were able to analyse problems in applying design for quality (DFQ) while devising and conducting continued education courses on DFQ for designers in an automotive company.

A large share of research projects involving TQM are concerned with small and medium-sized enterprises. Zink (1995) reports experiences originating from the introduction of TQM in these enterprises, while Christofolini and Wolf (1997) describe the introduction of a total quality approach in a small Italian company. In addition Dobberkau and Rauch-Geelhaar (1999) report the results of a project with the objective of transferring quality methods to small and mid-size industrial companies. Several studies concerning the introduction of TQM are based on surveys. Andersson (1993) carried out a series of interviews in Swedish industry, investigating the perception by industry of the DFQ methods. In contrast, a large sample of quality professionals was surveyed by Tamimi and Sebastianelli (1998) in order to investigate the degree to which various barriers hinder TQM initiatives in organisations.

The exact order in which the steps of an implementation strategy are performed is not crucial; however, each of these steps should be included.
(Usher, 1996, on TQM)

Pfeifer and Lesmeister (1999) initiated a research project that aims to simplify preventive quality management methods for an application in industrial practice. In the scope of this project, 126 companies were questioned in order to reveal frequencies of usage and benefit potentials of methods. A similar research project, described by Mai (1999), aimed to implement a holistic quality management approach in industrial practice.

Model of method implementation

Usher (1996) notes that the exact order in which the steps of an implementation strategy are performed is not crucial; however, each of these steps should be included in the strategy. Consequently, Stetter (2000) proposed a model that distinguishes layers rather than phases in the method implementation process. Such a model proposes no procedure but classifies activities into five distinguishable layers, as shown in Figure 18.2. The five layers summarise activities that exhibit a strong interrelation in terms of content. The chosen distinction between the layers is based on the comparison of the models of method implementation presented, their accompanying literature, and insights gained in several case studies. The course of action can start at any layer, but it must include activities on every layer in order to increase the potential of a method implementation to succeed.

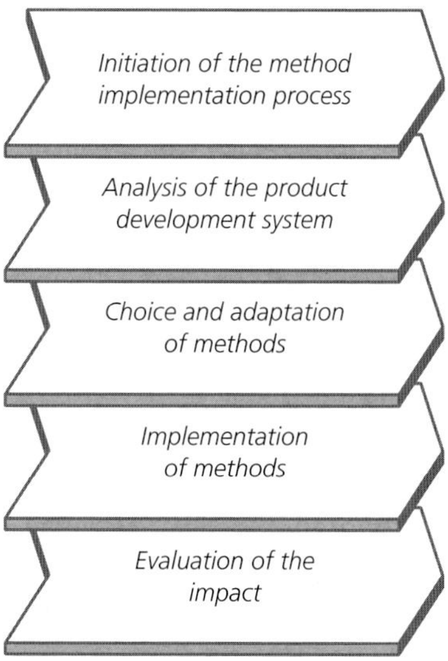

18.2 Five-layer model of method implementation (Stetter, 2000)

Strategies and guidance for method implementation

In this section, strategies and guidance for method implementation are presented according Stetter's (2000) model. For each layer, a general description is followed by a map of the key issues, and related literature that provides a rich resource of strategies and guidance for successful implementation.

Initiation of the method implementation process

This is the first layer of the method implementation model. The model does not prescribe a course of action. Nevertheless, this layer contains the activities that have to be performed (at least initially) in the very early stages of a method implementation process. They serve as a basis for the activities in the other layers and increase the potential for success.

The experience in several case studies agrees with insights in literature that certain activities are necessary for initiating the method implementation process. First, it needs to be emphasised that strengths and improvement potential have to be identified, e.g. by means of self-assessment or benchmarking, and have to be explicitly communicated to the stakeholders. Second, this improvement potential has to be transformed into operational objectives. It is necessary to consider method implementation as a project and consequently to plan a method implementation process for it.

The consideration of moderation and team building issues is mandatory. Furthermore, a number of organisational measures need to be carried out in method implementation. The common characteristic of the activities presented is that they serve to pave the way for a method implementation process. As a consequence of these activities, the organisation should be prepared for the change which inevitably accompanies it. Several appropriate approaches for the different activities and their origin are summarised in Figure 18.3.

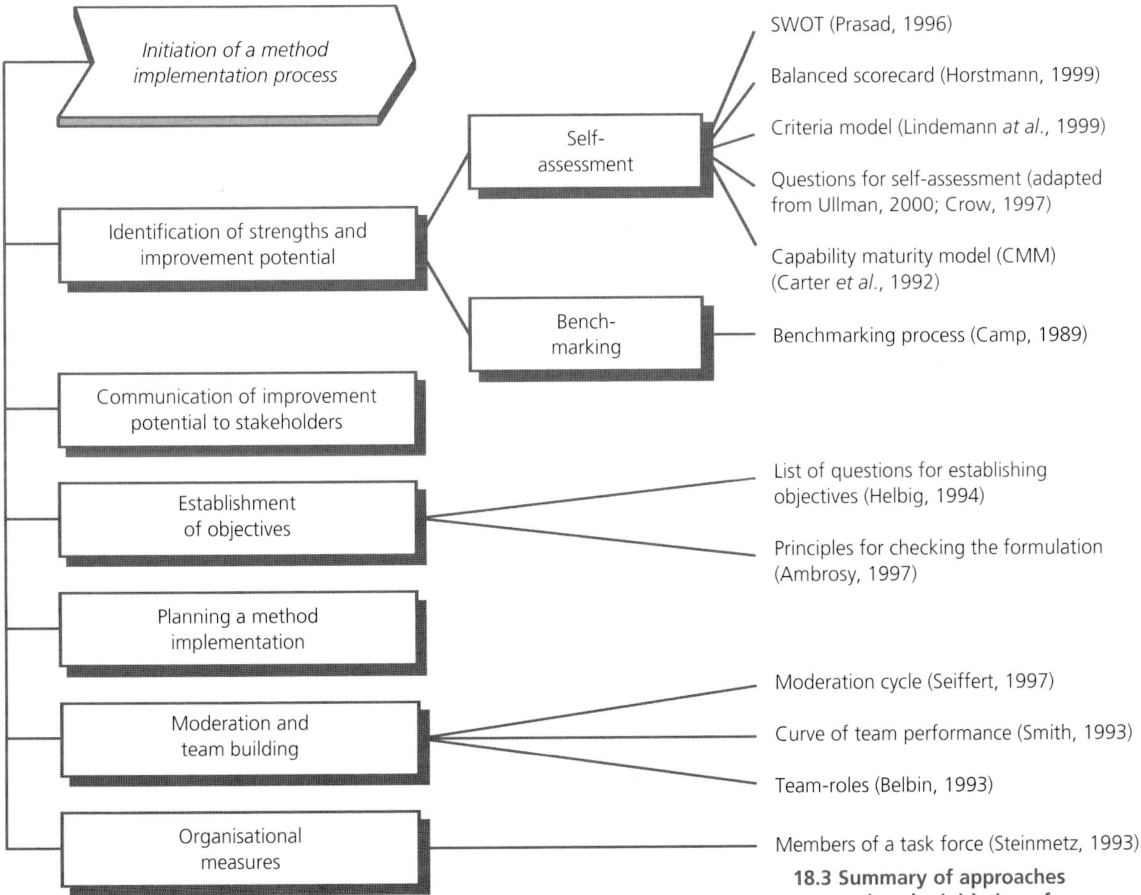

18.3 Summary of approaches concerning the initiation of a method implementation process and their origins

An intensive, unbiased analysis of the product development process is one of the cornerstones for successful method implementation.

Analysis of the product development system

An intensive, unbiased analysis of the product development process is one of the cornerstones for successful method implementation. In the first layer of the model of method implementation, 'initiation of a method implementation process', a project team is assembled and steps necessary for the acceptance of the analysis are initiated. Similarly, the activities in this layer create a basis for the activities in the other layers of the model.

Method implementation can only be successful if actual inadequacies in product development processes are addressed. Consequently, these inadequacies need to be identified, as the basis of information collected by multiple means. Commonly, document analysis, structured and semi-structured interviews, workshops with the designers involved and observation and protocol analysis can be employed. Process models as a depiction of the product development process can be used as underlying structures for identifying inadequacies. Moreover, process models may serve for determining the input and output information for tasks that are to be supported by methods.

The analysis of the product development systems should identify a collection of inadequacies in product development that should be addressed by method implementation and detailed analyses of these inadequacies, e.g. by means of performance measurement or cause–effect analyses. Strategies, methods and tools for analysing product development processes and their origin are summarised in Figure 18.4.

Choice and adaptation of methods

The choice and adaptation of methods are critical elements in every method implementation process. The choice of the appropriate method can be understood as being based on its essential, invariable characteristics. The adaptation is the altering of certain, more variable characteristics of a method in order to suit the specific needs of the industrial company.

The choice and adaptation of methods has to be based on the results of the activities in the layer 'analysis of the product development system' and serves as input for the core of the process, the actual 'implementation of methods'. Hence, it is vital to distinguish between the invariable core aspects of a method and the aspects that can be adapted to suit each application.

A conscious evaluation on different levels (function level, information level, procedure level, and applicability level) is the best approach for choosing appropriate methods. Three general recommendations can be proposed as to why we should adapt the more concrete aspects of a method:

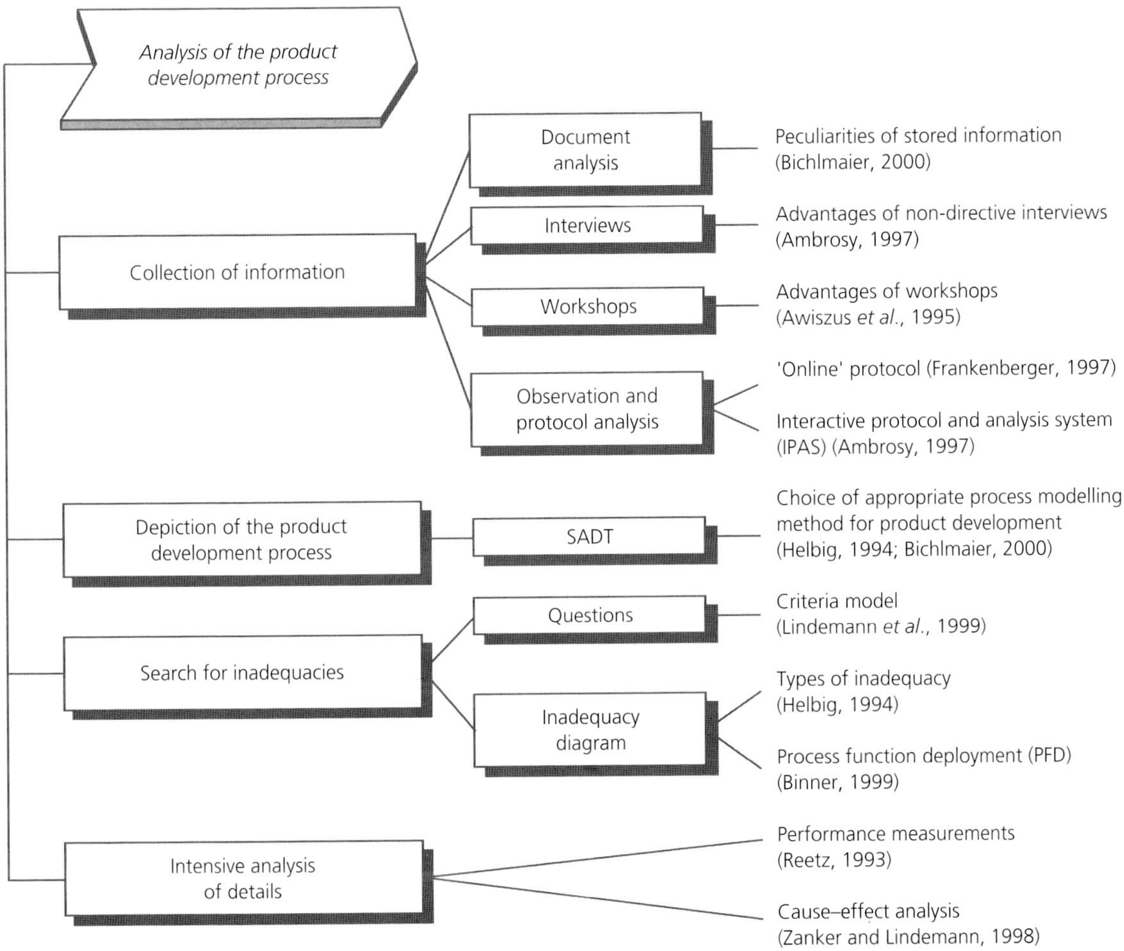

18.4 Strategies, methods and tools for the analysis of the product development system and their origins

- by adapting methods, immediate benefits can be achieved that will increase the chance that the methods will be adopted enthusiastically by designers;
- by enhancing the flexibility of methods, the designers' individual working styles can be supported;
- by adapting methods to the boundary conditions of the company, the perception that the methods to be implemented are foreign can be avoided and improved integration in the company processes can be achieved.

These approaches for choosing and adapting methods, and their origin, are summarised in Figure 18.5.

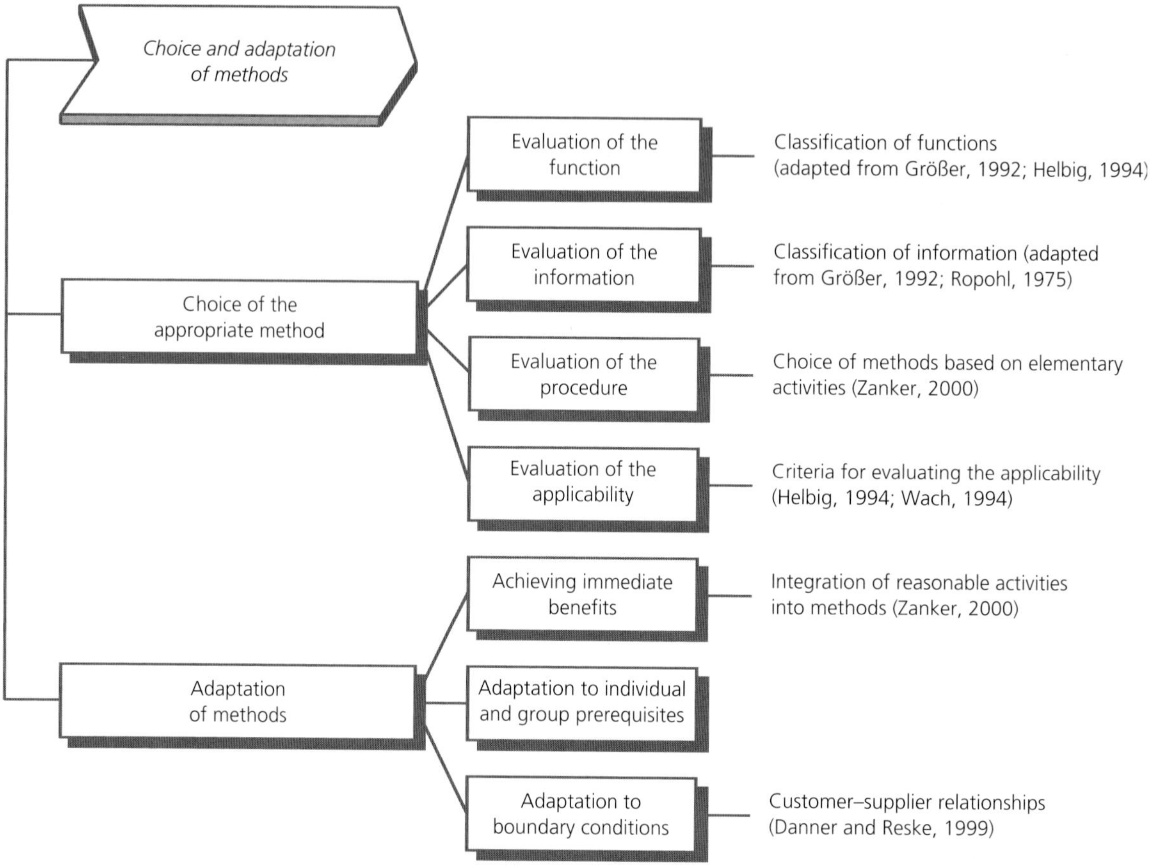

18.5 Procedure and strategies for the choice and adaptation of methods and their origins

Implementation of methods

The core of any method implementation endeavour is the actual 'implementation of methods'. If the implementation is carried out with anything less than full commitment, it will fail and all previous work will have been wasted. Essentially, the other layers of the model of method implementation consist of activities which prepare for or support the actual implementation.

In this phase it is necessary to mediate the objectives and the underlying logic of a method as well as the skills to apply it effectively. In order to do so, the method needs to be taught. In addition, the involvement of the designers in the implementation process is an essential ingredient for dealing with possible resistance. For the initial introduction of a method, several strategies and pilot projects can be utilised. During further application, methods need to be anchored in an organisation to ensure their continued use. They should also be subject to a continuous improvement cycle.

It is extremely important to refine and update methods, and especially the accompanying tools. In the literature, several approaches for the improvement and replacement of methods are described. These approaches and their origin are summarised in Figure 18.6.

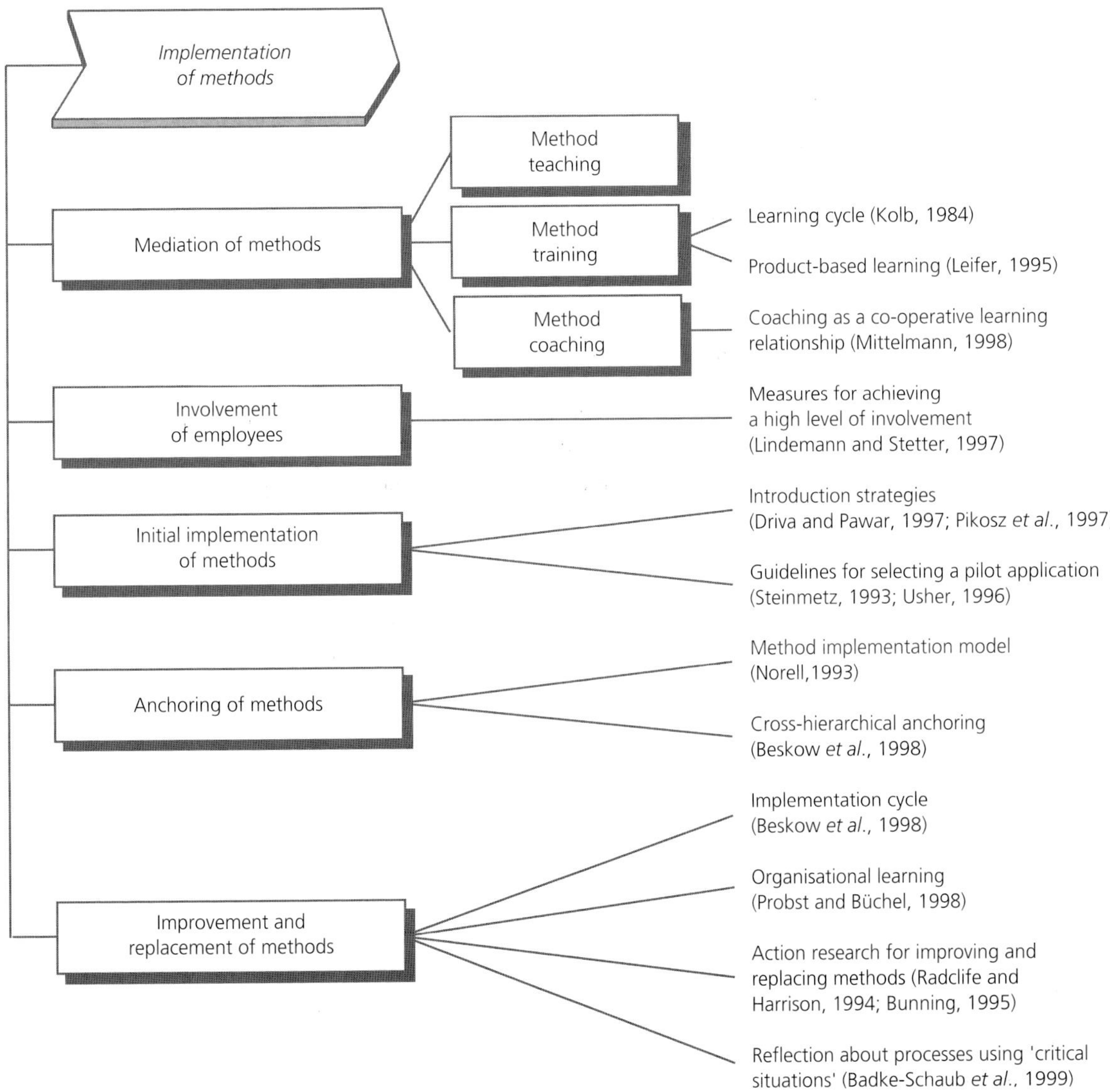

18.6 Summary of approaches concerning the introduction of methods and their origins

Evaluation of the impact

Evaluating the impact of a new method is important, but also difficult. To evaluate the impact of a method implementation essentially means to determine the effect of the methods, tools and strategies on the product development process. This evaluation is required to provide information for many activities in the other layers of the model of method implementation, e.g. for the improvement and replacement of methods.

An improvement cycle for method implementation can only be established if the organisation is capable of evaluating its impact. The basis of this evaluation has to be an assessment of the product development productivity before and after a method implementation. In general, several problems affect the assessment of the product development productivity, such as a possible disguising of the effects of a method implementation by probabilistic effects.

An improvement cycle for method implementation can only be established if the organisation is capable of evaluating its impact.

Two promising approaches to these problems can be found: the 'networked efficiency thinking' and the goal/question/metric (GQM) approach. Based on these, the impact of method implementation should be evaluated by using indicators, *i.e.* quantitative measurements or qualitative criteria correlated to the product development productivity.

Stetter (2000) presents a preliminary concept for evaluating the impact of a method implementation that incorporates the aspects described. The approaches developed and their origins are summarised in Figure 18.7.

Case studies of method implementation

The case studies of method implementation described in this section are based on 3 years of experience in the Seating Development Department of AUDI AG. In this timespan a systematic product development process for seat surfaces was developed, introduced, and applied in two seating development projects.

The main challenge of the systematic seat surface project was to achieve a synthesis of the different, sometimes contradictory requirements of product design and seat comfort (Siegmüller *et al.*, 2003). Figure 18.8 shows, for example, the seating system of the Audi A3.

In the same department, two connected tools were introduced to improve the management of the product development process. The first, 'project monitor', supports the representation of the most important information concerning the seating system development project. The second, 'information platform', makes the contents of the 'project monitor' available to all members of a product development team, distributed over different locations as well

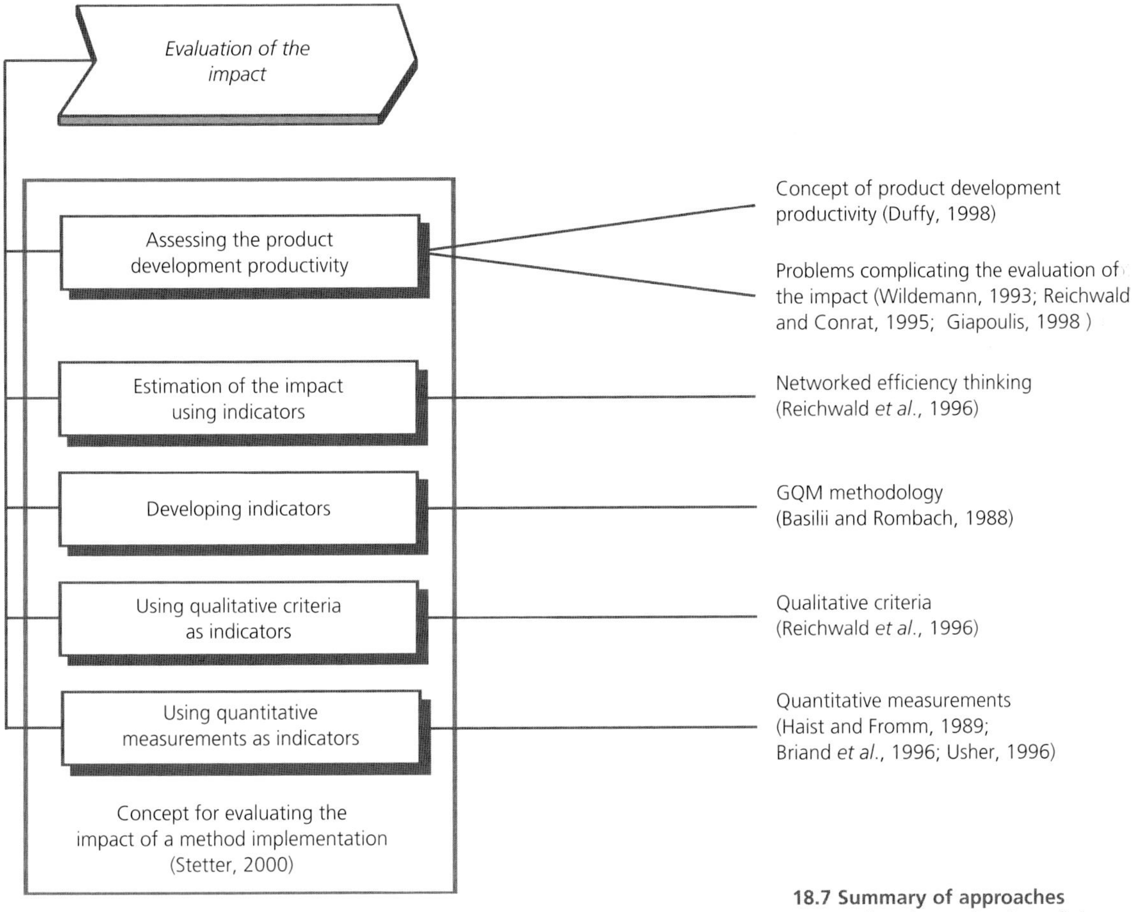

18.7 Summary of approaches concerning the evaluation of the impact and their origins

as different companies (Stetter, 2003). During the implementation of these tools, many insights were gathered. As a result, a number of 'success factors' were identified, which are presented in this section arranged according to the five-layer model of method implementation (Figure 18.2).

During the 'early phase of the method implementation process', there are three important success factors:

- the need for an enthusiastic and powerful method champion;
- a realistic distinction between variable and invariable characteristics;
- trust in external method sources.

During all phases of the method implementation process, it was observed that the chance that a method or tool would be used in the longer term was

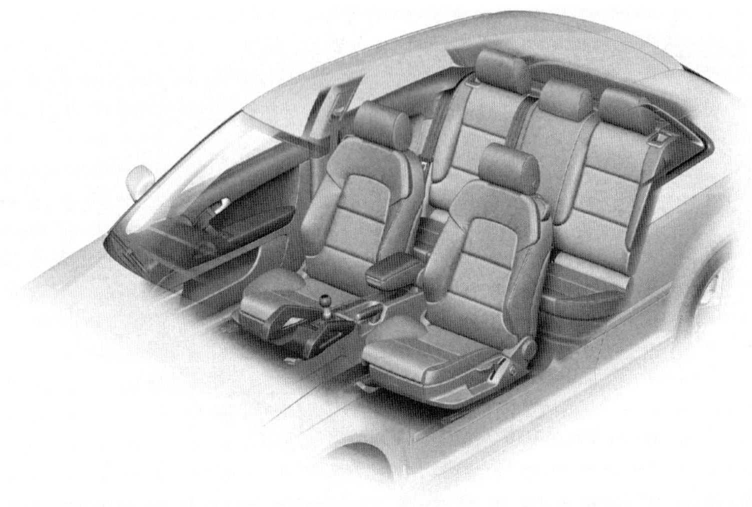

18.8 The seating arrangement of the AUDI A3 (reproduced from Siegmüller *et al.*, 2003)

greatly increased if it was initially promoted by someone who was deeply convinced of the benefits of the method.

A further observation was that method implementation was only successful where the variable characteristics were addressed, *i.e.* the characteristics of the product development system that actually could be changed. Attempts to change characteristics which lay outside the scope of change of the method implementation team were most often doomed to failure.

Typically, methods and tools were promoted by staff departments, internal and external consultants, or academics. A central success factor was trust – the designers in the product development department needed to trust the external method sources. They needed to believe that the proposed methods were suited to their actual situation and problems, that they were easy to use, and that they would have a positive impact.

Usually, this form of trust can only be established on the basis of a long-term partnership between product designers and method implementers. In this case, AUDI AG's seating system development department had a long established link (over 10 years) with the Institute of Product Development at the Technical University of Munich. This greatly enhanced the method implementation process.

Within the complex product development process of the automotive industry, the 'analysis of the product development system' is made more difficult by the fact that products are developed at different locations and at different companies. It is important to note that a product development process can only be fully understood if the process segments which take place at the suppliers and engineering consultant companies are analysed. This requires a certain degree of collaboration, not only on the product development, but also on the process improvement level.

It became obvious during the 'choice and adaptation of methods' that one of the most prominent barriers to successful method implementation can result from an overambitious attempt to address the weaknesses of the product development system. It is important to choose and develop methods that fulfil their main purpose.

In addition, it is frequently observed that during the adaptation of a method it becomes more and more complex in response to emerging requirements. The method champion needs to be aware of this development and, when necessary, make conscious decisions to focus on the most important requirements and to neglect others in order to keep the method and tools simple and appropriate.

With regard to the core of the process, the actual 'implementation of methods', little can be added to the conclusions and guidance presented in the earlier part of this chapter. Experience of the implementation process underlines the fact that people, and especially their knowledge of and their attitude towards new methods, play the major role in the success of any implementation process. The main challenge in this phase is to achieve consensus with all the key people involved.

People, and especially their knowledge of and their attitude towards new methods, play the major role in the success of any implementation process.

The 'evaluation of the impact' of the method implementation processes proved to be, as expected, the most difficult part of the job. An objective evaluation of product development productivity is nearly impossible, as a consequence of the ongoing development of greater functional and economic requirements with every new seating system.

Objective indicators of project performance are equally difficult to compare. However, the observation of the product development system supports the inference that a method improves the efficiency and effectiveness of the product development process if the method is used over a longer period of time and if there is a consensus that it is helpful.

The success factors arising from the case studies of method implementation observed in the seating system development department at AUDI AG are summarised in Figure 18.9. They are intended to support the application of the strategies and guidance presented in the earlier parts of this chapter.

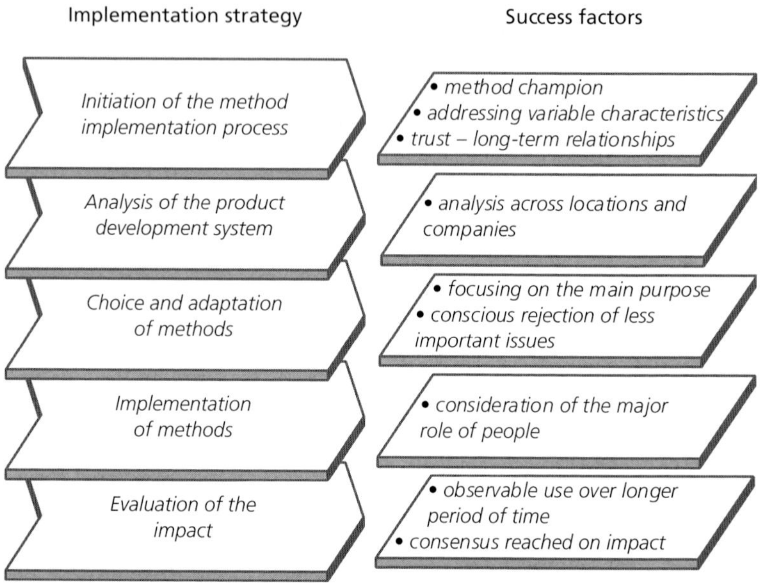

18.9 Success factors for method implementation

Conclusions

The potential success of companies developing products in an ever more competitive and dynamic environment is critically dependent upon their product development processes. Such processes need to be able to respond to the external market pressures, minimising risk in ever shorter development cycles. As a result, new methods and tools to support all aspects of product development are increasingly in demand. Their implementation within existing product development processes is a non-trivial activity and there are many stories of attempts to introduce excellent ideas that have ended in failure. The purpose of this chapter has been to highlight the key issues that determine success when implementing new methods and tools.

A five-stage implementation process has been presented which shows the important roles that the method and tool suppliers and the recipient company must play in any implementation. Numerous references are provided to studies related to process improvement, focusing on the selection and implementation of new methods and tools. These provide a rich resource for those wishing to enhance their product development processes.

The implementation of new product development methods and tools is an important activity. In order for it to be successful, it must be a carefully planned and executed collaborative effort between supplier and customer.

References

Ahrens G, Beitz W (1997) A method for the introduction of simultaneous engineering in a product development process. ICED'97, Tampere, Finland

Albin S, Crefeld P (1994) Getting started: concurrent engineering for a medium-sized manufacturer. Journal of Manufacturing Systems, 13: 48–58

Ambrosy S (1996) Methoden und Werkzeuge für die integrierte Produktentwicklung. PhD thesis, TU München, Germany

Andersson P (1993) Design for quality – as perceived by industry. ICED'93, The Hague, The Netherlands

Andreasen M, Hein L (1987) Integrated product development. IFS

Antill L (1986) Action research in information systems design. Design Studies, 7(4) 192–198

Araujo C, Duffy A (1997) Assessment and selection of product development tools. ICED'97, Tampere, Finland

Awiszus B, Brüning J, Goebel D, Hanewinckel F (1995) Analyse und Optimerung unternehmensübergreifender Geschäftsprozesse. VDI-Z, 137(5): 40–43

Badke-Schaub P, Wallmeier S, Dörner D (1999) Training for designers: a way to reflect design processes and cope with critical situations in order to increase efficiency. ICED'99, Munich, Germany

Basili V, Rombach D (1988) The TAME project: towards improvement-oriented software environments. IEEE Transactions on Software Engineering, SE-14(6): 758–773

Beitz W, Birkhofer H, Pahl G (1992) Konstruktionsmethodik in der Praxis. Konstruktion, 44: 391–397

Belbin M (1993) Team roles at work. Butterworth-Heinemann

Beskow C (2000) Towards a higher efficiency: studies of changes in industrial product development. PhD thesis, Royal Institute of Technology, Stockholm, Sweden

Beskow C, Johansson J, Norell M (1998) Changing the product development process: a study of QFD implementations in Swedish industry. Royal Institute of Technology, Stockholm, Sweden

Beskow C, Johansson J, Norell M (1999) Changing the product development process: a study of four QFD implementations. ICED'99, Munich, Germany

Bichlmaier C (2000) Methoden zur flexiblen Gestaltung von integrierten Entwicklungsprozessen. PhD thesis, TU München, Germany

Binner H (1999) Ein Vorgehensmodell zur Durchsetzung der Prozeßorientierung. ZWF, 94(6): 345–349

Birkhofer H (1991) Methodik in der Konstruktionspraxis: Erfolge, Grenzen und Perspektiven. ICED'91, Zürich, Switzerland

Bonaccorsi A, Manfredi E (1999) Design methods in practice: a survey on their adoption by the mechanical industry. ICED'99, Munich, Germany

Briand L, Young-Mi K, Melo W, Seaman C, Basili V (1996) Q-MOPP: qualitative evaluation of maintenance organisations, processes, and products. Fraunhofer ISERN-Report 96–02

Bullinger H-J, Warschat J eds (1996) Concurrent simultaneous engineering systems. Springer

Bunning C (1995) Professional development using action research. Look Forward Ask Questions Learn Virtual Conference. MCB University Press

Camp R (1989) Benchmarking: the search for industry best practices that lead to superior performance. ASQC Quality Press

Cantamessa M (1997) Design best practices at work: an empirical research upon the effectiveness of design support tools. ICED'97, Tampere, Finland

Cantamessa M (1998) Investigating productivity in engineering design: a theoretical and empirical perspective. In: The design productivity debate. Springer

Cantamessa M, Duffy A, Hein L, Rimmer D (1999) Design co-ordination in new product development. ICED'99, Munich, Germany

Carlson-Skalak S, Kemser H-P, Allen K (1997) Concurrent engineering applied to product development in small companies. ICED'97, Tampere, Finland

Carter D, Stilwell E, Baker B (1992) Concurrent engineering: the product development environment for the 1990's. Addison-Wesley

Christofolini I, Wolf G (1997) Total quality approach to design and manufacturing. DMMI'97, Portoroz, Slovenia

Crow K (1997) Improving the product development process. http://www.npd-solutions.com/pdforum.html

Danner S, Reske M (1999) Systematic design in practice: target finding for product and process. ICED'99, Munich, Germany

Dobberkau K, Rauch-Geelhaar C (1999) Zwischen Anstoß und Vision. Gestaltung und Einführungsmanagement von Qualitätsmethoden für kleine und mittelständische Unternehmen. QZ, 44(5): 605–610

Dooley K, Subra A, Anderson J (2002) Adoption rates and patterns of best practices in new product development. Journal of Innovation Management, 6(1): 85–103

Driva H, Pawar K (1997) Overview of PACE from conceptual model to implementation methodology. PACE'97, Marinha Grande, Portugal

Duffy A (1998) Design productivity. In: The design productivity debate. Springer

Eckert CM, Clarkson PJ (2003) The reality of design process planning. ICED'03, Stockholm, Sweden

Eneström P, Ritzén S, Norell M (1999) Co-operation between supplier – buyer in integrated product development: a pilot study. ICED'99, Munich, Germany

Ernzer M, Grüner C, Birkhofer H (2002) Implementation of DFE in the daily design work – an approach derived from surveys. ASME DTM'02, Montreal, Canada

Frankenberger E (1997) Arbeitsteilige Produktentwicklung – Empirische Untersuchung und Empfehlungen zur Gruppenarbeit in der Konstruktion. PhD thesis, TH Darmstadt, Germany

Giapoulis A (1998) Modelle für effektive Konstruktionsprozesse. PhD thesis, TU München, Germany

Glen J, Lord M (1995) Small scale, low volume new product development within the UK medical device industry. ICED'95, Praha, Czech Republic

Gouvinhas RP, Corbett J (1999) A discussion on why design methods have not been widely used within industry. ICED'99, Munich, Germany

Griffin A (1992) Evaluating QFD's use in US firms as a process for developing products. Journal of Product Innovation Management, 9(3): 171–187

Griffin A (1998) PDMA research on new product development practices: updating trends and benchmarking best practices. Journal of Product Innovation Management, 14(6): 429–458

Größer H (1992) Systematische rechnerunterstützte Ermittlung von Produktanforderungen. PhD thesis, TH Darmstadt, Germany

Haist F, Fromm H-J (1989) Qualität im Unternehmen: Prinzipien – Methoden – Techniken. Hanser

Helbig D (1994) Entwicklung produkt- und unternehmensorientierter Konstruktionsleitsysteme. PhD thesis, TU Berlin, Germany

Hindson G, Kochhar A, Cook P (1998) Procedures for effective implementation of simultaneous engineering in small to medium enterprises. IMechE, Part B, 212(4): 251–258

Horstmann W (1999) Der Balanced Scorecard-Ansatz als Instrument der Umsetzung von Unternehmensstrategien. Controlling, (4/5): 193–199

Katzenbach J, Smith D (1993) Teams: Der Schlüssel zur Hochleistungsorganisation. Ueberreuter

Kolb D (1984) Experiential learning: experience as the source of learning and development. Prentice Hall

Leifer L (1995) Evaluating product-based-learning education. Osaka'95, Japan

Lettice F, Evans S, Smart P (1998) Understanding the concurrent engineering implementation process – a study using focus groups. In: The design productivity debate. Springer

Lewis W, Samuel A (1991) An analysis of designing for quality in the automotive industry. Design Studies, 12(4): 252–260

Lindemann U (2003) Methods are networks of methods. ICED'03, Stockholm, Sweden

Lindemann U, Stetter R (1997) Restrukturierungsprozesse in Entwicklung und Konstruktion. Beteiligte und Betroffene. In: Unternehmenserfolg durch Restrukturierung von Entwicklungs- und Konstruktionsprozessen, VDI Bericht 1338

Lindemann U, Wulf J (2001) Action orientation in design methodology. ICED'01, Glasgow, UK

Lindemann U, Bichlmaier C, Stetter R, Viertlböck M (1999) Integrierte Produktentwicklung in der industriellen Anwendung – Eine Vorgehensweise zur verbesserten Umsetzung der Methodik. Konstruktion, 9: 30–34

Lindemann U, Heßling T, Hutterer P (2003) Introduction and adaptation of methods for sustainable product development in small and middle-sized companies – results and experiences. ICED'03, Stockholm, Sweden

Lohse G (1993) Methodical product development – experience of a practical study. ICED'93, The Hague, The Netherlands

Mai Ch (1999) Schneller – Höher – Weiter. Mit umfassenden Qualitätsmanagement zu Spitzenleistungen. wt Werkstattstechnik, 89(4): 145–148

Merte J, Ganio G, Schiller H, Tegel O, Beitz W (1999) Application of the German approach to systematic product development at a US automotive supplier. ICED'99, Munich, Germany

Mittelmann A (1998) Organisationales Lernen und Geschäftsprozess-management. Institutsbericht. Institut für Wirtschaftsinformatik

Norell M (1993) The use of DFA, FMEA and QFD as tools for concurrent engineering in product development processes. ICED'93, The Hague, The Netherlands

Norell M (1998) Competitive industrial product development needs multidisciplinary knowledge acquisition. In: The design productivity debate. Springer

Pfeifer T, Lesmeister F (1999) Präventive QM-Methoden einfacher gestalten. ZWF, 94(11): 642–645

Pikosz P, Malmström J, Malmqvist J (1997) Strategies for introducing PDM systems in engineering companies. CE'97, Rochester, MI, USA

Prasad B (1996) Concurrent engineering fundamentals. Prentice Hall

Probst J, Büchel B (1998) Organisationales Lernen: Wettbewerbsvorteil der Zukunft. Gabler

Radcliffe D, Harrison P (1994) Transforming design practice in a small manufacturing enterprise. ASME DTM'94, New York, USA

Reetz U (1997) Performance measurements – a key method for a guided implementation of concurrent engineering principles into product development processes. PACE'97, Marinha Grande, Portugal

Reichwald R, Conrat J-I (1995) Integrationslösungen für die Produktentwicklung. VDI-Z, 137(5): 58–60

Reichwald R, Höfer C, Weichselbaumer J (1996) Erfolg von Reorganisationsprozessen. Leitfaden zur strategieorientierten Bewertung. Schäffer-Poeschel

Ritzén S, Beskow C, Norell M (1999) Continuous improvement of the product development process. ICED'99, Munich, Germany

Rohatynski R (2001) Diagnosing the gap between methodology of engineering design and industrial practice. ICED'01, Glasgow, UK

Ropohl G (1975) Systemtechnik – Grundlagen und Anwendung. Hanser

Schneider M, Birkhofer H (1999) How to increase efficiency of design processes by design methods – an evaluation of successful co-operation projects among industry and universities. ICED'99, Munich, Germany

Seiffert J (1997) Visualisieren – Präsentieren – Moderieren. Gabal

Siegmüller C, Stetter R, Brandes B, Kirchhoff R, Mauter G (2003) Audi A3. Der Innenraum – Einklang von Ergonomie, Funktion und Design. In: Der neue Audi A3. ATZ/MTZ Sonderheft

Steinmetz O (1993) Die Strategie der Integrierten Produktentwicklung. Softwaretechnik und Organisationsmethode zur Optimierung der Produktentwicklung im Unternehmen. Vieweg

Stetter R (2000) Method implementation in integrated product development. PhD thesis, TU München, Germany

Stetter R (2003) Information demands of design managers in automotive industry. ICED'03, Stockholm, Sweden

Tamimi N, Sebastianelli R (1998) The barriers to total quality management. Quality Progress, 31(6): 57–60

Tebay R, Atherton J, Wearne S (1984) Mechanical engineering design decisions: examples of practice compared with theory. ImechE, Part B, 198(6): 87–96

Tiggesbäumker P, Pingel H (1992) Voraussetzungen für die Einführung der Konstruktionsmethodik in einem Industrieunternehmen. In: Praxiserprobte Methoden erfolgreicher Produktentwicklung. VDI Bericht 953

Ullman D (2000) Measuring and improving your concurrent capabilities (a questionnaire assessing concurrent engineering – demonstration). http://www.engr.orst.edu/~ullman/assessment

Usher J (1996) Implementing concurrent engineering in small manufacturing enterprises. Engineering Management Journal, 8(1): 33–43

Viertlböck M (2000) Modell der Methoden- und Hilfsmitteleinführung im Bereich der Produktentwicklung. PhD thesis, TU München, Germany

Wach J (1994) Problemspezifische Hilfsmittel für die integrierte Produktentwicklung. Hanser München

Weber F, Pawar K, Barson R, Santoro R (1999) Approach and concepts for a methodology and software system for the implementation and improvement of concurrent engineering in small and medium enterprises in the aeronautics industry. ICE'99, The Hague, The Netherlands

Whybrew K, Shaw A, Aitchison D, Raine J (2001) Use of design tools and methodologies for rapid product development in the New Zealand manufacturing industry. ICED'01, Glasgow, UK

Wildemann H (1993) Optimierung von Entwicklungszeiten: Just-In-Time in Forschung and Entwicklung und Konstruktion. München TCW Transfer Centrum

Zanker W (1999) Situative Anpassung und Neukombination von Entwicklungsmethoden. Aachen: Shaker 2000. PhD thesis, TU München, Germany

Zanker W, Lindemann U (1998) Weaknesses of today's development processes. International Congress of Project Engineering, Cordoba, Spain

Zink K (1995) Konzepte zur Umsetzung von Qualitätswissen. Praktische Erfahrungen in kleinen und mittelständischen Unternehmen. VDI-Z, 137 (11/12): 59–61

Design research

In the second part of this book each chapter provides a description of the research currently being undertaken at a leading design research centre. Their purpose is to encourage design practitioners to explore the benefits of collaboration with the research community. Whilst we realise that chapters describing research activities will, with the passage of time, be less representative of actual research, their main purpose is to identify centres of research excellence in the field of design process improvement.

Updated versions of the chapters will be posted from time to time on the book web-site (see back cover) and we would welcome electronic contributions from other leading centres.

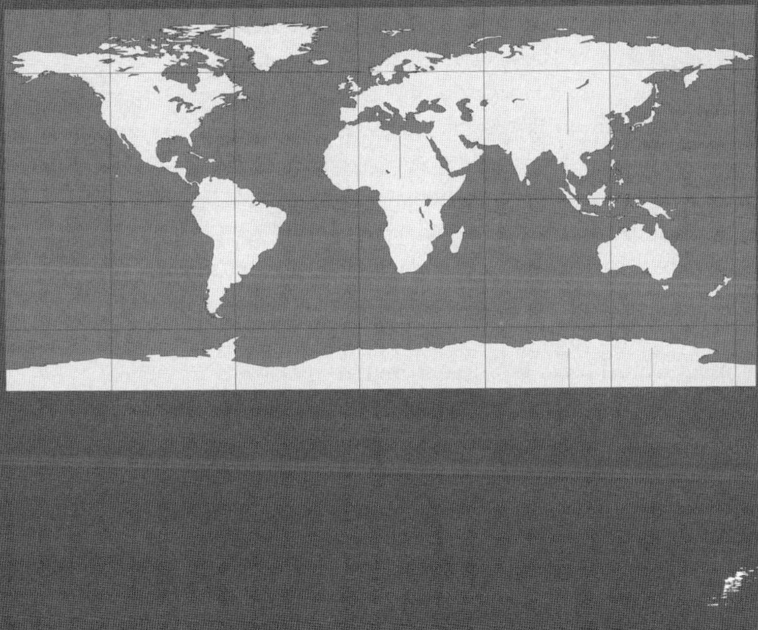

Chapter 19

Institute of Theoretical Psychology, University of Bamberg

Petra Badke-Schaub, Dietrich Dörner and Joachim Stempfle

19.1 Institute of Theoretical Psychology

Psychology, in general, aims to understand human thinking, feeling, and behaviour. The Institute of Theoretical Psychology is mainly concerned with the description, analysis and explanation of human action regulation in complex, dynamic and partially opaque situations. The behaviour of humans dealing with complex tasks can be decomposed into several cognitive processes, such as information gathering, planning, building a mental representation of the situation, decision making and self-organisation. Complex problem-solving behaviour, however, cannot be restricted to cognitive processes, but must also take into account motivation, emotion, and the social context. Therefore, the overall aim of the institute is to establish a theory for explaining human behaviour dealing within complex environments, thereby integrating cognition, motivation, emotion and the social context in a comprehensive framework.

In order to investigate human behaviour, extensive research is being conducted by the institute in the field as well as in laboratory settings. In the course of this research, subjects from diverse areas such as design, management, medicine, and also from different cultures such as India and Brazil are being investigated. In experimental settings we employ a variety of computer-simulated 'microworlds', which model the features of specific complex environments and allow us to analyse behavioural strategies in detail. Understanding the logic of human failure in problem solving provides us with the means to support managers, policymakers and designers in improving their problem-solving skills.

Since 1986 the institute has collaborated with engineers from the Universities of Technology in Darmstadt and Münich. The research programme was started by Professors Ehrlenspiel, Pahl and Dörner. In the course of this ongoing collaboration, now with Professors Birkhofer and Lindemann, several empirical studies on individual and team problem solving in engineering design have been conducted with the aim of penetrating the design process from a psychological point of view.

Research topics: design as complex problem solving

Individual strategies

One important research topic in the institute refers to the strategies designers use during the design process. In several laboratory studies engineering designers were assigned different tasks, e.g. designing a wall-mounted swivelling mechanism for an optical enlarger. The questions posed in Figure 19.2 characterise this aspect of the research.

A current challenging topic, relevant to education and industry, is the attempt to distinguish the approaches of experienced and non-experienced designers to different requirements during the design process.

Teamwork and design processes in industry

Another topic guiding the institute's research refers to the question of success and failure in design teams. Investigating design teams in industry means dealing with a complex technical and social environment. Individual-level characteristics of the designers (such as experience, competence, skills and problem-solving abilities), group-level characteristics (such as style of communication and group climate) and organisational-level characteristics (such as the coordination of activities between different divisions) combine to form a holistic network of influencing factors (Figure 19.3). We have observed and analysed a total of 10 projects in four organisations in order to understand the group design process. An important step in the analysis of the data was the development of the critical situation methodology that distinguishes between routine work and so-called critical situations, which have an important impact on the whole project. Building on the results of our empirical investigations, we have developed a training programme for designers that concentrates on coping with and reflecting on critical situations.

Leadership in design

In another project, research focuses on the analysis of leadership activity in product development. In the course of an extensive field study, we have observed managers in the context of their teams in three engineering companies over the course of several weeks. The data obtained in these studies has been

Designer
What individual prerequisites are related to which behavioural strategies?

Design methodology
Are the properties and methods of design methodology (VDI 2221) a reasonable foundation to formally describe and explain the empirically observed design process?
What are the differences in strategies of designers with methodological and without methodological education?

Design process
How do designers come to successful solutions?
What kind of errors and problems occur during the design process?
How can designers' strategies be improved?

19.2 Questions to characterise design process strategies

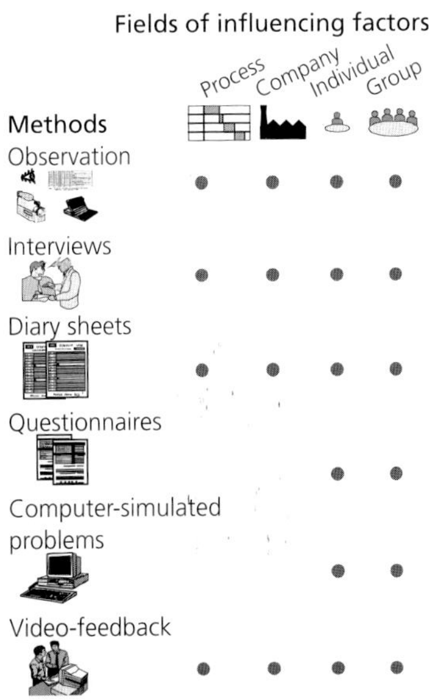

19.3 Research methods related to the design process, the company, the individual and the group

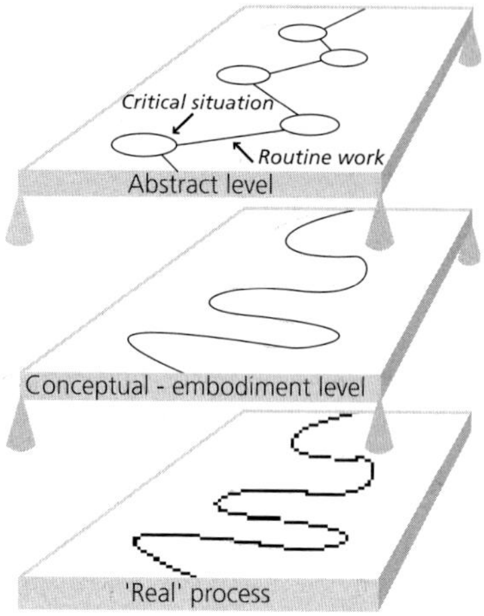

19.4 Critical situations as choice points in the design project

evaluated with regard to similarities and differences in the behaviour of the observed leaders. Furthermore, a typology of critical leadership situations (Figure 19.4) has been developed which comprehensively describes leadership requirements in design teams. This typology has been used in both a descriptive and a prescriptive way in order to explain observed leadership behaviour and to derive recommendations for successful leadership from situational characteristics. Conclusions based on these analyses will be elaborated in order to enhance existing design methods and to support design education as well as design practice.

Communication processes

Trying to analyse the thinking and reasoning process of designers is a difficult undertaking, as we have no direct means to inspect the process inside the designer's brain. Designers working in groups, however, communicate their thinking in the group setting and thus provide us partial access to their thought processes. In several investigations we have used protocol analysis in order to capture and analyse the thought processes of designers in the group. Based on the assumption that communication provides a prime access to the thought and problem-solving processes of design teams, we have developed a multi-level coding system for the analysis of the recorded data. A study conducted with design teams in the laboratory has yielded important insights into their problem-solving process, which can be characterised as a constant interweaving of task-oriented and process-oriented activity cycles.

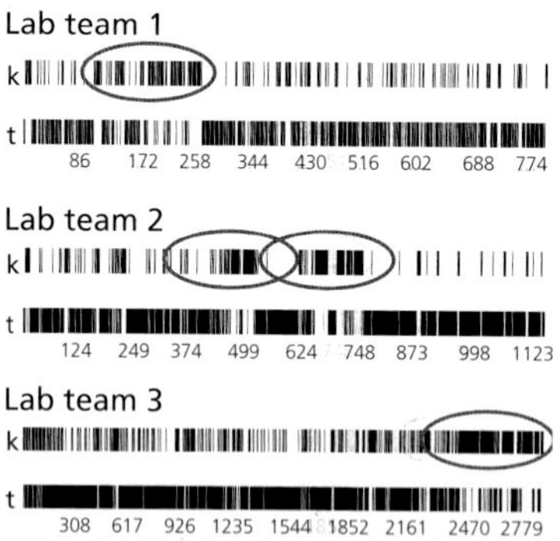

19.5 Activity focuses over the course of design work k= process, t= content

Transferability of results to other complex fields of work

Besides design there are other complex work environments, such as aviation, the petrochemical industry and medicine, with similar requirements, namely the handling of complex technological systems embedded in a social context. In all of these work environments, the results of unsuccessfully coping with critical situations, commonly called *human error*, may have a major impact on the achievement of goals and occasionally on the wellbeing of humans. Therefore, the findings from research in complex problem solving in one domain should be transferred to other domains in order to enhance the potential of human beings to cope effectively with complex environments. In 1998, members of the Institute of Theoretical Psychology founded a non-profit association, the so-called 'Platform – People in Complex Working Environments' in order to foster the exchange of experience between researchers and practitioners interested in human behaviour in complex environments.

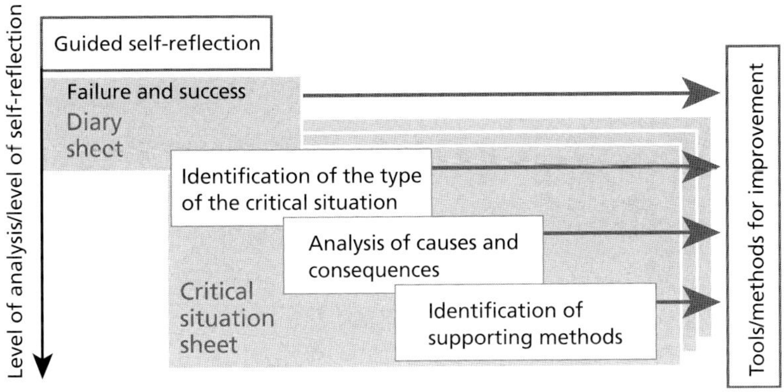

19.6 Training – identifying and reflecting on critical situations

Further reading

Badke-Schaub P, Frankenberger E (1999) Analysis of design projects. Design Studies, 20(5): 481–494

Badke-Schaub P, Frankenberger E (2003) Management Kritischer Situationen. Produktentwicklung erfolgreich gestalten. Springer

Dörner D (1996) The logic of failure. Holt

Frankenberger E, Badke-Schaub P, Birkhofer H eds (1998) Designers – the key to successful product development. Springer

Pahl G, Badke-Schaub P, Frankenberger E (1999) Résumé of 12 years interdisciplinary empirical studies of engineering design in Germany. Design Studies, 20(5): 481–494

Stempfle J (in press) Facilitating cooperation in multidisciplinary project teams in the pharmaceutical industry: an experience-centered approach. In: Facilitation group communication: innovations and applications with natural groups. Hampton Press

Stempfle J, Badke-Schaub P (2002) Thinking in design teams – an analysis of team communication. Design Studies, 23(5): 473–496

Wallmeier S (2001) Potenziale in der Produktentwicklung. Möglichkeiten und Grenzen von Tätigkeitsanalyse und Reflexion. VDI

For further information, see
http://www.bavarian-universities.de/Bamberg_eng.html

Chapter 20
Innovative Manufacturing Research Centre, University of Bath

Chris McMahon

20.1 The CPTMC Building - home of the Bath IMRC

The Bath Engineering Innovative Manufacturing Research Centre (IMRC) is a broad-based research centre located in the Department of Mechanical Engineering of the University of Bath, with interests across the spectrum of design and manufacturing. The focus of the centre's work is integrated product development. The centre is built upon broad foundations of research in design and in manufacturing processes and systems. The acknowledged strength in design research can be traced to the internationally recognised Engineering Design Centre in Fluid Power Systems (EDC) complemented by the leading-edge research in machine design and design information management. The manufacturing research is based upon established strengths in process and system modelling and simulation and is integrated into the design research through work on design for changeover. The centre's work is widely supported by industry, with strong support from the aerospace and packaging sectors, and with emerging strengths in the medical and electronic sectors, both of which build on the core skills in the centre.

Research within the IMRC currently follows three main themes: manufacturing processes and systems, design technologies and design information and knowledge. The ability to compete in today's world of globally distributed engineering teams rests on the ability of engineering organisations to respond to change. Central to this is the need to combine an understanding of manufacturing processes, design technologies and human issues, all supported by reliable and up-to-date information and knowledge. Two of the research themes have a strong design focus and are described in the following sections.

Design technologies

This theme involves research into techniques for the design, analysis and modelling of machine systems. Recent research has been based upon the team's previous work in the area of constraint modelling and design of mechanisms and machine systems. A design methodology has been created for assessing existing designs of machine systems and identifying their capability for improvement. This is based on comparing computational models of the design with experimental results derived from the use of high-speed video and instrumentation. The computational modelling methodology develops an understanding of the constraints that limit the action of a given design. Optimisation techniques are used to attempt to resolve constraints even when these are in conflict. Creation of a constraint model and subsequent exploration of the constraint space significantly aids in the understanding of an existing or proposed design. The work is looking at the generic areas of machine, product and system flexibility, with a particular focus on barriers to flexibility.

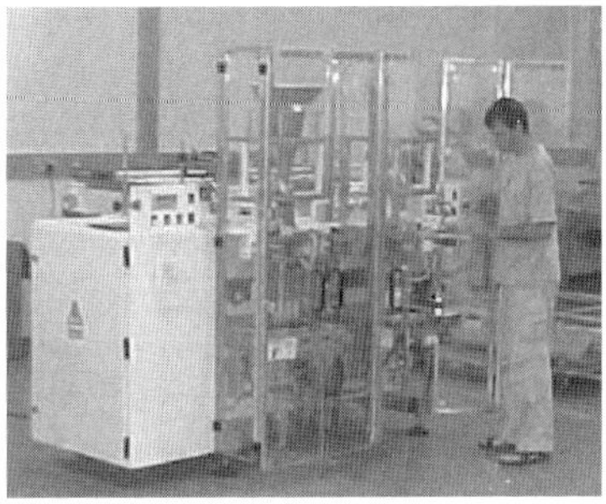

20.2 Research into packaging machine

The work of the design technology theme goes beyond machine performance to consider the interaction between machines and materials. A major project on such interactions that occur during packaging operations has allowed material aspects to be incorporated into the design methodology so that interactions can be considered. The above techniques have been applied, in particular, to the design of machinery for packaging and other production processes, but recent work has extended the application of the techniques to modelling the human body and this has led in turn to projects in the area of health care.

Design information and knowledge

Engineering design and manufacture are activities that are crucially dependent on timely access to relevant knowledge and information. The continued commercial success of companies and their ability to respond to change in a world in which design and manufacture are globally distributed is dependent on them implementing the most effective knowledge and information management approaches. The work of the second research theme, design information and knowledge, is focused on developing an understanding of the information needs of engineers and of approaches to information organisation and management. The specific objective of this theme is to develop approaches to design information and knowledge management that enable:

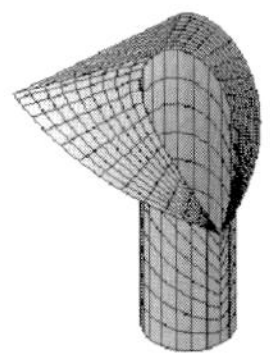

20.3 Modelling of a forming shoulder from a packaging machine

- rapid and reliable access to high-quality design information and knowledge;
- improvements in product quality and design lead time.

The research team undertakes a variety of work, ranging from short-term projects that exploit the earlier development by the group of faceted classification information systems, especially through the spin-out company Adiuri Systems Ltd, to more blue-skies consideration of such issues as information quality and the way in which engineering might function without conventional documents. All of the work is grounded in empirical study of the way engineers use information. Examples of current research programmes in this theme include:

- Application of techniques from the Semantic Web to document mark-up and to information retrieval.
- Identification of design rationale from discourse, in particular in design transactions such as meetings and conversations.
- Agent-based identification of the activities being undertaken by engineers, and their working contexts, and use of this identification for information push.

Two areas in which information and knowledge for design are particularly important concern evaluation of risk and uncertainty in the design process and understanding of the implications of the designer's work for people impacted by the design. With the increasing use of analysis and simulation techniques in engineering, it is very important for engineers to understand the uncertainties and risks inherent in their use of such techniques. The centre is working with the University of Bristol to identify a framework for the accumulation of knowledge on such uncertainties based on design process models. The centre is also working on the development of tools to help engineering designers reason about how people use engineered artefacts and work with engineered systems. In the course of this work, research issues such as risk assessment, risk perception, human error and distributed cognition have been addressed.

Conclusion

The Engineering IMRC at Bath involves about 30 academic staff and researchers in the Department of Mechanical Engineering. One of the centre's key strengths is its strong links with both small- and medium-sized enterprises and larger multinationals. This ensures the relevance of research work to current and future industrial needs. The current research programme incorporates 12 projects involving over 40 commercial organisations and trade associations. The Centre has access to the latest CAD systems, a virtual reality suite, rapid prototyping systems and machining systems. Global collaboration is facilitated by dedicated videoconferencing facilities. The centre is located in the recently completed Centre for Power Transmission and Motion Control (CPTMC) building that includes excellent laboratory facilities for machine systems research.

Further reading

Crossland R, McMahon CA, Sims Williams JH (2003) The practical application of design risk assessment models. IMechE, Part B, 214: 227–234

Darlington MJ, Culley SJ (2002) Current research in the engineering design requirements. IMechE, Part B, 216(3): 375–388

Hicks BJ, Bowler C, Medland AJ, Mullineux G (2002) A redesign methodology for analysing and improving the performance capability of packaging machinery. International Journal of Industrial Engineering, 9(4): 389–398

Hicks BJ, Medland AJ, Mullineux G (2003) A constraint-based approach for the optimum redesign of a packaging operation. Journal of Packaging Technology and Science, 16(4): 135–148

McMahon CA, Crossland R, Lowe A, Shah T, Sims Williams JH, Culley, SJ (2002) 'No zero match' browsing of hierarchically categorised information entities. AI EDAM, 16: 243–256

Potter S, Culley SJ, Darlington MJ, Chawdhry PK (2003) Automatic conceptual design using experience-derived heuristics. Research in Engineering Design, 14(3): 131–144

For further information, see
http://www.bath.ac.uk/imrc/

Chapter 21
Engineering Design Centre, University of Cambridge

John Clarkson

21.1 The Cambridge Engineering Design Centre

The Cambridge Engineering Design Centre (EDC) is a research centre for the development, validation and dissemination of advanced design understanding, methods and tools for technical systems. The research programme reflects UK industry's need for the best design methods and tools to achieve economic competitiveness, and to improve the quality of life through wealth creation and environmentally sustainable technology.

The overall aim of the Cambridge EDC is to improve the effectiveness and efficiency of engineering designers and design teams by undertaking research into the theories that will underpin the design methods of the future. These methods will be embodied in software tools, workbooks and publications that support the creation of reliable, high-quality, cost-effective products. This is being achieved through:

- *Research*, focusing on high-quality generic research to provide the theories and methods that will underpin engineering design in the future.
- *Technology transfer and exploitation*, transferring the research results into industry.
- *Education*, contributing to design education at both undergraduate and postgraduate levels to help create a pool of well-educated design engineers and designers.

21.2 Using computers to design products
Photo Kristi Shea and Janet Fan

Research

We currently have a team of over 40 people with expertise in aerospace, healthcare, architecture, engineering and construction (AEC), andgeneral product design. The research programme is split into a number of projects, these being grouped into eight closely linked themes:

- Design synthesis – using computers to design products.
- Design optimisation – searching for the best design.
- Design evaluation – proving that the design works.

- Materials selection – selecting the appropriate materials.
- Knowledge management – sharing knowledge in design.
- Process improvement – designing better products faster.
- Inclusive design – designing for the older or disabled user.
- Research methodology – ensuring high research standards.

Process improvement

There is always room for improvement in design. Maybe there is need for a better product, or for a better, more effective and economic design process. Late delivery of new products has been shown to be the single largest contributor to the loss of company profits. Our own experience of working with a number of companies, such as Lotus, GKN Westland Augusta and Rolls-Royce, has shown that effective communication, management of change and process planning are essential ingredients for a good product development process. We are continuing to develop our understanding of these issues and creating software tools to facilitate design process improvement.

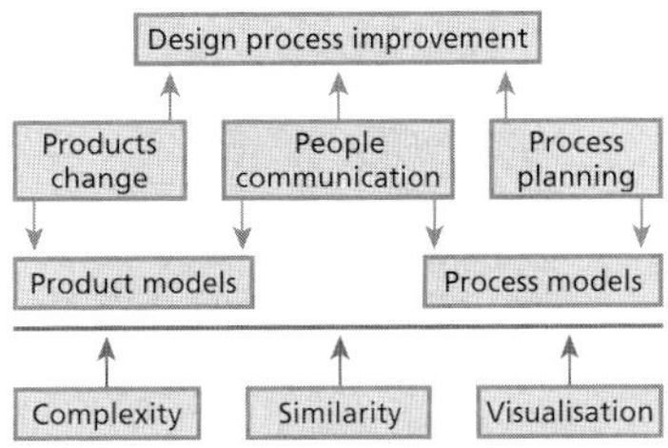

21.3 The research programme for process improvement

The main challenges arise from the complex nature of design and the interaction between products and processes. Practical design questions are closely related to such fundamental issues as the nature of complexity or similarity, as well as the visualisation and hierarchical ordering of complex data sets. We take a holistic approach in our design research, focusing on three aspects of design process improvement: communication, engineering change, and planning and scheduling. Our research is driven by empirical studies of design practice through interviews and complemented by rigorous theoretical investigations drawing on the theory of many different sciences. We foster cross-fertilisation between our case studies and have began systematic research into comparisons between design domains.

Drawing on methods from computer science, we base our tools for design process support on identified industrial needs and are sensitive to potential detrimental effects on the processes we are trying to assist. At the moment we are addressing the challenge of how to build data-rich models of processes and products without burdening the industrial user.

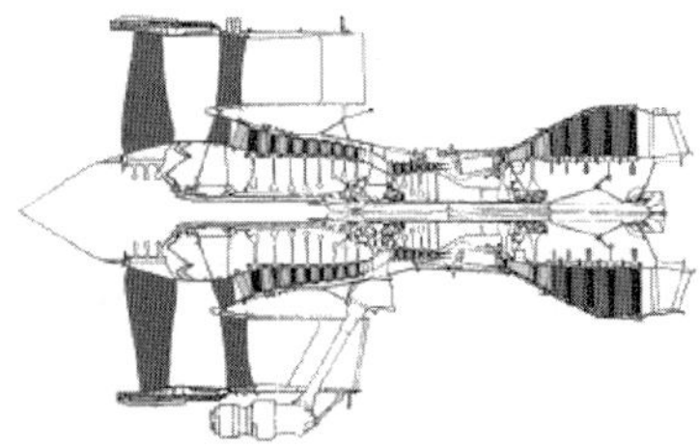

21.4 'Signposting' jet engine compressor design
Reproduced with the kind permission of Rolls-Royce plc

Planning and scheduling

We model design processes through signposting models, which capture tasks with their input and output parameters. The parameters have confidence levels associated with them, which reflect the maturity of the value and drive the sequence of the process. This enables us to suggest suitable tasks at any time during the design process, based on the design knowledge available and what would advance it. By modelling the probabilities of task failure and costs/durations, we can establish the lowest risk route through the design process. Additional information on resources allows simulation of likely processes outcomes.

21.5 The Perkins 1000 series engine
© Perkins Engines Company Limited

Engineering change

Design never starts completely from scratch, but reuses or adapts parts of existing designs. Even the design of variants in which many parts are kept is far from easy, because change can spread throughout. We are studying how companies carry out change processes and capture the links between components, systems or functions. These can be expressed in dependency matrices and we are developing probabilistic methods to predict the spread of change. The integration of change prediction and process planning will further support the selection of change routes.

Communication

Communication in industry can be problematic for many reasons, ranging from organisational to interpersonal. This research aims to explain communication behaviour in companies and build tools to support information flow and communication. These tools will utilise the connectivity models generated for process planning and change prediction.

21.6 Complex products demand complex design processes
Reproduced with the kind permission of Rolls-Royce plc

Complexity

Engineering products are inherently complex and the organisation of large companies and supply chains exacerbates this. We aim to understand the effect this complexity has on the design process and devise methods to reduce it.

Similarity

Products are designed based on similar products; process planning makes use of experiences from similar processes. We all talk about design with reference to similar products. We investigate what similarity means in these cases and address its formal structure.

Visualisation

Expressing process and product models for complex designs is difficult, because of the size of the models. We are looking at techniques for the display of complex models. This involves addressing the hierarchical grouping of design information.

Comparative design

We are developing a framework for comparison between design processes and design domains by identifying causal links between the characteristics of processes, such as safety, criticality or degree of innovation, and observed behaviour. This enables us to gain insights into processes by identifying hidden factors that are more easily observed in other processes or industry. It also gives us insights into the nature of design and will support the transfer of best practice in design across industry sectors.

21.7 The EH101 – a study of change by design
© AgustaWestland

Further reading

Clarkson PJ, Hamilton JR (2000) Signposting: a parameter-driven task-based model of the design process. Research in Engineering Design, 12(1): 18–38

Clarkson PJ, Simons CS, Eckert CM (2004) Predicting change propagation in complex design. ASME Journal of Mechanical Design

Eckert CM (2001) The communication bottleneck in knitwear design: analysis and computing solutions. Computer Supported Cooperative Work, 10(1): 29–74

Eckert CM, Clarkson PJ, Zanker W (2004) Change and customisation in complex engineering domains. Research in Engineering Design, 15(1): 1–21

Eckert CM, Stacey MK (2000) Sources of inspiration: a language of design. Design Studies, 21(5): 523–538

Jarratt TAW, Eckert CM, Clarkson PJ, Stacey MK (2004) Providing an overview during the design of complex products. DCC'04, MIT, Cambridge, MA, USA

Jarrett JP, Clarkson PJ (2002) The surge–stagnate model for complex design. Journal of Engineering Design, 13(3): 189–196

21.8 Design studies in automotive engineering

For further information, see
http://www-edc.eng.cam.ac.uk/

Chapter 22
A culture of design research and teaching, Carnegie Mellon University

Jonathan Cagan

22.1 Several Buick front ends generated from Buick brand shape grammar

Carnegie Mellon University has a rich and long history in design research and education. Arguably the root of the field of formal design research goes back to 1969 and the late Herb Simon's *The Sciences of the Artificial*. Simon, a Nobel Laureate, co-founder of the field of artificial intelligence, psychologist and economist, believed that a formal approach to synthesis, creativity and design methodology was critical and available as a field to researchers. Most, if not all, research today in design, be it computer or cognitively based, has influence from Simon's thoughts developed at Carnegie Mellon.

The culture at Carnegie Mellon goes back to the Carnegie Plan for Professional Education, developed by Carnegie Mellon President Robert E. Doherty in 1939–40. A portion of the plan stated that students must acquire "creativity and intellectual playfulness, moving beyond established knowledge and practice to create imaginative ideas and artifacts...We are committed to bring together the traditions of liberal and professional education".

Today, Carnegie Mellon University exemplifies interdisciplinary collaboration in research and education. There is a culture of design. Every college and every department has a faculty directly or indirectly thinking about and contributing to design theory, methodology, application and practice. The focus in this chapter is the engineering college (called Carnegie Institute of Technology) and a selection of the research areas from the college are discussed. The rich focus on design has also led to leadership and innovation in design education. For nearly two decades, the college has offered a wide selection of interdisciplinary design courses, and most recently introduced a professional Master's in Product Development.

Electrical and computer engineering

At Carnegie Mellon, there are numerous research thrusts in microelectromechanical systems (MEMS) covering microsensor systems integrated with electronics, microrobotics, RF-MEMS and microfluidic systems. One leader in the field is Electrical and Computer Engineering Professor Gary Fedder. In his work, development of such microsystems benefits from a hierarchical design approach, a concept borrowed directly from digital and analogue circuit areas. Prior research has evolved libraries of composable parametric MEMS elements for rapid design through schematic construction, parameter sizing, simulation and iteration.

Daniel P. Siewiorek, Professor of Electrical and Computer Engineering and Computer Science, and a world expert in wearable computers, has developed and employed an interdisciplinary concurrent design methodology over the past dozen years to design, fabricate and deploy over 20 dedicated wearable computer systems. Coining the term Wearable Computer in the early 1990s, interdisciplinary groups of students have created systems for a variety of applications, including vehicle inspection, used by the US Marines, and aircraft inspection, by Boeing.

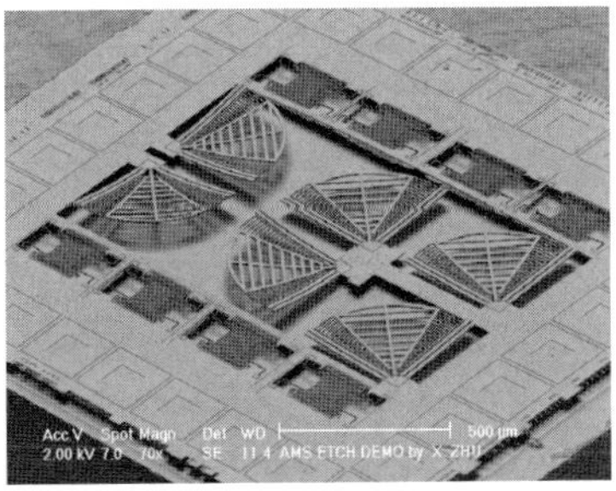

22.2 SEM of MEMS structures used for material characterisation
An array of "fan" structures for crack failure analysis, and on the periphery are crab-leg resonator structures for characterising material stiffness

Chemical engineering

Through the effort of Professors Larry Biegler, Ignacio Grossmann and Art Westerberg, Chemical Engineering at Carnegie Mellon has been a leader in pursuing mathematical programming approaches for synthesising process flowsheets of continuous and batch processes, and subsystems such as reactor networks, complex distillation systems, heat exchanger networks and utility plants. These approaches rely on the use of superstructure representations that systematically embed the alternative designs. The simultaneous optimisation of the topology and operating conditions is performed with large-scale mixed-integer nonlinear programming techniques which have been applied in the petroleum and chemical industry through the Center of Advanced Process Decision-making.

22.3 Belt-worn spot computer with wireless connectivity and head-mounted display

Civil and environmental engineering

In the late 1980s, Susan Finger, Professor of Civil and Environmental Engineering, began working with colleagues at the Institute for Complex Engineered Systems (then the Engineering Design Research Center) on concurrent engineering, exploring functional and behavioural representations of

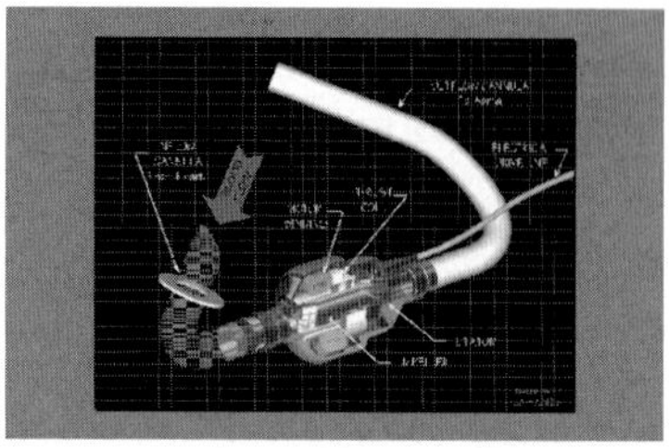

22.4 Streamliner artificial heart showing components and blood flow directions

designs, as well as organisational issues. They share an interest in integrating design and manufacturing and have worked together on systems as diverse as wearable computers and bone-tissue engineering. Current work includes modelling support for the early stages of design and the use of design tools and methods for collaborative learning.

Led by Professors Lester Lave, Chris Hendrickson and Fran McMichael, Carnegie Mellon is a leader in design for the environment, an increasingly important corporate and social goal. Reducing energy use, avoiding pollutant emissions or enabling reuse of products can often be accomplished through design changes rather than adding on control or remediation processes. Faster analysis methods for lifecycle assessments are critical to allow integration with design processes, such as the economic input–output lifecycle assessment model.

Biomedical engineering

The Biomedical Engineering Department has an emphasis on the design of artificial organs. Led by Professor Jim Antaki, the group has focused on improving the formalism of this field by employing design processes based on predictive mathematical models and quantitative objective functions. The most recent example is a novel magnetically levitated heart-assist pump, called 'the Streamliner'. Initial sizing and optimisation of components were enabled by a set of coupled closed-form models for the fluid dynamics, actuation, magnetic suspension and rotordynamics. A combination of computational fluid dynamics and finite-element electromagnetic analyses was used to fine-tune the design.

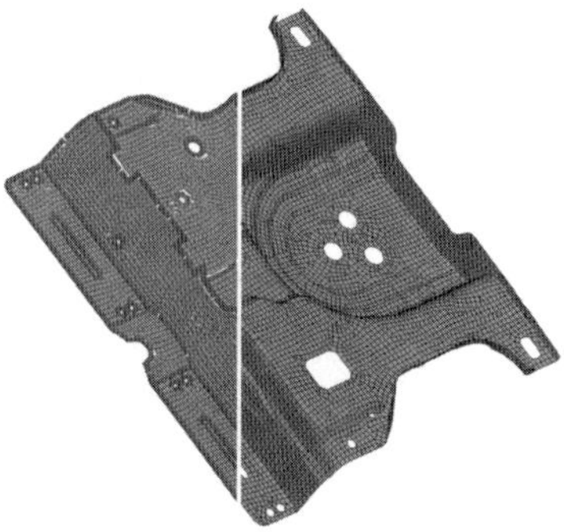

22.5 Rectangular cells, or bubbles, are tightly packed on the surface (left), and a high-quality quadrilateral mesh is created by connecting the centres of the bubbles (right)

Mechanical engineering

Kenji Shimada, Professor of Mechanical Engineering, focuses on physically based models for design, analysis and visualisation. For example, his approach to finite-element meshing, called bubble mesh, was the first to be physically based. The method was originally inspired by the observation of the regular hexagonal pattern of soap bubbles floating in liquid. This pattern is the geometric dual of an ideal triangular mesh. The bubble mesh method packs spherical or rectangular cells, or bubbles, tightly in a geometric domain by dynamic simulation, and the centres of the bubbles are then connected into a well-shaped mesh with controlled mesh size, directionality and anisotropy. The method has been extended to quadrilateral meshing, tetrahedral meshing and hexahedral meshing.

Jonathan Cagan, author of this article, and a Professor in Mechanical Engineering, focuses on the early stages of product development. Work in collaboration with Professors Craig Vogel and Laurie Weingart focuses on understanding and developing tools to improve interdisciplinary product development teams. Work on formal design synthesis explores three main areas: languages of design and their implementation, such as shape grammars to model functions or product brands; automated intelligent packaging or layout of products to optimise space and design goals; and, in collaboration with Psychology Professor Ken Kotovsky, cognitive-based models of design creativity including agent-based approaches applied, for example, to electro-mechanical synthesis problems. In the Carnegie Mellon tradition, much of this work has found application in industrial settings.

Further reading

Antaki JF, Ghattas O, Burgreen GW, He B (1995) Computational flow optimization in rotary blood pump components. Artificial Organs, 19(7): 608–615

Antonsson EK, Cagan J eds (2001) Formal engineering design synthesis. Cambridge University Press

Biegler LT, Grossmann IE, Westerberg AW (1999) Systematic methods for chemical process design. Prentice Hall

Cagan J, Vogel CM (2002) Creating breakthrough products: innovation from product planning to program approval. Financial Times Prentice Hall

Fedder GK, Jing Q (1999) A hierarchical circuit-level design methodology for microelectromechanical systems. Circuits and Systems II: Analog and Digital Signal Processing, 46(10): 1309–1315

Finger S, Konda S, Subrahmanian E (1995) Concurrent design happens at the interface. AI EDAM, 9: 89–99

Lave L, Hendrickson C, McMichael F (1995) Environmental implications of electric cars. Science, 268(19): 993–995

Shimada K, Liao J-H, Itoh T (1998) Quadrilateral meshing with directionality control through the packing of square cells. IMR'98, Dearborn, MI, USA

Siewiorek P *et al.* (1994) An interdisciplinary concurrent design methodology as applied to the navigator wearable computer system. Journal of Computer and Software Engineering, 2(3): 259–292

Simon HA (1996) The sciences of the artificial. MIT Press

For further information, see
http://www.cmu.edu/

Chapter 23

Product Development and Machine Elements, Darmstadt University of Technology

Herbert Birkhofer

The Product Development and Machine Elements (PMD) group is one of 23 groups within the faculty of Mechanical Engineering at the Darmstadt University of Technology in Germany. The research topics focus on a detailed understanding of the process of developing marketable and environmentally sound products and the sustained transfer of these practices into industry.

Design knowledge development

'Pingate' is a computer-based teaching, learning, and training system for product development which collects and consolidates existing knowledge with regard to design methodology.

Pingate produces specific documents adapted for various applications and different types of user. It is based on the concept of knowledge-modularisation. The knowledge-units are stored in a thematically structured database and can be accessed by a powerful navigation system.

Authors with different backgrounds and competencies in product development can add their contributions to a common pool of knowledge. The navigation system enables users like teachers and trainers to configure specific documents according to their needs and objectives in teaching and training courses. Direct access to the database could be useful for researchers and students alike, supporting independent learning via the Internet. Pingate is correctly used in preparing lessons and exercises at the university and for producing documents used in training courses in industry.

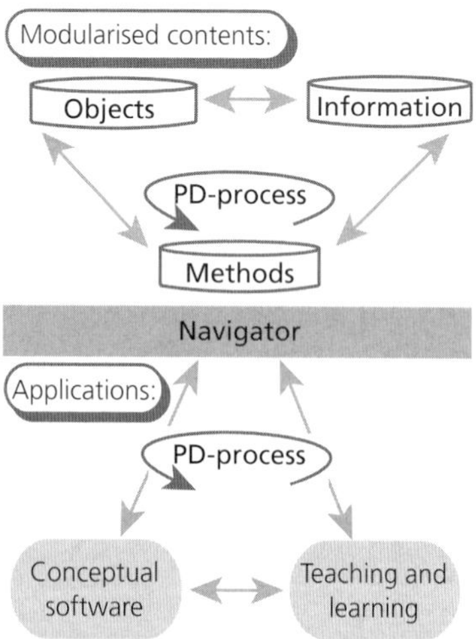

23.1 The 'pingate' approach

Collaborative project work

A key problem in industrial design is the lack of methodical work, especially in the early phases of design where functionality, costs and quality of the future product are substantially defined. Therefore, a project was established to undertake collaborative research with industrial partners, ensuring the successful transfer of existing methods into industry. The team adapts methods to the specific needs of the industrial partners and investigates and overcomes barriers to a successful transfer.

To improve the transfer of methods into industry, so-called 'Transfer-Workshops' are established, which are adapted to suit the particular design situation. Within these collaborative projects the industrial designers contribute specific product and process knowledge, whereas the members of the academic team act as trainers and coaches.

23.2 Collaborative project work

Collaborative project work meets the requirements of industry designers much more than standard seminars. The designers are able to learn methodical work practices while solving a specific problem in their own field of expertise.

Empirical design research

To support designers' work effectively, understanding how they think and act is indispensable. This subject is tackled by the cooperative research of the PMD group and cognition psychologists at Bamberg University.

Empirical design is carried out using detailed observations and enables the observers to describe phenomena in individual work, teamwork and leadership within industrial product development. A detailed classification and computer-based clustering system enables researchers to display the results of the studies, which serve as starting points for education and training courses.

The actual focus of research concentrates on improving the application of design methods in industry by creating highly adapted presentations of methods according to the specific design situation. They should enable designers to recognise the basic ideas of the method quickly, to develop a reliable understanding of the working steps and to get effective ideas for efficient use.

The generation of a specific presentation of methods may be supported by the Pingate approach with its own flexibility of access to and use of knowledge.

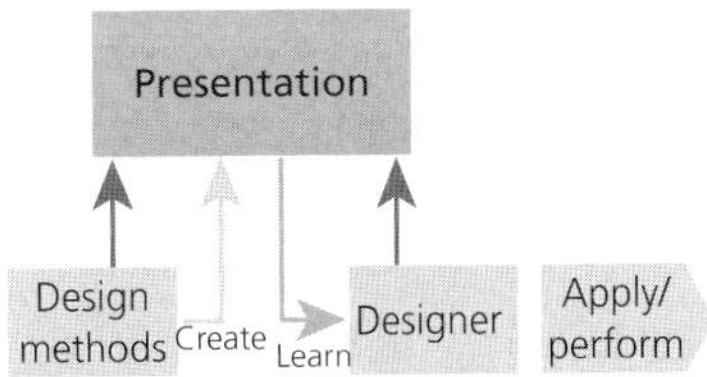

23.3 The role of presentation of method

Lifecycle design

Preventive environmental protection, avoiding or reducing resource consumption, waste production and emissions has to focus on all the design phases of a product. The EcoDesign group aims to support the holistic design of environmentally sound products by developing appropriate methods

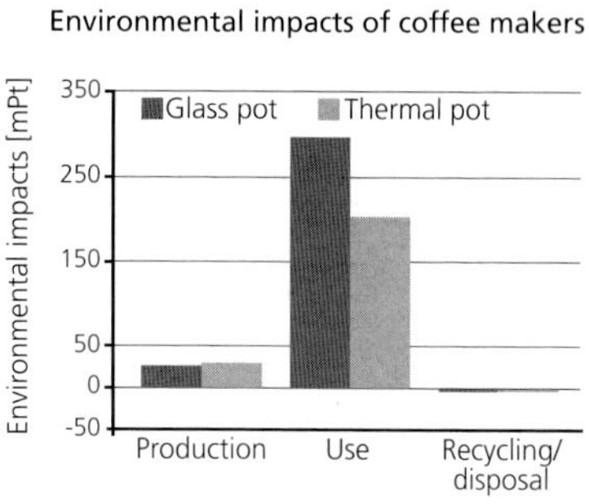

23.4 Environmental impacts of two product concepts

and tools. The research activities are integrated in the Collaborative Research Centre 392 'Design for Environment – Methods and Tools'.

The methodical support must enable the designers to recognise environmental impacts of their designs promptly, and to use these potentials preventively. EcoDesign is based on a modelling approach which simultaneously develops the product model in all design phases with the process models in all phases of the product lifecycle. The EcoDesign Group concentrates on three research issues:

- the environmentally friendly use of technical products;
- the integration of environmental knowledge in the design process;
- the management of information resources in design for environment.

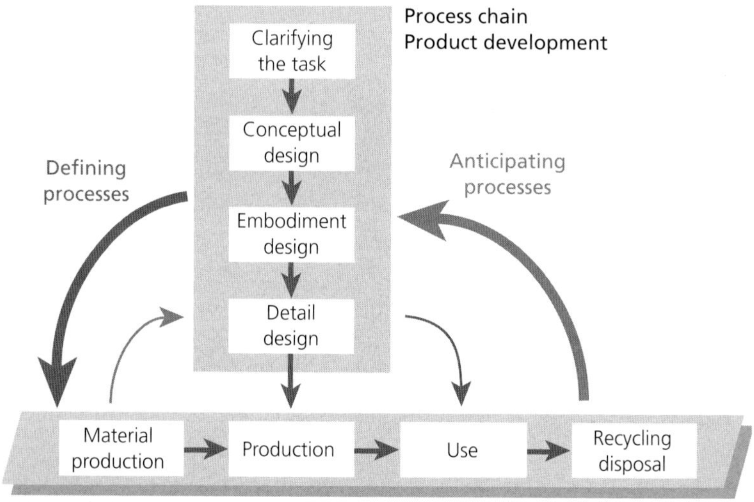

23.5 Holistic product and process development

Solid lubrication of ball bearings

The roller bearing technology group is working on a lifetime theory for solid lubricated roller bearings. Solid lubricants are used instead of oil or grease under extreme conditions, such as very high or very low temperatures, vacuum or radiation. The lubrication is achieved with thin layers of $MoS2$, graphite or soft metals, such as lead or silver, sputtered on the substrate.

The lifetime of a solid lubricant is limited by the wear of the lubricant coating, the amount of wear depending on the converted frictional energy within the tribological contact between ball and ring. A frictional energy model has been developed to calculate the extent of continuous erosion in the contact area.

A life-test for a solid lubricated roller bearing can take a few weeks. Hence, the roller bearing technology group developed a modified 'four-ball test rig' for rapid testing. Thus, one can analyse the quality of a solid lubricant coating under similar working conditions, e.g. in a roller bearing, but in considerably fewer hours.

It has been proven that the roller wear tests on the four-ball test rig are highly relevant for estimating the lifetime of solid lubricants in real roller bearings. The remarkable reduction in testing-time is a key factor in developing new and better solid lubricants, such as multi-layer or gradient-layer coats.

23.6 The four-ball test rig

Further reading

Birkhofer H, Berger B, Walter S (2002) Modularisation of knowledge – a new approach in the field of product innovation. Design 2002, Dubrovnik, Croatia

Birkhofer H, Kloberdanz H, Sauer T, Berger B (2002) Why methods don't work and how to get them to work. EDIProD'02, Zielona Gora, Poland

Ernzer M, Oberender C, Birkhofer H (2002) Methods to support EcoDesign in the product development process. CI'02, Vienna, Austria

Jänsch J, Sauer T, Walter S, Birkhofer H (2003) User-suitable transfer of design methods. ICED'03, Stockholm, Sweden

Li X, Heinz T, Schul C, Birkhofer H (2002) Lifetime calculation and analysis of solid lubricated ball bearings. 13th International Colloquium on Tribology, Esslingen, Germany

For further information, see http://www.tu-darmstadt.de/Welcome.en.html

Chapter 24

School of Industrial Design Engineering, Delft University of Technology

Henri HCM Christiaans

24.1 Delft School of Industrial Design Engineering

24.2 A low threshold virtual reality setup used for exploring and discussing sketchy design concepts

The School of Industrial Design Engineering at Delft University of Technology stresses its position as a 'designing' school with a strong research approach, directed by our vision that product development has to be considered an integrated, global, creative activity, driven by new information and communication technologies. We investigate those facets of product development that can allow designers and industry to fully exploit today's constantly changing markets and technologies. The research aims to develop methods to improve integration of user aspects in the product design process.

The school currently has a team of about 50 research fellows with expertise in disciplines such as product design and aesthetics, engineering and construction, product and systems ergonomics, innovation management, and marketing. There are another 35 PhD students who participate in research at the school.

The Delft Design Institute (DDI) is part of the school's coordinating and mediating research activities for industry. The school's research portfolio reflects the interdependence among these fields, promoting disciplinary strength whilst embracing interdisciplinary activities. It combines research-driven curiosity and societal and economic usefulness. Both are manifest in the two themes of the programme.

Design theory and support

The complexity of the design process is well recognised. Managing the design process to predict at an early stage the implications of a product concept and its usefulness is a critical task for a designer. In addition, designers need to integrate knowledge from different backgrounds into one single product concept. This research theme focuses on the exploration, development, validation and implementation of innovative methodologies, techniques and tools to support product designers in the creative and integrated development of new products.

Methodology, tools and techniques

The research of this program focuses on the theoretical fundamentals of the design process, as well as the methods, tools, and techniques that will be used by designers and design teams during the creative development of new product concepts. A proper understanding of the design process is fundamental to the effective development of methods, tools, and design support systems, as well as for design education. The broad range of traditional tools needs to be expanded with the new possibilities offered by digital technologies, such as (1) dynamic and interactive tools for expression, communication, experience and inspiration and (2) a collaborative virtual prototyping environment.

Lifecycle engineering: design for sustainability

For 10 years, ecological and energy-related aspects of products have been high on the societal agenda. Within the framework of the worldwide adopted policy concept of sustainable development, new knowledge is required on the ecological, economic and social–cultural impact of products over their lifecycle. Together with a radical reduction in products' environmental impact, and in the use of resources and space, the new product concepts should be competitive from a business perspective and fit into future social–cultural systems.

New materials (for example, polymers and material from fossil origin) are applied in designing strong but lighter products, resulting in less material and energy consumption. Such products are increasingly preferred in the transport and packaging industries. Renewable materials, aimed at replacing those of fossil origin, are also expected to contribute towards sustainability by reducing dependency on finite resources.

Supporting innovation in products and product-systems of superior quality with respect to sustainable development values (eco-design) is the central question of the subprogram. In addition, the program focuses on the business aspects of eco-design, based on current reality and expected developments in the near future.

Product functionality, aesthetics and experience

The ability to predict the implications of product concepts regarding

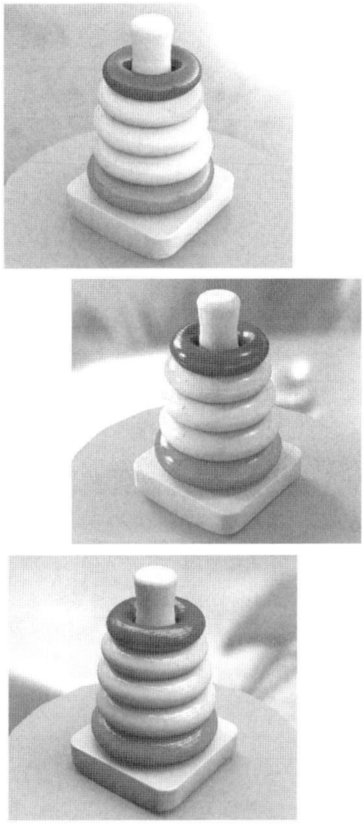

24.3 Using contextual influences in computer visualisations

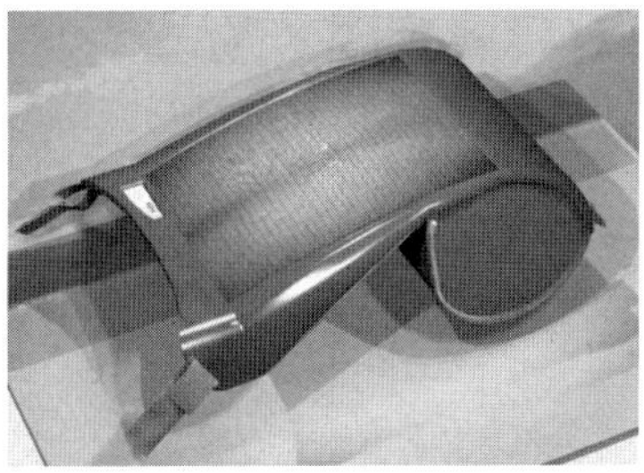

24.4 Solar energy on backpacks

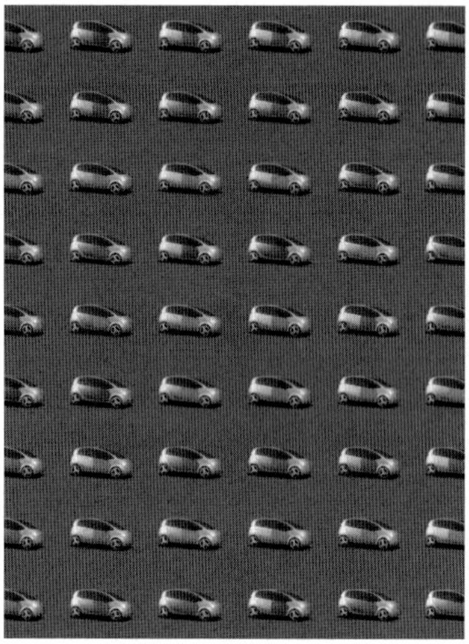

24.5 Designing emotions

usefulness, acceptance and experience by the user is a critical task for the designer. Designers and manufacturers are faced with deciding what kind of functionality to include in products and what experiences are to be elicited by products.

Although the functionality a product offers has always been, and will remain, an essential precondition for product satisfaction and market success, various developments point an increasing importance of product experience as a major driving force of product acquisition and use. In this subprogram, the focus is on the active consumer, with expectations and preferences, interacting with a product through all senses and within a particular context, and thereby undergoing a dynamic and multi-layered experience.

Design of future products

This research theme is directed at the development of future products using state-of-the-art and new technologies. The aim of this theme is to evaluate these technologies by developing new innovative products and solutions for existing problems and wishes of users. Special attention is paid to the inclusion of specific user groups with their own problems, wishes and preferences, taking into account the growing diversity reflected in current demographic changes. *Research through design*, preferably in a natural context, is a basic attitude of this theme.

24.6 Mitka

Product intelligence

Present-day technological advances foreshadow a world where a large variety of consumer and (semi-) professional products will contain powerful, intelligent hardware, inter-device communication via intelligent telecommunication networks and advanced user-input and display technologies. These upcoming products will be capable of processing information relating to the user's desired tasks and the environment in which the product is being used.

It is envisaged that users will interact with such products in an intuitive way. Therefore it is important to make the user interface as transparent or unobtrusive as possible, to enable the user to engage in the task or content at hand, rather than be bothered by how to control or interact with the product. In order to get an optimal user-product fit, multimodal channels of user-product communication, such as graphics combined with gesture input, speech-driven dialogues and auditory feedback are envisioned. The research also deals with the general problem of how marketing and technology innovators can communicate and evaluate new product functionality.

Design for all

This subprogram is focused on understanding the dynamic aspects of product use. The approach has a focus on the way the musculoskeletal system and the skin behave in human–product interaction. Special attention is given to the design of healthy environments. Problems for the user, such as pressure sores, RSI, low backpain, discomfort and lipoatrophia semicircularis, are being studied. Understanding the underlying medical aspects of these complaints and the anatomical deviations of the different users form the basis for bio-mechanical modelling. Verification experiments in real-life situations help to improve the dynamic model of the human–product interaction and lead to design guidelines for new products that prevent musculoskeletal complaints.

24.7 Design for all

Further reading

Christiaans HHCM (2002) Creativity as a design criterion. Creativity Research, 14(1): 41–54

De Jong AM, Vink P (2002) Participatory ergonomics applied in installation work. Applied Ergonomics, 33(5): 439–448

Horvath I, Vergeest JSM, Medland AJ (2002) Computer-based tools and methods of competitive engineering. Computer-Aided Design, 34: 869–870

Lloyd P, Busby J (2001) Softening up the facts: engineers in design meetings. Design Issues, 17(3): 67–82

Muller W (2001) Order and meaning in design. Lemma

Roozenburg NFM, Eekels J (1995) Product design: fundamentals and methods. John Wiley

Schifferstein HNJ, Blok ST (2002) The signal function of thematically (in) congruent ambient scents in a retail environment. Chemical Senses, 27: 539–549

For further information, see
http://www.io.tudelft.nl/index.php

Chapter 25

The Design Group, Technical University of Denmark

Mogens Myrup Andreasen

25.1 The Department of Mechanical Engineering on DTU's campus

The Design Group at the Technical University of Denmark (DTU) consists of researchers in the field of mechanical engineering. The group is currently part of the Section of Engineering Design and Product Development in the Department of Mechanical Engineering and part of the cross-departmental research and teaching initiative, Design and Innovation.

The group has more than 30 years of experience in creating a comprehensive school of engineering design and supplying Danish industry with approaches, models and terminology for innovative, competitive and efficient design. Its research is based on two fundamental theories:

- Theory of technical systems, originally brought to the group by Vladimir Hubka, and further developed into a productive theory of structure and behaviour of technical products, illustrated in the applications below.
- Theory of design processes, aimed at understanding individual, team, organisational and management aspects of designing, the role of externalisation, communication and coordination, and the role of digital product models for complex design.

The latter part of the research is supported by the group's sister organisation, the Institute for Product Development, a foundation for consultative support for Danish industry, with 15 staff members. The Design Group benefits from its small size and international reputation. It can apply an approach driven by curiosity, experimentation and opportunism, which has led to innovative research results and new initiatives over the years.

The research mainly focuses on mechatronic products and on general aspects of designing such as:

- Methods application in practice.
- Mapping the design activity for complex understanding – synthesis, learning, management, knowledge management, documentation, *etc.*

- The sharing of design models and knowledge structures between designer and computer.
- Design coordination for dynamic, effective and efficient designing.
- Re- and pre-use of knowledge, experienced activities and artefact designs.
- Understanding the role of education and training in the development of skills, knowledge, attitudes and personal working methods.
- Performing 'design of design', *i.e.* designing or innovating the design organisation, resources, strategy and methodologies in accordance with the design opportunity.
- The role of socio-technical understanding in the designer's awareness, understanding of need, domestication of solutions and product life responsibility.

Design for X

Design for X (DFX) is a specific methodology, telling us how we are able to fit design to and create competitive edge in the area X at the design stage, where X can be an activity in the product's lifecycle or a strength parameter in the creation of business: cost, quality, risk, flexibility, environmental effects, etc.

Research has been completed on design for quality, reliability and environment, and publications have been produced on design for assembly and manufacture. The chain-phenomenon and integration have been explored by the so-called theory of dispositions.

Many DFX-tools have been recognised as powerful and quite easy to apply, but surprisingly they have poor penetration in industry. Today, we therefore need to reach a better understanding of the relationships of stakeholders, motivations and traditions to achieve proper implementation and use of the methods.

As a group we focus upon a total lifecycle approach, bringing the phases of product life and their visualisation into the conceptual design activity, and focusing upon the benefits to be provided by services.

25.2 First semester project students working on models for sales tents in a market place; a project combining social and technical aspects

Eco-design

This topic belongs to the DFX-field, but there are additional aspects to be taken into account: vision, strategy, management and methodology integrated into the product development activity. Our research focuses on both the analysis and understanding of eco-design synthesis and the actual integration of tools and techniques into product development projects. A challenge here is to obtain sustainable ecological effects, which calls for innovative and explorative approaches.

Product modelling

On the basis of the so-called domain theory, we have developed a product modelling philosophy for structuring the constituent characteristics of a product's accompanying activities such as realisation, use, and disposal, its functional aspects and its parts structure. This product model, the chromosome model, may be applied for several design purposes, e.g. the configuration of products from a product family, reuse and pre-use of knowledge, experiences from activities and designs, knowledge management related to design, etc.

One could say that this modelling approach proposes a way of describing the product, which both supports the synthesis, structural definition and design reasoning performed by the human designer, and allows the computer to capture the design and performance of logical and computational operations related to the product design. The formal design language used here has proved useful for modularisation and for design reasoning. Because the language has a common origin with DFX-methods, it also allows for integrated use of these methods.

25.3 Students on the International BEST Summer School

We see that designing on and overall basis with a product model as an approach can radically change the conditions for product development due to re-use, configuration possibilities, resource allocation, redefined lead time, capturing of product life data, etc. Our challenge is to formalise this way of designing.

Design methodologies

Design methods are mainly seen as procedural instructions and modelling techniques described in text and/or built into software. From the method designer's perspective it seems that this text is sufficient for proper use of the method. It is our experience, however, that several factors and elements go into the proper execution of a method. Mainly the designer's mindset, i.e. his/her proper understanding of the method, its basic mechanisms, validity, proper application, etc.

In our research we try to articulate and test the transfer of such mindsets for the enhancement of given methodologies, for identifying fundamental methods, and for the proper teaching of such methods. Our research has resulted in new insights concerning mindsets for evaluation and decision-making, for conceptualisation and for basic concepts behind DFX-methods.

Understanding designing

The design activity is traditionally modelled as structured modules of activities, showing the causal and experimental time sequence of these activities. It is realised among researchers that such models only vaguely mirror proper instructions or sound explanations for designing.

How can we do this better? It is well known that design is a structured, plannable chain of activities, but at the same time a pattern of areas of feedback. Designing is learning and a process of knowledge transformation and information processing. Designing is awareness, understanding needs and documentation of solutions, in a socio-technical pattern of actors.

In our research we try to find modelling means, mindsets and 'world maps' for the communication of this complex understanding of designing. The purpose is to give designers and managers a better understanding, allowing design of design and management of the process.

Further reading

Andreasen MM (2003) Design method's usability by a mindset approach. In: Human behaviour in design. Springer

Andreasen MM, Wognum N, McAloone TC (2002) Design typology and design organisation. Design 2002, Dubrovnik, Croatia

Hansen CT, Andreasen MM (2002) Two approaches to synthesis based on the domain theory. In: Engineering design synthesis. Springer

Hansen CT, Andreasen MM (2003) A proposal for an enhanced design concept understanding. ICED'03, Stockholm, Sweden

Hansen CT, Andreasen MM (2004) A mapping of design decision-making. Design 2004, Dubrovnik, Croatia

McAloone TC (2000) Industrial application of environmentally conscious design. Professional Engineering Publishing

Mortensen NH (1999) Design modelling in a designer's workbench – contribution to a design language. PhD thesis, Technical University of Denmark

Olesen J (1992) Concurrent development in manufacturing – based on dispositional mechanisms. PhD thesis, Technical University of Denmark

Olesen J, Wenzel H, Hein L, Andreasen MM (1996) Design for environment. Danish EPA

For further information, see
http://www.dtu.dk/index_e.htm

Chapter 26
The Systems Realization Laboratory, Georgia Institute of Technology

Janet K Allen, Bert Bras, Farrokh Mistree, Christiaan Paredis and David Rosen

26.1 SRL logo. We view design and engineering as the continuous interaction between the human and the computer. We represent this synergy with a wrench which requires both the human and computer in order to operate. When the wrench needs adjustment, the computer moves away from the human, signifying human-centred design. The wrench sits atop the world, signifying our focus on the global market place.

Strategic design is a comprehensive approach for safeguarding the economic viability of a company. It necessitates the design of products and processes that efficiently and effectively accommodate changing markets and technological innovations. Accordingly, our vision involves identifying, developing and understanding principles, tools, and technologies to establish and preserve strategic, sustainable development – for products, processes, industries and careers. Our core activities include:

- conceiving and verifying foundational theories for the realisation of engineered products, processes, systems and services;
- promoting scholarship in the form of discovery, analysis, synthesis and education;
- developing technologies that enable companies to conceive and produce customised products that service various market segments;
- promoting technology transfer;
- fostering growth of intellectual capital among all stakeholders, including industrial partners, faculty and students.

Research thrusts

We have worked on many different projects, but our current research thrusts are related to strategic design in the following engineering domains:

Information modelling and simulations for collaborative distributed design

To meet demands of a global marketplace, product innovation and time-to-market are crucial. Thus, we focus on methods and the theoretical underpinnings of collaborative, distributed design. Consistent capture and storage of information and knowledge about the design and manufacturing process can save significant resources by enabling reuse and sharing of information and data and possibly by automated computer processing. Within the Systems Realization Laboratory, we are developing information models for both products and processes. For modelling design concepts, our focus currently is on behavioural modelling: How should simulation models of components be represented so that they can be used for automated generation of system-level models? We are developing semantically rich information models to support and automate module composition operations. We are also developing information models for capturing the design process. We consider the design process to be a sequence of information transformations; this makes it convenient to focus both on the transformations and the interfaces between them.

26.2 The Systems Realization Laboratory is housed in Georgia Tech's Manufacturing Research Center

Design of next-generation product realisation technologies of multi-functional materials

Our vision of the future includes a world where layer-based, additive fabrication technologies (for example, rapid prototyping) are recognised as production manufacturing technologies. We want to leverage the unique capabilities of these additive fabrication technologies to produce unique geometries and material structures. Our current focus includes understanding and improving stereolithography processes, design methods for multi-material and multi-functional devices, and methods for rapid manufacturing.

Not only do we see the potential for designing and manufacturing new material structures, we also are developing the capabilities to design the materials themselves.

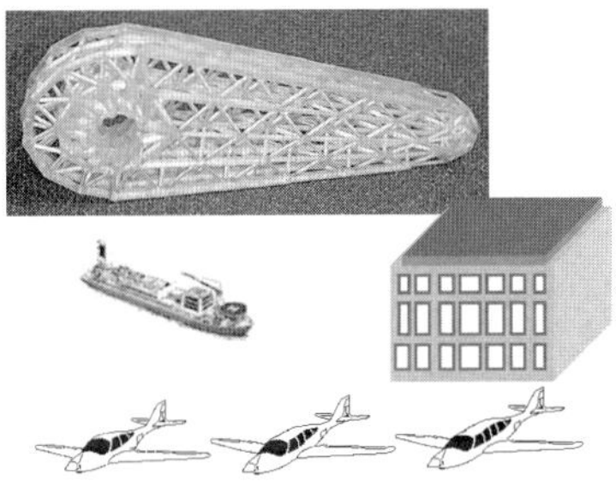

26.3 Applications from robot arms and heat exchangers for microprocessors to ships and aircraft

Design for sustainable development

Our work is anchored in the notion of sustainable development, *i.e.* development that does not compromise the needs of future generations. In this context, we pursue the design and realisation of (sustainable) technologies that not only increase industrial competitiveness, but also reduce the impact of our actions on the environment and enhance quality of life. In the Systems Realization Laboratory, we are researching new ways to assess design performance in terms of economic, environmental and social impact. We work with visionary companies to see how principles of sustainability

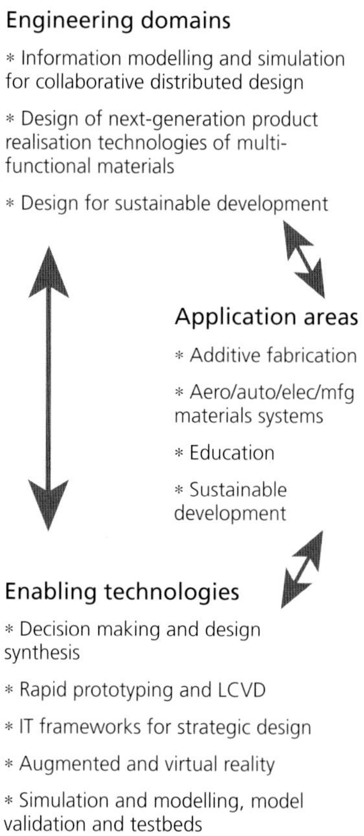

26.4 Strategic design in the Systems Realization Laboratory

can be cascaded downward from upper management to design engineers, and incorporated in company practices and tools. For strategic environmental and social impact assessments, we envision an integration of industrial models with ecological and social models and foresee an increased teaming with researchers in ecology and regional planning.

Applications

Our application areas are diverse and include: additive fabrication (stereolithography, laser chemical vapour deposition, LCVD); aircraft design (general aviation aircraft, high speed civil transport and the Boeing 727); design education; maintenance management (gas turbines); manufacturing and re-manufacturing; materials design; mechanical systems (aircraft and automobile engines); product families (consumer goods, automobiles); spacecraft (orbits, trajectories); ships (frigates and container ships); structural systems (ships and truss towers); sustainable development; and thermal systems (thermal-powered spacecraft, solar irrigation systems, air chillers).

The principal technologies we have developed which enable applications are anchored in: augmented and virtual reality; decision-based design and design synthesis; IT frameworks for distributed, collaborative design; and simulation and modelling, model validation and testbeds.

Foundation for success

The Systems Realization Laboratory was founded in 1992 by Janet Allen, Bert Bras, Farrokh Mistree and David Rosen; Farrokh Mistree was the Founding Director. Chris Paredis joined us 10 years later. Over the years we have had a series of remarkable students – people who choose to think outside the proverbial box – and who want to make a difference. We have graduated PhDs and more than 50 MS students; a third of the PhD students have pursued careers in academia. Together with our students we have published over 300 papers, half of them refereed.

The Systems Realization Laboratory is part of the George W. Woodruff School of Mechanical Engineering and is housed in Georgia Tech's Manufacturing Research Center. As part of an educational institution, our mission is to help everyone – students, faculty, staff and industrial colleagues – to rise to their full potentials.

In the belief that the combination of theory and application is more effective than either alone, we have sought extended partnerships with clients such as Ford, GM, Kvaerner, Lockheed-Martin, Carrier, Kodak, B.F. Goodrich,

Interface and Black and Decker. We are committed to technology transfer; we work with Georgia Tech's Rapid Prototyping and Manufacturing Institute and also with the Georgia Research Alliance to strengthen industry within the State of Georgia. We also build capability and scholarship with funding from national and international agencies and the military; in the last dozen years, we have received about $15,000,000 for research.

We seek collaborators who have a dream and a passion to change the world – those who wish to be the thought leaders of tomorrow and have a passion to make a significant difference.

26.5 Developing technologies anchored in virtual and augmented realities in the Systems Realization Laboratory

Further reading

Chen W, Allen JK, Mistree F (1997) The robust concept exploration method for enhancing concurrent systems design. Concurrent Engineering: Research and Applications, 5(3): 203–217

Emblemsvåg J, Bras BA (2000) Activity-based cost and environmental management. A different approach to the ISO 14000 compliance. Kluwer Academic

Mistree F, Smith WF, Bras B, Allen JK, Muster D (1990) Decision-based design: a contemporary paradigm for ship design in transactions. Society of Naval Architects and Marine Engineers, 98: 565–597

Mistree F, Bras BA, Smith WF, Allen JK (1996) Modeling design processes: a conceptual, decision-based approach. Engineering Design and Automation, 1(4): 209–221

Newcomb PJ, Bras BA, Rosen DW (1998) Implications of modularity on product design for the life-cycle. ASME Journal of Mechanical Design, 120: 483–490

Siddique Z, Rosen DW (2001) On discrete design spaces for the configuration design of product families. AI EDAM, 15: 1–18

For further information, see
http://www.srl.gatech.edu

Chapter 27
Engineering Design Research, University of Grenoble

Jean-François Boujut and Jean-Claude Léon

27.1 A view of the Grenoble campus

Grenoble has a long history in design research. However, engineering design research really started developing in the 1990s and is now articulated around two academic teams: the Integrated Design Centre (3S laboratory: soils, solids, structures) and the GILCO laboratory. The two teams work on various topics related to engineering design with strong interdisciplinary connections (*i.e.* industrial sociology, cognitive ergonomics, computer sciences, applied mathematics, etc.).

The main topics of the 3S laboratory are the product process integration and integration of downstream activities within the design process (manufacturing, assembly, recycling, etc.), CAD and geometric modelling, design cooperation and collaborative engineering involving new IT technologies.

The activity of the GILCO laboratory related to engineering design is focused on the various aspects of design information management (*i.e.* change management, product lifecycle management (PLM), modularity, informal information, etc.) and knowledge management.

All these developments aim at contributing to a better understanding of design and an improvement of the performance of design considered as a collective activity and a complex process. The research is performed by more than 20 permanent researchers and 30 PhD students.

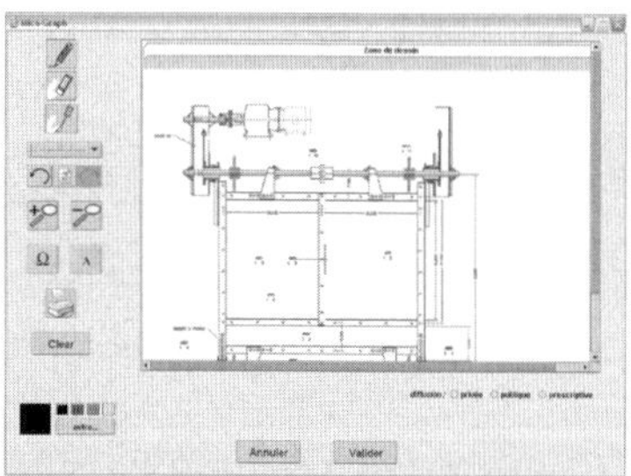

27.2 MICAGRAPH: a sketching and annotation support

Information management in design

GILCO is a laboratory in the industrial engineering school of the INP G (Grenoble Technical University). The overall aim of the laboratory is to improve industrial efficiency and, therefore, the research articulates two main themes: physical flows (logistics, supply chain management, etc.) and information flows (including product information and information management in design).

Knowledge management

Knowledge management is becoming a key issue in design. In fact, design is a highly cognitive activity that requires specific information supports. Current research is mainly centred on the research and development activities and upstream design phases. We develop information models and mediating supports in accordance with the concept of the learning organisation, enabling the management of knowledge within such organisations.

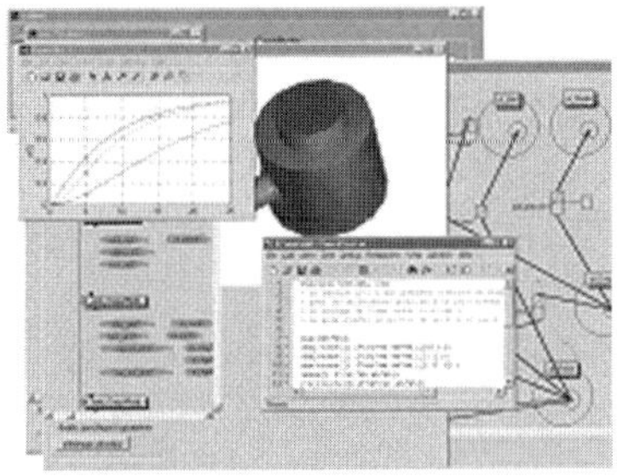

27.3 Product model to incorporate several views of the product

Information modelling in the early design phases

Information modelling (MOKA, UML, etc.) and workflow configuration. Having access to a unified version of the product (digital mock-up) at the same time everywhere in the company remains a problem. Our research aims at providing specific information models suitable for supporting various specific design phases within the frame of product data management (PDM) systems or more recently PLM systems. At the level of design interactions, however, the need to support the collective activity is not addressed by current PDM and PLM systems. Informal information structuring remains an important research area. We are developing sketching and annotation systems for structuring informal design information exchanges.

Change management and design modifications

The growing complexity of the industrial products and production architecture, and the growing diversity of the possible product configurations provide an opportunity for the development of specific tools for evaluating the impact of possible changes assisting decision making.

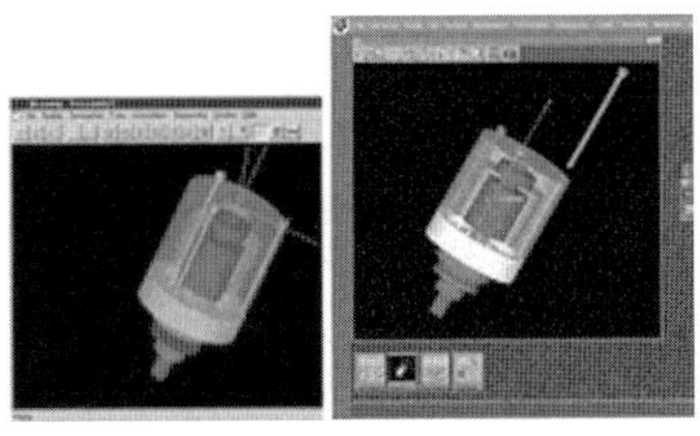

27.4 Product–process integration: the example of the assembly process

Modular design and design for the supply chain

Following the same idea, the growing complexity of the production systems and the necessity of managing the different delocalisation options highlight the concept of modularity. Our research aims at understanding the relations between product modularity and supply-chain configurations in highly changing environments.

Integrated design

The Integrated Design Centre of the 3S laboratory is attached to the INP G, the University Joseph Fourier and the CNRS. For more than 10 years, the activity of this team has focused on the improvement of the design and development of mechanical products/systems through new models, methods and computer tools dedicated to integrated design within the context of concurrent engineering.

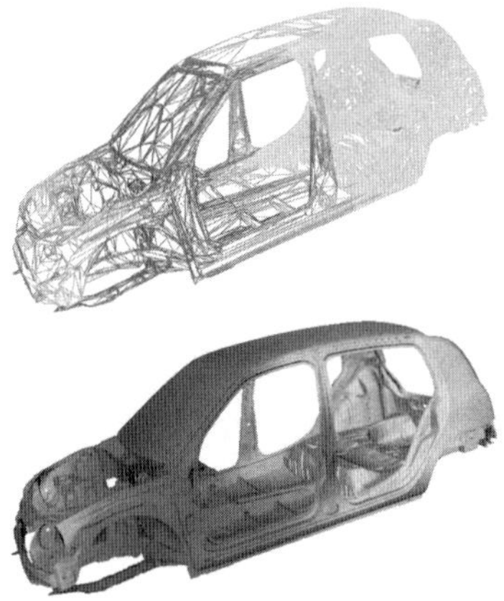

27.5 Digital mock-ups to work out models adapted to various stages of the design process (courtesy Renault)

Principles

As seen by the research team, design is a collaborative and distributed activity that covers the whole product lifecycle and incorporates analysis as well as synthesis activities. Such a context involves 'skilled actors' having points of view, knowledge and tools, interactions with an organisation, shared knowledge, decision-making processes, as well as technologies for cooperation activities, all for the multi-representation and the shape generation of a product. Thus, a multi-disciplinary research activity is conducted among mechanical engineering, human and social sciences (industrial sociology, cognitive ergonomics, and didactics), computer science and applied mathematics.

Methodologies for integrated design, innovation

This research topic is based on observations of the real design process. These include on-site studies in companies and experiments about the design activity, to set up methodologies for analysing the design process and lead to methods for incorporating innovative solutions in a design process. Software tools and models for defining the product model and an integrated design environment for multi-actors have also been set up.

Product–process integration in design

Formalisation of knowledge and methods related to process and production skills in mechanical design form the core of this field: manufacturing processes (forging, assembly, machining, process planning, aluminium extrusion, composite materials, etc.), processes for the end of life (disassembly, recycling, reuse), and tolerancing. Models for the dynamic behaviour of a high-speed machining system to improve the design process, and to set up new technologies for the drilling process are examples of detailed contributions, whereas concepts of product–process co-development of product lifecycle address a global level of the design process. Development of software demonstrators for product–process integration helps validate the proposed approaches.

Digital mock-ups for integrated design

Methods for performing shape changes on digital product models, like geometry simplification, adaptation for design and downstream processes, design data adaptation and idealisation for mechanical simulations, freeform shape parameterisation and deformation, shape optimisation, surface mesh generation and adaptation, are examples of research to produce such mock-ups. Methods for knowledge, know-how and service management around digital mock-ups are also addressed through the capitalisation and reuse of models concerning mechanical analyses of products. Thus, new concepts for the use of digital mock-ups in design can be evaluated through the development of software demonstrators and libraries for digital mock-ups of products.

Collaborative engineering and co-design

Characterised by methods and experiments for synchronous and asynchronous tasks among distant or collocated designers, concepts for providing common work environments between designers having different skills are proposed. Experiments for evaluating software tools in a collaborative context help validate these concepts.

Further reading

Blanco E, Gardoni M (2001) Supporting graphical non structured information in integrated design team. ICED'01, Glasgow, UK

Boujut J-F, Legardeur J (in press) ID^2: a new tool to foster innovation during the early phases of design projects. CERA Journal, 11(3)

Fine L, Rémondini L, Léon J-C (2002) A control criterion dedicated to detail removal for FEA geometry adaptation. In: Integrated design and manufacturing in mechanical engineering 2000. Kluwer

Noel F, Brissaud D, Tichkiewitch S (2003) Integrative design environment to improve collaboration between various experts. Annals of the CIRP, 52(1): 109–112

Pourroy F, Troussier N, Tollenaere M (2002) A method and a support for a better integration of mechanical simulation in the design process. IDMME'02, Clermont-Ferrand, France

For further information, see
http://www.inpg.fr/

Chapter 28

Institute of Machine Design and Automotive Engineering, University of Karlsruhe (TH)

Albert Albers

28.1 Institute of Machine Design and Automotive Engineering

The Institute of Machine Design and Automotive Engineering (mkl) at the University of Karlsruhe (TH) is a key centre for product development research in Germany. Its research is distributed over a very large span of the product development process, following the chain from market to product. The mkl was renamed as the Institute of Product Development – University of Karlsruhe (TH) in January 2004. However, this chapter will refer to the previous name.

The mkl team currently consists of 55 people, 31 of whom are scientific employees working on different high-quality scientific projects. The team is split into five closely linked research groups:

- Design Methodology and Design Management;
- Computer-Aided Engineering (CAE);
- Power Train Engineering;
- Mechatronics;
- Automotive Engineering.

The mkl team are involved in three of the 10 Collaborative Research Centers that are managed by the University of Karlsruhe (TH). Many important international companies trust in their efficient research and product development. The following describes their diverse research interests.

Contact and channel model

A central area of basic research at mkl is the Contact and Channel Model (C&CM). It describes the correlation between the design and functionality of technical systems. C&CM is the basis for all design research and education at mkl and has helped in solving many complex problems.

SPALTEN – an advanced problem-solving method

Practical product development is characterised by numerous problems for which there are often no solutions at the time of their discovery. The problem-solving methodology SPALTEN is a comprehensive one, which aims at shortening the problem-solving process for middle-sized and more complex problems by means of a systematic procedure, increasing the variety and the security of solutions.

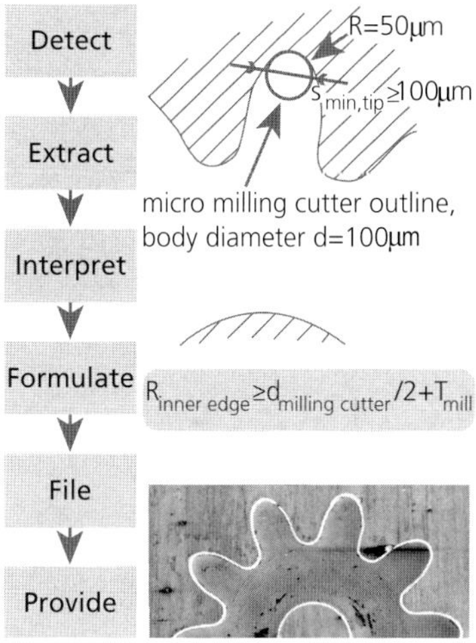

28.2 Detection of micro design rules

Design rules

In both micro-technology and macro-technology the early phases of product design are of considerable importance. Micro-technology design, however, is more technology driven than the conventional design process.

Design rules have been established to provide detailed instructions for micro-compatible design. By means of such rules technological boundary conditions and restrictions are derived and interpreted as relevant to design. They are made available via a knowledge-based design environment. In this sense, design rules can be understood as a methodical aid in order to make multidisciplinary knowledge available for one specific discipline.

Knowledge management

Design rules would be a great help to the designer if they could be applied automatically. The aim of this research work is to find a link between such design rules and a commercial CAD system that permits conventional design coupled with the automated application of the stored rules. Rule infringements are presented to the user in a dialogue and possibly corrected.

Multi-technology features

The 'multi-technology feature' research is concerned with the methodical support for a comprehensive cross-discipline development by employment of networked information units. This kind of approach is required for multi-technology products like humanoid robots, which exceed mere mechatronics since they include aspects of multi-modal interaction.

Expertise from different disciplines is collected and transformed to a methodologically standardised language. The extracted information units are made available in terms of multi-technology features. Using a special filter, the information units can be accessed assuming a particular expert point of view. As a result, design inconsistencies can be recognised at an early stage and a comprehensive cross-discipline development can effectively be supported.

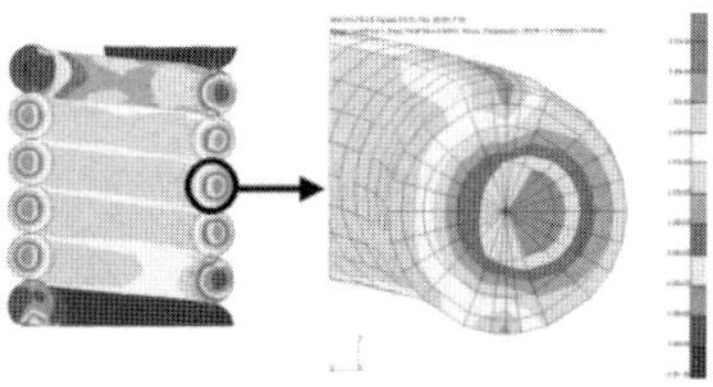

28.3 Optimisation of springs

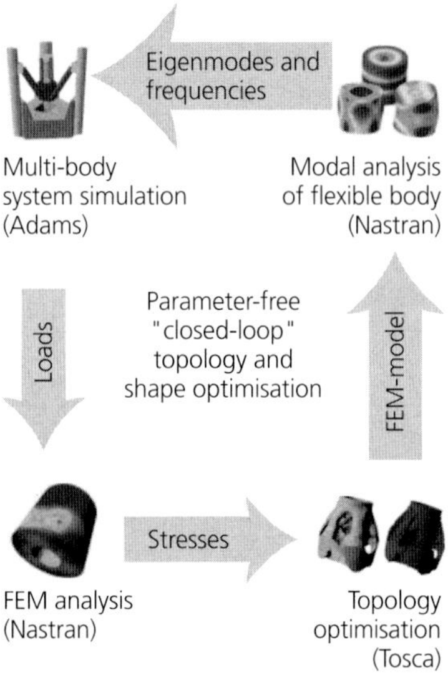

28.4 Parameter-free closed-loop topology and shape optimisation

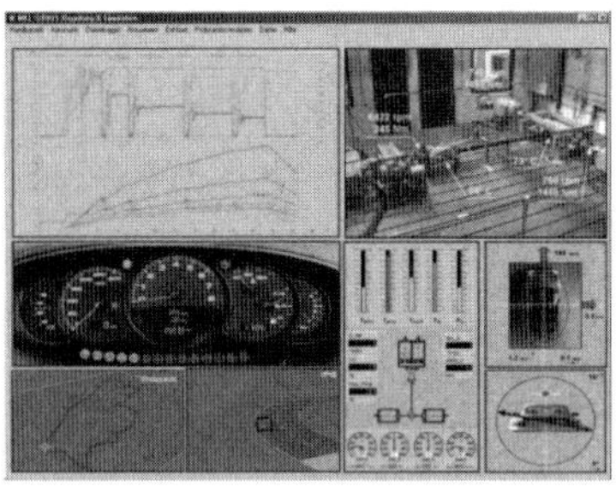

28.5 Integration of simulation and testing

Computer-aided engineering methods

The institute uses computational methods to simulate the mechanical behaviour of components and systems during different stages of the product development process. The finite element method (FEM) is used for the analysis of component stresses or deformation, as shown in Figure 28.3. Based on these results, the institute uses and develops software to optimise a component's shape for improved performance. This can be done as soon as the component's initial shape is determined.

Multi-body system (MBS) simulations may be used to simulate dynamic system behaviour. These give information about the complete dynamic behaviour of the system, such as positions, speeds and accelerations of any mechanical component.

Topology optimisation, based on the FEM, can provide basic design proposals and is often used to remove material in low-stressed regions of the structure in order to reduce the component's weight. This method is often used at a very early stage of the design process, when estimates of component stresses are known, and a first design proposal is needed.

Of particular interest is the combination of these two techniques, *e.g.* the FEM and the MBS simulation, for the optimisation of very dynamic systems. Topology optimisation is used to optimise a flexible component in the MBS, as shown in Figure 28.4, providing the opportunity to consider changes in system behaviour and their effects on the system components.

Chain of tribological testing

The optimisation of vehicle powertrains demands a holistic approach. This can be done by the investigation of friction systems at different levels of abstraction, where such results are linked to improve their significance. Numerical simulation can be used, as well as others tools like neuronal networks, in order to estimate a systems rating from measured data.

Mkl has set up a closed-loop testing facility and current research consists of adapting the links between the different levels of abstraction to maximise the prediction quality. This will provide tools and methods for time- and cost-saving development and optimisation of complex dynamic tribological systems such as vehicle powertrains.

Integration of simulation and testing

Future powertrains have to be developed with increasing use of virtual product models and supporting simulations. Owing to the recent progress in electric drive technology, the realistic replication of the torque characteristics and the non-uniform rotation of the internal combustion engine, as well as the slip of the tyres, is critically important. Recommended parameters for simulation are determined from field tests so that further investigations of the powertrain can be performed with real-time simulations on high-dynamic test benches. The future objective is to generate a 'virtual vehicle' which should enable the generation of robust performance data early in the development process.

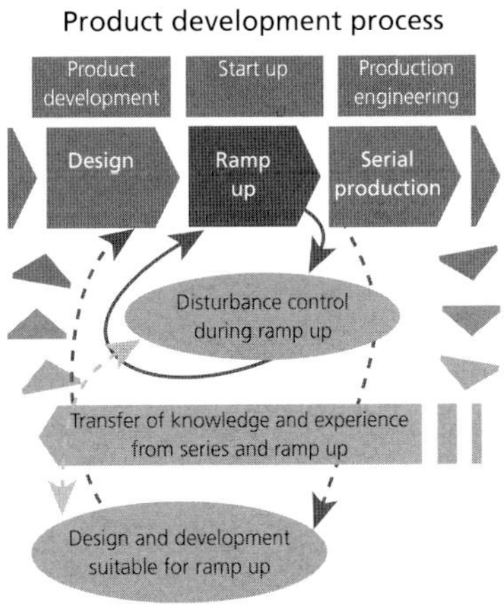

Small control circle for stabilisation of ramp up ——

Large control circle for sustainable long-term improvement of ramp up - - -

28.6 Fast ramp up

Ramp-up cybernetics

Ramp-up cybernetics is a method for sustainable support and knowledge integration from ramp-up into the development process with simultaneous provision of faster reaction mechanisms for dealing with emergency situations. The method assists consideration of specific ramp-up requirements, knowledge and restrictions during the product development process, improving process efficiency.

Further reading

Albers A, Marz J (2003) Knowledge-based design environment for primary shaped micro parts. HARMST'03, Monterey, CA, USA

Albers A, Matthiesen S (2002) Konstruktionsmethodisches Grundmodell zum Zusammenhang von Gestalt und Funktion technischer Systeme – Das Elementmodell. Konstruktio Zeitschrift für Produktentwicklung, 54(7/8): 53–59

Albers A, Albrecht M, Krüger A, Lux R (2001) New methodology for power train development in the automotive engineering – integration of simulation, design and testing. SAE Automotive and Transportation Technology Congress, Barcelona, Spain

Albers A, Burkardt N, Marz J (2003) Restrictions in the design of gear wheel components and drives for micro technology. Microsystem Technologies, 9(3): 192–196

Albers A, Matthiesen S, Ohmer M (2003) An innovative basic model in design methodology for analysis and synthesis of technical systems. ICED'03, Stockholm, Sweden

Ilzhöfer B, Müller O, Häußler P, Albers A (2001) Shape optimisation based on lifetime prediction measures. ICED'01, Glasgow, UK

For further information, see
http://www.mkl.uni-karlsruhe.de/

Chapter 29

Civil and Building Engineering, Loughborough University

Simon Austin

29.1 Civil and Building Engineering in Loughborough

The Construction Management Group has a well-established identity in the built environment with involvement in three university centres: the Centre for Innovative Construction Engineering, an engineering doctorate programme; the European Construction Institute, which supports client organisations; and the Innovative Manufacturing and Construction Research Centre, the largest IMRC in the UK, with a multi-disciplinary group of over 40 academic staff undertaking research to enhance the processes, products, and competitiveness of the UK's manufacturing and construction industries. Key objectives are to:

- undertake high-quality research that meets the needs of industry, its clients and customers, and enhances the knowledge base;
- forge close partnerships with industrial collaborators in addressing their core technical and business needs;
- disseminate research findings and engage in technology transfer.

Research can be grouped into four themes, which are related to the design and construction process.

- Advanced information and communication technologies: multi-media communications, artificial intelligence (AI), data exchange, product and process modelling, virtual and augmented reality and visualisation.
- Improved construction processes: supply-chain management, benchmarking, partnering, work process changes, collaborative working, briefing and strategic risk and design management.
- Innovative construction technologies: projects related to standardisation and pre-assembly, construction interfaces, automation and cladding.
- Human factors: health and safety, procurement and contractual relationships, learning organisations, simulation, knowledge management, project and performance management.

29.2 Change management at Glasgow Royal Infirmary

Process improvement

The importance of process in the construction industry has been at the forefront of government initiatives since the Latham and Egan reports of 1994 and 1998. Prior to this time the industry saw itself being fundamentally different from other engineering sectors involving mass production, in producing unique products without the advantages of prototyping. An understanding has emerged that the industry's processes, in both design and production, are highly repeatable, albeit that there is considerable scope for improvements in efficiency and effectiveness.

Our strategy revolves around the identification, modelling, integration, and improvement of construction business processes, which include: the commercial activities of construction organisations; their technical and managerial activities; and tools, techniques and cultural issues.

The research is undertaken with support from EPSRC, DTI, CIRIA, professional institutions, associations (including Collaboration for the Built Environment) and companies, such as AMEC, ARUP, BAA, and Sheppard Robson, who represent clients, designers, contractors and specialist suppliers. The main areas of our process research are: project-level phase/gate mapping, design process modelling, collaborative working, value management, design planning, control and change management, standardisation and pre-assembly, knowledge management and Web-based support.

29.3 ADePT was applied to University College London hospital project

Design planning and management

We developed an award-winning approach to the management of projects that involves planning the iterative flow of information, rather than simple activities and deliverables such as drawings. The Analytical Design Planning Technique (ADePT) offers opportunities to radically improve process and project management in a way similar to the improvements in sequential task scheduling brought about by the critical path method in the 1960s. The research is now being exploited through a spinout company, Adept Management, and its solution partner BIW Technologies who have developed PlanWeaver.

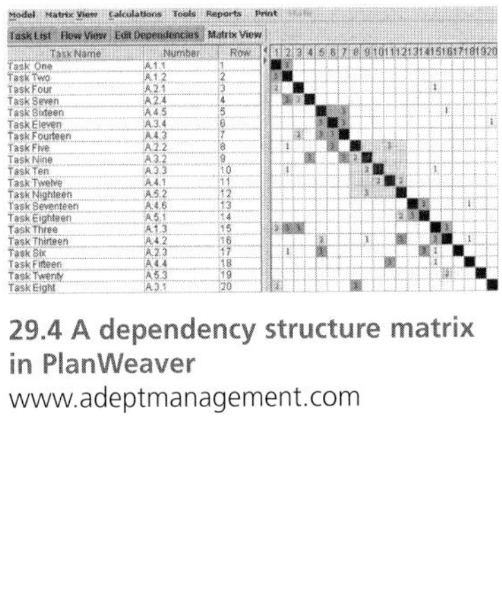

29.4 A dependency structure matrix in PlanWeaver
www.adeptmanagement.com

Standardisation and pre-assembly

IMMPREST aims to produce an interactive modelling tool that helps to evaluate the benefits of standardisation and pre-assembly. The model identifies factors to be considered in an assessment, the data required and where these reside in the supply chain.

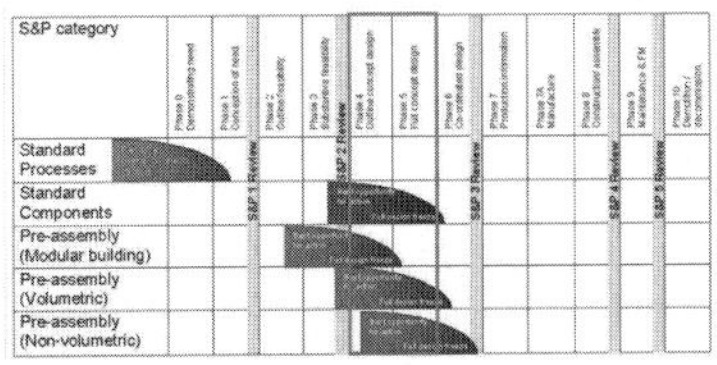

29.5 Standardisation and pre-assembly
www.immprest.com

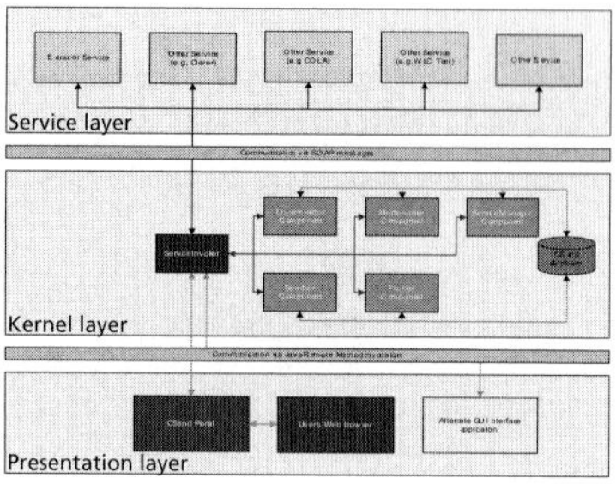

29.6 Knowledge management portal

29.7 ICD handbook
www.designchains.com

29.8 Telegenesis
www.telegenesis.org

Computer-aided collaborative conceptual design

Effective collaborative conceptual design in the AEC industry is essential for reducing lead times and improving design quality. We are investigating novel techniques to deal with the designers' needs to rapidly develop and assess ideas with computer-aided tools that facilitate collaborative distributed working, including visualisation, AI and knowledge management tools.

Knowledge management

Knowledge management is being established as one of the most important organisational assets in the construction industry. Techniques for the capture and reuse of knowledge are being investigated in a number of projects looking at information technologies and organisational aspects that influence how knowledge can be managed effectively.

Collaborative working

Integrated collaborative design (ICD) represents new thinking in supply chain management which places the design process and the management of design information at the centre of project management practice by creating suitable frameworks to ensure design information can be used effectively in both business and project-domain activities. It does this by focusing on the three principles of process management, supply-chain management and value framework and their application to the concept of the design chain. A handbook has been produced to help industry use the ICD approach and its 25 practices.

The Telegenesis project is identifying future scenarios for distributed design teams on complex products in aerospace and construction sectors and then make recommendations for innovation and improvement in the use of distributed design teams. This is partly being addressed by identifying the alignment of existing design processes and current design team characteristics for co-located and distributed working.

Value management

Value is a highly topical subject in the construction industry, with many industry bodies pressing for significant improvements in delivery to all key stakeholders. There is also a growing recognition that we must concentrate much more on what customers do with/in our facilities, rather than the products themselves. Our research aims to increase customer satisfaction through a better, shared understanding of appropriate value systems and standardised mechanisms and processes that map and measure the delivery of value within the design solution. Specific areas include a common value culture and language; communication of project values; their relation to design tasks; and monitoring effectiveness in value delivery.

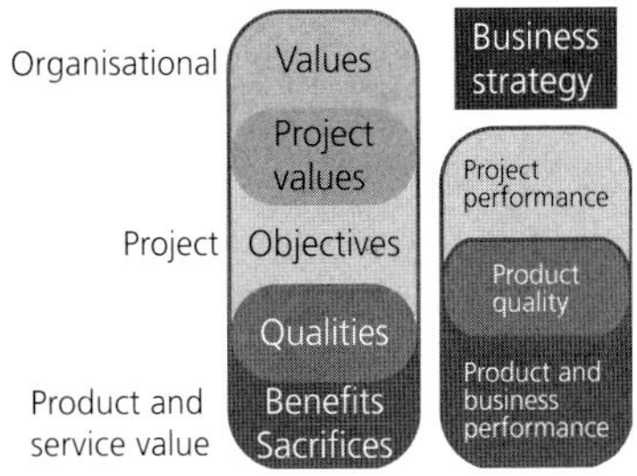

29.9 A framework of value
www.valueindesign.com

Project process

The generic design and construction process protocol project has been developed with Salford University to produce an industry-wide standard process map for the complete construction process, to carry out validation and testing, and to identify supporting IT systems. We have mapped sub-processes in eight zones, including development, project, resource, design, production facilities, and management. The process protocol has 10 phases, with a mixture of hard and soft gates, carried out in four stages. Using manufacturing principles, a framework of common definitions, documents and procedures has been developed to help integrate project participants working, together with a prototype process map creation tool.

29.10 The generic design and construction process protocol
www.processprotocol.com

Further reading

Austin SA, Baldwin AN, Li B, Waskett PR (2000) Application of the analytical design planning technique to construction project management. Project Management, 31(2): 48–59

Austin S, Baldwin A, Hammond J *et al.* (2001) Design chains – a handbook for integrated collaborative design. Thomas Telford

Austin S *et al.* (2001) Mapping the conceptual design activity of interdisciplinary teams. Design Studies, 22(3): 211–232

Bouchlaghem NM, Anumba CJ, Shang H *et al.* (2003) Concurrent engineering at conceptual design stage in the AEC industry. ISPE'03, Madeira, Portugal

Gibb AGF, Isack F (2003) Re-engineering through pre-assembly: client expectations and drivers. Building Research and Information, 31(2): 146–160

Kamara JM, Anumba CJ, Carrillo PM (2002) A CLEVER approach to selecting a knowledge management strategy. Project Management, 20(3): 205–211

Rezgui Y, Bouchlaghem D, Austin S (2003) An IT-based approach to managing the construction brief. IT in Architecture, Engineering and Construction, 1(1): 25–38

Thomson DS, Austin SA, Devine-Wright H *et al.* (2003) Managing value and quality in design. Building Research and Information, 31(5): 334–345

For further information, see
http://www.lboro.ac.uk/departments/cv/

Chapter 30
Information Technologies in Mechanical Engineering, University of Magdeburg

Sándor Vajna

30.1 Information Technologies in Mechanical Engineering

The main goals of the Information Technologies in Mechanical Engineering group is to contribute to the art of product development and to increase the permanent innovation ability of our partners, thus enabling high-valued, highly productive and dynamically acting product development.

- Basic research is focused on product development fundamentals.
- Application research is driven by projects with industrial partners, which foster quick and careful feedback on the success of the research in practice.
- In education, students are taught interdisciplinarily and in a holistic and project-driven way, thus assuring high practice relation – both within the graduate study course of integrated product development (IPD) and within the CAx applications cycle;
- The following leading CAx systems are applied in the work of the group: Unigraphics™, CATIA™ V4 and V5, Pro/Engineer™, SolidEdge™, Solid Works™, AutoCAD™, etc.
- In the CAx education cycle, it is our aim to create CAx generalists, but not a specialist of a particular system. Therefore, students are trained on the five leading CAx systems, which is unique at least in Germany.

We put a great emphasis on industrial cooperation. The smooth transfer of research and education results into practice is of high importance and an indispensable part of the work of the group.

Research areas

The research is concentrated on the improvement of product development procedures, processes, methods, and tools in product development, which lead to improved products and more reliable processes (Figure 30.2). Focus is on dynamic process management, on holistic process and organisation models (like integrated product development), on analogies to natural evolution, on computer-supported tools for product modelling, and on knowledge processing.

We observe a paradigm change in the application of computer-supported procedures and tools, because they, having become so powerful, go far beyond the capabilities needed by the classical design methods with their sequence-driven approaches. Today, these procedures and tools create radically enhanced application models and support procedures for product development (for example, consistent virtual worlds). In this context, we have been developing sequenceless design, which allows designers to start with any product development activity and to realise the product without losing consistency, since it is assured that all necessary activities to create the product are performed.

Dynamic process navigation

Processes in product development, due to their heuristic, spontaneous, and dynamic character, cannot be managed with traditional workflow systems that are mostly based on ERP approaches. Instead, these processes need an approach that does not direct the user strictly (like an ERP system), but offers him possible alternatives and allows him to select his own preferred way when working on the process. This approach should predict possible bottlenecks, should monitor the process in real time, and should evaluate potential process alternatives to overcome disturbances. This behaviour is best described with the term 'dynamic navigation', because it leaves the control and the decision competence with the user (whereas in ERP systems users are 'production means').

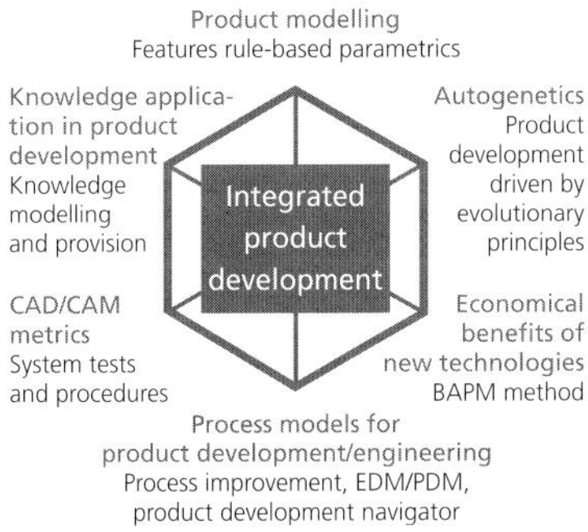

30.2 Research areas

A system named ProNavigator®, based on intelligent and generic process elements that are configured like building blocks, has been developed. It navigates engineers safely through any complex sets of processes. It ensures that the best possible process is always performed. It supplies knowledge, tools, and data at the right time and selects the process steps according to the current requirements and progress. Results are an increased transparency on active projects and on project statuses, a significantly shorter throughput time, a better balanced workload and the assured processing of all necessary working steps (see also www.pronavigate.com).

Integrated product development

IPD is a human-centred and interdisciplinary approach for the development of high-quality competitive products or services in an appropriate time, and at a reasonable price–performance ratio.

IPD has the following characteristics:

- The styling of the product (by industrial design), its ergonomics and its design (*i.e.* its shape to fulfil the functional requirements economically) form an inseparable unit.
- Within IPD, all participants, who are involved along the product lifecycle, cooperate and collaborate in a flexible way. IPD fosters all decisions to be taken at the right time.
- IPD includes the integrated application of holistic and multidisciplinary methods, procedures, and tools (both manual and computer aided) as well as process and organisation forms. IPD assures the minimised and sustainable use of production means and resources.

Within IPD, all characteristics and possible behaviour of a product are simulated during product development, in order to consider as many influences as possible from the product lifecycle at the earliest possible time.

Product modelling

The emphasis is on parametrics and enhanced features. In this view, a feature is an information unit representing a region of interest in a product model or a specific view on the product lifecycle. The enhanced feature is described by an aggregation of properties of a product, *i.e.* it contains a great part of the product knowledge. This information is separately stored in a database, which makes it easier to edit features or to exchange them between different CAD systems. With enhanced features, CAD systems become a modelling engine driven by the external feature description. Based on this definition, the feature-based plant design system IKA (Integriertes Konstruktionssystem fuer den Apparatebau) has been jointly developed with the Apparatus and Plant Design group of the Technical University of Munich. IKA provides modelling instructions for SolidEdge™ and AutoCAD™.

Autogenetic design theory (ADT)

The ADT describes the design process as an analogy to natural evolution. Terms and procedures from biology, evolution theory and (partly) chaos theory are transferred into product development. All actions in product development can be modelled unambiguously with the evolutionary operators replication, recombination, mutation, horizontal gene transfer, and selection. A partial application of ADT is product optimisation, since both changing an existing product (adaptation) or creating a new product (new design) can be described as the optimisation of an existing solution. This kind of optimisation is much faster than classical approaches. Furthermore, it evaluates the entire solution space for finding the best solution. We use our own optimising system NOA, which is based on genetic algorithms, and has been successfully applied to the design of automotive products and to process improvement.

Prediction of economical benefits

With standard approaches from controlling, it is only possible to calculate about 10% of the total benefits available of new approaches in engineering. The unique Benefit Asset Pricing Model (BAPM®) is able to predict and to calculate all possible benefits of engineering processes as well as costs and benefits of new technologies and computer-support systems. BAPM® applies analogies between the genesis of benefits in technical areas and prediction tools of the capital market (for example, balanced scorecard, portfolio theory of Markovitz). Its prediction accuracy is above 90%. This approach is as well suited to process simulation. BAPM has been used successfully in different companies for the above-mentioned aims (see also www.bapm.de).

Further reading

Freisleben D, Vajna S (2002) Project navigation – modelling, improving, and review of engineering processes. ASME'02, Montreal, Canada

Schabacker M (2002) Benefit evaluation of new technologies. ASME'02, Montreal, Canada

Vajna S, Burchardt C (1998) Dynamic development structures of integrated product development. Journal of Engineering Design, 2: 3–15

Vajna S, Weber C (2000) Teilmodelle im Konstruktionsprozeß – Bindeglied zwischen methodischer und rechnerunterstützter Konstruktion. Konstruktion, 52(4): 46–50

Vajna S, Weber C (2000) Sequenzarme Konstruktion mit Teilmodellen – Ein Beitrag zur Evolution des Konstruktionsprozesses. Konstruktion, 52(5): 35–38

Vajna S, Clement St, Jordan A (2002) Autogenetic design theory: an approach to optimise both the design process and the product. ASME'02, Montreal, Canada

Zirkel M, Pilhar S, Vajna S, Strohmeier K (2002) Continuous computation in apparatus engineering. ASME PVP'02, Vancouver, Canada

For further information, see
http://imk.uni-magdeburg.de/lmi.html

Chapter 31
Design Process Research, Massachusetts Institute of Technology

Daniel Whitney

31.1 MIT – the Great Dome

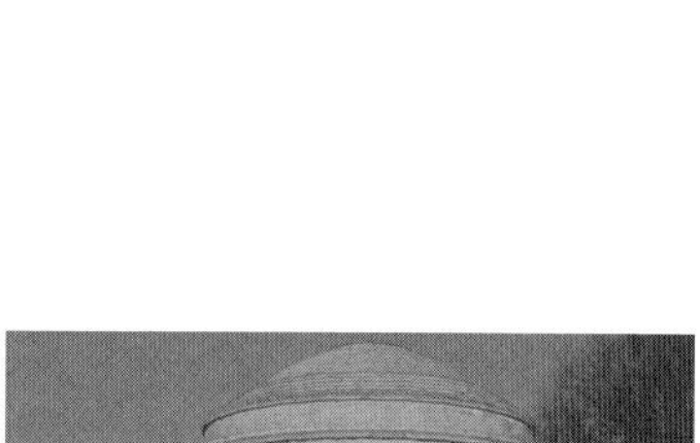

31.2 Multiobjective optimisation and isoperformance. A family of solutions can be generated that exhibits isoperformance: each design optimises the same function of the several objectives, but each objective individually is attained to different degrees.

This chapter provides a snapshot of design-process-related research currently being undertaken at the Massachusetts Institute of Technology in the Department of Mechanical Engineering, Engineering Systems Division and the Sloan School of Management. It is not possible in such a short chapter to do justice to the full range of activities. However, the following is representative of the strong tradition of research in this area.

Systems engineering

Professor Daniel Frey conducts research on system design methods, including robust design, design of experiments, probability, manufacturing and computational geometry. The overarching goal of his research is to identify principles and practices that improve the process of engineering design. In particular, Professor Frey is interested in strategies for dealing with uncertainty in design using experiments and simulations. One well-known approach to this challenge is known as 'robust design', in which factorial experiments are used to seek parameter settings that reduce sensitivity of the engineering system to variations.

The success of engineering systems is often determined during conceptual design, where quantitative methods are still in their infancy. Some high capital investment systems (such as new automobiles, aircraft or satellites) fail economically or technically due to an inadequate understanding of their underlying architecture. Professor de Weck's research program looks to build a rigorous methodological bridge between system architecture and multidisciplinary design optimisation, allowing systems to satisfy multiple criteria, while exhibiting desirable lifetime properties.

Design structure matrices

Professor Steven Eppinger's research addresses the management of complex engineering projects such as the development of an automobile, aircraft, or telecommunication system. This research has used the design structure matrix (DSM) method and has developed several analytical modelling extensions to the DSM approach. It is primarily conducted within the Center for Innovation in Product Development and the Leaders for Manufacturing and System Design and Management programs. Current projects include the application of DSM techniques to project management; understanding how product architecture drives communications in development organisations; metrics for product development process and organisational complexity; and spiral product development processes.

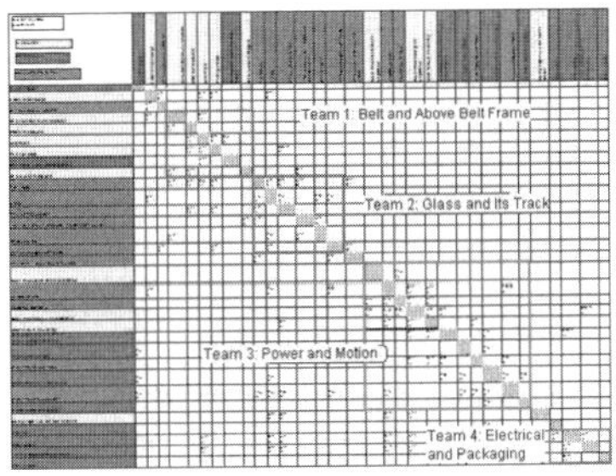

31.3 This design structure matrix reveals clusters of functional specialists that need to communicate in order to design the interior mechanisms of car doors

System modelling

Professor Dave Wallace's predictive integrated system modelling is now a pressing issue in the design of complex products ranging from home air conditioners, to automobiles and aircraft. While product development organisations have official top-down development processes, in practice individual participants perform their work in an informal marketplace, bartering service exchange relationships to get what they need to resolve their part of the problem. The informal, dynamic, heterogeneous and evolving characteristics of product development environments create many challenges to integrated system modelling.

Ideally, the naturally occurring informal activities of participants would also create a set of heterogeneous and distributed models representing the complete product, which could then be used for the rapid exploration of design trade-offs and global optimisation. The DOME (Distributed Object-based Modelling Environment) project-led by Professor Dave Wallace is developing a computational infrastructure for this purpose. DOME's simulation marketplace concept empowers participants to offer their capabilities digitally through simulation service interfaces instantiated by DOME object models accessible over the Internet. Participants can also independently negotiate and form local relationships between their simulation services and the services of other participants. The resultant service exchange network becomes an emergent distributed computational system with service state changes, rather than data models, propagating to predict the integrated behaviour of the emergent system rapidly.

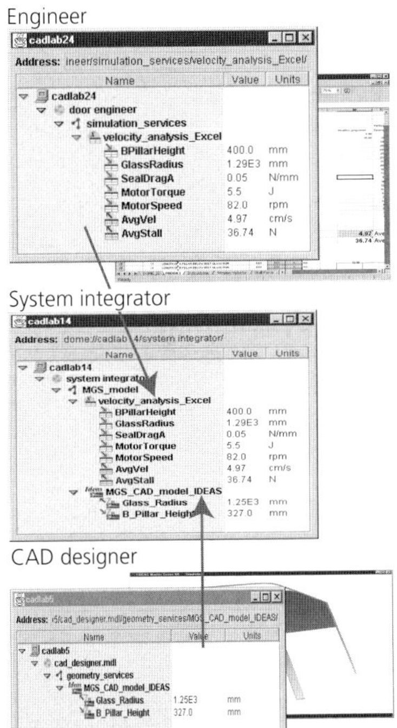

31.4 The DOME project has created software interface standards that permit proprietary software and data to interact so that a multi-function design can be created and optimised in a short time

Implementation dynamics

The history of management practice is full of process improvement innovations, e.g. job enrichment, quality circles and total quality management. Dr Nelson Repenning is investigating why innovations succeed in some organisations but fail in others.

The implementation dynamics initiative has sought to understand the dynamics of product development process improvement and the design of sustainable improvement programs. Working with Ford and Harley-Davidson, intensive case studies of successful and failed process improvement initiatives have been conducted and analysed to provide the basis for formal improvement models. Robust strategies have helped to eliminate the unanticipated side effects that routinely delay development. Management 'flight simulators' and learning labs have helped communicate these strategies to front-line managers and workers.

Axiomatic design

The overarching goal of Professor Nam Suh's research is to rationalise a design process such that the outcome of the design activity is assured to deliver its intended functions with maximum certainty and minimum resources. Professor Suh and members of the axiomatic design group have been applying the axiomatic design theory, which was established in the mid 1970, to a wide range of problems.

The motivation of this research comes from the underlying assumption that current design practice is both ineffective and inefficient, consequently failing to deliver an optimal result in many aspects. The goal of axiomatic systems design theory is to improve on the process for designing a large system by extending the scope of the axiomatic design approach to the large-scale system level. In particular, current research activities focus on investigating and applying these theories to space-vehicle system design, nano-manufacturing and systems biology.

Knowledge and design aids

Daniel Whitney is interested in advancing the knowledge and design aids applicable to product design. A major issue is to combine the usual issues of product performance with other equally important issues: manufacture, assembly, field use and repair, upgrading, even selling strategies.

Too often, product design has been sequential, with different designers doing their part (say function or manufacturing) one after the other, requiring later ones to live with the decisions made earlier. Since nearly all of the cost of making something is determined by early design decisions, the penalty for making these decisions incorrectly can be very large.

Therefore, there is increasing interest in improving design methodologies, understanding what questions to ask and when, what data are needed and to what accuracy, and how to embody the methodology and data into computer design tools and aids. Such aids will have to go well beyond mere geometric representations of the ideal shapes of parts, which is the current state of the art. New design aids will also have to extend beyond single parts to encompass whole assemblies. Factors that need to be modelled, in addition to traditional stress, weight, volume and thermal behaviour, are tolerances, cost vs. tolerance, ability to assemble, alternate assembly sequences, ability to test and ensure quality, ability to disassemble, and so on.

Further reading

De Weck O, Suh ES, Chang D (2003) Product family and platform portfolio optimization. ASME DETC'03, Chicago, IL, USA

Dong Q, Whitney D (2001) Designing a requirement driven product development process. ASME DETC'01, Pittsburgh, PA, USA

Eppinger SD, Whitney DE, Smith RP *et al.* (1994) A model-based method for organizing tasks in product development. Research in Engineering Design, 6(1): 1–13

Frey DD (2003) How one-factor-at-a-time experimentation can lead to greater improvements than orthogonal arrays. ASME DETC'03, Chicago, IL, USA

Repenning N (2002) A simulation-based approach to understanding the dynamics of innovation implementation. Organization Science, 13(2): 109–127

Senin N, Wallace DR, Borland N (2003) Distributed object-based modeling in design simulation marketplace. ASME Journal of Mechanical Design, 125(1): 2–13

Suh NP (2001) Axiomatic design: advances and applications. Oxford University Press

Ulrich KT, Eppinger SD (2000) Product design and development. McGraw-Hill

Wallace DR, Abraham S, Senin N *et al.* (2000) Integrated design in a service marketplace. Computer-Aided Design, 32(2): 97–107

For further information, see
http://www-me.mit.edu/

Chapter 32
Institute of Product Development, Technische Universität München

Udo Lindemann

32.1 Faculty of Mechanical Engineering, TUM

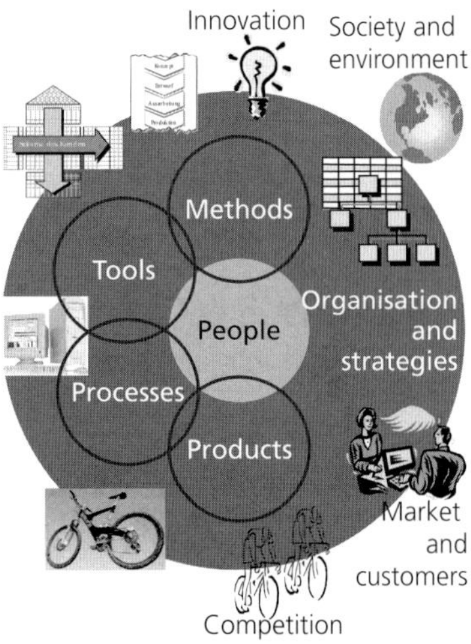

32.2 The institute's focus

The Institute of Product Development of the Technische Universität München (TUM) belongs to the faculty of mechanical engineering and consists of about 30 people of whom approximately 20 are researchers.

The development of competitive and innovative products and the optimisation of the respective product development processes are the main focus of the institute. The institute's aim is to support industrial product development by the generation and adaptation of effective methods, tools, and strategies (see Figure 32.2). This is complemented by competent and practical education of students and specific knowledge transfer into industry.

Research focuses on the flexible use, adaptation and implementation of methods in industry, on computer support of product development processes and the use of virtual reality, on cost management, distributed product development, knowledge management, functional analysis, innovation, variant management and mass customisation, sustainable development, empirical design research and rational design theory, system theory and interdisciplinary work.

Methods

Methods support engineers in product development by offering a regular and systematic proceeding. Product development, understood as problem solving, can be described by a flexible scheme called the Munich procedural model (Figure 32.3). Methods are described by the Munich model of methods (Figure 32.4), which helps with the selection, adaptation and use of the methods depending on boundary conditions, resources, etc. The methods are linked and implemented in a Web-based portal and knowledge base for developers from industry and academia, as well as for students.

Interdisciplinary collaboration

Product development takes place in interdisciplinary teams and processes. Either the expertise of other disciplines is needed or the product is built for a specific branch. The institute works on projects together with psychologists (design rationale, design thinking, for example, while sketching, elementary methods, tacit knowledge, etc.), medics (development of medical devices), civil engineers (method transfer), computer scientists and electrical engineers (for example, development of mechatronical products), business economists (for example, customer relationship management), sociologists (method implementation), as well as industrial designers or other engineers from mechanical engineering such as production engineering or logistics. The latter, especially, are important for industrial practice due to concurrent engineering strategies.

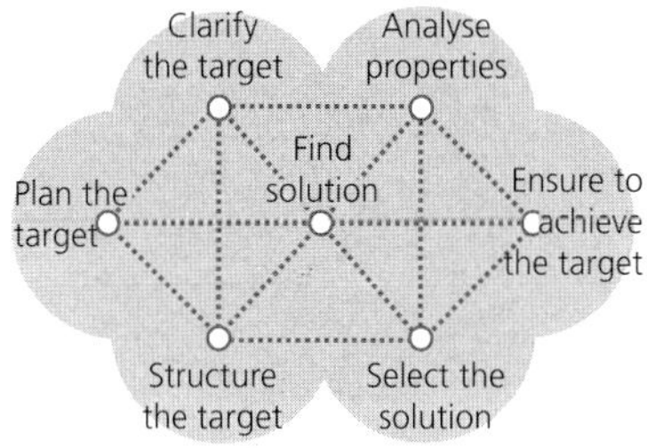

32.3 Munich procedural model

Tools

Tools, both physical tools and computer-supported tools, are based on methods or assist the use of methods. Examples of tools developed at the institute are IntraPAS, a tool to structure meetings and semi-automatically take the minutes, IntraPRO, a tool to add design information to CAD data, INKA, a designers' workbench documenting the product and process logic, or the idea database, resulting from empirical research is the 3D sketcher (Figure 32.5), which permits the drawing of three-dimensional sketches in a virtual environment.

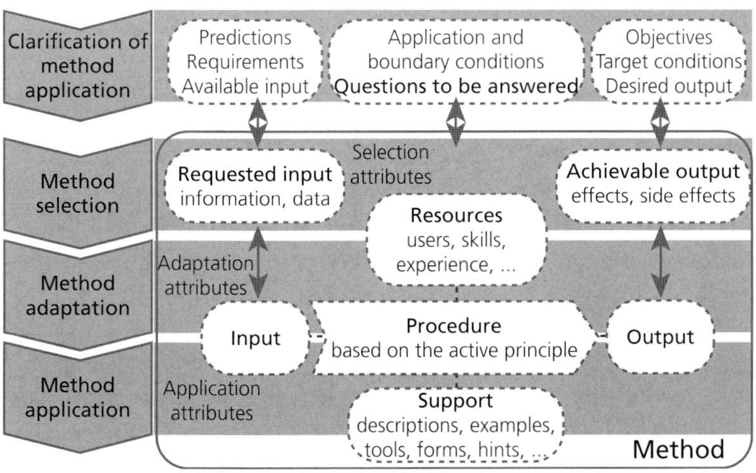

32.4 Munich model of methods

Individualised products

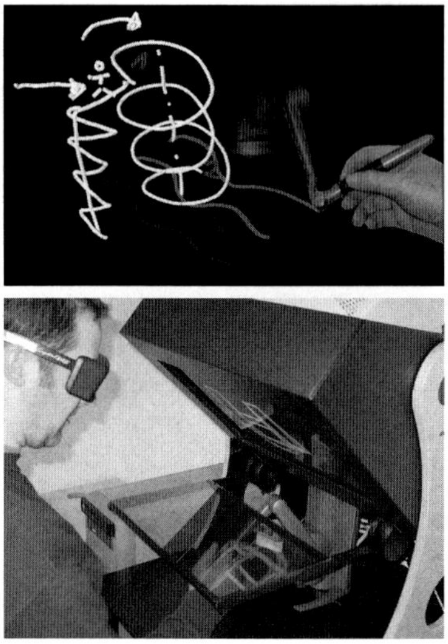

32.5 3D sketcher

Market pressure and customer demands force companies to offer more and more products or variants of their products. The growing complexity can hardly be handled by existing approaches. Within an interdisciplinary collaborative research centre (SFB 582) called 'Production of Individualised Products Close to the Market', the aim is to offer completely individualised products that perfectly fit the wishes and needs of the customers. The institute focuses on what the structure and architecture of an individualised product looks like, how processes can be modularised and standardised in order to develop varying products, and how an adequate cost management can be set up for individualised products. This topic is strongly connected to engineering change management, product configuration and specification, parametric design and artificial intelligence, as well as project management and system theory. A central characteristic is that this approach is implemented in a network of locally distributed miniature plants, so that aspects of distributed product development have to be considered, too.

Transfer of methods into industry

Concepts for implementing methods in industry and to develop specific methods for industry have been set up. A current project focuses on representing and optimising interdisciplinary development processes with the method of process building blocks as well as controlling and coordinating distributed development process with a large number of suppliers. The latter is supported by a tool which represents a project status including cost, weight, milestones, functions, tests, etc.

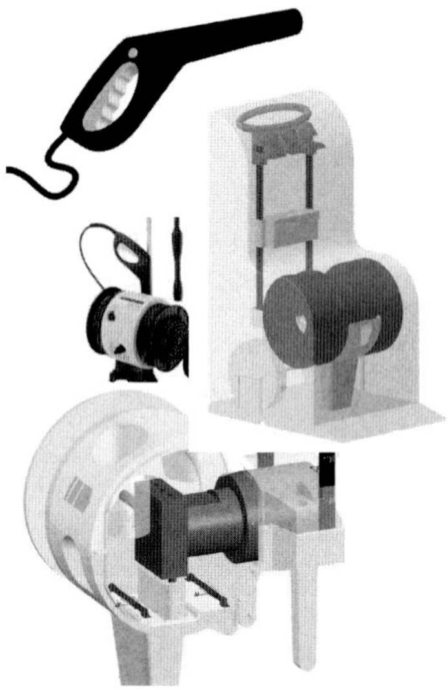

32.6 Individualised high-pressure cleaner

Strategies

The project 'Integrated product policy' aims at introducing and adapting methods for sustainable product design in practice by identification of effective, simple, and practical methods and tools as well as giving a guideline for small and medium-sized companies. The 'Strategic product and process planning' covers strategies and methods for identifying new and promising business segments, innovative products, and strategic measures.

Cost

Tools for estimating and controlling costs during the design process have been developed, including guidelines for cost-effective design and approaches for activity-based resource and time management.

Innovation

A central topic is methods and strategies for enhancing the designers' creativity. This concerns fundamental cognitive processes as well as applicable methods such as bionics. By use of these principles products such as a highly efficient bionic suction nozzle, an avalanche airbag, a rowing ergometer, a bicycle and bike equipment (anti-dive fork, wheel pressure control system, continuously adjustable gear), and a freezer working with solar energy, etc. have been developed.

32.7 Soft skills

Soft skills

A speciality offered by the faculty within the study courses is the so-called 'Tutor System Garching'. Elder students help young students beginning the study courses and teach them social soft skills such as conflict management, time management, presentations, negotiations, etc., which are required more and more in modern industry. Such soft skills are trained and deepened in product development seminars where students work in a project together with the industry and develop products such as those above.

Further reading

Baumberger C, Pulm U, Lindemann U (2003) Coordination and controlling of distributed product development processes. ICED'03, Stockholm, Sweden

Lindemann U (2003) Methods are networks of methods. ICED'03, Stockholm, Sweden

Lindemann U (2004) Methodische Entwicklung technischer Produkte – Methoden flexibel und situationsgerecht anwenden. Springer

Lindemann U, Wulf J (2001) Action orientation in design methodology. ICED'01, Glasgow, UK

Lindemann U, Maurer M, Pulm U (2003) Methods, tools, and processes for the design and development of individualised products. ICED'03, Stockholm, Sweden

Müller F, Pache M, Lindemann U (2003) Digital free-hand sketching in 3D – a tool for early design phases. ICED'03, Stockholm, Sweden

For further information, see
http://www.pe.mw.tu-muenchen.de/

Chapter 33

Engineering Design Centre, University of Newcastle

Bill Oliver and Pratyush Sen

33.1 Newcastle EDC

The Newcastle Engineering Design Centre undertakes fundamental and applied research into the design process for 'engineer-to-order' (ETO) products. Typically, these are 'one-off', multi-disciplinary products such as ships, power plants, military aircraft, offshore oil and gas platforms, etc. Increasingly complex in nature, ETO projects form a large sector of modern industry; there is a corresponding need for tailored research support in this area.

Working closely with leading industrial collaborators, the objective of the Newcastle EDC is to provide practical tools and methodologies which assist the designer with real problems such as design integration, sustainable life-cycle design, design option evaluation, selection and optimisation.

A variety of projects, aligned to these basic themes, have been undertaken by the research team at Newcastle over the past 12 years of EDC operation. There follows a synopsis of recent samples of such work.

Multiple-criteria design selection and synthesis

Engineering design often involves the balancing of potentially conflicting requirements. Classical optimisation deals with such problems by adjusting the requirements until feasible solutions emerge. An alternative view is to leave the requirements as they are and examine how a trade-off between requirements leads to candidate solutions.

There is a growing body of methodological tools to support this latter approach. Newcastle EDC continues to contribute to the development of such tools and their widespread application in areas such as spatial layout, design coordination and production scheduling for ETO products. Research is also progressing in the area of decisions under risk and incorporation of linguistic judgements.

Robust scheduling in production design

Competitiveness in the UK shipbuilding industry depends on building high value, complex specialist ships. There is an associated need for optimal use of production facilities within an environment that is unpredictable and susceptible to change.

The decision-support methodology developed in this research area aims to:

- Develop a formal, multiple-criteria schedule optimisation strategy and software (resource constrained).
- Examine the influence of variations in activity durations with a view to creating robust schedules that maintain performance over certain fluctuations.
- Assess the potential gains associated with specific improvement strategies using the new methodology, thus providing a management tool for continuous improvement.
- Identify higher level scheduling strategies on the basis of the above.

Industrial test applications of the methodology have yielded improved schedules that are not rendered inefficient in a changeable production process.

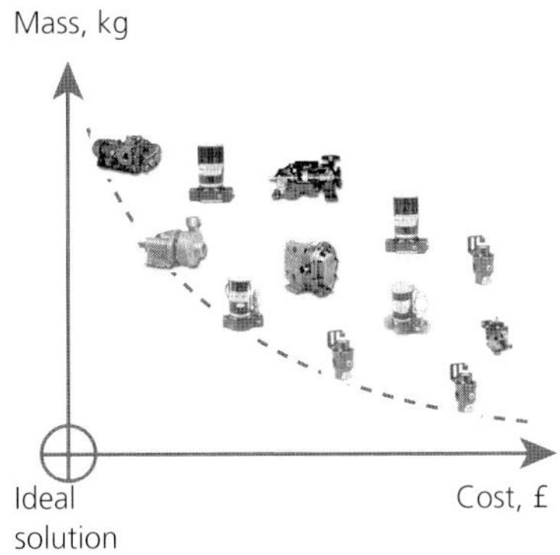

33.2 Example: conflicting criteria – aircraft fuel pumps?

Decision support for modelling and simulation

As our understanding of the world improves and as more complex products and processes are developed, effective computer modelling becomes increasingly difficult. Model acceptance is typically an *ad hoc* subjective procedure, but we are developing a methodology to capture the required reasoning formally and systematically through the use of Bayesian belief nets. Such an approach will enable resources for improving models of complex systems to be better targeted, thus promoting the use of modelling over costly systems testing.

Modelling framework representation

One of the great challenges in complex system development is the control of design properties throughout the lifecycle. Modelling plays a key role in this activity, but current approaches do not adequately support product integration. In particular, there is a failure to provide proper traceability of design properties throughout the product breakdown structure and an inadequate management of the impact of uncertainties in modelling activity throughout the lifecycle.

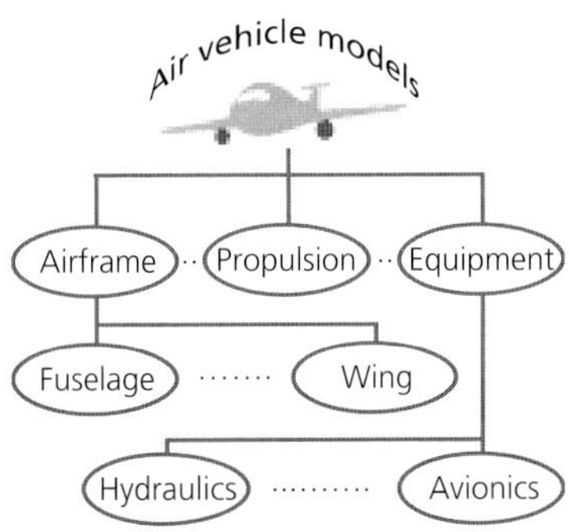

33.3 Aircraft model breakdown

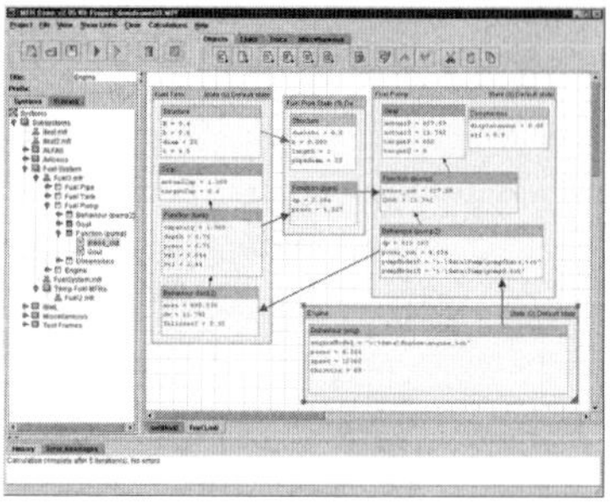

33.4 Cross-platform Java implementation of the modelling framework

This research project aims to develop an integrated modelling environment (IME) that can be fully deployed across a military aircraft supplier's operations. Benefits are:

- Increased understanding of modelling capability, leading to more extensive/effective use of modelling.
- More effective systems engineering achieved by increasing the awareness of interrelationships in the design community and product hierarchy.
- Wider dissemination, traceability, storage and reusability of best design practices.
- More effective targeting of modelling capability (specifically in development); identification of strengths and weaknesses in existing capabilities.

Traceability in design

Traceability is an important aspect of the design of complex systems. Object-oriented approaches to an integrated environment provide the basis for developing a traceability methodology on objects (such as properties) within a complex system design framework. This allows the design engineer to collect emerging information and target design effort early in the project. Traceability can also be used to measure and test the sensitivity of properties and estimate the lowest (and cheapest) fidelity of integrated models.

As part of the IME work, a generic methodology has been developed to trace a range of entity relationships within a variety of model environments, such as:

- information models;
- process models;
- documentation models;
- enterprise models.

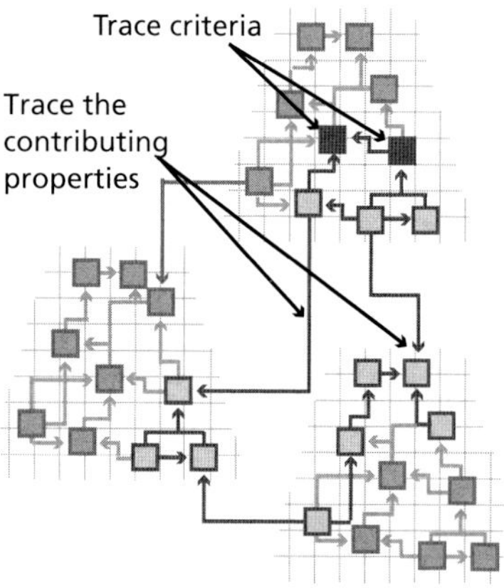

33.5 Tracing properties through a complex system

The application of margins in engineering design

Design margins are added to design variables to achieve a preventative compensation for parameter uncertainties in the models used for design. Typically, uncertainties are accounted for by introducing design margins based on hard-won experience. Increasingly, however, industry is interested in applying robust and probabilistic methods. It is clear that margins are potentially expensive to provide, but reducing them carries the risk of underperformance. There is, therefore, a trade-off involved. Present research is geared towards developing appropriate mathematical and statistical approaches to facilitate this trade-off.

With reference to Figure 33.6, the hashed area represents the region of underperformance if z is the specified level of performance. A systematic study of the methods of reducing the area of overlap is the main theme of this work.

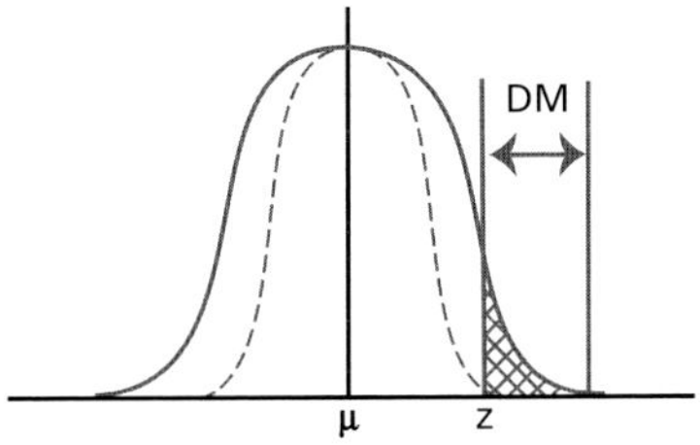

33.6 Design margin probability density function

Further reading

Dalton JA, Norman PW, Sen P, Whittle S (2002) Use of modelling frameworks in the design of complex engineering systems. EDC'02, King's College London, UK

Dalton JA, Norman PW, Whittle S, Rajabally E (2003) Design reuse of complex systems within an integrated modelling environment. CE'03, Madeira, Portugal

Dalton JA, Norman PW, Whittle S, Rajabally E (2003) The synthesis of complex engineering systems using an integrated modelling approach. VRAP2003, Leiria, Portugal

Rajabally E, Sen P, Whittle S (2002) A methodology for model dependability assessment. EDC'02, King's College London, UK

Rajabally E, Sen P, Whittle S (2002) Assessing uncertainty in models of complex systems. ESM'02, Fachhochsschule Darmstadt, Germany

Rajabally E, Sen P, Whittle S, Dalton JA (2003) Towards multi-criteria decision support for modelling and simulation. ISC'03, Universidad Politecnica de Valencia, Spain

Scott JS, Tolan C, Sen P (2003) A strategy for modelling the design development phase of a ship. IMDC'03, Athens, Greece

Sen P, Yang JB (1998) Multiple criteria decision support in engineering design. Springer

For further information, see
http://www.edc.ncl.ac.uk/

Chapter 34

Center for Design Research, Stanford University

Larry Leifer

The guiding question

34.1 Re-designing the future. Design as practised today has brought us to this posture. It is therefore our mission to re-design design. The drawing is an anonymous Web download.

Since its inception in 1984, the work of the center has been guided by one stimulus question and two corollary response questions. What do designers (Figure 34.1) do when they do design? How can we help them manage the process? How can information and communication technology support the process?

The product is the team

The Stanford Center for Design Research (CDR) is a collaborative faculty team doing empirical research on engineering design-process management, design-informatics, and design-education. It is situated within the Design Group (13 faculty) in Mechanical Engineering (42 faculty) within the School of Engineering (322 faculty), one of seven PhD-granting schools at Stanford University (Figure 34.2).

Our most important product is design-thinking education. Successful new product development companies (Figure 34.3) are our best proof evidence. To improve our own processes, thinking about design-thinking leads to our Doctor of Philosophy in Design Engineering (PhD). The experience at Stanford prepares our graduates for academic leadership positions, corporate research management, and new product innovation. As a rule, design researchers have reading committee members from across the university, with computer science, cognitive science, business administration, and education being amongst the most frequent interdisciplinary combinations.

34.2 Stanford University is in the heart of Silicon Valley

Our curriculum is driven by product innovation opportunities with corporate clients. Performance is team based, measured and graded. One course in particular has served as an important 'test-bed' for the research program, me310 'Team-Based Design Development with Corporate

Partners'. Within this simulator of corporate design practice, we endeavour to 're-design designers' (Figure 34.4)

The CDR team includes: four professors (Leifer, Cutkosky, Sheppard, and Gerdes); four researcher associates (Drs Mabogunje, Eris, Grossman, and Van der Loos); an average of 34 PhD candidates; 8 to 10 graduate research assistants; a dozen undergraduate researchers; and non-academic support staff. CDR is an affiliate of the Center for the Advancement of Scholarship in Engineering Education within the National Academy of Engineering, Washington, DC. Sheppard is a senior research fellow of the Carnegie Foundation for the Advancement of Teaching. Cutkosky is co-director of the Stanford Innovations in Manufacturing Program. Leifer is on the advisory boards of the Stanford Center for the Study of Language and Information and Media-X.

34.3 The product is the team. The image is a collage of logos representative of companies started by Stanford Designers.

Design team performance measurement

The guiding metaphor for team performance assessment and evaluation is process instrumentation (Tang, 1989) and control. The term is used in the sense of observing both independent and dependent variables in a control feedback environment similar to that found in aircraft flight simulators for crew training. The model asserts that there are several instrumentation requirements that must be satisfied to observe the input–output relationship between knowledge availability and decisions made in a design environment with noise and performance feedback (Figure 34.5). Mabogunje (1997) demonstrated that the incidenceunique noun-phrases in design documentation predicts product quality, a product knowledge variable. Eris (2003) showed that a team's question-asking rate predicts their performance, a human process variable.

34.4 We are re-designing designers through everyday interaction with corporate partners, including for 2003–2004, in our automotive studio: BMW, GM, TOYOTA, and VW

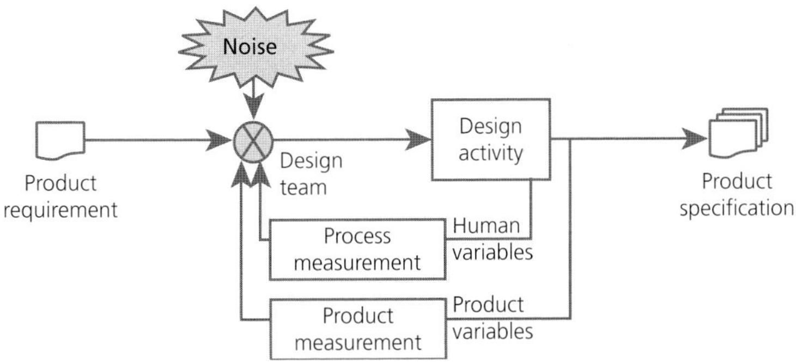

34.5 Design process management requires objective measurement of product knowledge variables and human design process variables in real-world environments with noise (that must also be measured)

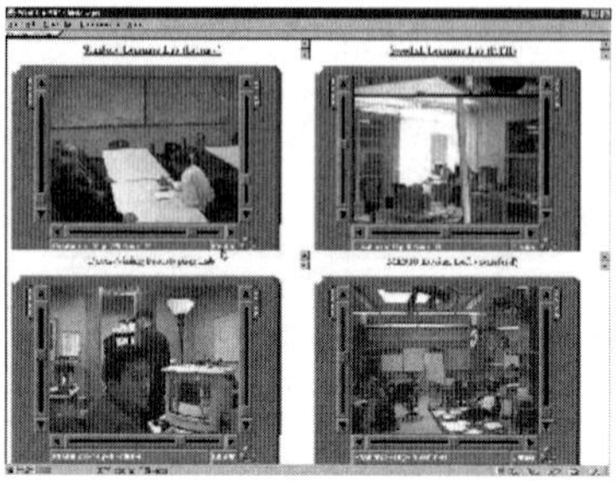

34.6 Workspaces for collaborative team-based design must accommodate people and information technology

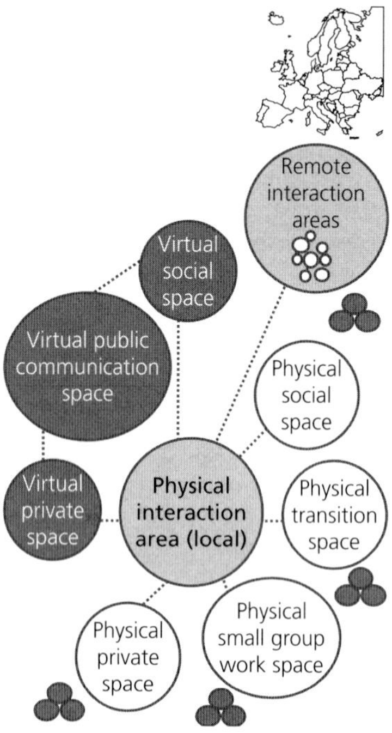

34.7 The performance of distributed innovation teams is our canonical challenge

Depth through breadth

Our working definition of 'design' is firstly broad, then deep. Comprehensiveness requires systems thinking. Competency and literacy are expected across the arts, sciences, and humanities with an overarching value proposition that demands 'design-for-wellbeing'. Our concerns are, in turn, driven today by rapid changes in the dynamics of transportation, work, information and new media systems (Figure 34.6).

Affiliated labs and programs

Depth in design at Stanford is best seen in the emphasis on graduate studies in affiliated laboratories, including the: Manufacturing Modeling Lab for lifecycle design (Professor Kosuku Ishii); Rapid Prototyping Lab for mezzo and nano fabrication (Professor Fritz Prinz); Dynamic Design Lab for drive-by-wire studies (Professor Chris Gerdes); Machine-Dissection-Lab for design education studies (Professor Sheri Sheppard); the Design-Observatory for video interaction analysis and the iLoft for distributed team interaction (Figure 34.7) studies (Professor Larry Leifer); Haptics and Grasp Lab for biomimetic design (Professor Mark Cutkosky); Human Computer Interaction Lab for interactive media design (Professor Terry Winograd, CS); Social Response to Technology Lab for emotional response to new media studies (Professor Clifford Nass, Communication); Interactive Tele-Robotics Lab for surgical robotics (Professor Gunter Neimeyer); Mobile Robotics Lab for autonomous robotics (Professor Kenneth Waldron, ME); Micro Sensors Lab for MEMS, nano, and bio sensor development (Professor Thomas Kenny). And finally, the Product Realization Lab is the locus for our intensely hands-on education program. Professor David Beach directs the lab and related teaching program.

Affiliated academic programs include: the Product Design Program, a joint venture between Mechanical Engineering and the Art Department (Professor David Kelley director (IDEO founder)): the Media-X program, a joint venture between the social sciences and engineering focused on interactive media (Professor Byron Reeves, Director, Communication); the engineering entrepreneurship program, a joint venture between the Management Sciences department in Engineering and the Graduate School of Business (Professor Thomas Byers, Director); and the Bio-Design program, a joint venture between engineering design medicine and biology (Professor Paul Yock, Director (Medicine)) (see Figure 34.8).

Further reading

Adams JD (2001) The transfer of scanning probe microscope research to the university classroom: lessons in distributed collaboration tasks. PhD thesis, Stanford University, CA, USA

Baya V (1996) Information handling behavior of designers during conceptual design: three experiments. PhD thesis, Stanford University, CA, USA

Brereton MF (1998) The role of hardware in learning engineering fundamentals: an empirical study of engineering design and product analysis activity. PhD thesis, Stanford University, CA, USA

Eodice MT (1999) A theory of requirements definition in engineering design. PhD thesis, Stanford University, CA, USA

Eris O (2003) Effective inquiry for innovative engineering design. Kluwer Academic

Liang AT (2001) Mapping experience: understanding socio-technical inter-team knowledge sharing in product development communities. PhD thesis, Stanford University, CA, USA

Mabogunje AO (1997) Measuring conceptual design process performance in mechanical engineering: a question based approach. PhD thesis, Stanford University, CA, USA

Minneman SL (1991) The social construction of a technical reality: empirical studies of group engineering design practice. PhD thesis, Stanford University, CA, USA

Rosenberg LB (1994) 'Virtual fixtures': perceptual overlays enhance operator performance in tele-presence tasks. PhD thesis, Stanford University, CA, USA

Tang JC (1989) Toward an understanding of the use of shared workspaces by design teams admin. PhD thesis, Stanford University, CA, USA

Yen SJ (1999) Capturing multimodal design activities in support of information retrieval and process analysis. PhD thesis, Stanford University, CA, USA

For further information, see
http://stanford.edu http://cdr.stanford.edu
http://soe.stanford.edu http://design.stanford.edu
http://www-cdr.stanford.edu/publications

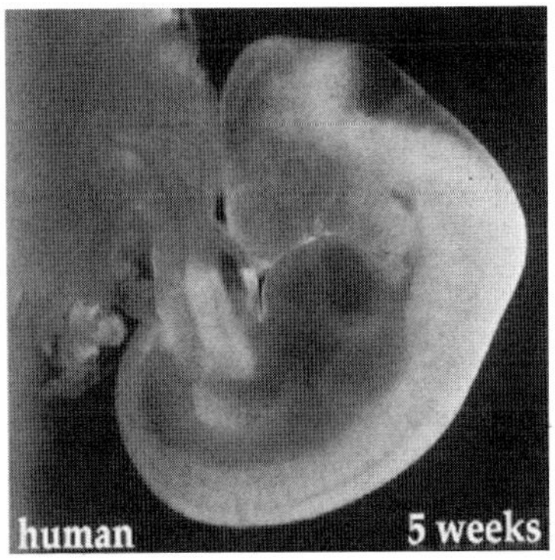

34.8 The bio-design program is a joint venture between engineering design, medicine and the life sciences

Chapter 35

Integrated Product Development, the Royal Institute of Technology, Stockholm

Margareta Norell and Sofia Ritzén

35.1 The Department of Machine Design

The development of society as a whole and the ability of industry to compete are highly dependent on our capability to innovate, design and develop new and better products. In Sweden, different research initiatives have been taken to meet these challenges. The largest during the last 5 years is the national ENDREA programme (www.endrea.sunet.se) finishing in 2003. The ENDREA program, founded by the Swedish Foundation for Strategic Research (SSF), followed by the ProViking Research School, aims to develop leading competence in product realisation.

The Department of Machine Design at KTH, the Royal Institute of Technology in Stockholm, performs research and education in product development. Our core scientific profile, designed to meet the requirements from society and industry, is best described as research concerning processes and technology for efficient design of complex physical products.

Product development is a strategic area of competence which needs a strong position in both education and research. The activities at the department treat, from a broad and multidisciplinary perspective, the design and development of modern products in which mechanical components are important for the overall product functionality. Our research and education is characterised by substantial industrial cooperation and several national and international collaborations, securing both industrial and scientific relevance.

Engineering design research at KTH

The Department of Machine Design currently has a team of over 100 people with expertise in engineering design, integrated product development, internal combustion engines, machine elements and mechatronics. The research is performed in areas concerning a number of key issues, some of which are:

- Tools, procedures and organisation for competitive performance of interdisciplinary product development.
- Modular and component-based architectures and technologies.
- Model-based development for mechatronic systems.
- Technology and competence integration.
- Industrial design, user adaptation and design for environment.

These issues are explored and concretised within the specific research divisions, aiming for a leading edge in particular research topics. Each specific area can be illustrated as the point of a star, as shown in Figure 35.3.

35.2 KTH hosted the ICED'03

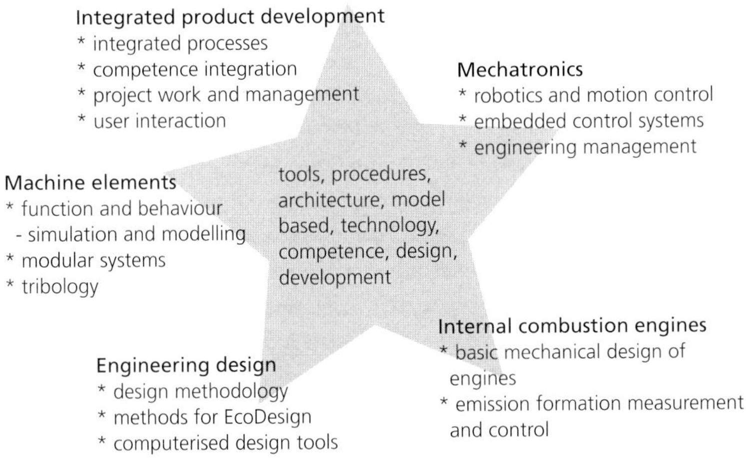

35.3 The research of the Department of Machine Design

Integrated product development

Integrated product development (IPD) is a core issue in engineering design and, as defined in the IPD research team, includes technology, tools, procedures and work organisation for increased efficiency, and learning in industrial product development processes. The research of the group is focused on key factors for competitive industrial product development in terms of utilisation of tools and procedures, organisational cooperation and parallel processes, as well as interdisciplinary teamwork. The work on change in industrial organisations is especially focused; the early phases of product development, as well as support for learning in the organisation, are of particular interest. The division cooperates closely with many industrial companies and researchers with behavioural science competence. The research approach chosen to

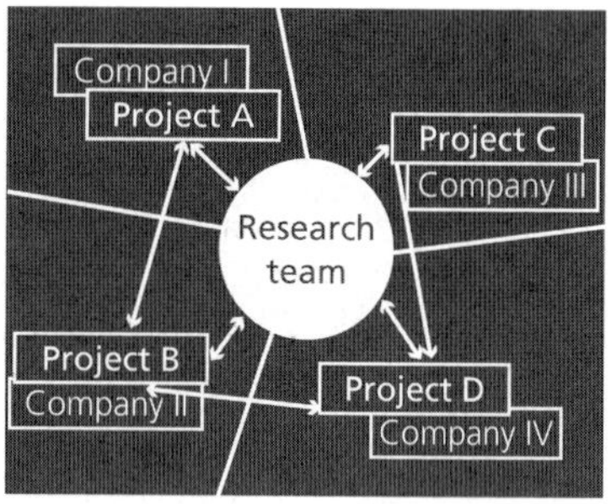

35.4 Empirical research depends on company collaboration – preferably in networks

35.5 Research results could typically be a management model, for example implementation of tools

35.6 First year students prototype sketching

meet the goal of more efficient and 'learning product' development processes is based on empirical studies combined with theoretical studies. Current research projects within IPD are:

- Engineering management in technical development work.
- Work and management in multi-project settings.
- Design for user satisfaction.
- Environmental concerns in product development when developing new business concepts.
- Integration of competences – development of mechatronic solutions.
- Innovation and creativity in product development.

In an attempt to deepen the research as a whole, and to further more basic research linked to industrial challenges, a broad research program was started in 2003. The program, 'Engineering management for integration – competitiveness and sustainability in industrial development processes', includes several companies and is driven by an action-learning approach combined with other empirical data-collection methods and analysis.

The IPD division coordinates the program, and is one of three research nodes; the other two are Operations Management and Industrial Ergonomics, the Department of Industrial Economics and Management at KTH and innovation and operations management at the Stockholm School of Economics. The aim of the program is to develop models for Innovative and Sustainable Leadership in a changing industrial setting. Models are to be developed by a synthesis of results from several research projects, some of which are defined above.

Education and training

The Department for Machine Design offers five specialisations in the undergraduate program: machine elements, mechatronics, engineering design, internal combustion engines and integrated product development. Besides the senior courses that come with the specialisations, the department offers courses at freshman level in industrial design, machine elements and machine design. In these areas, as well as fluid technology, internal combustion engines, mechatronics, and electrical engineering, courses are given to students not only from the study program for mechanical engineering, but also from industrial economics, metallurgy and materials technology. In total, the department offers more than 50 courses to over 2100 students.

The department is also involved in the provision of courses (continuing education) for professional engineers, to bring their knowledge up to date and to teach new topics.

The course on IPD is based on theory and experiential learning in practical product development projects in cooperation with industry. The students are trained in product development models, idea generation, project organisation, planning, project management, application of support methods, integration of environmental aspects, modelling, proto-typing and other areas related to development projects.

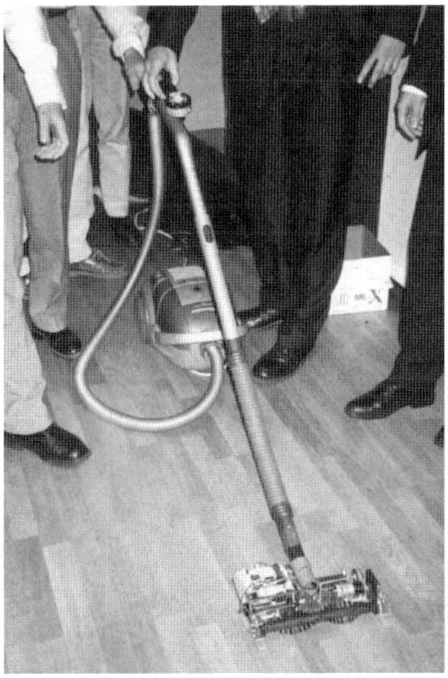

35.7 Appearance and functional models presented to a student project commissioner. In this case: Electrolux

Further reading

Beskow C, Ritzén S (2000) Performing changes in product development: a framework with keys for industrial application. Research in Engineering Design, 12: 172–190

Engwall M, Forslin J, Kaulio M, Norell M, Ritzén S (2003) Engineering management for integration. ICED'03, Stockholm, Sweden

Janhager J (2003) Utilization of scenario building in the technical process. ICED'03, Stockholm, Sweden

Ritzén S, Beskow C (2002) Actions for integrating environmental aspects into product development. Sustainable Product Design, 1: 91–102

Zika-Viktorsson A, Hovmark S, Nordqvist S (2003) Psychosocial aspects of project work: a comparison between product development and construction projects. Project Management, 21(8): 563–569

Ölundh G (2003) Environmental and developmental perspectives of functional sales. Licentiate thesis, KTH, Sweden

For further information, see
http://www.damek.kth.se

Chapter 36

The CAD Centre, University of Strathclyde

Alex Duffy

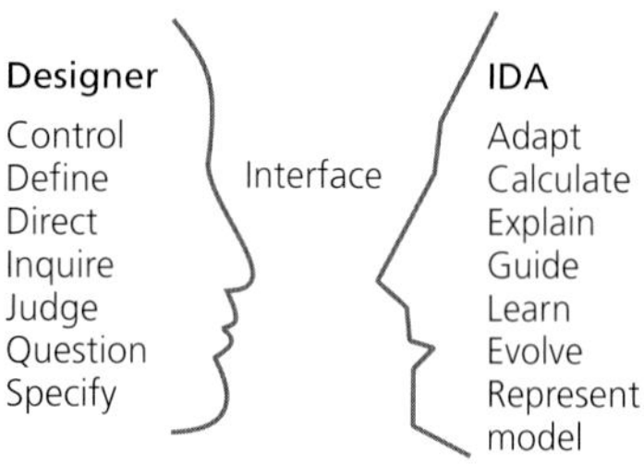

36.1 IDA

The CAD Centre was established in 1986 as a research and postgraduate unit now within the Department of Design Manufacture and Engineering Management. The aims of the centre are to develop the computing technology that supports a creative design partnership between man and machine, and to deliver the underlying technology, techniques and approaches to industry. The CAD Centre has an established track record in the fields of artificial intelligence and knowledge engineering applied at the early design stage. The centre has a long-term research goal to develop a fully integrated computing environment which supports design and its management based upon fundamental understanding and theories. Working towards this goal, in the mid-1980, the team originated and developed a long-term research strategy to realise a computer-based *intelligent design assistant* (IDA) which would be managed within an *integrated design environment* (IDE).

As a significant step towards achieving this goal, the CAD Centre has coordinated a 4year (2001–05) European Commission 5th Framework research project termed VRShips-ROPAX (Life-Cycle Virtual Reality Ship Systems). As VRShips-ROPAX (VRS) coordinators, the CAD Centre is responsible for directing 36 partners in 14 European countries towards the generation of an integrated, virtual, computer-based, generic platform in the marine domain. The platform supports distributed working practice in design, product modelling, performance analysis, simulation and process management between one or a number of VRS partners and supports any application domain or ship type. The VRS project encapsulates many of the previously developed theories, models, approaches and systems of the CAD Centre. In addition, based on the success of VRS, an extension of this research into a number of additional application domains is currently being sought.

Research approach

The CAD Centre has developed an integrated research approach where all aspects of design and CAD research are based upon industry needs and practice (reality) and performance improvement (envisaged reality).

Interdisciplinary collaboration occurs at a number of levels. Staff at the centre are from disciplines such as Computer Science, Electrical Engineering, Mechanical Engineering, Architecture (Building and Naval) and Product Design. In addition, the centre exists within a department whose expertise includes business process re-engineering, management of innovation, manufacturing management, manufacturing simulation, performance measurement, people and organisations, logistics and enterprise modelling.

Close links have also been established with research groups in other institutions through collaboration in the UK, industrial organisations, EU research projects and membership of standards and advisory committees.

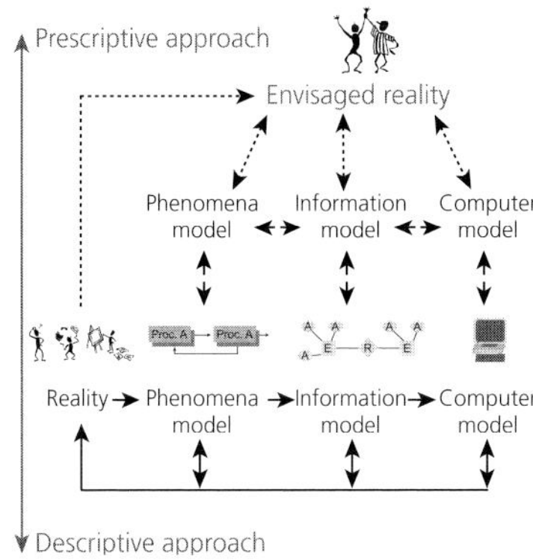

36.2 CAD Centre research approach

Research areas

Researchers at the CAD Centre have been active in the areas of artificial intelligence and advanced computational techniques, knowledge engineering, design management and coordination, design reuse, product modelling, process modelling, performance improvement and team-based design.

Knowledge engineering

A pioneering approach, encapsulated in a system named DENOTE, has been developed to model current working and domain knowledge over multiple viewpoints of the design solution, and manage its reuse and evolutionary nature.

A knowledge of lifecycle consequences model (KCM), based on knowledge-intensive component life-design synthesis, has been produced as a means to achieve lifecycle providence that design for multiple-X provides. A key conclusion from the work in knowledge modelling is that it is inextricably linked to the process or techniques being supported.

An engineering design reuse process model based upon the concept of the application, and not just the regurgitation, of knowledge has been proposed. The model, derived from examples in software engineering and engineering design practice, has acted as a base for the implementation of computational models within a prototype system.

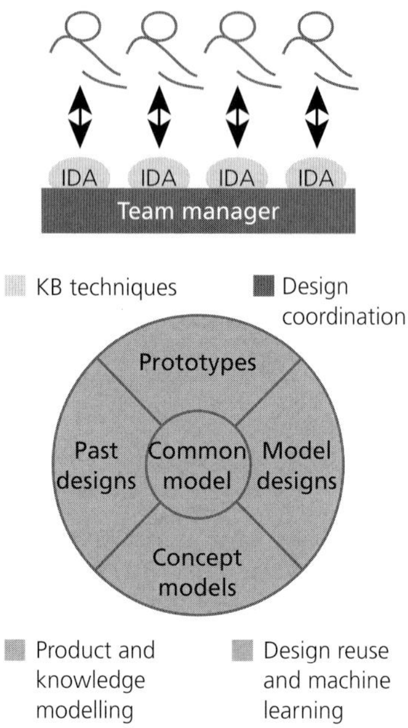

36.3 Intelligent Design Environment (IDE)

Design management and coordination

A design coordination framework has been developed through cross-institutional cooperation after recognising that coordination is a fundamental bottleneck in product development. It is also critical as the activity that requires successful management of the multi-disciplinary aspects of the design.

Shared workspaces for distributed design environments have been developed in order to support the multiple contributions which form a design process. These utilise existing technology to support synchronous application, video and voice communication and file transfer between distributed design teams. Furthermore, a role-interaction approach has been used to map the roles and their relationships within an organisation and thereby better understand and facilitate cooperation and automation within a diverse team.

Design reuse and machine learning

Formalisation of the learning activities in engineering design through addressing what, when and how designers learn, to provide a basis for further projects and a unifying theme in learning and design reuse. This activity has encompassed automated learning and generalisation of knowledge from past designs for use in the early conceptual design stages.

Product and knowledge modelling

Conventional CAD systems support geometric modelling that is well defined and precise. However, in the early stages, geometry is often ambiguous and uncertain in nature. Strathclyde have developed a 'vague geometric modeller' for supporting this early stage geometric modelling, where geometry is often vague and ambiguous in nature, allowing uncertain concepts to be built and analysed.

Methods and tools

Researchers are currently building on the above theories, methods, principles and systems to enhance the spectrum of the CAD Centre research and ensure its industrial applicability. A number of key areas are currently under investigation and returning encouraging results in application; these include:

- A 'multiple viewpoint modular design methodology' which capitalises on advances in knowledge engineering and product modelling. The approach has been applied in two design organisations resulting in a 26% increase in the modularity of the products.

- A process modelling and optimisation method and accompanying computational support tool, employing techniques including IDEF models, structure matrices, genetic algorithms and multiple criteria optimisation. The approach has been successfully applied to a number of processes relating to pre-contractual design work within a large shipyard and has resulted in between 51 and 70% reduction in the amount of iteration in the process.
- An E2 Model of performance management, focusing on the effectiveness and efficiency of processes, is currently being utilised to support design process performance management and has supported the definition of 3000 process performance metrics.
- A design for distribution (D4D) framework, based on the results of two live industrial case studies. The framework allows companies to analyse and prescribe solutions within their distributed design process.
- A virtual environment for presenting tolerance analysis data, applied to a gap and flush study on a Jaguar X200 glovebox assembly.

Further reading

Duffy AHB, O'Donnell FJ (1998) A design research approach. AID'98, Lisbon, Portugal

Fleming DA, Forbes GA, Hayfron LE, Duffy AHB, Ball PD (2003) Optimisation of the estimating and tendering process in warship refit – a case study. CAPE'03, Edinburgh, UK

MacGregor SP (2002) New perspectives for distributed design support. Design Research, special issue on Design as a Social Process, 2(2)

O'Donnell FJ (2000) A methodology for performance modelling and analysis in design development. CAD Centre

Smith JS (2002) A multiple viewpoint modular design methodology. PhD thesis, University of Strathclyde, Glasgow, UK

Wu Z, Duffy AHB (2002) Using protocol analysis to investigate collective learning in design. AID'02, Cambridge, UK

For further information, see
http://www.cad.strath.ac.uk/

Chapter 37

BAE SYSTEMS/Rolls-Royce University Technology Partnership for Design

Ken Wallace
University of Cambridge

37.1 The UTP at the Keynes House

Formed on 1 October 1998, the BAE SYSTEMS/Rolls-Royce University Technology Partnership (UTP) for Design is a long-term research partnership linking the two companies and the Universities of Sheffield, Southampton and Cambridge. The UTP is embedded within the Cambridge Engineering Design Centre (EDC), where it contributes significantly to the knowledge management research theme of the EDC.

BAE SYSTEMS and Rolls-Royce both supply complex products and services in aerospace and other industrial sectors. Essential market differentiators for the companies are performance, safety, reliability, short time to market, high quality and low cost of ownership. Continuous improvement of the product definition process is essential for the companies to maintain world-class performance against these measures.

The companies identified 3 key design areas that needed to be researched to enable further improvements to be made to the product definition process:

- role of innovation and people issues within the process (Sheffield);
- optimisation of the design taking into account all relevant factors (Southampton).
- management of the total knowledge needed for the design task (Cambridge).

The overall aim of the Sheffield research into human factors and innovation is to understand and improve the people and organisational aspects of the design process. The work adopts a broadly sociotechnical emphasis which lays stress on the interconnectedness of the social (people and organisational) and technical (methods, tools and techniques) issues.

The overall aim of the Southampton research into design search and optimisation is to understand, develop and improve the increasingly sophisticated search and optimisation software tools being used by both companies. These tools span commercial products, plug-ins to commercial codes and fully in-house capabilities.

The overall aim of the Cambridge research in engineering knowledge management (EKM) is to understand how to make more knowledge available to designers and engineers in a readily usable form. This includes both novices who are acquiring expertise in a particular area, as well as experienced staff who need to move into a new area to meet changing business requirements.

The UTP's overall research programme represents a novel approach based around the interaction of technologies, tools, processes and people.

37.2 Rolls-Royce Trent 700
Reproduced with the kind permission of Rolls-Royce plc

Engineering knowledge management

In the future, there are likely to be fewer opportunities to talk to the experienced designers and technology experts who were involved in previous projects. The specific aims of the EKM research are:

- to understand the *capture, storage and retrieval* of engineering design knowledge;
- to understand decision making in engineering design and the nature of design expertise;
- to develop theories that can form the basis of new methods and tools;
- to develop and test prototype methods and tools.

Engineering design is a knowledge-intensive activity. Knowledge exists in the heads of individuals and provides them with the capacity to make decisions and adopt courses of action. What is stored and transmitted externally is information and data. Knowledge is generated and evolves: (1) by observing; (2) by interpreting information; and (3) through reasoning.

Explicit knowledge can be articulated. Once articulated, it can be represented as information, e.g. written down, and thus stored externally and transferred. An example of explicit knowledge is the factual description of a process or product. Implicit knowledge cannot be articulated by the person possessing it, but it is possible for it to be elicited and articulated by others. An example of implicit knowledge is the strategy adopted by an experienced designer to undertake a particular task in the design process. Tacit knowledge cannot, by definition, be articulated, but its role in the design process can be investigated. An example of tacit knowledge is the intuitive feel that an experienced designer has for the correct shape of a component in a product.

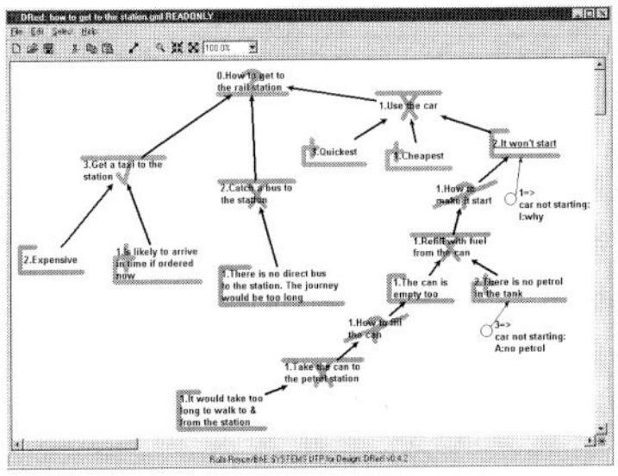

37.3 DRed screenshot

The intention is to articulate more process and product knowledge, *i.e.* transform it into information that can be stored and retrieved externally. Two particular research approaches are adopted: (1) observational studies in industry to obtain data; and (2) rapid development and testing of robust prototype software tools. The EKM research is organised under six projects for which the objectives are set out below.

37.4 Eurofighter
Reproduced with the kind permission of Rolls-Royce plc

Knowledge structure for design

- To understand how engineering design knowledge can be structured.
- To develop and evaluate an overall framework for the capture, storage and retrieval of engineering design knowledge.

Use of experience in design

- To understand the relationships between the strategies adopted by experienced designers and how designers retrieve information.
- To develop and evaluate an indexing structure for capturing and retrieving knowledge and experience.
- To understand the differentiation between design expertise and design experience.

Design rationale capture

- To understand how design rationale can be captured, stored and retrieved in a wide spectrum of design processes.
- To develop and evaluate a software tool for capturing design rationale.

Capturing and structuring design knowledge

- To understand the information requests of designers and establish what knowledge to capture.
- To undertake observational studies to answer questions such as: What triggers an engineer's need for knowledge? What types of knowledge do engineers require? How do engineers search for knowledge?

Retrieving and using design knowledge

- To understand how designers retrieve information from paper and electronic sources during the design process.
- To undertake case-study experiments to answer questions such as: Why do some designers prefer paper-based documentation? What differences are there when electronic documentation is used?

Information retrieval using design guidelines

- To understand how engineering designers intuitively structure, relate and store information in their heads.
- To undertake experiments to answer questions such as: Do engineers order information in recognisable patterns? Will a classification of design information based on such patterns improve retrieval?

37.5 Rolls-Royce Trent 800
Reproduced with the kind permission of Rolls-Royce plc

Further reading

Ahmed S, Wallace KM (2004) Identifying and supporting the knowledge needs of novice designers within the aerospace industry. Journal of Engineering Design, 15(5)

Aurisicchio M, Langdon PM, Ahmed S, Wallace KM (2003) Investigating knowledge searches in aerospace design. ICED'03, Stockholm, Sweden

Bracewell RH, Wallace KM (2003) A tool for capturing design rationale. ICED'03, Stockholm, Sweden

Del-Rey-Chamorro FM, Wallace KM (2003) A study of information retrieval in the aerospace domain. ICED'03, Stockholm, Sweden

Japikse R, Langdon PM, Wallace KM (2003) Structuring engineering design information using a model of how engineers' intuitively structure design information. ICED'03, Stockholm, Sweden

Wallace KM, Ahmed S (2003) How engineering designers obtain information. In: Human behaviour in design. Springer

For further information, see
http://www-edc.eng.cam.ac.uk/

Chapter 38
M.J. Neeley School of Business, Texas Christian University

Tyson R Browning

38.1 The M. J. Neeley School of Business

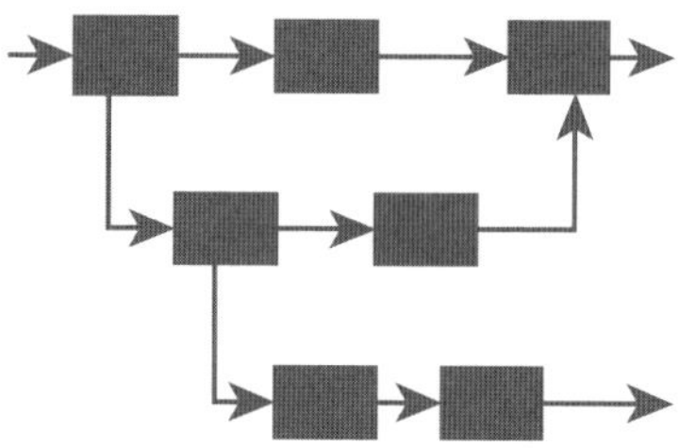

Original, typical process flowchart model

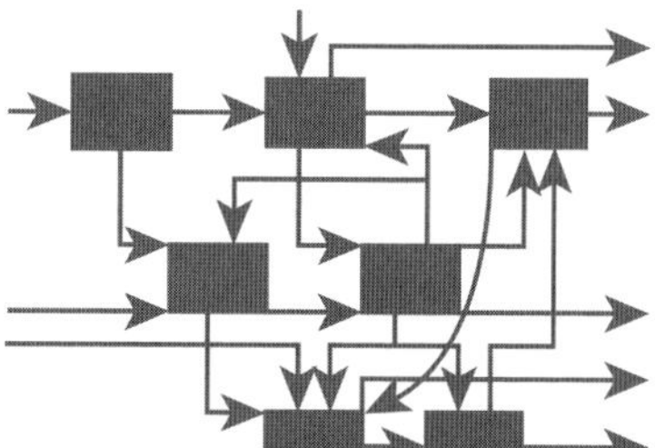

One view of an information-flow-based process model

38.2 There are more interactions in design processes than are typically captured in most process models

The M.J. Neeley School of Business at Texas Christian University has on its rolls 1470 undergraduate and 370 graduate students, a number of whom work with local companies such as Lockheed Martin, Bell Helicopter Textron, BNSF Railroad, Fidelity Investments, Motorola, Nokia, and many others. Because the design and development of timely and affordable products and services is a management issue as well as an engineering issue, the Neeley School is leading and collaborating on research to further the design, analysis, management and improvement of design processes.

Process modelling

Every organisation, team, or individual that does work and produces results has a process – a set of actions and interactions. That process may not be documented, modelled, effective, efficient, consistent, or understood, but it is the actual way the work occurs. A process model (or documentation, or map) is an abstract description of process reality. Our research seeks practical insights to five problems in process modelling:

1. How do we capture workers' tacit knowledge about reality – particularly in the context of large, complex design processes, where individuals do certain activities infrequently and intuitively?
2. A great amount of process modelling focuses on the activities or actions in a process and pays relatively little attention to the *interactions* (cf. many flowcharts that label the boxes but not the arrows), but the interactions are the primary driver of value in complex, iterative design processes. Our research has enhanced ways to build process models that capture 'system -level' process knowledge – i.e. that give emphasis to the interactions as well as the actions. Improved process models have first and foremost

enabled organisations to reach common understanding, vocabulary, and agreement about actions and interactions – thus enabling the establishment of realistic commitments, true empowerment and accountability.

3. Process models are built by a variety of constituencies within organisations and for a range of purposes, from project planning to compliance with standards. Our research addresses the diverse users of process models and their requirements. We have found that a single, rich process model with the power to satisfy a variety of purposes is preferable to having many disparate yet overlapping process models each targeting a particular purpose. Hence, we advocate process modelling on the basis of simple but highly 'enrich-able' components. The process models are object oriented, based on two fundamental objects representing actions and interactions, each with a myriad of potential attributes.
4. How realistic should a process model be? What is the necessary and sufficient set of attributes for a useful model? To what level should the actions and interactions be decomposed to understand and control the effects of their underlying structure on the process?
5. Many process models are paper based (even if stored as 'electronic paper') – making rapid and low-effort maintenance a problem – and not amenable to alternative representations. We are exploring ways to rapidly and inexpensively build, modify, update and enhance process models.

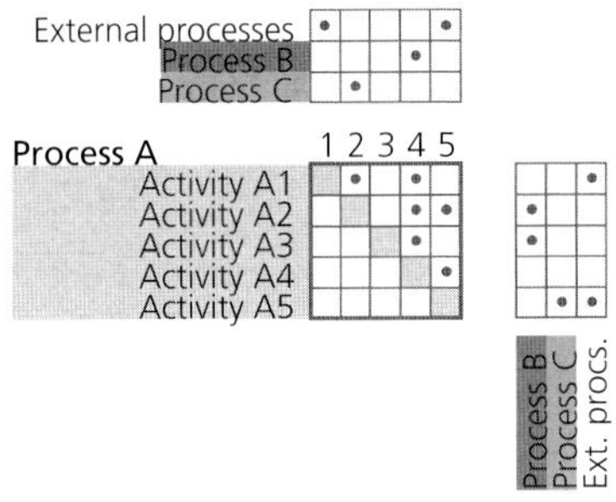

38.3 Design structure matrix. Representation of a process and its external interfaces

Process representation

A rich, integrated, object-oriented process model, capable of simultaneously satisfying the needs of a variety of users, resides in a database. No one representation or view is adequate to convey all of the attributes of each object or the emergent properties of the integrated set of objects. A system architecture is best represented through a number of views. Different users, querying the model for diverse purposes, will filter the model's information through one or more views. Our research explores views of process model information such as Gantt charts, flowcharts, narratives, design structure matrices, value stream maps, etc. For design process models, special emphasis is given to views that highlight activity interactions and their emergent effects on time, cost, quality and risk.

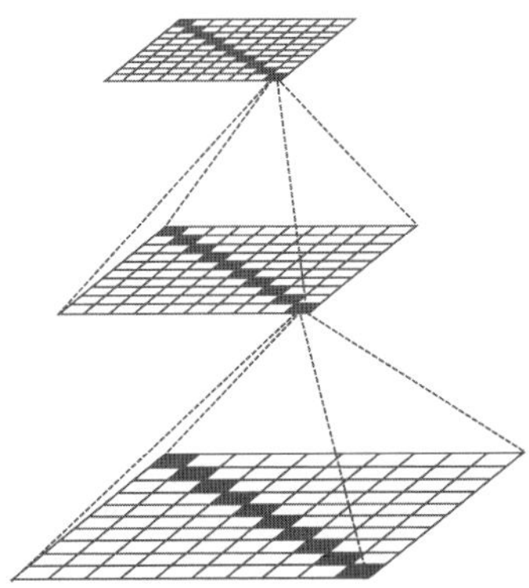

38.4 Process decomposition represented using design structure matrices

Process model usage

Project planning and scheduling

Unlike many business processes, such as volume manufacturing and order

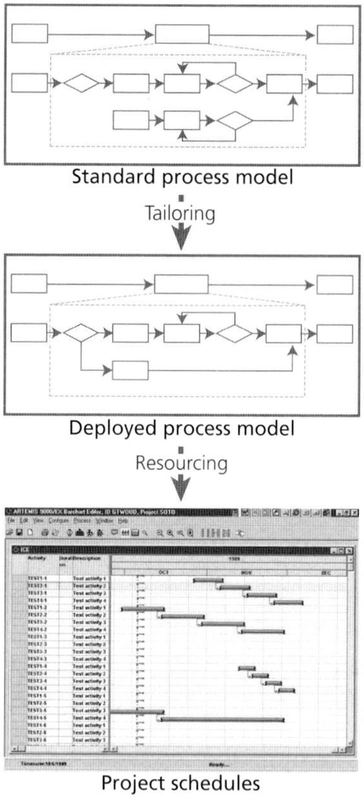

38.5 Project execution and control

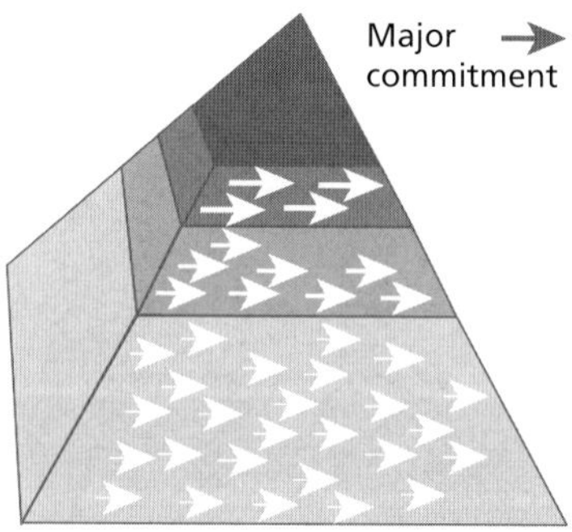

38.6 A stylised hierarchy of commitments

fulfilment, that seek to do exactly the same thing many times, the goal of product design is to do something new, once. Yet even product design projects have some repeatable structure from one instance to the next. Multi-project organisations seek to take advantage of this to benefit from economies of scale. Thus, many organisations, some prompted to comply with external standards and capability maturity models, keep a set of standard processes for use (to some extent) on all projects. However, since each project is somewhat different, the standard processes must be tailored and/or scaled in each instance. Our research explores guidelines for and ways to automate 'process (model) deployment' and process-based project planning and scheduling.

Project execution and control

The planned interactions in a design process signify agreements and commitments. Thus, the design process may be viewed as a holonic network of commitments. Breakdowns in low-level commitments provide a 'leading indicator' of risks in meeting higher-level commitments. With a 'nervous system' that quickly communicates changed commitments throughout a project, project managers can take appropriate corrective actions and re-plan in near real time. Our work on information-driven project management and deliverable-oriented project management explores this perspective.

Risk and opportunity management

Managing a product design project is a risk management exercise. A product design is hypothesised early in the project; the rest of the project seeks to detail that design and confirm that it indeed satisfies expectations. Design activities seek to reduce uncertainties. Choosing which activities to do, and thus what information to create and when, and hence which uncertainties and risks to reduce, is the project planning problem. Ensuring that risks in fact go down according to plan is the project execution problem. This perspective motivates the risk value method, which we are studying as a promising approach to improved management of product design projects. The method enables trade-offs among areas of risk to maximise the overall value of the product design.

The actions and interactions forming the structure of the design process determine what information will be created when, and thus what work must be done based on assumptions instead of firmer information. We studied the impact of varying the design process architecture (through alternative project plans) on cost and schedule risk; ongoing work investigates the effects of design actions and interactions on project cost, duration, quality, and risk. We are also looking at the effects of resource constraints on the outcomes of iterative design processes.

Consequences or impacts may be positive. Opportunity is the opposite of risk – the expected rewards – a function of the probability of desirable outcomes and their rewards. Our research considers ways to manage opportunities and risks, both separately and jointly, to maximise the overall value of a product design.

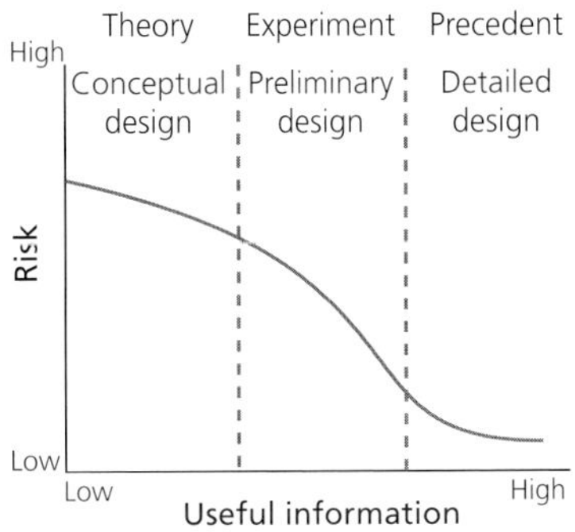

38.7 In the design process, risk decreases as activities produce useful information

Knowledge management

As an organisation plans and executes projects based on a process model, the model can be improved by incorporating lessons learned. We use a simple but highly extendible process modelling framework with unlimited ability to grow as the organisation learns. Ease of maintenance is emphasised. This improves the likelihood that knowledge will be captured. Using an improved process model as the basis for planning and controlling the next project forces closure of the organisational learning loop. A rich process model 'maps the genome' of a project, and the knowledge captured can be used to comprehend and improve processes in ways beyond our current understanding.

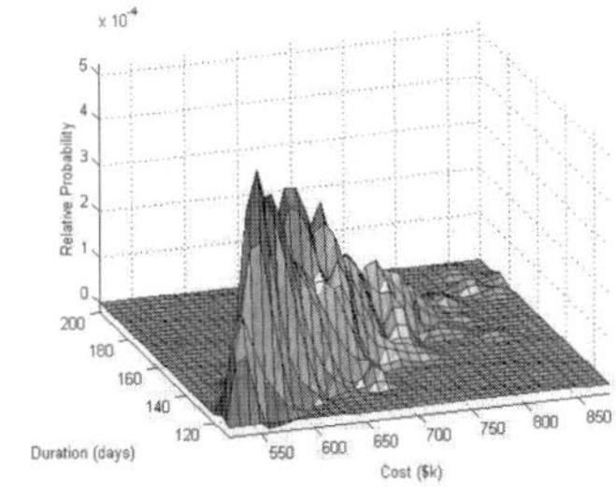

38.8 Joint probability distribution of cost and duration outcomes for a preliminary design process

Further reading

Browning TR (2001) Applying the design structure matrix to system decomposition and integration problems: a review and new directions. IEEE Transactions on Engineering Management, 48(3): 292–306

Browning TR (2002) Process integration using the design structure matrix. Systems Engineering, 5(3): 180–193

Browning TR (2003) On customer value and improvement in product development processes. Systems Engineering, 6(1): 49–61

Browning TR, Eppinger SD (2002) Modeling impacts of process architecture on cost and schedule risk in product development. IEEE Transactions on Engineering Management, 49(4): 428–442

Browning TR, Deyst JJ, Eppinger SD, Whitney DE (2002) Adding value in product development by creating information and reducing risk. IEEE Transactions on Engineering Management, 49(4): 443–458

For further information, see http://www.neeley.tcu.edu/

Chapter 39
Technological Innovation Research Group, Politecnico di Torino

Laura Alessi and Marco Cantamessa

39.1 Technological Innovation Research Group
Luigi Amerio © Politecnico di Torino

Technological Innovation Research Group (GRIT) approaches the topic of technological innovation from a multidisciplinary perspective that encompasses economic, engineering and management issues. The group focuses on sustaining innovation, new product development and the organisational change required to exploit the capabilities offered by innovation, especially when related to information and communication technology (ICT).

GRIT's location in Turin within the Politecnico di Torino, which is Italy's second largest technical university, is a key component to the group's success. This is due to close links with both academic researchers in different branches of engineering and with local industry, where the latter is currently undergoing significant structural change.

The approach

GRIT's research is based on the notion that technological innovation has nowadays become a complex phenomenon. Successful innovation depends on the convergence of efforts made by different stakeholders, ranging from business managers across the supply chain to government policy-makers. This calls for a multidisciplinary approach in which these multiple perspectives can simultaneously be taken into account.

GRIT operates according to the conceptual scheme depicted in the diagram in Figure 39.2. The boxes on the right-hand side represent a generic 'innovation value chain' in which results from research are blended with, and transferred, to industry, which then incorporates them in marketable goods and services. The research themes indicated in white coincide with priorities set by the Italian National Research Plan (Infoscience, Nanoscience, Bioscience) and are therefore currently receiving (or soon will) special attention within GRIT projects. GRIT covers the 'innovation value chain' according to the following three main themes:

Innovation policy

The theme deals with innovation dynamics in the context of the economic system. The goal of the approach is to define instruments for strategic planning, policy-making and for science and technology. Research projects include empirical research and the development of analytical models. The lead researcher is Mario Calderini, Associate Professor of Economics of Innovation.

Business and product development

The theme covers the management of the technological innovation process, ranging from strategic issues such as technology assessment, project portfolio management and product architecture design to operational problems associated with the product development process. Research is based on empirical instruments (case studies and surveys) as well as on the development of decision-support tools. Research is targeted at a variety of industries, with a special attention given to engineering and ICT. The lead researcher is Marco Cantamessa, Associate Professor of Technology Management and Product Development.

Industry transformation

The theme studies the transformation that product and process innovations induce on companies, considering both their internal assets and their relations in the value chain. Research is mainly based on empirical studies in which the impact delivered by ICT receives special attention. The lead researcher is Emilio Paolucci, Associate Professor of Business and Information Systems.

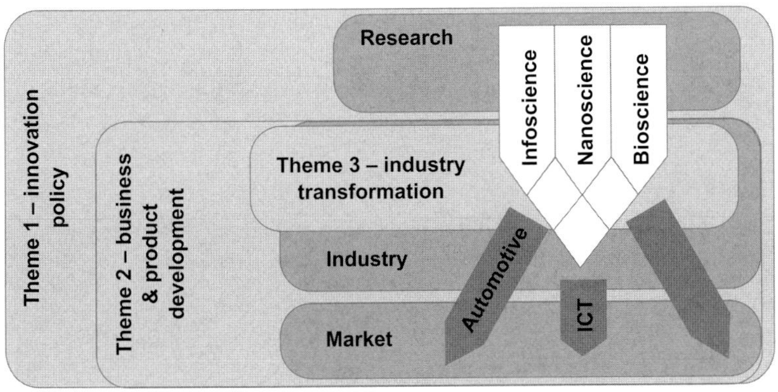

39.2 Three main themes within the GRIT

The group

GRIT is managed by three leading experts, each of whom is responsible for one of the themes described above. The group also includes eight researchers and doctoral students with engineering, business and economics degrees who constantly interact in their research projects, thus allowing implementation of a truly multidisciplinary approach. In addition to the Politecnico di Torino's researchers, GRIT involves a number of research affiliates that work on related projects with the Mario Boella Institute (www.ismb.it, specialising in ICT) and the Fondazione Rosselli (www.fondazionerosselli.it, specialising in economics of innovation).

Research projects are generally externally funded, with grants awarded by both private firms and government bodies (the European Union, the Italian government, national regulation authorities and the regional government of Piedmont).

Research in product design and development

Research in product design and development is central to the second theme. As mentioned, there is a strong interplay between empirical and normative research projects with respect to the product development process.

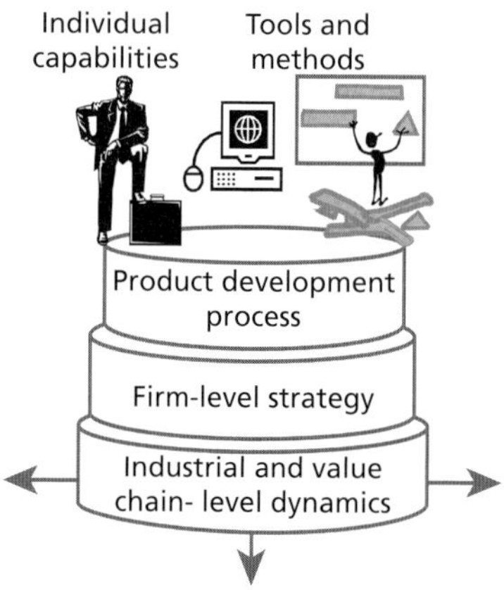

39.3 Reference model for empirical research on the product development process

Empirical research is primarily concerned with understanding the way in which technological innovation, together with novel design support methods and tools impact the design function of the firm. The outcome of this strand of research is a set of best practices for design management at a strategic level taking to account the roles of industry, firm-level dynamics and designers' individual capabilities (Figure 39.3).

Normative research is currently concerned with the relationship between organisation, product architecture and technological innovation. This research is currently shifting the unit of analysis from the individual firm to inter-firm relationships in the supply chain and deals with design coordination, negotiation support in new product development and representations of distributed knowledge on product architecture.

Further reading

Calderini M, Garrone P (2001) Liberalisation, industry turmoil and the balance of R&D activities. Information Economics and Policy, 13: 199–230

Calderini M, Garrone P, Sobrero M eds (2003) Market structure, corporate governance and innovation. Edward Elgar

Calosso T, Cantamessa M, Vu T, Villa A (2003) Production planning and order acceptance in business to business electronic commerce. International Journal of Production Economics, 85: 233–249

Cantamessa M (2003) An empirical perspective upon design research. Journal of Engineering Design, 14(1): 1–15

Cantamessa M, Valentini C (2000) Planning and managing manufacturing capacity when demand is subject to diffusion effects. International Journal of Production Economics, 66(3): 227–240

Cantamessa M, Villa A (2000) Product and process design effort allocation in concurrent engineering. International Journal of Production Research, 38(14): 3131–3147

Clark W, Paolucci E (2001) Commercial development of environmental technologies for the automotive industry; towards a new model of technological innovation. International Journal of Technology Management, 21(5/6): 565–585

Corso M, Paolucci E (2001) Fostering innovation and knowledge transfer in product development through information technology. International Journal of Technology Management, special issue on Implementation of Business Process Innovations, 22(1/2/3): 126–148

Corso M, Martini A, Paolucci E, Pellegrini L (2001) Knowledge management in product innovation: an interpretative review. International Journal of Management Review, 3(4): 341–352

Corso M, Martini A, Paolucci E, Pellegrini L (2003) Technological and organizational configurations for knowledge management. An empirical analysis in Italian SMEs. Integrated Manufacturing Systems, 14(1): 46–56

For further information, see
http://www.polito.it/research/grit

Chapter 40
Intelligent Interactive Distributed Systems, Vrije Universiteit, Amsterdam

Frances MT Brazier and Niek JE Wijngaards

40.1 Vrije Universiteit, Amsterdam

Design, and especially distributed design, is a typical application of dynamic agent systems, involving a highly interactive, heterogeneous agent population in which cooperation is of great importance. The focus of the design research at our Intelligent Interactive Distributed Systems (IIDS) research group is on distributed design of dynamic artefacts, with an emphasis on management of coordination processes, researching both theory and implementation. Dynamic artefacts include self-configuring computer systems (cf. IBM's autonomic computing), autonomous self-adapting artefacts (for example, software agents), and artefacts designed to be re-designed (for example, office buildings, PC-hardware). The IIDS group is part of the Computer Science Department of the Faculty of Sciences at the Vrije Universiteit Amsterdam, and combines expertise from artificial intelligence (AI) and computer systems (see http://www.iids.org/ for more information).

Design

The IIDS group builds on a decade of theoretical and prototypical research in AI & Design. The initial generic design model, introduced in 1994, is based on the premise that design involves exploration and reflection. Within this model, manipulation of the description of the design object is explicitly separated from manipulation of sets of qualified requirements, which is in turn separated from coordination of the overall design process: a structure which facilitates acquisition and explication of design domain and design process knowledge, including strategic knowledge, design rationale and conflict management. The generic design model has been formalised and tested in a number of domains, including aircraft emergency exit design, self-adapting software agents, environmental inventory model design, automated Web-service configuration, aircraft toilet unit design and elevator configuration, yielding a number of prototype implementations and increasing insight into the types of strategy used in design.

Design process improvement

40.2 Do all chimneys count as chimneys?
© Michiel Wijnbergh

The generic design model's usefulness has been proven in both elicitation and acquisition of design-related knowledge, as well as in modelling and implementation of (automated) design systems and design support systems. The generic design model has been used as a shared model between knowledge engineers and designers. In one project the generic design model was used as the graphic metaphor to control a design-support system for the design of environmental policies. In another project the generic design model was used as a metaphor for the design of a decision-support system for an environmental inventory of Dutch brick and tile fabrication. Both individual designers and teams of designers have used this model to explicate and to improve their design process. For example, in both projects the shared design process ontology and shared process model's explicit incorporation of design process coordination, design strategies, conflict resolution and rationale have facilitated the design and development of support systems.

Design and the Internet

The Internet is a dynamic environment in which software agents and Web services appear, disappear, roam, interact and cooperate. This large-scale ever-changing environment requires agents and web services to be flexible and adaptive, at this scale an unmanageable process for humans. Unsupervised automated (re-) design of agents (Brazier *et al.*, 2001b) and (re-) configuration of Web services (van Splunter *et al.*, 2003) are domains of application involving the study of dynamic (software) artefacts which are designed to be re-designed.

Distributed design

One of the greatest benefits of the Internet is the ease of communication between (human or automated) agents: a natural environment to investigate automated support of design teams (involving both human and automated participants) on the basis of an understanding of distributed design processes. Distributed design involves many participants, each with their own expertise, experience, and goals, requiring participants to deliberate about coordination (Brazier *et al.*, 1997), trust, reflection (Brazier *et al.*, 2001c), and the design

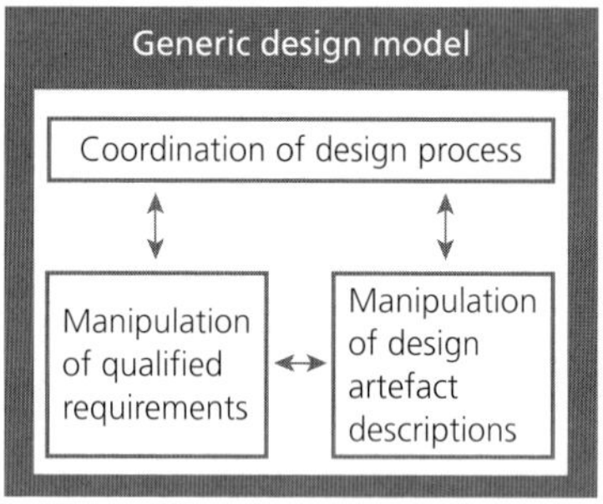

40.3 Generic design model

process at hand. As such, the area of distributed design provides a research setting in which results from studies of teams of human designers can be applied to, and structured by, models of automated design agents.

By combining the generic design model with the generic agent model, a model of a generic design agent (Brazier *et al.*, 2001a) resulted: a basis for studying the knowledge and behaviour of individual agents in distributed design processes. A design agent model explicitly distinguishes communication, service interactions, and self-management from its design capabilities, providing a structure to model aspects of (human) designers involved in cooperative design tasks. Each designer has his or her own view of the world and other agents, and their environments, including assessments of their expertise, reliability, experience, etc. The role of trust in distributed design is of special interest, as trust is often left implicit in design studies, yet it determines the way in which members of a design team assess and incorporate each others' designs, objectives, and evaluations.

Initial coordination models (Brazier *et al.*, 1997), acquired in the context of aircraft design, form the basis for our research in coordination models for distributed design. Task delegation is only one aspect of these coordination models: existing models and heuristics for the resolution of conflicts, distribution of design strategies, and sharing of design process objectives, sets of qualified requirements, and (partial) design object descriptions are aspects that also need to be considered. In addition, coordination models need to include distributed design histories and rationale.

Design challenges

The current overall progress of task-based design-process coordination is promising, yet needs to increase the understanding of distributed design to be able to provide automated support for such processes. The Internet has been shown to be a suitable domain to demonstrate theories and investigate emergent effects. It is also a domain that itself could do with support from the design community with respect to agent and Web-service (re-) configuration processes. The real proof of viability is, however, to be found in design practice.

40.4 Fokker airplane
Supporting design of complex artefacts

Further reading

Brazier FMT, van Langen PHG, Ruttkay ZS, Treur J (1994) On formal specification of design tasks. AID'94, Lausanne, Switzerland

Brazier FMT, Jonker CM, Treur J (1997) Formalization of a cooperation model based on joint intentions. ATAL'96, Budapest, Hungary

Brazier FMT, Jonker CM, Treur J, Wijngaards NJE (2001a) Compositional design of a generic design agent. Design Studies, 22(5): 439–471

Brazier FMT, Jonker CM, Treur J, Wijngaards NJE (2001b) Deliberative evolution in multi-agent systems. Software Engineering and Knowledge Engineering, 11(5): 559–581

Brazier FMT, Moshkina LV, Wijngaards NJE (2001c) Knowledge level model of an individual designer agent in collaborative distributed design. Artificial Intelligence in Engineering, 15(2): 137–152

Van Splunter S, Sabou M, Brazier FMT, Richards D (2003) Configuring web services using structurings and techniques from agent configuration. WI'03, Halifax, Canada

For further information, see
http://www.iids.org/

Chapter 41

The Center for Engineering Learning and Technology, University of Washington

Robin S Adams, Jennifer Turns and Cynthia J Atman

The College of Engineering at the University of Washington established the Center for Engineering Learning and Teaching (CELT) in 1998 to close the gap between engineering education and engineering practice. CELT is the nation's first engineering education center with a dual purpose of conducting rigorous research on engineering student learning while providing instructional development services for engineering faculty. In just a few years CELT's methodical, data-driven work across research and teaching practice has transformed students' understanding about their own lives as engineers, faculty perceptions about adopting new approaches to teaching, and corporate expectations of entry-level engineers. CELT's efforts have attracted broad attention and much of our work is funded through grants from national organisations.

Working with interdisciplinary colleagues across the University of Washington campus, CELT pursues activities in three strategic areas:

- *Research* – understanding what engineering graduates need to know and how they can best master the skills and knowledge necessary for success as engineering professionals.
- *Instructional development services* – working with faculty to shift their focus from teaching to learning and incorporate active-learning and student-centred practices in their courses.
- *Advancing the CELT model* – developing programmes and organisational models for replicating the CELT approach across the nation, accelerating rates of change and expanding the impact of reform in engineering education.

Results from these efforts have been presented at national meetings and published in major journals in the field.

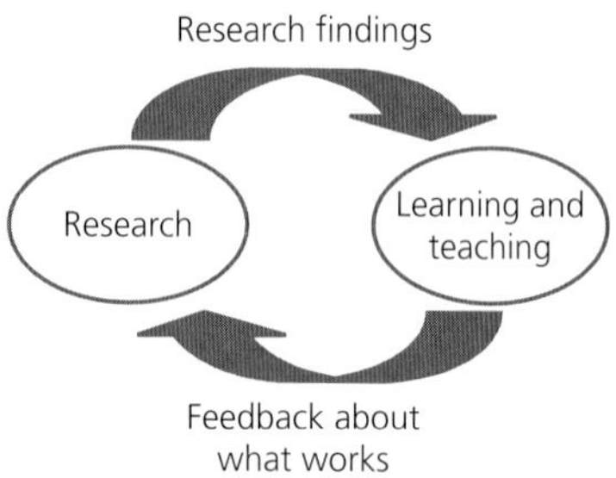

41.1 The CELT model
Synergistic activities in research on engineering student learning and programmes to improve teaching in engineering classrooms

Because design is central to engineering practice, a significant CELT emphasis has been promoting a research-informed approach to design education. In such an approach, teaching is influenced by existing research on engineering student learning. Our CELT model illustrates this relationship, highlighting the important role of research in engineering student design learning and the teaching of engineering design. The model also illustrates two kinds of synergistic cross-talk: instances of teaching giving rise to research and instances of research informing teaching practice. The remainder of this chapter provides examples of CELT's research-informed approach to advancing engineering design education in 4 areas:

- Empirical characteristics of design process knowledge – design processes, iteration, problem scoping behaviour, communication practices, and use of design representations.
- Empirical characteristics of design content knowledge – conceptions of design and professional engineering practice.
- Metrics – metrics and methodologies for characterising the design process and content knowledge.
- Bridging research and teaching practice – research-based instructional activities, assessment tools, and workshops for design educators.

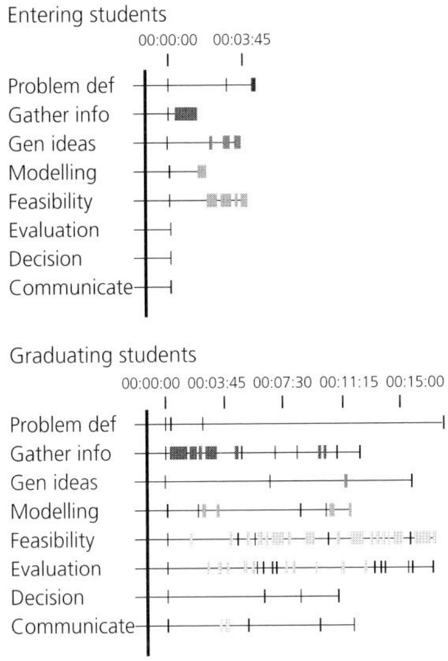

41.2 Individual design processes become more complex as students gain experience

Design process

Because little is known about how engineering students approach design problems and the difficulties they face, a core research focus has been on providing empirical accounts of student design activity. We have conducted multiple empirical studies building on three datasets to comprehensively document and understand engineering student design processes. We are also working with industry partners to extend this work with studies of practising engineering design experts. Particular foci are documenting expert behaviours and knowledge, comparing experts with the students in our earlier studies to identify appropriate learning targets and educational experiences in the teaching of design, and developing a continuum for describing the learning of design. Our research in the area of design processes are described in the following paragraphs.

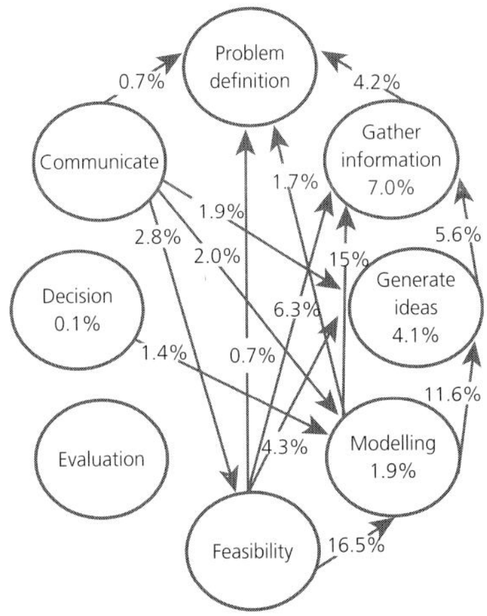

41.3 Effective designers iterate frequently revisiting both problem and solution elements (Reprinted from Adams *et al.*, 2003)

Iteration

Design problems are frequently ambiguous and ill-structured and can have multiple solutions. As a result, a designer's understanding of a problem or possible solutions evolves through a process of iteration. Although iteration is widely considered a key aspect of engineering design behaviour, little research has focused on how designers engage in iterative activity or how iterative activity supports design performance. A significant part of our work has been the development and use of coding schemes to analyse iterative activity across levels of performance and experience.

Problem scoping

Problem scoping refers to the portion of the design process where designers define the nature of the design problem and the space in which they will search for design solutions. Our focus has been on developing strategies for documenting problem scoping behaviours and using these strategies to analyse patterns in problem scoping behaviour.

Communication

Because design is often a team-based activity, we are working towards understanding communication practices in design teams, particularly interdisciplinary teams of engineering students.

Design representations

Design representations are multifunctional – they serve to capture a designer's current understanding of a problem, provide feedback to a designer about strengths and weaknesses of a particular solution approach, and communicate both prototype ideas and a final design to others. A recent CELT research focus is on analysing the ways in which designers' creation and manipulation of representations support their design activity.

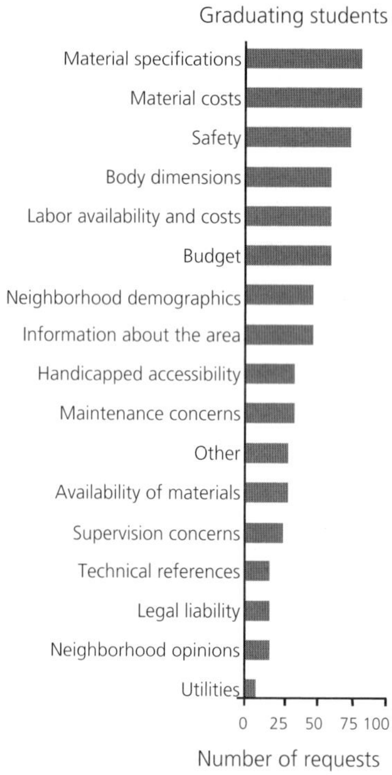

41.4 Engineering designers gather broad categories of information encompassing technical, environmental, social, and economical issues (Reprinted from Adams *et al.*, 2003) © 2003 Elsevier

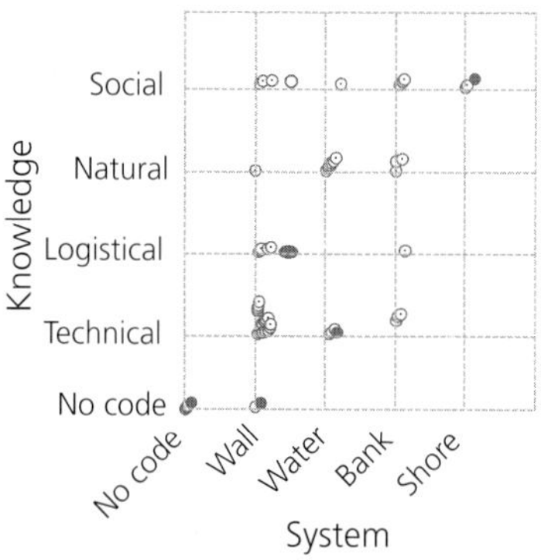

41.5 Engineers utilise their design knowledge in relation to a problem's frame of reference (Reprinted from Adams *et al.*, 2003) © 2003 Elsevier

Empirical characteristics of design content knowledge

Beliefs about the nature of design and design processes can shape a designer's strategy in solving design problems. For example, we have found that engineers who consider iteration as central to design are more likely to spend time iterating and use a wider variety of iterative strategies to work towards a high-quality solution.

In interdisciplinary work, professionals need to be able to translate their knowledge to others, understand the contributions of other professionals, and operate in the areas between professions. They also need an integrated understanding of their own discipline, an ability to integrate their skills in the context of a given project, and an ability to evolve their conceptions as they acquire new knowledge.

Metrics

We have identified and validated a set of measures for capturing various aspects of design activity. These include measures of design processes, design content, design performance, and conceptions of professional practice. We have also developed representational formats such as process timelines and problem space grids for illustrating differences in design behaviour and inquiring into the nature of design ability. Some of these metrics and methodologies have been transformed from research tools into instructional tools for improving design education.

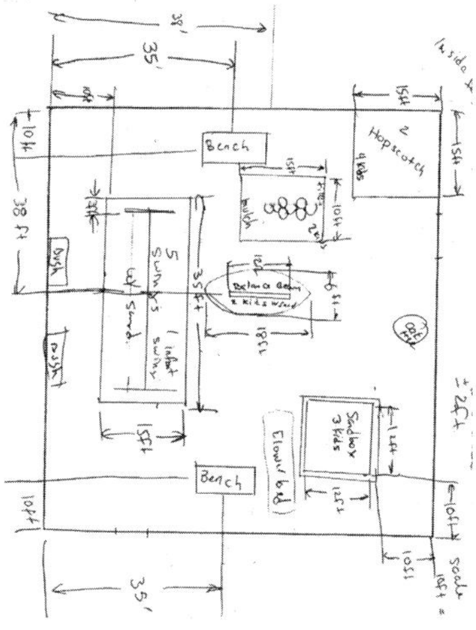

41.6 Sketching supports engineering design activity in a variety of ways

Bridging research and practice

A critical feature of CELT's research-informed approach is making robust links between what we know about how engineers design and how to educate effective designers. Because of the complexity of real education practice, we have been developing and demonstrating multiple strategies to bridge research and practice. These all serve one goal: enhancing engineering student learning.

Further reading

Adams RS (2002) Understanding design iteration: representations from an empirical study. London Design Research Society: Common Ground

Adams RS, Turns J, Atman CJ (2003) Educating effective engineering designers: the role of reflective practice in design studies. Design Studies, 24(3): 275–294. Figures reprinted with permission from Elsevier

Atman CJ, Turns J (2001) Studying engineering design learning: four verbal protocol analysis studies. In: Design learning and knowing. Lawrence Erlbaum

Atman CJ, Chimka JR, Bursic KM *et al.* (1999) A comparison of freshman and senior engineering design processes. Design Studies, 20(2): 131–152.

Bursic KM, Atman CJ (1997) Information gathering: a critical step for quality in the design process. Quality Management, 4: 60–75

Safoutin MJ, Atman CJ, Adams R *et al.* (2000) A design attribute framework for course planning and learning assessment. IEEE Transactions on Education, 43(2): 188–199

Turns J, Atman CJ, Cardella M, Adams R (2002) Do we see within subject change? Four cases of engineering student design processes. London Design Research Society: Common Ground

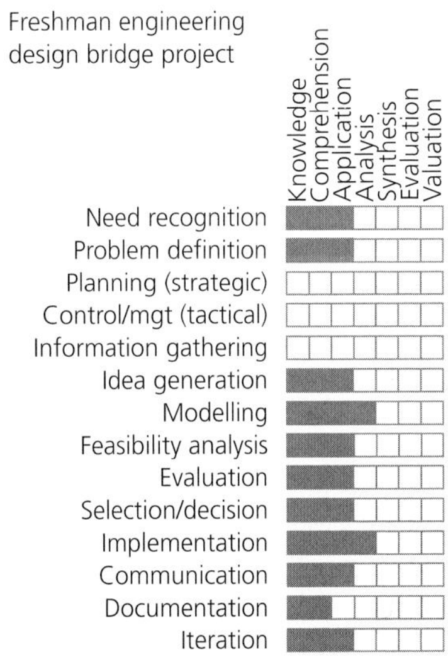

41.7 Frameworks for characterising design activity can be used to assess design education experiences (Safoutin *et al.*, 2000) © 2000 IEEE

For further information, see
http://depts.washington.edu/celtweb/

Index and authors

Keyword index

This keyword index cites the first page of each chapter in which there is a significant discussion of the term.

Authors, biographies

Editors

John Clarkson

Cambridge Engineering Design Centre, University of Cambridge. Reader in Engineering Design and director of the Cambridge Engineering Design Centre. Interests include inclusive design, design evaluation, design optimisation and design process improvements.

Claudia Eckert

Cambridge Engineering Design Centre, University of Cambridge. Senior Research Associate and assistant director of the Cambridge Engineering Design Centre. Interests include all aspects of design process improvement, design cognition and comparisons between design domains.

Contributing authors

Saeema Ahmed

Cambridge Engineering Design Centre, University of Cambridge. Senior Research Associate and deputy co-director of the BAES/RR University Technology Partnership for Design. Interests include product design, the use of experience and expertise in design, indexing structures for capturing engineering design knowledge and empirical design research methods.

Rob Bracewell

Cambridge Engineering Design Centre, University of Cambridge. Senior Research Associate and assistant director of the Engineering Design Centre. Interests include design research methodology, design rationale capture, design synthesis, knowledge-based engineering and rapid development of usable prototype engineering software tools.

David C Brown

Artificial Intelligence in Design Group, Worcester Polytechnic Institute, MA, USA. Professor of Computer Science and of Mechanical Engineering. Interests: computational models of engineering design and the applications of artificial intelligence to engineering and manufacturing. Editor in Chief, Cambridge UP journal "AI EDAM: AI in Engineering, Design, Analysis and Manufacturing".

Tyson R Browning

Department of Information Systems and Supply Chain Management, M.J. Neeley School of Business, Texas Christian University. Assistant Professor of Enterprise Operations. Research interests include product development, systems engineering, process modelling, project management and enterprise engineering.

Peter Buckle

Robens Centre for Health Ergonomics, University of Surrey. Professor of Health Ergonomics and director of the RCHErg. Interests include patient safety, participatory ergonomics, systems design and evaluation of interventions.

Jerry Busby

University of Lancaster. Senior Lecturer in Management Science. Interests include systemic failure, human and organisational error, risk analysis and risk management, design organisations and design processes.

Jonathan Cagan

A Professor in the School of Design at Carnegie Mellon University, with an appointment in the Biomedical Engineering Program. His research, teaching and consulting are in the areas of product development and design history.

Marco Cantamessa

Department of Manufacturing and Economics, Politecnico di Torino. Associate Professor of Management of Innovation and Product Development. Interests cover management and organisational issues associated with design (management of the design process, implementation of design-support methods and tools, product architecture).

Martina Dündorfer
Managing director of the Management Perspective (Augsburg, Munich). Expertise: process management, human resources process engineering, value-based management and human capital management.

Chris Earl
Professor of Engineering Product Design and Head of the Department of Design and Innovation at the Open University. Research interests include generative design, planning for design and manufacture and shape interaction.

Timothy Jarratt
Former research student and Research Associate at the Cambridge Engineering Design Centre, University of Cambridge. Now working as an associate consultant at Bain & Company, Inc., London. Interests: engineering change management, product modelling and product design process.

Jeffrey Johnson
Professor of Complexity Science and Design, Department of Design and Innovation, Open University. Research interests include the foundations of complexity science and applications in the design and management of complex physical and socio-technical systems. Applications include the design process, organisational dynamics; urban and regional planning; traffic dynamics; robotics; machine vision and self-organising machines.

Michael Koch
Scientific assistant at the Institute for Engineering Design, University of Erlangen-Nuremberg, Germany. Interest: digital mock-up, early stages of design and function structures.

Kristina Lauche
A lecturer at the University of Aberdeen and a co-director of the Industrial Psychology Research Centre. Interests include team interaction and non-technical skills in design and other complex work.

Udo Lindemann
Head of the Institute and Professor of Mechanical Engineering, Institute of Product Development, Technische Universität München, Germany. Interests include methods of product development, design thinking, interdisciplinary research, design process improvement, cost engineering and computer support.

Pericles Loucopoulos
Professor of Information Systems Engineering, School of Informatics, University of Manchester. Interests include requirements engineering, enterprise knowledge modelling and information systems development approaches. Co-editor-in-chief of the Journal of Requirements Engineering. Advisor to various European governmental and industrial institutions.

Anja Maier
Research student, Cambridge Engineering Design Centre, University of Cambridge. Interests: engineering design communication, communication audits, philosophy of design.

Chris McMahon
Innovative Manufacturing Research Centre, University of Bath. Reader in Engineering Design and engineering director of the Bath IMRC. Interests include information and knowledge management, risk and uncertainty in design and computer-aided design.

Harald Meerkamm
Professor of Engineering Design and director of the Institute for Engineering Design, University of Erlangen-Nuremberg, Germany. Interests include integrated product development, methodical and computer-supported design, design process, product and process modelling and ecodesign.

Brendan O'Donovan
Former research student at the Cambridge Engineering Design Centre, University of Cambridge. Now working as a consultant at L.E.K. Consulting, London. Interests: computer assisted design optimisation, CFD applications and design methodology in aerospace and automotive fields.

Martin Stacey
School of Computing, De Montfort University. Senior Lecturer. Interests include psychology of design, artificial intelligence and human computer interaction aspects of computer tools for designers and multiple-level analyses of design processes.

Ralf Stetter
Professor for Design and Development in Automotive Technology at the Fachhochschule Ravensburg-Weingarten. Previously he worked as a team coordinator "seating comfort" at AUDI AG. Interests: implementation of product development methods. Member of the advisory board "Zentrum für Entwicklungsmethoden", München, Germany.

Graham Thompson
Professor of Engineering Design and Head of the Department of Mechanical, Aerospace and Manufacturing Engineering at the University of Manchester Institute of Science and Technology. Alex F. Osborn Visiting Professor at the International Centre for Creative Studies, Buffalo, NY, and an Adjunct Professor at Luleå University of Technology, Sweden. Interests: concept design, especially multi-disciplinary system design with respect to reliability and maintainability.

Sándor Vajna
Otto von Guericke University Magdeburg, Germany. Professor of Information Technologies in Mechanical Engineering, previously 12 years of industrial experience. Interests include integrated product development, design methodology based on biological evolution, dynamic process navigation, CAx applications and PLM (former CIM).

Craig M Vogel
Professor in the School of Design, Carnegie Mellon University, and Associate Dean of the College of Fine Arts. A Fellow and former President of the Industrial Designers Society of America (FIDSA). His areas of expertise include product design, product aesthetics, design history, team management and design patent litigation.

Ken Wallace
Cambridge Engineering Design Centre, University of Cambridge. Professor of Engineering Design, chairman of the Engineering Design Centre and co-director of the BAES/RR University Technology Partnership for Design. Interests include systematic engineering design methods and engineering knowledge management.

David Wynn
Research student, Cambridge Engineering Design Centre, University of Cambridge. Interests: knowledge elicitation, visualisation of complex models, signposting.